Wireless Power Transfer
2nd Edition

RIVER PUBLISHERS SERIES IN COMMUNICATIONS
Volume 45

Series Editors

ABBAS JAMALIPOUR
The University of Sydney
Australia

MARINA RUGGIERI
University of Rome Tor Vergata
Italy

HOMAYOUN NIKOOKAR
Delft University of Technology
The Netherlands

The "River Publishers Series in Communications" is a series of comprehensive academic and professional books which focus on communication and network systems. The series focuses on topics ranging from the theory and use of systems involving all terminals, computers, and information processors; wired and wireless networks; and network layouts, protocols, architectures, and implementations. Furthermore, developments toward new market demands in systems, products, and technologies such as personal communications services, multimedia systems, enterprise networks, and optical communications systems are also covered.

Books published in the series include research monographs, edited volumes, handbooks and textbooks. The books provide professionals, researchers, educators, and advanced students in the field with an invaluable insight into the latest research and developments.

Topics covered in the series include, but are by no means restricted to the following:

- Wireless Communications
- Networks
- Security
- Antennas & Propagation
- Microwaves
- Software Defined Radio

For a list of other books in this series, visit www.riverpublishers.com
http://riverpublishers.com/series.php?msg=Communications

Wireless Power Transfer
2nd Edition

Johnson I. Agbinya

Melbourne Institute of Technology,
Australia

LONDON AND NEW YORK

Published 2016 by River Publishers
River Publishers
Alsbjergvej 10, 9260 Gistrup, Denmark
www.riverpublishers.com

Distributed exclusively by Routledge
4 Park Square, Milton Park, Abingdon, Oxon OX14 4RN
605 Third Avenue, New York, NY 10017, USA

First issued in paperback 2023

Wireless Power Transfer, 2nd Edition / by Johnson I. Agbinya.

Routledge is an imprint of the Taylor & Francis Group, an informa business

Publisher's Note
The publisher has gone to great lengths to ensure the quality of this reprint but points out that some imperfections in the original copies may be apparent.

While every effort is made to provide dependable information, the publisher, authors, and editors cannot be held responsible for any errors or omissions.

ISBN 13: 978-87-7022-977-7 (pbk)
ISBN 13: 978-87-93237-62-9 (hbk)
ISBN 13: 978-1-003-34007-2 (ebk)

Contents

2   Efficient Wireless Power Transfer based on Strongly Coupled Magnetic Resonance   73

Fei Zhang and Mingui Sun

9   Technology Overview and Concept of Wireless Charging Systems   347

Pratik Raval, Dariusz Kacprzak and Aiguo Patrick Hu

10   Wireless Power Transfer in On-Line Electric Vehicle   385

Preface

Since the publication of the first edition of Wireless Power Transfer a couple of years ago significant technical developments in relation to the topic have taken place prompting the writing of the second edition. As its publication explained the analysis and applications of frequency splitting of the frequency response of inductive systems, design of inductive systems using bandpass filter approach and impedance matching have been presented. New applications of these techniques in addition to existing systems already in the public domain have become prevalent.

Until now, the common view is that electric power is transmitted and distributed using transmission lines. As true as this is, we have lived all along with electromagnetic signals from radio, television (TV) and mobile communication systems. They all use wireless power transfer in the form of modulated radio frequencies. Existing methods of generating electrical power and distribution place severe demands on the environment. Coal-powered electric power sources create gases which contribute to global warming. Nuclear power sources are not only dangerous as witnessed in the nuclear accidents in Chenobyl and in Japan but are also costly and come with political conflicts between nations. Wireless energy sources and distribution systems are relatively new and safe but have limited range, a problem being addressed by several chapters of this book.

Two laws of electrical engineering lay the theoretical foundation for the near-field magnetic flux coupling form of wireless power transfer. These laws are termed Ampere's law and Faraday's law and form parts of Maxwell's equations. In simplified terms, Ampere's law shows that a varying electric current produces a varying magnetic field. This magnetic field may be determined using Ampere's circuital law which relates the integrated magnetic field around a closed loop to the changing electric current passing through the loop. The second law credited to Michael Faraday also shows that an electromotive force (EMF) is induced in any closed circuit through a time-varying magnetic flux through the circuit. In fact, the EMF generated is proportional to the rate of change of the magnetic flux. In essence, there is need for one or more parameters of the system to be varying. These two laws are the main foundations of NFMI technology.

This edition contains twenty chapters and includes the eleven chapters of the first edition. The new chapters complement the basic concepts in wireless power transfer. These are design of wireless power transfer systems using bandpass filters, impedance matching, frequency splitting and its applications, multi-dimensional wireless power transfer using cellular concepts and inductive heating and cooking.

Impedance matching chapters present how to reduce power losses in inductive systems through techniques for cancelling the reactive power component. The chapter on multi-dimensional wireless power transfer shows how to design wide area wireless power transfer by using the cellular concept, a technique that was made popular in cellular communication networks. Frequency splitting was shown to provide a means for optimum power transmission at two frequencies concurrently. Its application to inductive communication using frequency modulated continuous wave (FMCW) or chirp signals is demonstrated in this edition. A chapter on induction cooking and heating has been added. Induction ovens and cook tops benefit from the technology covered in this chapter.

Collectively, the chapters of the book have provided techniques on how to extend the range of the energy system, how to increase efficiency of power conversion and induction and also how to control the power systems. Furthermore, they detail techniques for power relay and also some of the applications of wireless energy transfer. The book is written in a progressive manner and the first chapters make it easier to understand later chapters.

Objectives and Prerequisites

A new book on a new technology ought to be written in a form which makes it understandable to not only establish practioners and researchers on the topic, but also to postgraduate and where necessary undergraduates in the field of study. This edition has done that. Most students of electrical or electronic engineering with basic circuit theory should understand most of the chapters without much difficulty. However, there are also new concepts which are covered in these chapters to enhance the learning process and lay foundations on the topic. The three approaches for designing wireless power transfer technologies are covered in Chapters one (coupled mode theory), five (circuit theory) and sixteen (bandpass filter theory). The rest of the chapters are individually focused on specific concepts in wireless power transfer.

Many applications of wireless power transfer require strong coupling of the receiver to the magnetic field created by the transmitter. This is discussed in Chapter two. The Chapter three also explains new concepts which address the limited range of inductive power sources and also control techniques in wireless energy transfer. Inductive antennas are discussed in Chapter three. Chapter four then discusses wireless power generation and distribution. Recent advances in wireless power transfer are discussed in Chapter six.

The two chapters, seven and eight discuss problems which normally impact optimum wireless power transfer and how system parameters may be optimized to achieve higher power transfer. Misalignment and tuning of system parameters are covered.

Some of the most popular applications of wireless power transfer are discussed in Chapters nine, ten, eleven, twelve, nineteen and twenty. Medical applications, wireless power chargers, electric vehicles, inductive ovens and heating are the main areas covered. Other applications in telecommunications are discussed in Chapters seventeen and eighteen. Association between splitting frequency and power transfer at two or more frequencies are discussed in this chapter. Conditions that must be met for frequency splitting to happen are analyzed in great detail.

In a nutshell, the chapters deal with:

- Power Transfer by Magnetic Induction Using Coupled Mode Theory
- Wireless Power Transfer with Strongly Coupled Magnetic Resonance
- Low-Power Rectenna Systems for Wireless Energy Transfer
- Inductive Wireless Power Transfer Using Circuit Theory
- Magnetic Resonant Wireless Power Transfer
- Techniques for Optimal Wireless Power Transfer Systems
- Directional Tuning/Detuning Control of Wireless Power Pickups
- Technology Overview and Concept of Wireless Charging Systems
- Wireless Power Transfer in On-Line Electric Vehicle
- Design and optimization of resonant power transfer systems using bandpass filter theory
- Frequency splitting
- Impedance Matching Concepts, Techniques and Circuits
- Wireless Powering and Propagation of Radio Frequencies Through Tissue
- Microwave Propagation and Inductive Energy Coupling in Biological Skin for Body Area Network Channels
- Inductive cooking and heating

Organization of the Book

The first chapter focuses exclusively on the coupled mode theory that is central to the Physics of wireless energy transfer. The physical principles are based on and rely on concepts including near and far fields and radiative propagation related to the antenna. In the *near field region of an antenna, energy transfer is mostly non-radiative and takes place* close to the antenna at a distance smaller than one wavelength and decays very fast. The *far field energy transfer is mostly radiative and is considered in the later chapters of the book.* The energy *coupling in the near field region is through magnetic resonance. Hence* coupled systems work at their resonance frequencies. In principle, the method relies on *resonant objects exchanging energy efficiently* while the non-resonant objects interact weakly. High-power radiation poses safety concerns and also requires a complicated tracking system, while the current non-radiative types only work within a short distance between the energy transmitter and receiver.

The inadequacy of weak coupled systems is demonstrated in Chapter two. The chapter emphasizes the coverage in Chapter one and covers both the coupled mode and the circuit theories of energy transfer. To overcome problems associated with the traditional inductive methods, 'wiTricity'. This technique is based on non-radiative strongly coupled magnetic resonance and can work over the mid-range distance, defined as several times the resonator size. Strongly coupled magnetic resonance is the core focus of this chapter. It also covers relay effect which extended the original wiTricity from the source–device(s) scenario to the source-relay-device scenario to enable more flexible and distant wireless power transfer. The chapter demonstrates by both theoretical treatments and experimental studies that this scheme succeeds beyond the limit of mid-range, allowing much longer and more flexible power transmission without sacrificing efficiency. The latter part of this chapter extended the concept to a network architecture involving a number of resonators and a spectral method to assist in the design and evaluation of wireless power transfer network is described.

There is a massive development of a wide range of portable electronic devices, consumer devices like smartphones and also industrial applications, like wireless sensor networks with increased miniaturization. The objective is to facilitate their portability and their integration in the environment. The use of electrical wires for powering them is in many cases impossible due to the "wireless" nature of such devices. Also wide reliance on batteries for power and subsequent disposal of depleted batteries is costly and pollute the environmentally. The natural evolution of power transfer seems to take the same path as information transfer: cutting the cord and stepping in the wireless era. This chapter dwells on this general theme and describes techniques for converting RF to DC power. It also studies different circuit topologies for wireless energy transfer. Models of rectenna (receiving antenna) are given.

For readers desiring less extensive mathematics including differential equations, the circuit theory approach to modeling and design of wireless power transfer is presented in Chapter four. The content presented in this chapter provides a concise overview of the field and it is most suited as a first introduction to the topic at an undergraduate level from the circuit theory point of view. The chapter dwells on the loosely coupled system where a weak magnetic link is adopted for wireless power transfer. In other words, the coupling coefficient is a lot smaller than unity.

Chapter five provides the circuit theory model.

Two port network concepts are used in Chapter six. Thus discussions on impedance mismatch, signal reflection, reflection coefficient, travelling wave theory, scattering parameter (S-parameter) matrix and impedance inverters/ matching are covered in this chapter. The contention between maximum efficiency and maximum power to the load are explained in the chapter. The use of the quality factors of the power transmitter and receiver circuits is introduced. The approach helps to easily quantify sources of power delivery and losses in the system and hence the efficiency of the energy transfer system.

The use of arrays of inductive loops has become prominent within the last few years. Chapter seven provides coverage of the techniques for deriving the power transfer functions of such systems. The power transfer function equation for an inductive array of N loops including split ring arrays is provided. Nearest neighbor interaction concept is employed and magneto-inductive wave transmission is assumed. The easy to use algorithm for the transfer function equations assumes also low power coupling approximation which applies to many current applications of inductive methods. The assumptions lead to the power transfer equation for any N coils. Correction terms are suggested for larger coupling coefficients and quality factors. Interpretation of the overall system of loops based on the approximation is suggested and shown to be a very reasonable approach for explaining what takes place in such systems from the electronic communication points of view.

The chapter concludes with proposals and analysis of a new framework for wireless feedback control systems. The feedback control problem is modeled as inductive flux link from a coil to its neighbors. This provides a foundation for system control of wireless Internet of things and an easier path to the design, analysis, control and performance analysis of such systems.

Chapter eight presents an algorithm named directional tuning/detuning control (DTDC) for power flow regulation of wireless power transfer (WPT) systems. The controller regulates the power being delivered to the pickup load by deliberately tuning/detuning the centre frequency of the pickup tuning circuit with the operating frequency according to the load demand and circuit parameter variations. This is essentially achieved by controlling the duty cycle of a switch-mode tuning capacitor in the pickup resonant tank, which in turn gives a desired equivalent tuning capacitance. The controller allows the wireless pickup to operate with full-range tuning and eliminates the tedious fine-tuning process associated with traditional fixed tuning methods. This, therefore, eases the components selection of WPT system design and allows the system to have higher tolerance in circuit parameter variations.

One of the most attractive applications of wireless energy transfer is wireless charging of devices and is covered in Chapter ten. Several products by several manufacturers have been in the market for several years now. They are mostly for low proximity and low-power charging of electronic devices. For this purpose, this edition provides magnetic structures which have been developed to demonstrate the concept of two-dimensional and three-dimensional wireless low-power transfer systems. The presented development is aided by state of the art finite-element-based simulation software packages.

Magnetic levitation has been with us for quite a while and is used in fast trains in many countries. The basis for magnetic levitation is shared with WPT. It is an application of magnetic force. This chapter provides a general overview of wireless power transfer system using magnetic field and the application to online electric vehicles. The magnetic field shape design and shielding technologies are introduced and discussed. The simulation and measurement results supporting the design methods are explained.

Chapters eleven and twelve is an overview of some of the applications of wireless powering and focuses on far-field remote powering for implantable applications using radiative electromagnetic fields. Analytical and simulation models are presented to help illustrate and gain an intuitive understanding of RF propagation through tissue. *Ex vivo* experiments and *in vivo* studies are performed to access the validity and accuracy of the models, and these empirical results are discussed. Lastly, the electrical components, circuit design, and system integration required for transcutaneous wireless power transfer are described. There are significant benefits of the radiative wireless powering technique, as compared to its inductive coupling counterpart, but with these advantages comes extremely complex challenges. In Chapter eleven, we perform a rigorous analysis of the effects of body tissues and specifically skin, fat and muscle on biomedical systems where energy is propagated or coupled from a source to a receiver. The two scenarios of electric field and magnetic flux coupling are described. Models of tissues including the skin are given and analyzed in great detail. The effects of tissues on the induced magnetic field are shown to be mainly threefold, Ohmic heating in the tissue at both low and high frequencies which reduces the induced voltage available to the receiver and a phase change resulting from the relaxation time (characteristic frequency) of the tissue. At mid-frequency range, a gain factor is also manifested in the voltage at the output of the tissue impedance transfer function. We show that absorption of E-field in tissue can be greatly minimized by using protective materials with left-handed materials which maximizes reflection. We also show that for the same system and tissue characteristics, the specific magnetic field absorption rate is always smaller than the specific absorption rate for the electric field suggesting preference for inductive coupling in embedded medical applications.

Some applications of inductive systems will need more than one frequency for transmission. Applications in telecommunications, inductive transceivers, payment cards and magnetic sensors may need to operate optimally at two frequencies. Chapter thirteen lays the foundation for them through analysis of the frequency response of typical inductive arrays. Coupling regimes including weak, strong and critical coupling are discussed and distinguished from each other. At critical coupling or more the frequency response of inductive systems presents several frequencies at which power and data transfers are optimum. This chapter also lays the foundation for discussions in Chapters fourteen and fifteen on impedance matching.

Impedance matching concepts including conjugate matching and conditions for maximum power transfer for resistive, capacitive and inductive circuits are discussed in Chapter fourteen and analysis of reflection coefficients, standing wave ratio are given. The notion of ideal impedance transformers is introduced. Short circuit stubs, open-circuit stubs and Q-sections are analyzed with examples.

In Chapter fifteen, formal discussions on impedance matching circuits are presented. Series-parallel transformations, using L-sections (low-pass and high-pass sections) and equivalent circuits have been analyzed in the chapter. Furthermore, impedance matching networks specifically pi-networks, T-networks and tunable networks are discussed with examples. The relationship between impedance matching and maximum power transfer is demonstrated.

Recently, several groups have demonstrated the bases for multi-dimensional wireless power transfer. In one example given in Chapter seventeen cubical inductive nodes combined with resonant arrays and cellular concept are presented and explained. These techniques lead to wide area wireless power transfer and thus extend the range of wireless power transmission not only in one direction but also to any desirable directions concurrently. Several connection configurations are given in the chapter.

Chapter eighteen is a formal presentation and analysis of split frequency techniques in magnetic inductive systems. Examples of applications in short range data communications are demonstrated.

The second edition is concluded with two applications. The first is a chapter on advances in wireless powering of medical applications with particular focus on implantable medical devices. Product design processes are discussed.

Lastly, Chapter twenty completes the edition through discussion of induction cooking and heating. Induction cooking ware have become ubiquitous in recent times. Therefore, the chapter is eminently relevant as it presents the design process with examples.

Acknowledgment

This book is the third in the series of books by the author on inductive near field communications. It is also the second edition of a well-received book titled Wireless Power Transfer. I acknowledge the contributions of all the authors including my postgraduate students who have contributed a significant proportion of the book. Their contributions have made this edition comprehensive and technically relevant to applications and system developments based on inductive systems.

List of Contributors

Johnson I. Agbinya *Melbourne Institute of Technology, Australia*

Bruno Allard *Ecole Centrale de Lyon, INSA Lyon, Ampère UMR 5005, INL UMR 5270, France*

William J. Chappell *Center for Implantable Devices, Weldon School of Biomedical Engineering, Purdue University, School of Electrical and Computer Engineering, Purdue University, USA*

Eric Y. Chow *Cyberonics Inc., Houston, TX 77058 USA*

Alessandra Costanzo *DEI, University of Bologna, Italy*

Marco Dionigi *DI, University of Perugia, Italy*

Lucia Dumitriu *Electrical Engineering Department Politehnica University of Bucharest, Romania*

Brian Flynn *Institute for Integrated Micro and Nano Systems, The University of Edinburgh, UK*

Kyriaki Fotopoulou *Institute for Integrated Micro and Nano Systems, The University of Edinburgh, UK*

Dohyuk Ha *School of Electrical and Computer Engineering, Purdue University, USA*

Aiguo Patrick Hu *Department of Electrical and Computer Engineering, The University of Auckland, New Zealand*

Mihai Iordache *Electrical Engineering Department Politehnica University of Bucharest, Romania*

Pedro P. Irazoqui *Center for Implantable Devices, Weldon School of Biomedical Engineering, Purdue University, School of Electrical and Computer Engineering, Purdue University, West Lafayette, IN 47907 USA*

Dariusz Kacprzak *Department of Electrical and Computer Engineering, The University of Auckland, New Zealand*

Vlad Marian *Ecole Centrale de Lyon, INSA Lyon, Ampère UMR 5005, INL UMR 5270, France*

Franco Mastri *DEI, University of Bologna, Italy*

Henry Mei *Center for Implantable Devices, Weldon School of Biomedical Engineering, Purdue University, USA*

Nagi F. Ali Mohamed *Department of Electrical Engineering, Faculty of Engineering Technology, Libya*

Mauro Mongiardo *DI, University of Perugia, Italy*

Giuseppina Monti *DII, University of Salento, Italy*

Hoang Nguyen *Department of Electronic Engineering, La Trobe University, Australia*

Dragos Niculae *Electrical Engineering Department Politehnica University of Bucharest, Romania*

Pratik Raval *Department of Electrical and Computer Engineering, The University of Auckland, New Zealand*

Mingui Sun *Departments of Neurosurgery and Electrical Engineering, University of Pittsburgh, USA*

David L. Thompson *Cyberonics Inc., Houston, TX 77058 USA*

Jacques Verdier *Ecole Centrale de Lyon, INSA Lyon, Ampère UMR 5005, INL UMR 5270, France*

Christian Vollaire *Ecole Centrale de Lyon, INSA Lyon, Ampère UMR 5005, INL UMR 5270, France*

Xiyao Xin *Department of Electrical and Computer Engineering at University of Houston, Houston, TX 77204*

Chin-Lung Yang *Department of Electrical Engineering in National Cheng Kung University, Tainan 70101, Taiwan*

Fei Zhang *Departments of Neurosurgery and Electrical Engineering, University of Pittsburgh, USA*

List of Figures

List of Tables

List of Abbreviations

AC	Alternating Current
A4WP	Alliance For Wireless Power
BPF	Bandpass Filter
CT	Circuit Theory
CMT	Coupled-mode Theory
EM	Electromagnetic
ECG	Electrocardiogram
EMG	Electromyogram
EMP	Electromagnetic Powering
HF	High Frequency
IFF	Identification Friend or Foe
ISM	Industrial Scientific Medical
LF	Low Frequency
LBS	Biot Savart Law
MF	Medium Frequency
MRC	Magnetic Resonance Coupling
MNE	Modified Nodal Equations
NFC	Near-Field Communication
PMA	Power Matters Alliance
PTE	Power Transfer Efficiency
PW	Power waves
RF	Radio Frequency
RFID	Radio Frequency Identification
RX	Receiver
RX	Receiving
SMD	Smart Medical Devices
SoC	System on Chip
SVE	State Variable Vector
SESYMGP	State Equation Symbolic Generation Program
SPICE	Simulation Program with Integrated Circuit Emphasis
V2i	Vehicular to infrastructure

TX	Transmiter
TX	Transmitting
VNA	Vector Network Analyzer
wBSN	Wireless body sensor networks
WET	Wireless Energy Transfer
WITRICITY	WIreless elecTRICITY
WPC	Wireless Power Consortium
WPT	Wireless Power Transfer
WPTN	Wireless Power Transfer Network
WREL	Wireless Resonant Energy Link

1

Power Transfer by Magnetic Induction Using Coupled-Mode Theory

Mihai Iordache, Lucia Dumitriu and Dragos Niculae

Electrical Engineering Department Politehnica University of Bucharest, Romania
E-mail: {mihai.iordache; lucia.dumitriu; dragos.niculae}@upb.ro

Abstract

This chapter presents the concepts behind wireless power transfer by induction and describes the phenomenon both by Circuit Theory (CT) and by Coupled-mode Theory (CMT). The two theories are compared, outlining their advantages and disadvantages. Methods to compute the resonator parameters based on the Neumann formula and using the professional ANSOFT Q3D EXTRACTOR 6 are also developed and compared. This chapter concludes with optimization of wireless power transfer.

1.1 Introduction

The electric signals are basic concepts of the electromagnetic field. They are carriers of energy and information, and the applications take this into account.

The study of the physical principle of these applications is based on the concepts like **near** and **far fields** and **radiative propagation** related to the antenna.

The *near field* is referred to as a *non-radiative* type that occurs close to the antenna at a distance smaller than one wavelength and decays very fast ($\sim 1/r^3$) [1].

The *far field is considered to be of a radiative* type. It propagates starting from a distance equal to two wavelengths from the antenna up to infinity. This type of radiation decays much slower than the near field ($\sim 1/r$). The emitted power decays with the square of the distance [1].

Wireless Power Transfer 2nd Edition, 1–72.

There is a transition zone starting from a distance of one wavelength from the antenna up to two wavelengths in which the combined effects of the near and far fields occur.

The modern applications in telecommunications are based on the propagation of electromagnetic waves (on far field), but the **antenna radiation technology** is not suitable for power transfer. The main reason is that the radiated electromagnetic power is small (a vast majority of the energy is wasted by dispersion into the free space) making this technology more suitable to transfer information rather than power.

Wireless energy transfer—Witricity (WIreless elecTRICITY)—is different from wireless transmission of information. In this new technology (useful in cases where instantaneous or continuous energy is needed, but interconnecting wires are inconvenient, hazardous, or impossible), the transfer is made over distances at which the electromagnetic field is strong enough to allow a reasonable power transfer. This is possible if both the emitter and the receiver achieve magnetic resonance.

The physical principle behind the Witricity concept is based on the *near field* in correlation with the resonant inductive coupling.

The *coupling through magnetic resonance* implies that the coupled systems work at their resonance frequency. The principle on which the method relies is that ***the resonant objects exchange energy efficiently,*** while the non-resonant objects interact weakly.

The Witricity system consists of two resonators—source and device—which theoretically could be in one of the following connections: series–series, parallel–parallel, series–parallel, and parallel–series.

The source resonator emits a lossless non-radiative magnetic field oscillating at MHz frequencies, which mediates an efficient power exchange between the source and the device resonators.

Some remarks have to be made [2]:

- The interaction between source and device is strong enough so that the interactions with no resonant objects can be neglected, and an efficient wireless channel for power transmission is created.
- Magnetic resonance is particularly suitable for applications because, in general, the common materials do not interact with magnetic fields.
- It seems [3] that the power transfer is not visibly affected when humans and various objects, such as metal, wood, and electronic devices, are placed between the two coils at more than a few centimeters from each

of them, even in cases where they completely obstruct the line of sight between source and device.

- Some materials (such as aluminum foil and humans) just shift the resonant frequency, which can in principle be easily corrected with a feedback circuit.

The principle behind the antenna radiation technology was described by the CMT, which recently was extended to describe wireless power transfer [4–13], but, because *RLC* resonators are used, the Circuit Theory can be applied as well.

Considering for multiple reasons two series–series resonators, we shall describe the behavior of the inductively coupled system used in Witricity by the two theories: coupled-mode and circuit.

1.2 Series–Series Resonators Inductively Coupled

1.2.1 Analysis by the Circuit Theory

Let us consider two series–series resonators inductively coupled as in Figure 1.1, where L_3 and L_4 are two coaxial identical coils represented in Figure 1.2 [13, 14]. The parameters of the coils are as follows: the radius

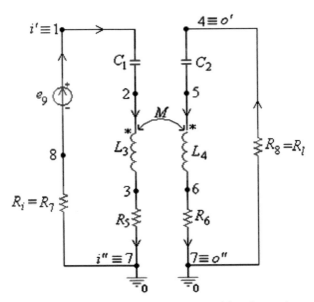

Figure 1.1 System of two series–series resonators driven by a voltage source.

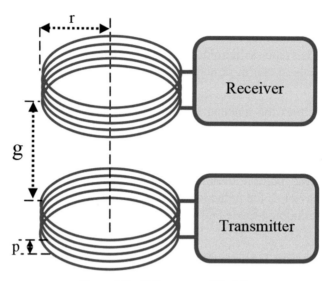

Figure 1.2   Coil geometry [13, 14].

$r = 150$ mm, the pitch $p = 3$ mm, the wire size $w = 2$ mm, the distance between the coils $g = 150$ mm, and the number of the turns $N = 5$. Using the ANSOFT Q3D EXTRACTOR program [15], we get the following numerical values for the parameters of the system of the two inductively coupled coils:

$$C_1 = C_2 = C = 1.0404 \, \text{nF}, \quad L_3 = 16.747 \, \mu\text{H}, \quad L_4 = 16.736 \, \mu\text{H},$$
$$M = 1.4898 \, \mu\text{H}, \quad R_5 = 0.12891 \, \Omega, \quad R_6 = 0.12896 \, \Omega$$

The source circuit has the resistance $R_7 = 5.0 \, \Omega$, and the load is $R_8 = 5.0 \, \Omega$.

Performing the SESYMGP program [16], we get the state equations of the circuit in full-symbolic normal form:

$$\frac{dv_{C_1}}{dt} = \frac{1}{C_1} i_{L_1} \tag{1.1}$$

$$\frac{dv_{C_2}}{dt} = \frac{1}{C_2} i_{L_2} \tag{1.2}$$

$$\frac{di_{L_1}}{dt} = -\frac{L_4}{L_3 L_4 - M^2} v_{C_1} + \frac{M}{L_3 L_4 - M^2} v_{C_2} - \frac{L_4(R_5 + R_7)}{L_3 L_4 - M^2} i_{L_1}$$
$$+ \frac{M(R_6 + R_8)}{L_3 L_4 - M^2} i_{L_2} + \frac{L_4}{L_3 L_4 - M^2} e_9 \tag{1.3}$$

$$\frac{di_{L_2}}{dt} = -\frac{M}{L_3 L_4 - M^2} v_{C_1} + \frac{L_3}{L_3 L_4 - M^2} v_{C_2} - \frac{M(R_5 + R_7)}{L_3 L_4 - M^2} i_{L_1}$$
$$- \frac{L_3 (R_6 + R_8)}{L_3 L_4 - M^2} i_{L_2} + \frac{M}{L_3 L_4 - M^2} e_9 \tag{1.4}$$

where $i_{L_1} = i_{L_3}$ is the current through the first circuit, $i_{L_2} = i_{L_4}$ is the current through the second one, and v_{C_1}, v_{C_2} are the capacitor voltages of the two circuits.

Considering null initial conditions, that is,

$$v_{C_1}(0) = 0.0 \text{ V}, \; v_{C_2}(0) = 0.0 \text{ V}, \; i_{L_1}(0) = 0.0 \text{ A and}$$
$$i_{L_2}(0) = 0.0 \text{ A} \tag{1.5}$$

and taking $e_9(t) = 100.0 \cdot \cos(\omega_0 t)$ V, with $\omega_0 = 1/\sqrt{((L_3 + L_4)/2.0) \cdot C}$, if we integrate the state Equations (1.1)–(1.4) for the numerical values of the circuit parameters given above, we get the solution shown in Figures 1.3 and 1.5. For comparison, we present in Figures 1.4 and 1.6 the SPICE simulation results.

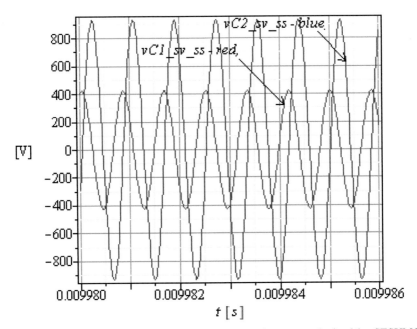

Figure 1.3 Time variation of the voltages $v_{C1\_ss\_red}$ and $v_{C2\_ss\_blue}$, obtained by SESYMGP simulation.

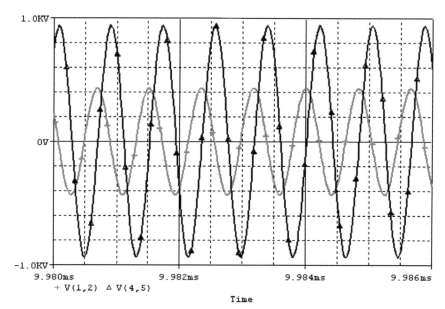

Figure 1.4 Time variation of the voltages $v_{C1\_Spice\_ss\_green}$ and $v_{C2\_Spice\_ss\_red}$, obtained by SPICE simulation.

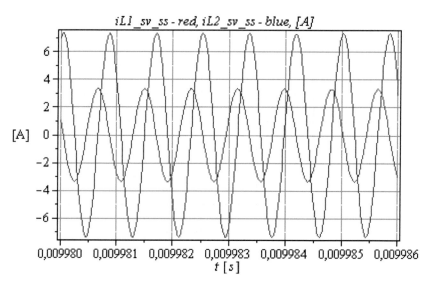

Figure 1.5 Time variation of the currents $i_{L1\_ss\_red}$ and $i_{L2\_ss\_blue}$, obtained by SESYMGP simulation.

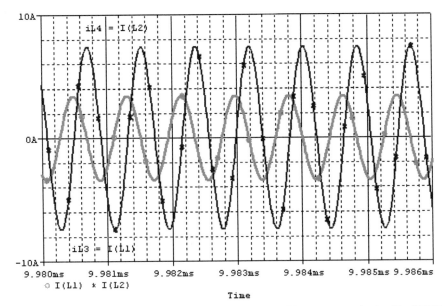

Figure 1.6 Time variation of the currents $i_{L1\_Spice\_ss\_green}$ and $i_{L2\_Spice\_ss\_red}$, obtained by SPICE simulation.

We note that the same solution can be obtained much more easily using the Laplace transform.

1.2.2 Analysis by the Coupled-Mode Theory

In order to obtain the differential equations of the coupled-mode amplitudes, the Equations (1.2) and (1.4) are multiplied by $\sqrt{C/2}$ and the Equations (1.3) and (1.5) by $j\sqrt{L_3/2}$ and $j\sqrt{L_4/2}$, respectively. Adding the first and the third equations so modified, we get

$$\frac{\mathrm{d}}{\mathrm{d}t}\left(\sqrt{\frac{C}{2}}v_{C_1} + j\sqrt{\frac{L_3}{2}}i_{L_1}\right)$$

$$= -\frac{j}{\sqrt{L_3 C}}\left(\frac{L_3 L_4}{L_3 L_4 - M^2}\sqrt{\frac{C}{2}}v_{C_1} + j\sqrt{\frac{L_3}{2}}i_{L_1}\right)$$

$$-\frac{L_4\left(R_5 + R_7\right)}{L_3 L_4 - M^2}\left(j\sqrt{\frac{L_3}{2}}i_{L_1}\right) + j\frac{M\left(R_6 + R_8\right)}{L_3 L_4 - M^2}$$

$$\cdot \left(\frac{1}{R_6} \sqrt{\frac{L_3}{C}} \sqrt{\frac{C}{2}} v_{C_2} + (-j) j \sqrt{\frac{L_3}{L_4}} \sqrt{\frac{L_4}{2}} i_{L_2} \right)$$

$$+ j \sqrt{\frac{L_3}{2}} \frac{L_4}{L_3 L_4 - M^2} e_9 \tag{1.6}$$

which can be written in compact form as

$$\frac{\mathrm{d}a_1(t)}{\mathrm{d}t} = -j\omega_1 a_1''(t) - \gamma_1 a_1' + k_{12} a_2''(t) + k_{1e} e_9 \approx -j\omega_0 a_1(t) - \gamma_1 a_1(t)$$

$$+ k_{12} a_2(t) + k_{1e} e_9 \tag{1.7}$$

where

$$a_1'' = \frac{L_3 L_4}{L_3 L_4 - M^2} \sqrt{\frac{C}{2}} v_{C_1} + j\sqrt{\frac{L_3}{2}} i_{L_1} \approx a_1 \stackrel{\mathrm{d}}{=} \sqrt{\frac{C}{2}} v_{C_1} + j\sqrt{\frac{L_3}{2}} i_{L_1}$$

$$a_1' = j\sqrt{\frac{L_3}{2}} i_{L_1} \approx a_1 \stackrel{\mathrm{d}}{=} \sqrt{\frac{C}{2}} v_{C_1} + j\sqrt{\frac{L_3}{2}} i_{L_1}$$

$$\gamma_1 = \frac{L_4 (R_5 + R_7)}{L_3 L_4 - M^2}; \quad k_{12} = j\sqrt{\frac{L_3}{C}} \frac{M}{L_3 L_4 - M^2};$$

$$k_{1e} = j\sqrt{\frac{L_3}{2}} \frac{L_4}{L_3 L_4 - M^2} \tag{1.8}$$

$$a_2'' = \sqrt{\frac{C}{2}} v_{C_2} + (-j) j (R_6 + R_8) \sqrt{\frac{C}{L_4}} \sqrt{\frac{L_4}{2}} i_{L_2}$$

$$\approx a_2 \stackrel{\mathrm{d}}{=} \sqrt{\frac{C}{2}} v_{C_2} + j\sqrt{\frac{L_4}{2}} i_{L_2}$$

$$\omega_1 = \frac{1}{\sqrt{L_3 C}}$$

Adding the modified second equation and the modified fourth one, we get

$$\frac{\mathrm{d}}{\mathrm{d}t} \left(\sqrt{\frac{C}{2}} v_{C_2} + j\sqrt{\frac{L}{2}} i_{L_2} \right)$$

$$= -\frac{j}{\sqrt{L_4 C}} \left(\frac{L_3 L_4}{L_3 L_4 - M^2} \sqrt{\frac{C}{2}} v_{C_2} + j \sqrt{\frac{L}{2}} i_{L_2} \right)$$

$$- \frac{L_3 R_6}{L_3 L_4 - M^2} \left(j \sqrt{\frac{L_4}{2}} i_{L_2} \right)$$

$$\overset{MR_c}{+ j \sqrt{\frac{L_4}{C}} \frac{M}{L_3 L_4 - M^2}} \left(\sqrt{\frac{C}{2}} v_{C_1} + (-j) R_5 \sqrt{\frac{C}{L_4}} j \sqrt{\frac{L_4}{2}} i_{L_1} \right)$$

$$- j \sqrt{\frac{L_4}{2}} \frac{M}{L_3 L_4 - M^2} e_9 \tag{1.9}$$

which can be written in compact form as

$$\frac{\mathrm{d}a_2(t)}{\mathrm{d}t} = -j\omega_2 a_2''(t) - \gamma_2 a_2' + k_{21} a_1'(t) + k_{2e} e_9 \approx -j\omega_0 a_2(t) - \gamma_2 a_2(t)$$

$$+ k_{21} a_1(t) + k_{2e} e_9 \tag{1.10}$$

where

$$a_2'' = \frac{L_3 L_4}{L_3 L_4 - M^2} \sqrt{\frac{C}{2}} v_{C_2} + j \sqrt{\frac{L}{2}} i_{L_2} \approx a_2 \overset{\mathrm{d}}{=} \sqrt{\frac{C}{2}} v_{C_2} + j \sqrt{\frac{L_4}{2}} i_{L_2}$$

$$a_2' = j \sqrt{\frac{L_4}{2}} i_{L_2} \approx a_2 \overset{\mathrm{d}}{=} \sqrt{\frac{C}{2}} v_{C_2} + j \sqrt{\frac{L_4}{2}} i_{L_2}$$

$$a_1' = \sqrt{\frac{C}{2}} v_{C_1} + (-j) j R_5 \sqrt{\frac{C}{L_3}} \sqrt{\frac{C}{2}} \sqrt{\frac{L_3}{2}} i_{L_1} \approx a_1 \overset{\mathrm{d}}{=} \sqrt{\frac{C}{2}} v_{C_1} + j \sqrt{\frac{L_3}{2}} i_{L_1}$$

$$\gamma_2 = \frac{L_3 R_6}{L_3 L_4 - M^2}; \quad k_{21} = j \sqrt{\frac{L_4}{C}} \frac{M}{L_3 L_4 - M^2};$$

$$k_{2e} = -j \sqrt{\frac{L_4}{2}} \frac{M}{L_3 L_4 - M^2}$$

$$\omega_2 = \frac{1}{\sqrt{L_4 C}} \tag{1.11}$$

The solutions of the differential Equations (1.7) and (1.10) can be obtained by integration and they have the general form:

$$a(t): \quad = a_0 e^{-\gamma t} e^{j\omega_0 t} \tag{1.12}$$

For small loss, because $|a|^2$ (which is the circuit energy) decays as $\exp(-2\gamma t)$, we can express the dissipated power as a function of the square of the mode amplitude:

$$P_{d\_cmt} = -\frac{dW(t)}{dt} = 2\gamma W(t) = f(|a(t)|^2) \tag{1.13}$$

1.2.3 Transfer Power Computation

We can compute the power dissipated on the load resistance as Joule power

$$P_{R8\_sv\_ss} = R_8 \left(i_{L2\_sv\_ss}\right)^2 \tag{1.14}$$

If we use the mode amplitude, we can express this power either by the equation

$$P_{R8\_cmt\_aprox\_ss} = abs(-2 \cdot \gamma_2 \cdot a_{2\_cmt\_ss} \cdot a^*_{2\_cmt\_ss}) \tag{1.15}$$

or by

$$P_{R8\_cmt\_d\_ss} = \frac{d}{dt} \left(a_2(t) \cdot a_2^*(t)\right) \tag{1.16}$$

We can also compute the current through the load circuit $i_{L2\_cmt\_ss}$ from the expression of a_{2_ss} by identifying its imaginary part. With this value, we compute the Joule power dissipated on load:

$$P_{R8\_cmt\_ss} = R_8 \left(i_{L2\_cmt\_ss}\right)^2 \tag{1.17}$$

The graphical representation of these instantaneous powers is done in Figure 1.7.

We can remark a significant difference between the power computed in CT (blue) and those obtained by the CMT formulas. This is a consequence of the approximations made in the expressions of a_1 and a_2.

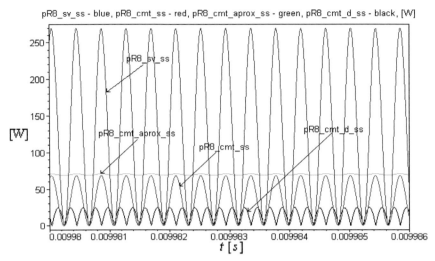

Figure 1.7 Time variation of the transferred power in the four approaches: $P_{R8\_sv\_ss\_blue}$, $P_{R8\_cmt\_ss\_red}$, $P_{R8\_cmt\_aprox\_ss\_green}$, and $P_{R8\_cmt\_d\_ss\_black}$.

The time-averaged transferred powers on a period $T_2 = 2\pi/\omega_0$, with $n = 1,00,000$, have the following values:

$$n = 1,00,000; T_2 = 8.28814 \cdot 10^{-7}\,s; P_{R8\_med\_sv\_ss} = 134.78914\,\text{W};$$

$$P_{R8\_med\_cmt\_ss} = 34.2224\,\text{W}; P_{R8\_med\_cmt\_aprox\_ss} = 69.7736\,\text{W};$$

$$P_{R8\_med\_cmt\_d\_ss} = 15.796511\,\text{W} \tag{1.18}$$

1.2.4 Remark

As expected, the difference between the instantaneous powers (Figure 1.7) is projected over the time-averaged powers.

Using CT and either *Laplace transform* or *complex representation*, and *symbolic analysis*, we can obtain information regarding the variation of the power transferred on the circuit load with respect to different parameters needed in the optimized design.

Using a symbolic simulator [16], we get the symbolic expression of the active power

$$
\begin{aligned}
PR8\_f\_ss := 0.2453679569 10^{14}\, R8\, f^6\, C1^2\, E\_c^2\, M^2\, C2^2 \Big/ \big(&-0.3155072000010^{11}\, L4\, f^2\, C2 - 0.3155072000010^{11}\, f^2\, C1\, L3 \\
&- 0.1244309920 10^{13}\, f^4\, C1\, C2\, M^2 + 0.2488619836 10^{13}\, f^4\, C1\, L4\, L3\, C2 + 0.9676919856 10^{15}\, f^8\, C1^2\, C2^2\, M^4 \\
&+ 0.3155072000010^{11}\, f^2\, C1^2\, R7\, R5 + 0.6221549578 10^{12}\, f^4\, C1^2\, L3^2 + 0.6221549578 10^{12}\, L4^2\, f^4\, C2^2 \\
&+ 0.1577536000010^{11}\, f^2\, C2^2\, R8^2 + 0.1577536000010^{11}\, f^2\, C2^2\, R6^2 + 0.1577536000010^{11}\, f^2\, C1^2\, R5^2 \\
&+ 0.4907359235 10^{14}\, f^6\, C1^2\, C2^2\, M^2\, R5\, R6 + 0.4907359235 10^{14}\, f^6\, C1^2\, C2^2\, M^2\, R7\, R8 - 0.1935838397 10^{16}\, f^8\, C1^2\, C2^2\, M^4\, L4\, L3 \\
&+ 10048.\, f^4\, C1^2\, L3\, R7\, C2\, R8 + 10048.\, f^4\, C1^2\, L3\, R5\, C2\, R6 + 10048.\, f^4\, C1^2\, L3\, R5\, C2\, R8 + 10048.\, f^4\, C1^2\, L3\, R7\, C2\, R6 \\
&+ 10048.\, L4\, f^4\, C2^2\, C1\, R7\, R8 + 10048.\, L4\, f^4\, C2^2\, C1\, R5\, R6 + 10048.\, L4\, f^4\, C2^2\, C1\, R5\, R8 + 10048.\, L4\, f^4\, C2^2\, C1\, R7\, R6 \\
&+ 0.6221549578 10^{12}\, f^4\, C1^2\, R7^2\, C2^2\, R8^2 + 0.2488619831\, 10^{13}\, f^4\, C1^2\, R7\, C2^2\, R8\, R5\, R6 + 0.1244309916\, 10^{13}\, f^4\, C1^2\, R7\, C2^2\, R8^2\, R5 \\
&+ 0.1244309916\, 10^{13}\, f^4\, C1^2\, R7^2\, C2^2\, R8\, R6 + 0.6221549578\, 10^{12}\, f^4\, C1^2\, R5^2\, C2^2\, R6^2 + 0.1244309916\, 10^{13}\, f^4\, C1^2\, R5^2\, C2^2\, R6\, R8 \\
&+ 0.1244309916\, 10^{13}\, f^4\, C1^2\, R5\, C2^2\, R6^2\, R7 + 0.6221549578\, 10^{12}\, f^4\, C1^2\, R5^2\, C2^2\, R8^2 + 0.6221549578\, 10^{12}\, f^4\, C1^2\, R7^2\, C2^2\, R6^2 \\
&- 0.2488619811 10^{13}\, f^4\, C1^2\, L4\, R7\, C2\, R5 + 0.2453679569 10^{14}\, f^6\, C1^2\, C2^2\, R6\, L3^2\, R8 - 0.2488619811 10^{13}\, f^4\, C1\, C2^2\, R6\, L3\, R8 \\
&- 0.1244309960 10^{13}\, f^4\, C1\, C2^2\, R6^2\, L3 + 0.4907359138 10^{14}\, f^6\, C1^2\, C2^2\, R6\, L3^2\, R8 + 0.3155072000010^{11}\, f^2\, C2^2\, R8\, R6 \\
&- 0.1244309960 10^{13}\, f^4\, C2^2\, R8^2\, C1\, L3 + 0.2453679569 10^{14}\, f^6\, C1^2\, L4^2\, R5^2\, C2^2 - 0.1244309960 10^{13}\, f^4\, C1^2\, L4\, R5^2\, C2 \\
&+ 0.2453679569 10^{14}\, f^6\, C1^2\, C2^2\, R8^2\, L3^2 + 0.4000000000010^9 \big)
\end{aligned}
$$

$$\text{(1.19)}$$

Keeping as variables only the frequency and the mutual inductance, we can study the influence of these parameters on the transferred power value.

This is shown in Figures 1.8 and 1.9.

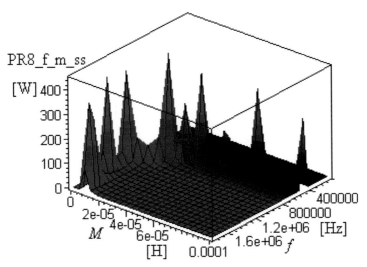

Figure 1.8 Power variation in respect of frequency and mutual inductance.

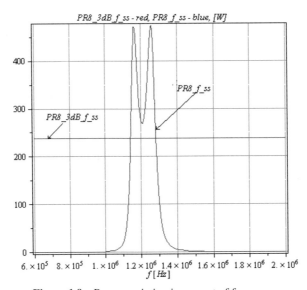

Figure 1.9 Power variation in respect of frequency.

We can see that the power has two maxima: $P_{R8\_f\,\text{max}\,1\_ss} = 472.7066$ W and $P_{R8\_f\,\text{max}\,2\_ss} = 476.7677$ W at the frequencies $f_{PR8\_\text{max}\,1\_ss} = 1.16183 \cdot 10^6$ Hz and $f_{PR8\_\text{max}\,2\_ss} = 1.25798 \cdot 10^6$ Hz, respectively, and a minimum $P_{R8\_\text{min}\_ss} = 268.9381$ W at the frequency $f_{PR8\_\text{min}\_ss} = 1.20798 \cdot 10^6$ Hz.
    This situation is called frequency splitting.
    The quality factor of the circuit is

$$Q_{c\_ss1} = \frac{f_{0\_ss}}{f_{s\_ss} - f_{i\_ss}} = \frac{1.21 \cdot 10^6}{1.28 \cdot 10^6 - 1.14 \cdot 10^6} = 8.643 \qquad (1.20)$$

where $f_{0\_ss} = 1.21 \cdot 10^6$ Hz, while f_{i_ss} and f_{s_ss} are identified by the intersection of the curve $P_{R8\_f\_ss}$ with $y = \max{(P_{R8\_f\_ss})}/2.0$ and have the values $f_{i\_ss} = 1.14$ MHz and $f_{s\_ss} = 1.28$ MHz.

1.3 Mutual Inductance Computation

In order to compute the mutual inductance for different situations occurring in practice, many formulas have been developed.
    The first and the most important, dedicated to the coaxial circles, is the formula involving elliptic integrals given by [16–18]:

$$M = 4\pi\sqrt{Rr}\left\{ \left(\frac{2}{k} - k\right)F - \frac{2}{k}E \right\} \qquad (1.21)$$

in which R and r are the radii of the two circles,

$$k = \frac{2\sqrt{Rr}}{\sqrt{(R+r)^2 + h^2}} \qquad (1.22)$$

where h is the distance between their centers. F and E are the complete elliptic integrals of the first and second kind. In general, elliptic integrals cannot be expressed in terms of the elementary functions. They can be expressed as a power series.
    Weinstein [17] gives an expression for the mutual inductance of two coaxial circles in terms of the complementary modulus k':

$$\begin{aligned} M = 4\pi\sqrt{Rr}\Bigg\{ &\left(1 + \frac{3}{4}k'^2 + \frac{33}{64}k'^4 + \frac{107}{256}k'^6 + \dots\right)\left(\log\frac{4}{k'} - 1\right) \\ &- \left(1 + \frac{15}{128}k'^4 + \frac{185}{1536}k'^6 + \dots\right)\Bigg\} \end{aligned} \qquad (1.23)$$

where $k' = \sqrt{1 - k^2}$.

Nagaoka [18] proposes formulas for the calculation of the mutual inductance of coaxial circles without using elliptic integral tables. These formulas use Jacobi's q-series [19]. The first is to be used when the circles are not too close to each other and the second one when they are close to each other. The first formula has the following expression:

$$M = 16\pi^2 \sqrt{Rr} q^{\frac{3}{2}} (1 + \varepsilon), \qquad (1.24)$$

where

$$\varepsilon = 3q^4 - 4q^6 + 9q^8 - 12q^{10} + \ldots$$

$$q = \frac{l}{2} + 2\left(\frac{l}{2}\right)^5 + 15\left(\frac{l}{2}\right)^9 + \ldots \qquad (1.25)$$

$$l = \frac{1 - \sqrt{k'}}{1 + \sqrt{k'}}$$

Nagaoka's second formula [17] has the form:

$$M = 4\pi\sqrt{Rr} \cdot \frac{1}{2(1 - 2q_1)^2} \left\{ \left[1 + 8q_1 - 8q_1^2 + \varepsilon_1\right] \log \frac{1}{q_1} - 4 \right\}$$

$$q_1 = \frac{l_1}{2} + 2\left(\frac{l_1}{2}\right)^5 + 15\left(\frac{l_1}{2}\right)^9 + \ldots \qquad (1.26)$$

$$l_1 = \frac{1 - \sqrt{k}}{1 + \sqrt{k}}$$

$$\varepsilon_1 = 32q_1^3 - 40q_1^4 + 48q_1^5 - \ldots$$

Other formulas use different mathematical tools: Havelock's formulas are based on certain definite integrals of Bessel functions, and E. Mathy's formulas work with elliptic integrals of the third kind; formulas have also been developed by Coffin, Rowland, and Rayleigh [17].

To facilitate the analytical calculation of the mutual inductances between different magnetically coupled coils, an efficient procedure based on the Neumann formula was implemented in MATLAB [20].

Let C_1 and C_2 be two circuits placed into a homogeneous medium (μ_0), with $i_1 \neq 0$ and $i_2 = 0$.

The magnetic flux through the circuit C_2 produced by the circuit C_1 is

$$\Phi_{21} = \int_{S_{C2}} B_1 \mathrm{d}A = \int_{S_{C2}} \mathrm{rot} A_1 \mathrm{d}A = \int_{C_2} A_1 \mathrm{d}l_2 \qquad (1.27)$$

and taking into account the magnetic potential vector expression

$$A_1 = \frac{\mu_0}{4\pi} \cdot i_1 \cdot \int_{C1} \frac{\mathrm{d}l_1}{R_{12}} \qquad (1.28)$$

the magnetic flux expression becomes

$$\Phi_{21} = \frac{\mu_0}{4\pi} i_1 \int_{C1} \int_{C2} \frac{\mathrm{d}l_1 \mathrm{d}l_2}{R_{12}} \qquad (1.29)$$

Thus, starting from the definition, the mutual inductance is given by the formula:

$$M = \frac{\Phi_{21}}{i_1} = \frac{\mu_0}{4\pi} \int_{C1} \int_{C2} \frac{\mathrm{d}l_1 \mathrm{d}l_2}{R_{12}} \qquad (1.30)$$

The MATLAB implementation for computing the above expressions uses two grids for each winding—one inside and another outside (Figure 1.10). In this

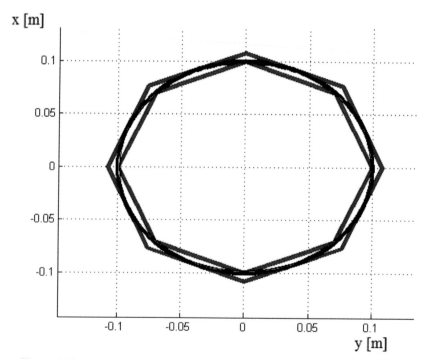

Figure 1.10   Description of the grids corresponding to each coil (Winding) [20].

manner, we can limit the mutual inductance value between a maximum and a minimum. The final value of the mutual inductance will be the average of two values, corresponding to the two grids.

For a better comparison, we have chosen from the literature the four cases described in Figure 1.11.

We denote by M_p the mutual inductivity value computed by the above procedure implemented in MATLAB and by M_t—the mutual inductivity value according to the tables for elliptic integrals [20]. The results, given as a function of the grid density, are presented in Figure 1.12.

We have also simulated the configurations in Figure 1.12 using ANSOFT Q3D EXTRACTOR program. For modeling, we used a coil with square-section wire of side 1 mm. The simulation results are shown in Figure 1.13.

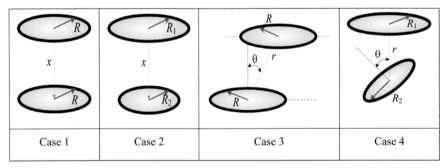

Figure 1.11 Geometry of the four cases which were studied.

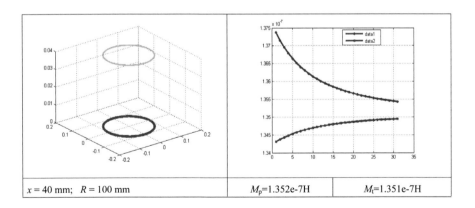

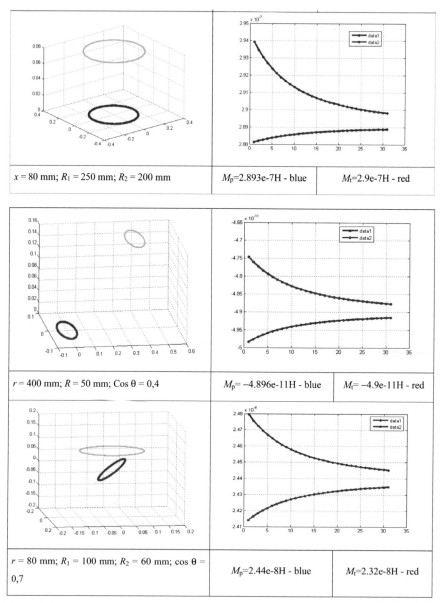

Figure 1.12   Results obtained by the integration procedure implemented in MATLAB (first column—the device geometry; second column—values of the mutual inductance for each grid type, function of the grid density).

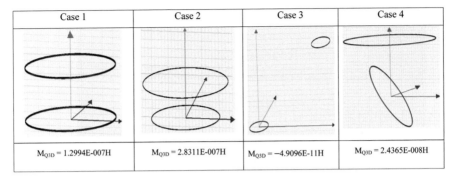

Case 1	Case 2	Case 3	Case 4
$M_{Q3D} = 1.2994E\text{-}007H$	$M_{Q3D} = 2.8311E\text{-}007H$	$M_{Q3D} = -4.9096E\text{-}11H$	$M_{Q3D} = 2.4365E\text{-}008H$

Figure 1.13 Results obtained by ANSOFT Q3D EXTRACTOR.

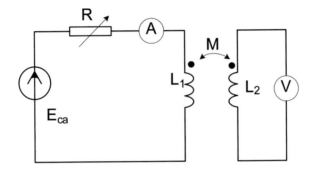

Figure 1.14 The experimental circuit scheme.

Using two handmade coils and the scheme in Figure 1.14, we have measured the mutual inductivity value $M = M_m$ in the four configurations.

The coils have 16 turns of insulated solid copper wire arranged in two rows with 8 turns in series (Figure 1.15). The geometric parameters are as follows: coil radius 100 mm, pitch 1.9 mm, and rectangular coil section 3.2 × 1.8 mm.

An initial mesh is chosen to obtain the most accurate solutions in a short time; later the program will refine the mesh to meet the convergence criteria (Figure 1.16).

The coil parameters, determined by ANSOFT Q3D EXTRACTOR simulation, are presented in Table 1.1, where

- M_{Q3D} is the mutual inductance value obtained by ANSOFT Q3D EXTRACTOR;
- L_1, L_2 are self-inductance values;

Figure 1.15 Experimental model.

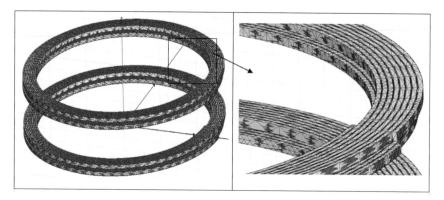

Figure 1.16 The mesh net.

Table 1.1 Results of ANSOFT Q3D EXTRACTOR simulation for 50 Hz

	L_1 (μH)	L_2 (μH)	M_{Q3D} (μH)	R_1 (mΩ)	R_2 (mΩ)	C (pF)	Number of Mesh Elements
Case 1	99.32	98.897	47.483	23.531	23.31	14.831	77,934
Case 2	99.435	98.907	14.823	22.146	21.884	9.5873	55,230
Case 3	98.467	98.246	15.734	23.401	23.664	11.312	77,752
Case 4	99.977	98.933	16.799	21.999	23.976	10.467	79,790

- R_1, R_2 are coil resistances.
- C is the capacitance ($C = C_1 \approx C_2$)

The values of the mutual inductance obtained by measurements (M_m), those computed by integration in MATLAB (M_p), and the ones obtained by ANSOFT Q3D EXTRACTOR simulation (M_{Q3D}), for different values of x, are presented in Table 1.2.

The analysis is done for a 50 Hz frequency, and the surface current density is shown in Figure 1.17.

If the ANSOFT Q3D EXTRACTOR simulations are performed at a 10 MHz frequency, the values of the self-inductances, the capacitances, and the mutual inductance remain unchanged, but the resistance values are higher due to the skin effect, as it is shown in Table 1.3.

The surface current density is represented in Figure 1.18.

Table 1.2   Experimental results and their comparison with the computed ones

Case 1	
$x = 30$ mm	$M_m = 51{,}97$e-6H
	$M_p = 50{,}33$e-6H
	$M_{Q3D} = 47.483$e-6H
Case 2	
$x = 100$ mm	$M_m = 16{,}19$e-6H
	$M_p = 15{,}92$e-6H
	$M_{Q3D} = 14.823$e-6H
Case 3	
$x = 30$ mm	$M_m = 17{,}61$e-6H
$\theta = \text{arctg}(105/30)$	$M_p = 17{,}01$e-6H
	$M_{Q3D} = 15.734$e-6H
Case 4	
$x = 100$ mm	$M_m = 14{,}73$e-6H
$\theta = 45^0$	$M_p = 16{,}52$e-6H
	$M_{Q3D} = 16.799$e-6H

Table 1.3   Results obtained by Ansoft Q3D Extractor at the 10 MHz frequency

	L_1 (μH)	L_2 (μH)	M_{Q3D} (μH)	R_1 (Ω)	R_2 (Ω)	C (pF)	Number of Mesh Elements
Case 1	99.320	98.897	47.483	10.523	10.425	14.813	77,934
Case 2	99.983	99.354	14.886	9.9168	9.8068	9.5873	77,766
Case 3	98.467	98.246	15.735	10.465	10.583	11.312	77,752
Case 4	99.977	98.933	16.799	9.8382	10.722	10.467	79,790

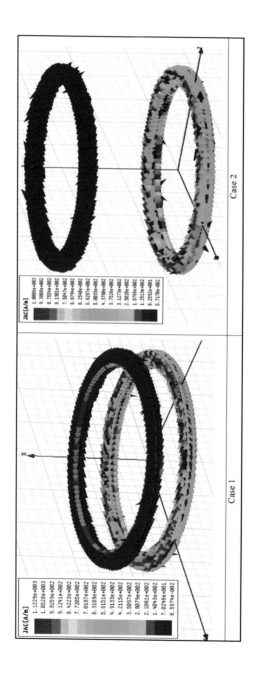

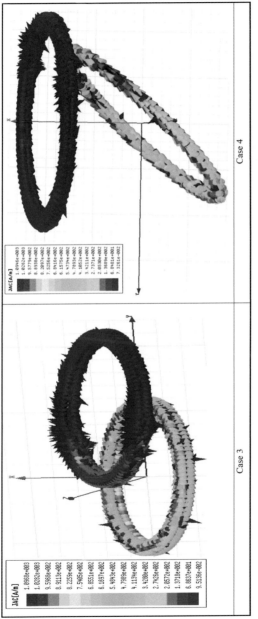

Figure 1.17   Surface current density at 50 Hz frequency.

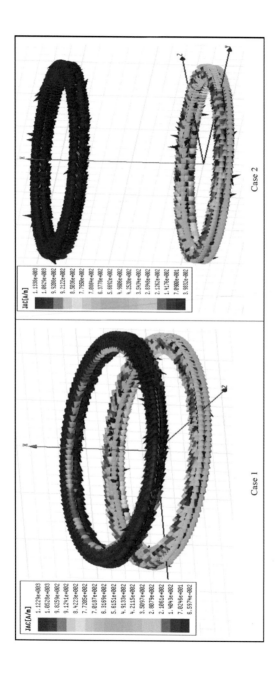

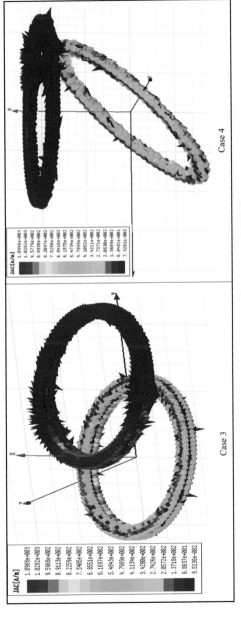

Figure 1.18   Surface current density at 10 MHz frequency.

1.4 Efficiency of the Active Power Transffer

1.4.1 Scattering Parameters $\underline{S}$

The scattering parameters $\underline{S}$ are especially useful in the evaluation of the active power transferred to the load, taking into account the frequency as well. The existence of some efficient techniques for the measurement of $\underline{S}$ parameters [vector network analyzer (VNA)] is a good reason to use them [13, 14].

In order to define the scattering parameters of a two-port structure, we consider the circuit in Figure 1.19.

We make a mathematical change of variables from the pairs $(\underline{U}_1, \underline{I}_1)$, $(\underline{U}_2, \underline{I}_2)$ to the pairs $(\underline{a}_1, \underline{b}_1)$, $(\underline{a}_2, \underline{b}_2)$, by the following relations:

$$
\begin{aligned}
\underline{U}_1 = \sqrt{Z_0}\,(\underline{a}_1 + \underline{b}_1), \quad \underline{I}_1 = \frac{1}{\sqrt{Z_0}}\,(\underline{a}_1 - \underline{b}_1), \\
\underline{U}_2 = \sqrt{Z_0}\,(\underline{a}_2 + \underline{b}_2), \quad \underline{I}_2 = \frac{1}{\sqrt{Z_0}}\,(\underline{a}_2 - \underline{b}_2)
\end{aligned}
\tag{1.31}
$$

where Z_0 is the magnitude of a positive real variable, called *reference impedance*.

By similarity with the wave equation, the solution $\underline{a}_1$ ($\underline{a}_2$) represents the incident wave to the port $i' - i''$ ($o' - o''$), and $\underline{b}_1$ ($\underline{b}_2$) is the reflected wave to the same port.

For the linear circuits, the variables associated with each port can be considered as a superposition of incident (direct) and reflected (indirect) waves.

The magnitudes of the new variables have the dimension $\sqrt{AV}$ and that is why the square of these magnitudes are powers.

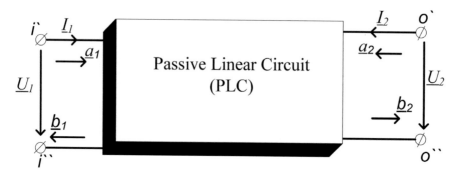

Figure 1.19 Scheme for scattering parameter definition.

Usually, the reference impedance is equal to the magnitude of the load impedance.

From Equation (1.31), it results

$$\underline{a}_1 + \underline{b}_1 = \frac{1}{\sqrt{Z_0}}\underline{U}_1, \quad \underline{a}_1 - \underline{b}_1 = \sqrt{Z_0}\underline{I}_1$$
$$\underline{a}_2 + \underline{b}_2 = \frac{1}{\sqrt{Z_0}}\underline{U}_2, \quad \underline{a}_2 - \underline{b}_2 = \sqrt{Z_0}\underline{I}_2 \tag{1.32}$$

and solving for the new variables, we get

$$\underline{a}_1 = \frac{1}{2}\left(\frac{1}{\sqrt{Z_0}}\underline{U}_1 + \sqrt{Z_0}\underline{I}_1\right) = \frac{1}{2\sqrt{Z_0}}(\underline{U}_1 + Z_0\underline{I}_1)$$
$$\underline{b}_1 = \frac{1}{2}\left(\frac{1}{\sqrt{Z_0}}\underline{U}_1 - \sqrt{Z_0}\underline{I}_1\right) = \frac{1}{2\sqrt{Z_0}}(\underline{U}_1 - Z_0\underline{I}_1) \tag{1.33}$$

and

$$\underline{a}_2 = \frac{1}{2}\left(\frac{1}{\sqrt{Z_0}}\underline{U}_2 + \sqrt{Z_0}\underline{I}_2\right) = \frac{1}{2\sqrt{Z_0}}(\underline{U}_2 + Z_0\underline{I}_2)$$
$$\underline{b}_2 = \frac{1}{2}\left(\frac{1}{\sqrt{Z_0}}\underline{U}_2 - \sqrt{Z_0}\underline{I}_2\right) = \frac{1}{2\sqrt{Z_0}}(\underline{U}2 - Z_0\underline{I}_2) \tag{1.34}$$

The scattering parameters $\underline{S}$ of a two-port structure satisfy the following equations between the incident signals and the reflected ones:

$$\underline{b}_1 = \underline{S}_{11}\underline{a}_1 + \underline{S}_{12}\underline{a}_2$$
$$\underline{b}_2 = \underline{S}_{21}\underline{a}_1 + \underline{S}_{22}\underline{a}_2 \tag{1.35}$$

They are dimensionless quantities.

The quantity $\underline{S}_{11} = \frac{\underline{b}_1}{\underline{a}_1}\Big|_{\underline{a}_2=0}$ defines the scattering coefficient at the port 1 ($i' - i''$). The condition $\underline{a}_2 = 0$ is realized when an adapted load is connected to the port 2 ($o' - o''$), which absorbs the signal $\underline{b}_2$ and does not generate any signal. Obviously, from 1 to 34, the condition $\underline{a}_2 = 0$ implies $\underline{U}_2 = -Z_0\underline{I}_2$. On the other hand, if at the port 2, the load impedance $\underline{Z}_L$ is connected, then $\underline{U}_2 = -Z_L\underline{I}_2$. Consequently, $\underline{a}_2 = 0$ implies $\underline{Z}_L = Z_0$, meaning a purely resistive load, the resistance being equal to the magnitude of the reference impedance.

$|\underline{S}_{11}|^2$ is the scattering coefficient at the port 1.

The quantity $\underline{S}_{21} = \frac{\underline{b}_2}{\underline{a}_1}\Big|_{\underline{a}_2=0}$ defines the transfer coefficient from the port 1 to the port 2. It is in correlation with the power–frequency dependence and with the efficiency of the power transfer from the first circuit to the second one.

$|\underline{S}_{21}| < 1$, and $|\underline{S}_{21}|^2$ is the transfer coefficient from the port 1 to the port 2.

For a two-port structure (Figure 1.19) with two resonators magnetically coupled in harmonic behavior, the $\underline{S}$ parameters are computed as follows:

- **The parameter $\underline{S}_{11}$ is**

$$
\underline{S}_{11} = \frac{b_1}{a_1}\Big|_{a_2 = 0 \Leftrightarrow \underline{U}_2 = -Z_c \underline{I}_2} = \frac{\underline{U}_1 - Z_c \underline{I}_1}{\underline{U}_1 + Z_c \underline{I}_1}\Big|_{a_2 = 0 \Leftrightarrow \underline{U}_2 = -Z_c \underline{I}_2}
$$

$$
\underline{Z}_l = Z_c, \underline{U}_1 = \underline{E}_i - \underline{Z}_c \underline{I}_1 \qquad\qquad \underline{Z}_i = Z_c, \underline{U}_1 = \underline{E}_i - \underline{Z}_c \underline{I}_1
$$

$$
= \frac{\underline{E}_i - 2Z_c \underline{I}_1}{\underline{E}_i} = 1 - 2\underline{A}_{1i}
$$

$$
(1.36)
$$

where $\underline{A}_{1i} = \dfrac{Z_c \underline{I}_1}{\underline{E}_i}\Big|_{\underline{U}_2 = -Z_c \underline{I}_2,\ \underline{Z}_i = Z_c}$ is the voltage gain;

- **The parameter $\underline{S}_{12}$ is**

$$
\underline{S}_{12} = \frac{b_1}{a_2}\Big|_{\substack{a_1 = 0 \Leftrightarrow \underline{U}_1 = -Z_c \underline{I}_1 \\ \underline{Z}_i = Z_c, \underline{Z}_o = Z_c}}
$$

$$
= \frac{\underline{U}_1 - Z_c \underline{I}_1}{\underline{U}_2 + Z_c \underline{I}_2}\Big|_{\substack{a_1 = 0 \Leftrightarrow \underline{U}_1 = -Z_c \underline{I}_1 \\ \underline{Z}_i = Z_c, \underline{Z}_o = Z_c}}
$$

$$
= \frac{-2Z_c \underline{I}_1}{\underline{E}_o} = -2\underline{A}_{io} \qquad\qquad (1.37)
$$

where $\underline{A}_{io} = \dfrac{Z_c \underline{I}_1}{\underline{E}_o}\Big|_{\underline{U}_1 = -Z_c \underline{I}_1,\ \underline{Z}_o = Z_c}$ is the voltage gain, computed if an impedance Z_c is connected to the input port $i' - i''$ ($\underline{E}_i$ being zero) and an impedance $\underline{Z}_o = Z_c$ is connected to the output port $o' - o''$ in series with $\underline{E}_o$;

- **The parameter $\underline{S}_{21}$ is**

$$
\underline{S}_{21} = \frac{b_2}{a_1}\Big|_{a_2 = 0 \Leftrightarrow \underline{U}_2 = -Z_c \underline{I}_2}
$$

$$
\underline{Z}_i = Z_c, \underline{Z}_l = Z_c, \underline{E}_i = \underline{U}_1 + Z_c \underline{I}_1
$$

$$= \frac{U_2 - Z_c I_2}{U_1 + Z_c I_1} \bigg|_{\underline{a}_1 = 0} \Leftrightarrow \underline{U}_2 = -Z_c \underline{I}_2$$

$$\underline{Z}_i = Z_c, \underline{Z}_l = Z_c, \underline{E}_i = \underline{U}_1 + Z_c \underline{I}_1 \qquad (1.38)$$

$$= \frac{-2 Z_c \underline{I}_2}{\underline{E}_i} = -2 \underline{A}_{oi}$$

where $\underline{A}_{oi} = \frac{Z_c \underline{I}_2}{\underline{E}_i} \bigg|_{\underline{U}_2 = -Z_c \underline{I}_2, \underline{Z}_i = Z_c}$ is the voltage gain;

- **The parameter $\underline{S}_{22}$ is**

$$\underline{S}_{22} = \frac{b_2}{a_2} \bigg|_{\underline{a}_1 = 0} \Leftrightarrow \underline{U}_1 = -Z_c \underline{I}_1$$

$$\underline{Z}_i = Z_c, \underline{Z}_o = Z_c, \underline{U}_2 = \underline{E}_o - Z_c \underline{I}_2$$

$$= \frac{U_2 - Z_c \underline{I}_2}{U_2 + Z_c \underline{I}_2} \bigg|_{\underline{a}_1 = 0 \Leftrightarrow \underline{U}_1 = -Z_c \underline{I}_1} \qquad (1.39)$$

$$\underline{Z}_i = Z_c, \underline{Z}_o = Z_c, \underline{U}_2 = \underline{E}_o - Z_c \underline{I}_2$$

$$= \frac{\underline{E}_0 - 2 Z_c \underline{I}_2}{\underline{E}_o} = 1 - 2 \underline{A}_{2o}$$

where $\underline{A}_{2o} = \frac{Z_c \underline{I}_2}{\underline{E}_o} \bigg|_{\underline{U}_1 = -Z_c \underline{I}_1, \underline{Z}_o = Z_c}$ is the voltage gain, computed if an impedance Z_c is connected to the input port $i' - i''$ ($\underline{E}_i$ being zero) and an impedance $\underline{Z}_o = Z_c$ is connected to the output port $o' - o''$ in series with $\underline{E}_o$. The reflection $\underline{S}_{11}$ and the transmission $\underline{S}_{21}$ can be measured by the VNA [13].

1.4.2 Efficiency Computation

The most important scattering parameters are the reflection $\underline{S}_{11}$ and the transmission $\underline{S}_{21}$, because the ratio of power reflection is $\eta_{11} \stackrel{d}{=} S_{11}^2 \times 100$ (%), while the ratio of power transmission is $\eta_{21} \stackrel{d}{=} S_{21}^2 \times 100 (\%)$ [14, 15].

The transmission $\underline{S}_{21}$ is related to the efficiency of the power transfer according to the relation

$$\eta_{21\_S21\_f\_ss} = \underline{S}_{21} \cdot \underline{S}_{21}^* \times 100 \qquad (1.40)$$

The variation of the efficiency η_{21} with respect to the frequency and the mutual inductance for the circuit in Figure 1.19 is represented in Figure 1.20, while in Figure 1.21, we can see the variation of η_{21} with respect to the frequency.

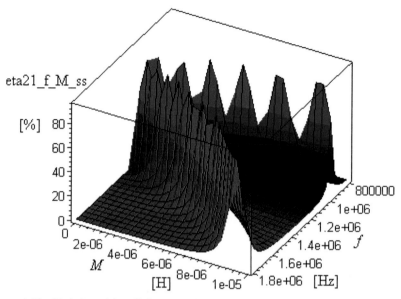

Figure 1.20   Variation of the efficiency η_{21} in respect of frequency and mutual inductance.

The efficiency $\eta_{21\_sv\_ss}$ of the power transferred from the resonator one to the resonator two can be computed in CT using the relation:

$$\eta_{21\_sv\_ss} - \frac{P_{R8\_f\_ss}}{P_{Rit\_f\_ss} + P_{Rlt\_f\_ss}} \times 100 \qquad (1.41)$$

where $P_{R8\_f\_ss}$, $P_{Rit\_f\_ss}$, and $P_{Rlt\_f\_ss}$ are the powers dissipated in the resistor R_8, respectively, $(R_5 + R_7)$ and $(R_6 + R_8)$.

The variation of the two quantities computed with formulas Equations (1.40) and (1.41) is represented in Figure 1.22. One can notice the difference between the values at the resonance frequency. The efficiency computed by the mean of the powers is significantly higher than the one computed by the transmission coefficient.

The magnitude variations of the reflection $\underline{S}_{11}$ and the transmission $\underline{S}_{21}$ with respect to the frequency, for different distances (g) between the two magnetic-coupled coils (i.e., different mutual inductances M), are represented in Figure 1.23a–f. For all these cases, the values of M_p (computed by MATLAB implementation of the integration procedure presented in Section 1.3) and M_{Q3D} (obtained by ANSOFT Q3D EXTRACTOR simulations) are given in the mentioned figures.

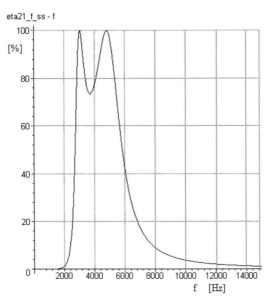

Figure 1.21 Variation of the efficiency η_{21} in respect of the frequency.

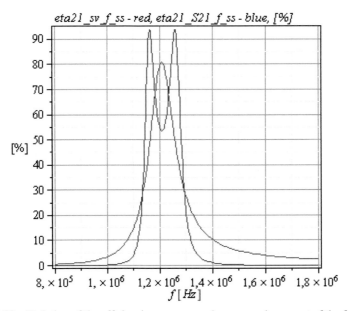

Figure 1.22 Variation of the efficiencies $\eta_{21\_sv\_f\_ss}$ and $\eta_{21\_S21\_f\_ss}$ in respect of the frequency.

$S11\_f\_ss\_P$ - *blue-l*, $S11\_f\_ss\_Q3D$ - *blue-p*, $S21\_f\_ss\_P$ - *black-l*,
$S21\_f\_ss\_Q3D$ - *black-p*, g = 100 mm, [%]

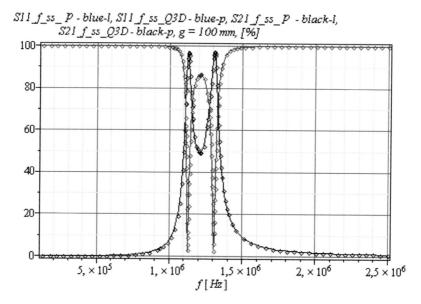

(a) $g = 100$ mm, $M_p = 2.5$ µH, $k_p = 0.14933$, $M_{Q3D} = 2.468$ µH, $k_{Q3D} = 0.14742$

g = 150 mm, [%]

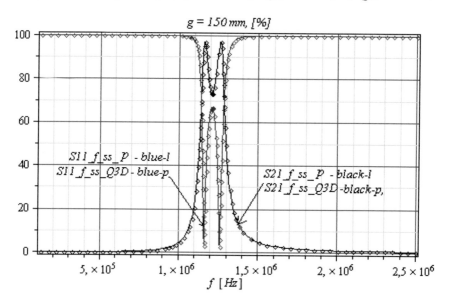

(b) $g = 150$ mm, $M_p = 1.5$ µH, $k_p = 0.094376$, $M_{Q3D} = 4.4898$ µH,
$k_{Q3D} = 0.088988$

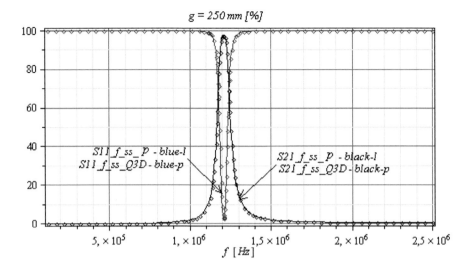

(c) g = 250 mm, M_p = 0.65 µH, k_p = 0.038825, M_{Q3D} = 0.64191 µH, k_{Q3D}= 0.03834

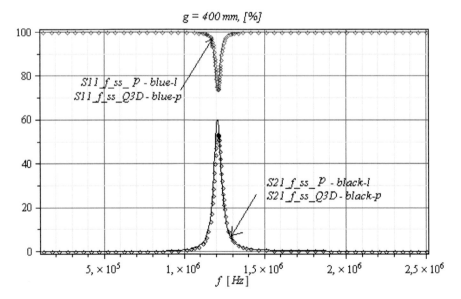

(d) g = 400 mm, M_p = 0.24 µH, k_p = 0.014335, M_{Q3D} = 0.23558 µH, k_{Q3D} = 0.01407

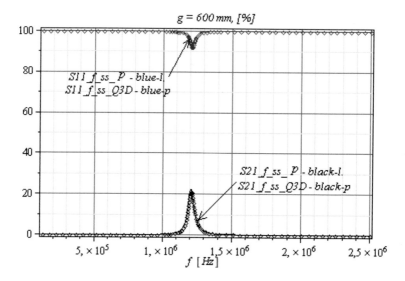

(e) $g = 600$ mm, $M_p = 0.088$ µH, $k_p = 0.005256$, $M_{Q3D} = 0.08279$ µH, $k_{Q3D} = 0.004945$

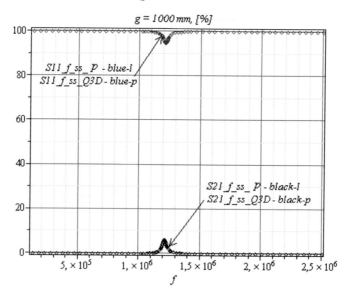

(f) $g = 1000$ mm, $M_p = 0.022$ µH, $k_p = 0.001314$, $M_{Q3D} = 0.0203$ µH, $k_{Q3D} = 0.001215$

Figure 1.23   Magnitude variations of the reflection $\underline{S}_{11}$ and the transmission $\underline{S}_{21}$.

1.5 Some Procedures for Optimal Wireless Energy Transfer Systems

In this section a set of constraints applied to a wireless power-supply system (emitter and receiver) for operating in optimal conditions is described. These constraints regard both geometrical aspects and operational conditions of the system. The performance optimization in the power wireless transfer is also presented. Starting from the equivalent scheme of a wireless power transfer system sinusoidal behavior, we generate the system performances (the delivered active power or the system efficiency) in full symbolic and/or symbolic-numeric form, obtaining an appropriate frequency space representation based on the complex or Laplace modified nodal equations (MNE) or/and Laplace state variable equations (SVE). In order to obtain the optimal values of the system parameters, which provide the maximum performances, it is used the MATLAB procedures which minimize these objective functions. These equations are combined with some measurements performed on the real system and an unconstrained minimization algorithm for some scalar and/or vector functions of multiple variables provided by MATLAB Optimization Toolbox. The algorithm is suitable to compute optimal circuit parameters which guarantee the minimum and maximum values of the performance quantity. Linear or small-signal nonlinear circuits can be treated in this manner. This section has also highlighted some aspects regarding the conditions that have to be accomplishing by a power supply system that uses wireless technologies, for an optimal transfer of energy. The proposed optimization techniques was tested and validated with simulation data on some illustrative examples.

1.5.1 Indroduction

In the last decade, advances in wireless communication and semiconductor technology have produced a wide variety of portable consumer electronic, medical and industrial devices. However, the mobility degree of these devices is strong relied on how often you have to manually plug in for recharging their batteries. Furthermore, as portable devices shrink, connectors become a larger fraction of system size. Thus, a growing interest of researchers, focused on implementation of wireless technologies in batteries recharging, was emerged. Wireless power offers the possibility of connector-free electronic devices, which could improve size and reliability, [11, 22–24, 30–34].

Radiative transfer, although perfectly suitable for transferring information, poses a number of difficulties for power transfer applications: the efficiency of power transfer is very low if the radiation is omni-directional, and unidirectional radiation requires an uninterrupted line of sight and sophisticated tracking mechanisms [11, 24–31].

In recent papers, various researchers have been trying to transfer energy using wireless technologies like: laser beam, piezoelectric principle, radio waves and microwaves, inductive coupling [22]. Based on last principle, the technology Witricity (from WIreless elecTRICITY) provides large air gaps and large amounts of power. The physical principle behind the Witricity concept is based on the *near-field* in correlation to the resonant inductive coupling.

The *coupling through magnetic resonance* implies that the coupled systems work at their resonance frequency.

In practice it is generally difficult to set-up the circuit parameters of two magnetic coupled resonators to obtain the desired performances. For a pair of magnetic coupled resonators, the power transfer efficiency is the high interest as performance quantity. For this reason, the section is focused on a novel optimization technique of two magnetic coupled resonators where the power transfer efficiency is chosen as objective function.

This purpose imposes exploiting parameter estimation techniques based on iterative computation [34]. The success of these techniques requires appropriate mathematical models and it depends strongly on the starting point of the iteration process.

To minimize or maximize scalar objective functions with multiple variables we propose using an unconstrained nonlinear optimization algorithm function *fminunc* provided by the Optimization Toolbox of MATLAB [34, 35].

This MATLAB function covers a wide spectrum of methods for unconstrained optimization, depending on the chosen computation options. The simplest are the search methods that use only function evaluations, being most suitable for problems that are not smooth or have discontinuities. On the other hand, gradient methods use information about the slope of the function to establish the direction of the search. Higher order methods, such as Newton's method, are most suitable when the second-order derivatives of the objective function, as the Hessian matrix, can be calculated with decent effort [13, 26–34].

1.5.2 Optimal Parameter Computing Performance Optimization of Magnetic Coupled Resonators

We use some algorithms developed to analyze any analog circuit in full symbolic form starting from the equivalent diagram of a linear or linearized lumped circuit [33]. The network function in the Laplace domain $H(s)$ allows expressing the complex network function in the frequency domain $\underline{H}(j\omega)$ or $\underline{H}(f)$, if necessary. Their coefficients depend on the circuit parameters and on the complex or real frequency.

Assuming that one or more circuit parameters are unknown and need to be estimated, a wide spectrum of algorithms based on network functions can be exploited to solve such a problem. Independently of the used algorithm, the magnitude and the phase of the complex network function must be known at one or at more working frequencies. These parameters can be measured directly by supplying the circuit with a variable frequency sinusoidal voltage.

Assuming that the system parameters which follow to be optimized and/or estimated are: x_1, x_2, ..., x_p (p being the number of unknown parameters), others $n - p$ parameters have the nominal (catalog) values. We consider k frequency samples at the network function (the system performance) are measured (simulated).

The following objective function is considered:

$$f_j\left(x_1, x_2, ..., x_p, f_j\right) = \left(|\underline{H}\left(f_j\right)| - |\underline{H}\left(f_j, x_1, x_2, ..., x_p\right)|\right)^2, \; j = \overline{1, k},$$

(1.42)

In which the objective function is a vector having k components or

$$f\left(x_1, x_2, ..., x_p\right) = \sum_{j=1}^{k}\left(|\underline{H}\left(f_j\right)| - |\underline{H}\left(f_j, x_1, x_2, ..., x_p\right)|\right)^2,$$

(1.43)

when the objective function is a scalar. In this case, k can be equal to 1 or great than 1.

If the objective function is the load active power $P_2 = P_L$ or the transfer efficiency η_{21}, $\eta_{21} = \frac{P_2}{P_1} \cdot 100$, the objective function has the structure:

$$f\left(x_1, x_2, ..., x_p\right) = \sum_{j=1}^{k} P_2\left(x_1, x_2, ..., x_p, f_j\right).$$

(1.44)

respectively,

$$f\left(x_1, x_2, ..., x_p\right) = \sum_{j=1}^{k} \eta_{21}\left(x_1, x_2, ..., x_p, f_j\right). \tag{1.45}$$

Let us consider two series-series resonators inductively coupled as in Figure 1.24, where $L_3 = L_1$ and $L_4 = L_2$ are two coaxial identical coils.

The parameters of the coils are: the radius $r = 150$ mm, the pitch $p = 3$ mm, the wire size $w = 2$ mm, the distance between the coils $g = 150$ mm, and the number of the turns $N = 5$. Using the Q3D EXTRACTOR program [37], we get the following numerical values for the parameters of the system of the two inductively coupled coils (see Figure 1.2): $C_1 = 13.8974$ pF; $C_2 = 13.8915$ pF; $C_{10} = 27.51354$ pF; $L_1 = 16.748$ μH; $L_2 = 16.735$ μH; $M = 1.4899$ μH; $R_1 = R_{L1} = 1.352$ Ω and $R_2 = R_{L2} = 1.3524$ Ω.

The source circuit has the resistance $R_7 = R_i = 5.0$ Ω, and the load is $R_8 = 5.0$ Ω.

We want to find some optimal resonator parameters which provide the minimum and maximum values of the power transfer efficiency η_{21}. The study will be focused on the parameters: L_1, L_2, M, R_1, and R_2.

To generate the power transfer efficiency η_{21} in full symbolic form we use the SYMNAP – SYmbolic Modified Nodal Analysis Program [35], based on

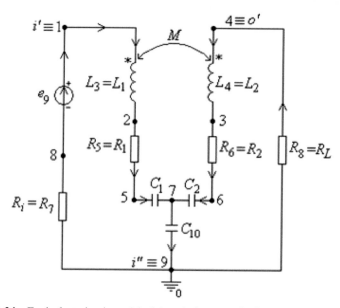

Figure 1.24 Equivalent circuit model of the wireless transfer for electromagnetic energy.

the complex or Laplace modified nodal equations. Running the SYMNAP the following expression for the power transfer efficiency is obtained:

$$
\begin{aligned}
\eta_{21} = {} & 100.^*C2^\wedge 2^*(1. - 2.^*\omega^\wedge 2^*C10^*M + \omega^\wedge 4^*C10^\wedge 2^*M^\wedge 2) \\
& ^*RL/((C10^\wedge 2^*C2^\wedge 2^*Rl^*M^\wedge 2 + C10^\wedge 2^*L2^\wedge 2^*C2^\wedge 2^*Ri \\
& + C10^\wedge 2^*C2^\wedge 2^*R2^*M^\wedge 2 + C10^\wedge 2^*L2^\wedge 2^*C2^\wedge 2^*R1) \\
& ^*\omega^\wedge 4 + (2.^*C10^\wedge 2^*C2^\wedge 2^*RL^*Ri^*R22.^*C10^*C2^\wedge 2 \\
& ^*RL^*M + C10^\wedge 2^*C2^\wedge 2^*R1^*RL^\wedge 2 + C10^\wedge 2^*C2^\wedge 2 \\
& ^*R1^*R2^\wedge 2 + C10^\wedge 2^*C2^\wedge 2^*Ri^*R2^\wedge 2 + C10^\wedge 2^*C2^\wedge 2 \\
& ^*Ri^*RL^\wedge 2 - 2.^*C10^*C2^\wedge 2^*R2^*M + 2.^*C10^\wedge 2^*C2^\wedge 2 \\
& ^*RL^*R1^*R2 - 2.^*C10^*L2^*C2^\wedge 2^*Ri - 2.^*C10^\wedge 2 \\
& ^*L2^*Ri^*C2 - 2.^*C10^\wedge 2^*L2^*R1^*C2 - 2.^*C10^*L2 \\
& ^*C2^\wedge 2^*R1)^*\omega^\wedge 2 + C10^\wedge 2^*Ri + C10^\wedge 2^*R1 + R2^*C2^\wedge 2 + R1 \\
& ^*C2^\wedge 2 + RL^*C2^\wedge 2 + 2.^*C10^*Ri^*C2 + 2.^*C10^*R1^*C2 \\
& + Ri^*C2^\wedge 2);
\end{aligned}
\tag{1.46}
$$

With the frequency of 12.46 MHz (identical to the resonance frequency), at which the efficiency has the maximum value (when the resonator parameters have the nominal values), we obtain the objective function for the procedure *fminunc* used to determine the minimum and maximum values of the power transfer efficiency.

The initial variation intervals for the five parameters are:

$$
\begin{aligned}
& L_{1\_0} \in [1.6e - 05, 2.0e - 05], \, L_{2\_0} \in [1.6e - 05, 2.0e - 05], \\
& M_{\_0} \in [1.4e - 06, 1.8e - 06], \, R_{1\_0} \in [1.0, 3.0], \\
& \text{and} \quad R_{2\_0} \in [1.0, 3.0]
\end{aligned}
\tag{1.47}
$$

Running the program with the initial variation intervals Equation (1.47), we obtained the following results:

1. Optimal (maximum) parameters: L_1, L_2, M, R_1, and R_2 for the maximum value of the efficiency η_{21}:

$$
\begin{aligned}
& \eta_{21\_max} = 83.3146[\%]; \, L_{1\_max} = 0.00002[\text{H}]; \\
& L_{1\_max} = 0.0000177[\text{H}]; \, M_{\_max} = 0.00000118[\text{H}]; \\
& R_{1\_max} = 1.99999[\Omega]; \, R_{2\_max} = 0.99999[\Omega]; \\
& k_{\_max} = 0.06272.
\end{aligned}
\tag{1.48}
$$

2. Optimal (minimum) parameters: L_1, L_2, M, R_1, and R_2 for the minimum value of the efficiency η_{21}:

$$\eta_{21\_min} = 62.471376[\%]; L_{1\_min} = 1.8e - 005[H];$$
$$L_{2\_min} = 1.7680161e - 005[H]; M_{\_min} = 1.57627e - 006[H];$$
$$R_{1\_min} = 2.99999[\Omega]; R_{1\_min} = 2.99999[\Omega];$$
$$R_{2\_min} = 2.99999[\Omega]; k_{\_min} = 0.08836. \tag{1.49}$$

Substituting the five parameter values from Equations (1.48) and (1.40) in the expression (1.46) we obtain the same maximum and minimum values for the efficiency η_{21} as ones in Equations (1.48) and (1.49).

In Figure 1.25 (Figure 1.26) the transfer active power P_L and the transfer efficiency η_{21} vs frequency for the optimal parameter values L_1, L_2, M, R_1, and R_2 which maximize (minimize) the transfer efficiency are presented.

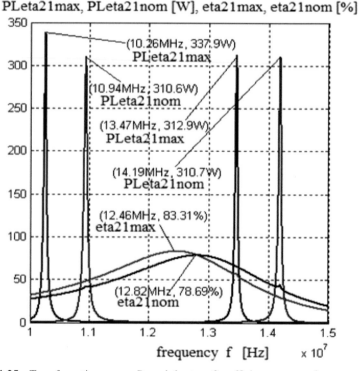

Figure 1.25   Transfer active power P_L and the transfer efficiency η_{21} vs frequency for the optimal parameter values L_1, L_2, M, R_1, and R_2 which maximize the transfer efficiency.

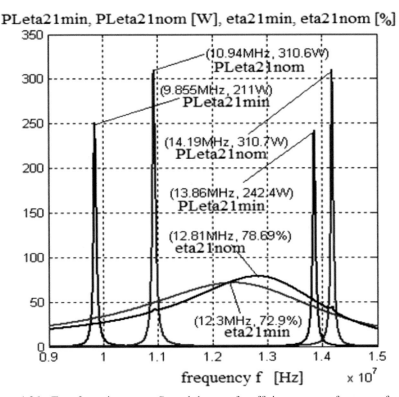

Figure 1.26 Transfer active power P_L and the transfer efficiency η_{21} vs frequency for the optimal parameter values L_1, L_2, M, R_1, and R_2 which minimize the transfer efficiency.

If we take into account the parameters: C_1, C_2, C_{10}, R_i, and R_L, and we consider the following initial variation intervals for the five parameters:

$$C_{1\_0} \in [1.0e - 11, 2.0e - 11], C_{2\_0} \in [1.0e - 11, 2.0e - 11],$$
$$C_{10\_0} \in [1.5e - 11, 3.0e - 11], R_{i\_0} \in [4.0, 8.0] \text{ and } R_{L\_0} \in [4.0, 8.0],$$

$$(1.50)$$

and running the program with these initial variation intervals, we obtained the following results: optimal (minimum) parameters: C_1, C_2, C_{10}, R_i, and R_l for the minimum value of the efficiency η_{21}:

- Optimal (maximum) parameters: C_1, C_2, C_{10}, R_i, and R_L for the maximum value of the efficiency η_{21}:

$$\eta_{21\_max} = 85.4915[\%]; C_{1\_max} = 1.8e - 011[F];$$
$$C_{2\_max} = 1.4e - 011[F]; C_{10\_max} = 2.7e - 011[F];$$
$$R_{i\_max} = 7.0[\Omega] \text{ and } R_{L\_max} = 8.0[\Omega]. \tag{1.51}$$

- Optimal (minimum) parameters: C_1, C_2, C_{10}, R_i, and R_L for the minimum value of the efficiency η_{21}:

$$\eta_{21\_min} = 32.1[\%]; C_{1\_min} = 2.0e - 011[F];$$
$$C_{2\_min} = 2.0e - 011[F]; \tag{1.52}$$

In Figure 1.27 (Figure 1.28) the transfer active power P_L and the transfer efficiency η_{21} vs frequency for the optimal parameter values L_1, L_2, M, R_1, and R_2 which maximize (minimize) the transfer efficiency are shown.

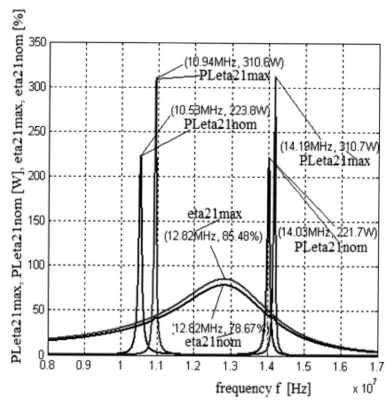

Figure 1.27   Transfer Active power P_L and the transfer efficiency η_{21} vs frequency for the optimal parameter values C_1, C_2, C_{10}, R_i, and R_L which maximize the transfer efficiency.

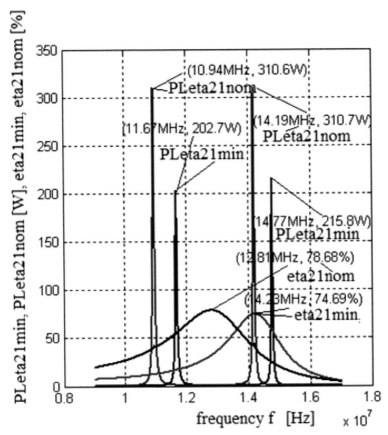

Figure 1.28   Transfer active power P_L and the transfer efficiency η_{21} vs frequency for the optimal parameter values C_1, C_2, C_{10}, R_i, and R_L which minimize the transfer efficiency.

Substituting the five parameter values from Equations (1.51) and (1.52) in the expression (1.46) we obtain the same maximum and minimum values for the efficiency η_{21} as ones in Equations (1.51) and (1.52).

Other method for the improving of the power transfer efficiency consist in a system of two series-series resonators driven by a voltage source having three magnetic couplings, as in Figure 1.29, [22].

The magnetic couplings k_{12} and k_{34} are very strong and the magnetic coupling k_{23} is weakly. In this way, for some numeric values for the system parameters (the ones above), it is obtained efficiency bigger than the one obtained by a system with only two magnetic couplings (see Figure 1.30).

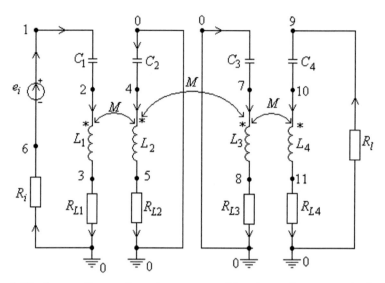

Figure 1.29   System of two series-series resonators driven by a voltage source having three magnetic couplings.

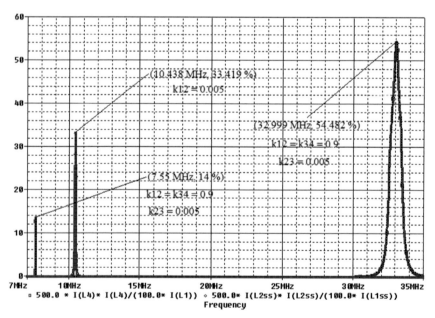

Figure 1.30   Transfer efficiency variation vs frequency.

A parameter very important, in order to maximize the power transfer efficiency, is the mutual inductance. We have implemented in MATLAB a very efficient procedure that determines the mutual inductance between two circular coils with different geometrical parameter and relative positions based on Neumann (see relation Equation (1.30)), [22, 23, 28].

If we keep the mutual inductance constant while the receiver is drift away from the emitter, we have to enlarge the receiver, growing the radius of the second coil following the field lines of the emitter coil (see Figure 1.31b).

In practice, the two devices (emitter and receiver) are not always in parallel planes (Figure 1.31a). For distances between the two coils smaller than their diameter (the radius is 50 cm) a variation about 40 degrees of the angle α improves the value of the mutual inductance (see Figure 1.32, and Figure 1.33a).

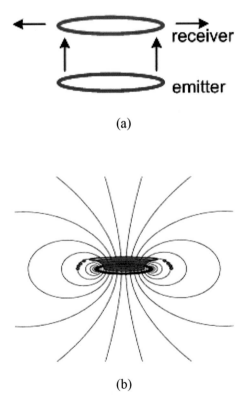

(a)

(b)

Figure 1.31   The receiver dimensions related to a constant value of the mutual inductance.

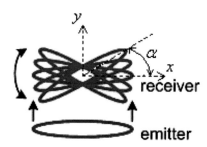

Figure 1.32   The receiver positions related to the emitter coil.

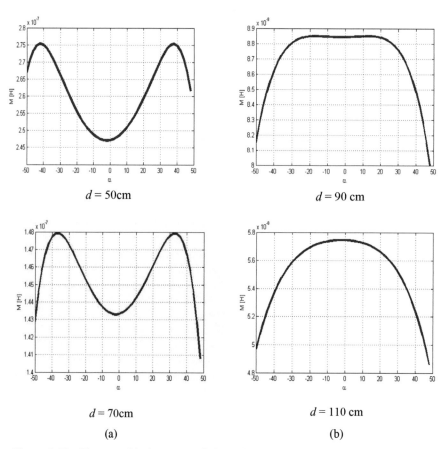

Figure 1.33   The mutual inductance variation related to the angle between the two coils.

1.5.3 Remarks

For distances bigger than the diameter, in order to obtain a maximum value of the mutual inductance, the two coils have to be in parallel planes (Figure 1.33b);

The paper highlights some aspects regarding the conditions that have to be satisfied by a wireless system for an optimal transfer of energy.

Using the facilities provided by the Optimization Toolbox of MATLAB and taking as objective function the power transfer efficiency, the circuit parameters which affects this performance were identified and the values which maximize the efficiency were obtained;

The two parameters which control the amount of the transferred power by magnetic coupling and the transfer efficiency as well are the frequency and the coupling coefficient. The analysis proves that a more efficient power transfer is obtained in a wireless power system with three magnetic couplings: two strong couplings in the source, respectively the load circuits and one weak coupling on the wireless power transfer path;

The simulations show that on the power supply side, the working frequency of the emitter has to be near the resonance frequency of the two magnetic coupled circuits and on the receiver side, the geometrical size and the relative position (defined by the angle α) are the most significant parameters for an efficient transfer of energy using wireless technologies.

1.6 Conclusions

The study realized in this chapter allows us to make the following remarks:

1. Wireless power transfer is a new technology, useful when electrical energy is needed but interconnecting wires are, for certain reasons, impractical or impossible; the transfer can be made over distances at which the electromagnetic field is strong enough to allow a reasonable power transfer. This is possible if both the emitter and the receiver achieve magnetic resonance, because the resonant objects exchange energy more efficiently than the non-resonant ones;
2. The modern applications in telecommunications (information transfer) are based on the propagation of electromagnetic waves, but the antenna radiation technology is not suitable for power transfer because of the weak efficiency (a vast majority of the energy is wasted by dispersion into free space);
3. The principle behind the information transfer can be described either by Maxwell's equations or by the Coupled-Mode Theory (CMT). In wireless

power transfer, because *RLC* resonators are used, the Circuit Theory can be successfully applied;

4. The main advantage of CMT is reducing the number of differential equations which describe the circuit behaviour by a half. In the case of cascade resonators this involves a significant decrease in the computational effort.

5. The assumptions made in CMT regarding

   - the frequency interval on which the study is performed, which has to be small enough so that the coefficients which arise in the mode amplitude equations are constant (independent of frequency)
   - the coupling factor k, which needs to be small enough so that the resonance frequency splitting doesn't occur, have to be considered when one intends to apply this analysis.

6. The power transferred on the load resistance can't be accurately computed by means of the mode amplitudes;

7. CT offers an important analysis tool for linear or nonlinear circuits which is the state variable approach [21]. The analysis performed by this method gives exact information about the power transfer, which is the great advantage of the Circuit Theory by comparison with Couple-mode Theory;

8. Using a suitable symbolic simulator [16], we can get useful information for optimizing the design of the wireless power transfer system;

9. The efficiency of the wireless transfer power deeply depends on the resonator parameters (self-inductances, mutual inductances, capacitances, and ohmic resistances of the two magnetically-coupled coils). Consequently, parameter identification is a very important objective in the automatic design of such a system. In this chapter an efficient procedure to compute the mutual inductance is developed and the results of its implementation in MATLAB are compared with those obtained by using the well performing program ANSOFT Q3D EXTRACTOR. We note the agreement between the two;

10. The measurements made using a device with two copper coils in various positions give similar results. The differences between the measured values and the ones computed could be explained by apparatus accuracy and by the fact that the integration procedure neglects the repartition of coil turns, considering a single medium turn;

11. The scattering parameters of a two-port were defined and the power transfer efficiency was studied in terms of the transmission parameter. Significant differences between the efficiency values computed in CMT

and CT, respectively, appear in the graphical representation with respect to the frequency;

12. Using the facilities provided by the Optimization Toolbox of MATLAB and taking as objective function the power transfer efficiency, the circuit parameters which affects this performance were identified and the values which maximize the efficiency were obtained;

13. The two parameters which control the amount of the transferred power by magnetic coupling and the transfer efficiency as well are the frequency and the coupling coefficient. The analysis proves that a more efficient power transfer is obtained in a wireless power system with three magnetic couplings: two strong couplings in the source, respectively the load circuits and one weak coupling on the wireless power transfer path;

14. The simulations show that on the power supply side, the working frequency of the emitter has to be near the resonance frequency of the two magnetic coupled circuits and on the receiver side, the geometrical size and the relative position (defined by the angle α) are the most significant parameters for an efficient transfer of energy using wireless technologies.

15. The efficiency of the wireless transfer power deeply depends on the resonator parameters (self inductances, mutual inductances, capacitances, and ohmic resistances of the two magnetically-coupled coils). Consequently, parameter identification is a very important objective in the automatic design of such a system. In this chapter an efficient procedure to compute the mutual inductance is developed and the results of its implementation in MATLAB are compared with those obtained by using the well performing program ANSOFT Q3D EXTRACTOR. We note the agreement between the two;

16. The measurements made using a device with two copper coils in various positions give similar results. The differences between the measured values and the ones computed could be explained by apparatus accuracy and by the fact that the integration procedure neglects the repartition of coil turns, considering a single medium turn;

17. The scattering parameters of a two-port were defined and the power transfer efficiency was studied in terms of the transmission parameter. Significant differences between the efficiency values computed in CMT and CT, respectively, appear in the graphical representation with respect to the frequency;

18. Using the facilities provided by the Optimization Toolbox of MATLAB and taking as objective function the power transfer efficiency, the circuit

parameters which affects this performance were identified and the values which maximize the efficiency were obtained;

19. The two parameters which control the amount of the transferred power by magnetic coupling and the transfer efficiency as well are the frequency and the coupling coefficient. The analysis proves that a more efficient power transfer is obtained in a wireless power system with three magnetic couplings: two strong couplings in the source, respectively the load circuits and one weak coupling on the wireless power transfer path;

20. The simulations show that on the power supply side, the working frequency of the emitter has to be near the resonance frequency of the two magnetic coupled circuits and on the receiver side, the geometrical size and the relative position (defined by the angle α) are the most significant parameters for an efficient transfer of energy using wireless technologies;

21. The simulations show that on the power supply side, the working frequency of the emitter has to be near the resonance frequency of the two magnetic coupled circuits and on the receiver side, the geometrical size and the relative position (defined by the angle α) are the most significant parameters for an efficient transfer of energy using wireless technologies.

1.7 Problems

P1.1: The system of two parallel-parallel (P-P) resonators driven by a voltage source is shown in Figure P1.1.

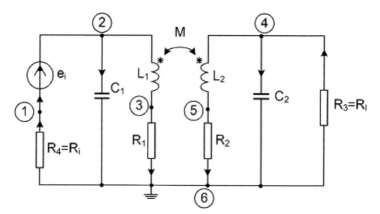

Figure P1.1   System of two parallel-parallel resonators driven by a voltage source.

For the parallel-parallel (P-P) resonator shown in Figure P1.1 find the state equations in the full-symbolic normal form.

P1.2: Set up the differential equations of the coupled-mode amplitudes for the circuit represented in Figure P1.1.

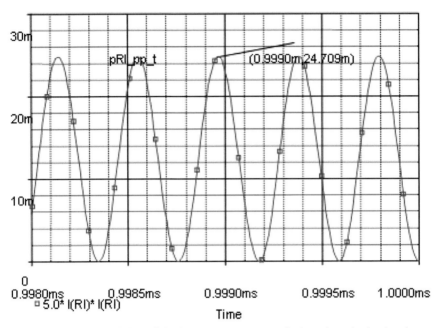

Figure P1.2 Time variation of the instantaneous power dissipated on the load resistance $p_{R1\text{-}pp}$, obtained by SPICE simulation.

P1.3: If we consider the numerical values of the circuit parameters presented in Section 1.2.1, plot the variation of the instantaneous power dissipated on the load resistance in respect of the time.

The time variation of the instantaneous power, dissipated on the load resistance, is plotted in Figure P1.2 (obtained by SPICE simulation).

P1.4: Keeping as independent variable only the frequency (in the sinusoidal behavior), study the influence of the frequency on the transferred power value and on the power transfer efficiency.

In the Figures P1.3, and P1.4 are represented the frequency variation of the active power $P_{R1\text{-}pp}$, delivered on the load resistance, and the frequency variation of the active power transfer efficiency $\eta_{21\text{-}ss}$ (obtained by SESYMGP simulation), respectively.

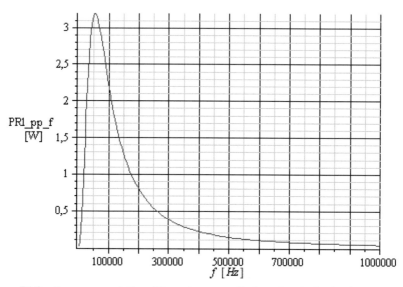

Figure P1.3   Frequency variation of the active power dissipated on the load resistance P_{R1_pp}, obtained by SESYMGP simulation.

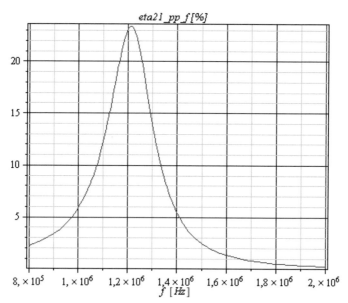

Figure P1.4   Frequency variation of the active power transfer efficiency η_{21_pp}, obtained by SESYMGP simulation.

P1.5: In Figure P1.5 is represented the system of two parallel-series (P-S) resonators driven by a voltage source.

Find, for the parallel-series (P-S) resonators in Figure P1.5, the state equations in the full symbolic normal form.

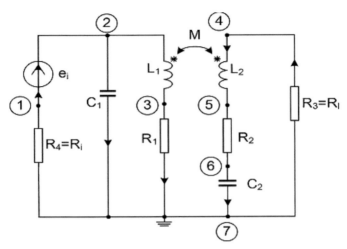

Figure P1.5   System of two parallel-series resonators driven by a voltage source.

P1.6: Formulate the differential equations of the coupled-mode amplitudes for the circuit represented in Figure P1.5.

P1.7: The numerical values of the circuit parameters (in Figure P1.5) are identical with the ones considered for the circuit in Figure 1.1 (Section 1.2.1). Plot the time variation of the instantaneous power dissipated on the load resistance.

The time variation of the instantaneous power, dissipated on the load resistance, is represented in Figure P1.6 (obtained by SESYMGP simulation).

P1.8: Considering as independent variable only the frequency (in the sinusoidal behavior), analyze the influence of the frequency on the transferred power value and on the power transfer efficiency.

In the Figures P1.7, and P1.8 are plotted the frequency variation of the active power P_{Rl_ps}, dissipated on the load resistance, and the frequency variation of the active power transfer efficiency η_{21_ps}, respectively (obtained by SPICE simulation).

P1.9: The system of two series-parallel (S-P) resonators driven by a voltage source is shown in Figure P1.9.

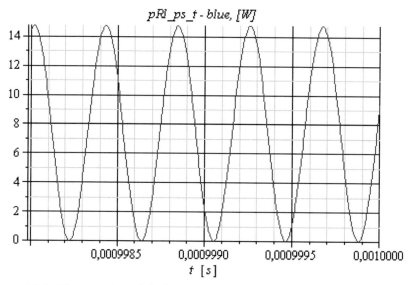

Figure P1.6   Time variation of the instantaneous power dissipated on the load resistance p_{Rl_ps}, obtained by SESYMGP simulation.

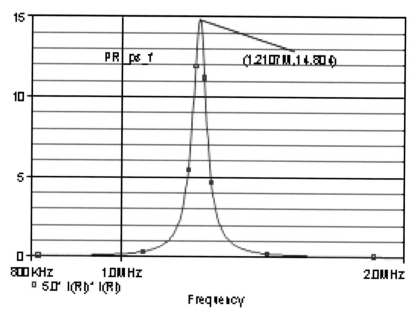

Figure P1.7   Frequency variation of the active power dissipated on the load resistance P_{Rl_ps}, obtained by SPICE simulation.

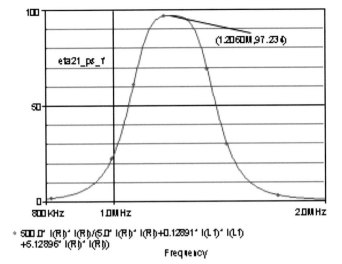

100

(1.2060M,97.23 %)

eta21_ps_1

50

0

800 KHz   1.0M Hz   2.0M Hz

∘ 500.0* I(Rb)* I(Rb)/(5.0* I(Rb)* I(Rb)+0.12891* I(L1)* I(L1)
+5.12896* I(Rb)* I(Rb))

Freqlency

Figure P1.8 Frequency variation of the active power transfer efficiency η_{21_ps}, obtained by SPICE simulation.

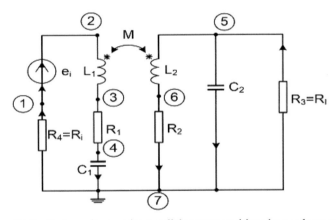

Figure P1.9 System of two series-parallel resonators driven by a voltage source.

For the series-parallel (S-P) resonators shown in Figure P1.9 set up the state equations in the full-symbolic normal form.

P1.10: Set up the differential equations of the coupled-mode amplitudes for the circuit represented in Figure P1.9.

P1.11: If we consider the numerical values of the circuit parameters presented in Section 1.2.1, plot the variation of the instantaneous power dissipated on the

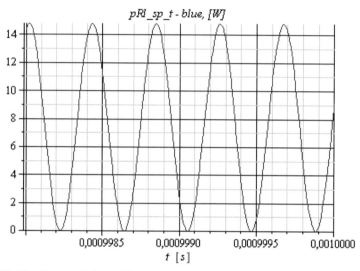

Figure P1.10   Time variation of the instantaneous power dissipated on the load resistance p_{R1_pp}, obtained by SESYMGP simulation.

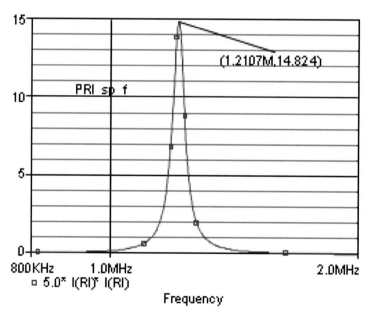

Figure P1.11   Frequency variation of the active power dissipated on the load resistance P_{R1_sp}, obtained by SPICE simulation.

load resistance in respect of the time. The time variation of the instantaneous power, dissipated on the load resistance, is plotted in Figure P1.10 (obtained by SESYMGP simulation).

P1.12: Keeping as independent variable only the frequency (in the sinusoidal behavior), study the influence of the frequency on the transferred power value and on the power transfer efficiency.

In the Figures P1.11 and P1.12 are represented the frequency variations of the active power P_{Rl_sp}, delivered on the load resistance, and the frequency variations of the active power transfer efficiency η_{21_sp}, respectively (obtained by SPICE Simulation).

P1.13: Performing the ANSOFT Q3D EXTRACTOR program at the 22 MHz frequency we obtain, for the system of two series-series resonators in Figure 1.22, the following parameter numeric values: $C_1 = 13.8974$ pF, $C_2 = 13.8915$ pF, $L_1 = 16.758$ μH, $L_2 = 16.735$ μH, $R_1 = 1.352$ Ω, $R_2 = 1.3524$ Ω and $C_{10} = 27.5135$ pF. The input resistance R_i and the load resistance R_l have the same value of 5.0 Ω. Determine the frequency variations

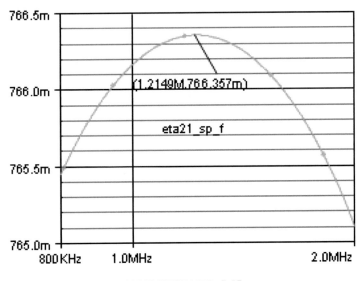

Figure P1.12 Frequency variation of the active power transfer efficiency η_{21_sp}, obtained by SPICE simulation.

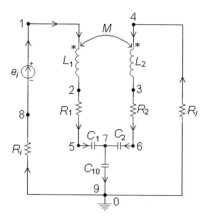

Figure P1.13   System of two series-series resonators driven by a voltage source.

of the transfer power dissipated in the load resistance and of the efficiency of this power transfer.

In the Figures P1.14 and P1.15 are plotted the frequency variations of the active power P_{Rl_ss}, delivered on the load resistance, and the active power transfer efficiency η_{21_ss}, respectively.

P1.14: Compare the results from Figures 1.23 and 1.24 with the ones in Figures 1.9 and 1.21.

P1.15: Using two of the mutual inductance formulas (Maxwell, Weistein or Nagaoka), find the mutual inductance values for the 2 handmade coils, with geometrical characteristics from Section 1.3, choosing one of the four layout cases.

P1.16: Let be the circuit from Figure 1.14. If you know the frequency of the power supply and you can measure the current and the voltage (according to positions of two apparatus from the figure), determine the expression of mutual inductance in function of above parameters.

P1.17: We wants to find some optimal resonator parameters which provide the minimum and maximum values of the load active power $P_2 = P_L = P_l$. The study will be focused on the parameters: L_1, L_2, M, R_1, and R_2 for the resonators in Figure P1.13. To generate the the load active power $P_2 = P_L = P_l$ in symbolic – numeric form we use the SYMNAP – SYmbolic Modified Nodal Analysis Program [35], based on the complex or Laplace modified nodal equations. Running the SYMNAP the following expression for the power transfer efficiency is obtained:

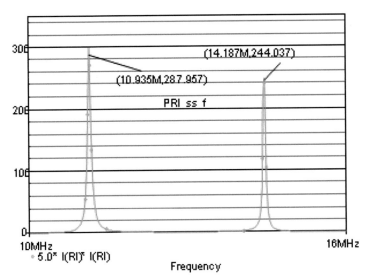

Figure P1.14 Frequency variation of the active power dissipated on the load resistance P_{R1_ss}, obtained by SPICE simulation.

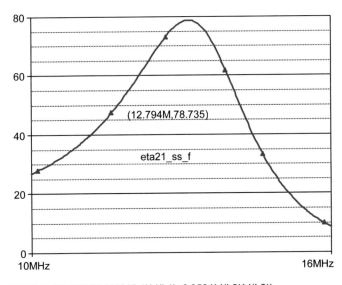

Figure P1.15 Frequency variation of the active power transfer efficiency η_{21_ss}, obtained by SPICE simulation.

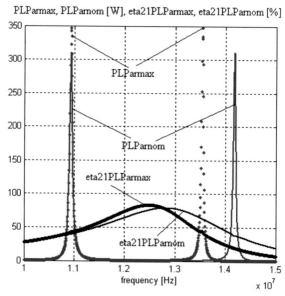

Figure P1.16 Transfer active power P_L and the transfer efficiency η_{21} vs frequency for the optimal parameter values L_1, L_2, M, R_1, and R_2 which maximize the load active power.

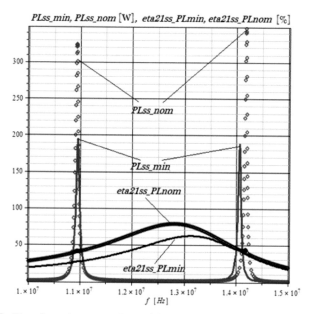

Figure P1.17 Transfer active power P_L and the transfer efficiency η_{21} vs frequency for the optimal parameter values L_1, L_2, M, R_1, and R_2 which minimize the load active power.

$PL\_L1L2MR1R2\_f = 0.7525834345 \ 10^{-34} f^2$

$(390625. + 0.4599298045 \ 10^{-12} f^4 M^4 - 0.0008477265594 \ f^8 M) / (0.1223343948 \ 10^{-11}$

$+ (-0.5460413004 \ 10^{-49} M^8 L1 L2 + 0.2730206474 \ 10^{-49} L1^2 L2^2 + 0.2730206474 \ 10^{-49} M^4) f^8 + ($

$-0.1499481986 \ 10^{-39} L1 L2^2 - 0.1006444247 \ 10^{-39} M^3 + 0.3461355472 \ 10^{-49} M^2$

$+0.1730677736 \ 10^{-49} L1^2 + 0.6922710943 \ 10^{-51} L1^2 R2^2 + 0.1499481986 \ 10^{-39} M^2 L2$

$+0.1499905118 \ 10^{-39} M^2 L1 + 0.1006444247 \ 10^{-39} L1 L2 M + 0.6922710945 \ 10^{-50} M^2 R2$

$-0.6800815260 \ 10^{-57} L2 L1 - 0.2720326103 \ 10^{-58} L1 L2 R1 R2 + 0.6922710945 \ 10^{-50} M^2 R1$

$-0.1499905118 \ 10^{-39} L1^2 L2 + 0.6922710943 \ 10^{-51} L2^2 R1^2 + 0.1730677736 \ 10^{-49} L2^2$

$+0.6922710805 \ 10^{-50} L2^2 R1 - 0.1360163052 \ 10^{-57} L1 L2 R1$

$+0.1384542189 \ 10^{-50} M^2 R1 R2 + 0.6922710805 \ 10^{-50} L1^2 R2) f^6 + (-0.6379849510 \ 10^{-40} M$

$-0.9507897711 \ 10^{-40} L2 - 0.9505215475 \ 10^{-40} L1 + 0.4388306120 \ 10^{-50} R1 + 0.4388306119 \ 10^{-50} R2$

$+0.7774007162 \ 10^{-30} L2 L1 - 0.2763792825 \ 10^{-30} L2 M + 0.1755322448 \ 10^{-50} R2 R1$

$-0.2764572727 \ 10^{-30} L1 M + 0.1755322476 \ 10^{-52} R1^2 R2^2 - 0.3802086221 \ 10^{-41} L1 R2^2$

$-0.3803159115 \ 10^{-41} L2 R1^2 + 0.1755322448 \ 10^{-51} R1 R2^2 + 0.1535122496 \ 10^{-48} R2 L2$

$-0.3802086207 \ 10^{-40} R2 L1 - 0.1275969902 \ 10^{-40} R2 M - 0.3803159101 \ 10^{-40} L2 R1$

$+0.1755322448 \ 10^{-51} R1^2 R2 + 0.1535555684 \ 10^{-48} R1 L1 - 0.1275969902 \ 10^{-40} R1 M$

$+0.3070244991 \ 10^{-49} R1 R2 L2 + 0.3071111369 \ 10^{-49} R1 R2 L1 - 0.2551939805 \ 10^{-41} R1 R2 M$

$+0.1097076548 \ 10^{-49} - 0.2727601789 \ 10^{-30} M^2 + 0.2060023133 \ 10^{-30} L1^2 + 0.2058861007 \ 10^{-30} L2^2$

$+0.4388306190 \ 10^{-51} R1^2 + 0.4388306190 \ 10^{-51} R2^2) f^4 + (0.5223394308 \ 10^{-32} R1^2$

$+0.5220447622 \ 10^{-32} R2^2 + 0.1175911605 \ 10^{-32} R2 R1 + 0.5808403422 \ 10^{-31} R2$

$-0.1004015307 \ 10^{-20} L1 + 0.6736995611 \ 10^{-21} M + 0.5811350112 \ 10^{-31} R1 + 0.2904938384 \ 10^{-30}$

$-0.1003732068 \ 10^{-20} L2) f^2)$

(P1.28)

With the frequency of 14.184 MHz (the frequency corresponding to the maximum value of the load active power, when the resonator parameters have the nominal values), we obtain the objective function for the procedure fminunc used to determine the minimum and maximum values of the load active power. The initial variation intervals for the five parameters are

$$L_{1\_0} \in [1.6e - 05, 2.0e - 05], L_{2\_0} \in [1.6e - 05, 2.0e - 05],$$
$$M_{\_0} \in [1.4e - 06, 1.8e - 06], R_{1\_0} \in [1.0, 3.0],$$
$$\text{and} \quad R_{2\_0} \in [1.0, 3.0] \tag{P1.29}$$

Running the program with the initial variation intervals (P1.29), we obtained the following results:

1. Optimal (maximum) parameters: L_1, L_2, M, R_1, and R_2 for the maximum value of the load active power P_L:

$$P_{L\_max} = 349.238[\text{W}]; L_{1\_max} = 0.000017566[\text{H}];$$
$$L_{2\_max} = 0.00001756[\text{H}]; M_{\_max} = 0.000002314[\text{H}];$$
$$R_{1\_max} = 1.0[\Omega]; R_{2\_max} = 1.0[\Omega]; k_{\_max} = 0.1317. \tag{P1.30}$$

2. Optimal (minimum) parameters: L_1, L_2, M, R_1, and R_2 for the minimum value of the load active power P_L:

$$P_{L\_min} = 149.09[\text{W}]; L_{1\_min} = 1.7893e - 005[\text{H}];$$
$$L_{2\_min} = 1.5924e - 005[\text{H}]; M_{\_min} = 1.7998e - 006[\text{H}];$$
$$R_{1\_min} = 2.99999[\Omega]; R_{1\_min} = 3.0[\Omega];$$
$$R_{2\_min} = 3.0[\Omega]; k_{\_min} = 0.10663. \tag{P1.31}$$

Substituting the five parameter values from (P1.30) and (P1.31) in the expression (P1.28) we obtain the same maximum and minimum values for load active power P_L as ones in (P1.30) and (P1.31).

In Figure P1.16 (Figure P1.17) the transfer active power P_L and the transfer efficiency η_{21} vs frequency for the optimal parameter values L_1, L_2, M, R_1, and R_2 which maximize (minimize) the load active power are presented.

1.8 Solutions to Problems

P1.1: Solution Because the circuit in Figure P1.1 has not excess elements, the state vector has the following structure:

$$x = [v_{C_1}\ v_{C_2}\ i_{L_1}\ i_{L_2}]^{\mathrm{t}} \tag{P1.1}$$

Performing the SESYMGP program [16], we get the state equations of the circuit in normal-symbolic form:

$$\frac{dv_{C_1}}{dt} = -\frac{1}{C_1 R_i} v_{C_1} - \frac{1}{C_1} i_{L_1} + \frac{1}{C_1 R_i} e_i \tag{P1.2}$$

$$\frac{dv_{C_2}}{dt} = -\frac{1}{C_2 R_l} v_{C_2} - \frac{1}{C_2} i_{L_2} \tag{P1.3}$$

$$\frac{di_{L_1}}{dt} = \frac{L_2}{L_1 L_2 - M^2} v_{C_1} - \frac{M}{L_1 L_2 - M^2} v_{C_2}$$

$$- \frac{L_2 R_1}{L_1 L_2 - M^2} i_{L_1} + \frac{M R_2}{L_1 L_2 - M^2} i_{L_2} \tag{P1.4}$$

$$\frac{di_{L_2}}{dt} = -\frac{M}{L_1 L_2 - M^2} v_{C_1} + \frac{L_1}{L_1 L_2 - M^2} v_{C_2}$$

$$+ \frac{M R_1}{L_1 L_2 - M^2} i_{L_1} - \frac{L_1 R_2}{L_1 L_2 - M^2} i_{L_2} \tag{P1.5}$$

P1.2: Solution The differential equation corresponding to the first coupled-mode amplitude, for the two parallel-parallel resonators in Figure P1.1, has the following compact form:

$$\frac{da_1(t)}{dt} = j\omega_1 a_{11}(t) - \gamma_1 a_{111}(t) + k_{12} a_{12}(t) + k_{1e_i} e_i \approx j\omega_1 a_1(t)$$

$$- \gamma_1 a_1(t) + k_{12} a_2(t) + k_{1e_i} e_i \tag{P1.6}$$

where

$$a_{11} = \frac{L_1 L_2}{L_1 L_2 4 - M^2} \sqrt{\frac{C_1}{2}} v_{C_1} + j\sqrt{\frac{L_1}{2}} i_{L_1} \approx a_1 \stackrel{\mathrm{d}}{=} \sqrt{\frac{C_1}{2}} v_{C_1} + j\sqrt{\frac{L_1}{2}} i_{L_1}$$

$$a_{111} = \sqrt{\frac{C_1}{2}} v_{C_1} + \frac{R_1 R_i L_2 C_1}{L_1 L_2 - M^2} j\sqrt{\frac{L_1}{2}} i_{L_1} \approx a_1 \stackrel{\mathrm{d}}{=} \sqrt{\frac{C_1}{2}} v_{C_1} + j\sqrt{\frac{L_3}{2}} i_{L_1}$$

$$\gamma_1 = \frac{1}{C_1 R_i}; \quad k_{12} = -j\sqrt{\frac{L_1}{C_2}} \frac{M}{L_1 L_2 - M^2}; \quad k_{1e_i} = \frac{1}{\sqrt{2 C_1 R_i}}$$

$$a_{12} = \sqrt{\frac{C_2}{2}} v_{C_2} + j R_2 \sqrt{\frac{C_2}{L_2}} j \sqrt{\frac{L_2}{2}} i_{L_2} \approx a_2 \stackrel{\text{d}}{=} \sqrt{\frac{C_2}{2}} v_{C_2} + j \sqrt{\frac{L_2}{2}} i_{L_2}$$

$$\omega_1 = \frac{1}{\sqrt{L_1 C_1}} \tag{P1.7}$$

The differential equation corresponding to the second coupled-mode amplitude, for the two parallel-parallel resonators in Figure P1.1, is:

$$\frac{da_2(t)}{dt} = j\omega_2 a_{22}(t) - \gamma_2 a_{222}(t) + k_{21} a_{21}(t) + k_{2e_i} e_i \approx j\omega_2 a_2(t)$$
$$- \gamma_2 a_2(t) + k_{21} a_1(t) + k_{2e_i} e_i \tag{P1.8}$$

where

$$a_{22} = \frac{L_1 L_2}{L_1 L_{24} - M^2} \sqrt{\frac{C_2}{2}} v_{C_2} + j \sqrt{\frac{L_2}{2}} i_{L_2} \approx a_2 \stackrel{\text{d}}{=} \sqrt{\frac{C_2}{2}} v_{C_2} + j \sqrt{\frac{L_2}{2}} i_{L_2}$$

$$a_{222} = \sqrt{\frac{C_2}{2}} v_{C_2} + \frac{R_2 R_l L_1 C_2}{L_1 L_2 - M^2} j \sqrt{\frac{L_2}{2}} i_{L_2} \approx a_2 \stackrel{\text{d}}{=} \sqrt{\frac{C_2}{2}} v_{C_2} + j \sqrt{\frac{L_2}{2}} i_{L_2}$$

$$\gamma_2 = \frac{1}{C_2 R_l}; \qquad k_{21} = -j \sqrt{\frac{L_2}{C_1}} \frac{M}{L_1 L_2 - M^2}; \qquad k_{2e_i} = 0$$

$$a_{21} = \sqrt{\frac{C_1}{2}} v_{C_1} + j R_1 \sqrt{\frac{C_1}{L_1}} j \sqrt{\frac{L_1}{2}} i_{L_1} \approx a_1 \stackrel{\text{d}}{=} \sqrt{\frac{C_1}{2}} v_{C_1} + j \sqrt{\frac{L_1}{2}} i_{L_1}$$

$$\omega_2 = \frac{1}{\sqrt{L_2 C_2}} \tag{P1.9}$$

P1.5: Solution The circuit in Figure P1.5 has not excess elements, thus the state vector has the following structure:

$$x = [v_{C_1} \ v_{C_2} \ i_{L_1} \ i_{L_2}]^{\text{t}} \tag{P1.10}$$

Performing the SESYMGP program [16], we obtain the state equations of the circuit in full-symbolic normal form:

$$\frac{dv_{C_1}}{dt} = -\frac{1}{C_1 R_i} v_{C_1} - \frac{1}{C_1} i_{L_1} + \frac{1}{C_1 R_i} e_i \tag{P1.11}$$

$$\frac{dv_{C_2}}{dt} = \frac{1}{C_2}i_{L_2} \tag{P1.12}$$

$$\frac{di_{L_1}}{dt} = \frac{L_2}{L_1 L_2 - M^2}v_{C_1} + \frac{M}{L_1 L_2 - M^2}v_{C_2}$$

$$-\frac{L_2 R_1}{L_1 L_2 - M^2}i_{L_1} + \frac{M(R_2 + R_l)}{L_1 L_2 - M^2}i_{L_2} \tag{P1.13}$$

$$\frac{di_{L_2}}{dt} = -\frac{M}{L_1 L_2 - M^2}v_{C_1} - \frac{L_1}{L_1 L_2 - M^2}v_{C_2}$$

$$+\frac{M R_1}{L_1 L_2 - M^2}i_{L_1} - \frac{L_1(R_2 + R_l)}{L_1 L_2 - M^2}i_{L_2} \tag{P1.14}$$

The differential equation corresponding to the first coupled-mode amplitude, for the two parallel-series resonators in Figure P1.5, has the following compact form:

$$\frac{da_1(t)}{dt} = j\omega_1 a_{11}(t) - \gamma_1 a_{111}(t) + k_{12}a_{12}(t) + k_{1e_i}e_i \approx j\omega_1 a_1(t)$$

$$- \gamma_1 a_1(t) + k_{12}a_2(t) + k_{1e_i}e_i \tag{P1.15}$$

where

$$a_{11} = \frac{L_1 L_2}{L_1 L_2 4 - M^2}\sqrt{\frac{C_1}{2}}v_{C_1} + j\sqrt{\frac{L_1}{2}}i_{L_1} \approx a_1 \overset{d}{=} \sqrt{\frac{C_1}{2}}v_{C_1} + j\sqrt{\frac{L_1}{2}}i_{L_1}$$

$$a_{111} = \sqrt{\frac{C_1}{2}}v_{C_1} + \frac{R_1 R_i L_2 C_1}{L_1 L_2 - M^2}j\sqrt{\frac{L_1}{2}}i_{L_1} \approx a_1 \overset{d}{=} \sqrt{\frac{C_1}{2}}v_{C_1} + j\sqrt{\frac{L_3}{2}}i_{L_1}$$

$$\gamma_1 = \frac{1}{C_1 R_i}; \qquad k_{12} = j\sqrt{\frac{L_1}{C_2}}\frac{M}{L_1 L_2 - M^2}; \qquad k_{1e_i} = \frac{1}{\sqrt{2C_1 R_i}}$$

$$a_{12} = \sqrt{\frac{C_2}{2}}v_{C_2} - j(R_2 + R_l)\sqrt{\frac{C_2}{L_2}}j\sqrt{\frac{L_2}{2}}i_{L_2} \approx a_2 \overset{d}{=} \sqrt{\frac{C_2}{2}}v_{C_2}$$

$$+ j\sqrt{\frac{L_2}{2}}i_{L_2}$$

$$\omega_1 = \frac{1}{\sqrt{L_1 C_1}} \tag{P1.16}$$

P1.6: Solution The differential equation corresponding to the second coupled-mode amplitude, for the two parallel-series resonators in Figure P1.8, has the following form:

$$\frac{\mathrm{d}a_2(t)}{\mathrm{d}t} = -j\omega_2 a_{22}(t) - \gamma_2 a_{222}(t) + k_{21} a_{21}(t) + k_{2e_i} e_i \approx j\omega_2 a_2(t)$$
$$- \gamma_2 a_2(t) + k_{21} a_1(t) + k_{2e_i} e_i$$

(P1.17)

where

$$a_{22} = \frac{L_1 L_2}{L_1 L_2 - M^2} \sqrt{\frac{C_2}{2}} v_{C_2} + j\sqrt{\frac{L_2}{2}} i_{L_2} \approx a_2 \stackrel{\mathrm{d}}{=} \sqrt{\frac{C_2}{2}} v_{C_2} + j\sqrt{\frac{L_2}{2}} i_{L_2}$$

$$a_{222} = j\sqrt{\frac{L_2}{2}} i_{L_2} \approx a_2 \stackrel{\mathrm{d}}{=} \sqrt{\frac{C_2}{2}} v_{C_2} + j\sqrt{\frac{L_2}{2}} i_{L_2}$$

$$\gamma_2 = \frac{L_1 (R_2 + R_l)}{L_1 L_2 - M^2}; \qquad k_{21} = -j\sqrt{\frac{L_2}{C_1}} \frac{M}{L_1 L_2 - M^2}; \qquad k_{2e_i} = 0$$

$$a_{21} = \sqrt{\frac{C_1}{2}} v_{C_1} + jR_1 \sqrt{\frac{C_1}{L_1}} j\sqrt{\frac{L_1}{2}} i_{L_1} \approx a_1 \stackrel{\mathrm{d}}{=} \sqrt{\frac{C_1}{2}} v_{C_1} + j\sqrt{\frac{L_1}{2}} i_{L_1}$$

$$\omega_2 = \frac{1}{\sqrt{L_2 C_2}}$$

(P1.18)

P1.9: Solution Because the circuit in Figure P1.9 has not excess elements, the state vector has the following structure:

$$x = [v_{C_1} \ v_{C_2} \ i_{L_1} \ i_{L_2}]^{\mathrm{t}}$$

(P1.19)

Running the SESYMGP program [16], we obtain the state equations of the circuit in normal-symbolic form:

$$\frac{\mathrm{d}v_{C_1}}{\mathrm{d}t} = \frac{1}{C_1} i_{L_1}$$

(P1.20)

$$\frac{\mathrm{d}v_{C_2}}{\mathrm{d}t} = -\frac{1}{C_2 R_l} v_{C_2} - \frac{1}{C_2} i_{L_2}$$

(P1.21)

$$\frac{di_{L_1}}{dt} = -\frac{L_2}{L_1 L_2 - M^2} v_{C_1} - \frac{M}{L_1 L_2 - M^2} v_{C_2} - \frac{(R_1 + R_i) L_2}{L_1 L_2 - M^2} i_{L_1}$$

$$+ \frac{M R_2}{L_1 L_2 - M^2} i_{L_2} + \frac{L_2}{L_1 L_2 - M^2} e_i \qquad \text{(P1.22)}$$

$$\frac{di_{L_2}}{dt} = \frac{M}{L_1 L_2 - M^2} v_{C_1} + \frac{L_1}{L_1 L_2 - M^2} v_{C_2} + \frac{(R_1 + R_i) M}{L_1 L_2 - M^2} i_{L_1}$$

$$- \frac{L_1 R_2}{L_1 L_2 - M^2} i_{L_2} - \frac{M}{L_1 L_2 - M^2} e_i \qquad \text{(P1.23)}$$

P1.10: Solution The differential equation corresponding to the first coupled-mode amplitude, for the two parallel-parallel resonators in Figure P1.9, has the following compact form:

$$\frac{da_1(t)}{dt} = -j\omega_1 a_{11}(t) - \gamma_1 a_{111}(t) + k_{12} a_{12}(t) + k_{1e_i} e_i \approx j\omega_1 a_1(t)$$

$$- \gamma_1 a_1(t) + k_{12} a_2(t) + k_{1e_i} e_i \qquad \text{(P1.24)}$$

where

$$a_{11} = \frac{L_1 L_2}{L_1 L_2 4 - M^2} \sqrt{\frac{C_1}{2}} v_{C_1} + j\sqrt{\frac{L_1}{2}} i_{L_1} \approx a_1 \overset{d}{=} \sqrt{\frac{C_1}{2}} v_{C_1} + j\sqrt{\frac{L_1}{2}} i_{L_1}$$

$$a_{111} = j\sqrt{\frac{L_1}{2}} i_{L_1} \approx a_1 \overset{d}{=} \sqrt{\frac{C_1}{2}} v_{C_1} + j\sqrt{\frac{L_3}{2}} i_{L_1}$$

$$\gamma_1 = \frac{L_1 (R_1 + R_i)}{L_1 L_2 - M^2}; \qquad k_{12} = -j\sqrt{\frac{L_1}{C_2}} \frac{M}{L_1 L_2 - M^2};$$

$$k_{1e_i} = j\sqrt{\frac{L_1}{2}} \frac{L_2}{L_1 L_2 - M^2}$$

$$a_{12} = \sqrt{\frac{C_2}{2}} v_{C_2} + jR_2 \sqrt{\frac{C_2}{L_2}} j\sqrt{\frac{L_2}{2}} i_{L_2} \approx a_2 \overset{d}{=} \sqrt{\frac{C_2}{2}} v_{C_2} + j\sqrt{\frac{L_2}{2}} i_{L_2}$$

$$\omega_1 = \frac{1}{\sqrt{L_1 C_1}} \qquad \text{(P1.25)}$$

The differential equation corresponding to the second coupled-mode ampli-tude, for the two parallel-parallel resonators in Figure P1.9, is:

$$\frac{da_2(t)}{dt} = j\omega_2 a_{22}(t) - \gamma_2 a_{222}(t) + k_{21} a_{21}(t) + k_{2e_i} e_i \approx j\omega_2 a_2(t)$$

$$- \gamma_2 a_2(t) + k_{21} a_1(t) + k_{2e_i} e_i \tag{P1.26}$$

where

$$a_{22} = \frac{L_1 L_2}{L_1 L_2 4 - M^2} \sqrt{\frac{C_2}{2}} v_{C_2} + j\sqrt{\frac{L_2}{2}} i_{L_2} \approx a_2 \overset{d}{=} \sqrt{\frac{C_2}{2}} v_{C_2} + j\sqrt{\frac{L_2}{2}} i_{L_2}$$

$$a_{222} = \sqrt{\frac{C_2}{2}} v_{C_2} + \frac{R_2 R_l L_1 C_2}{L_1 L_2 - M^2} j\sqrt{\frac{L_2}{2}} i_{L_2} \approx a_2 \overset{d}{=} \sqrt{\frac{C_2}{2}} v_{C_2} + j\sqrt{\frac{L_2}{2}} i_{L_2}$$

$$\gamma_2 = \frac{1}{C_2 R_l}; k_{21} = j\sqrt{\frac{L_2}{C_1}} \frac{M}{L_1 L_2 - M^2}; k_{2e_i} = -j\sqrt{\frac{L_2}{2}} \frac{M}{L_1 L_2 - M^2}$$

$$a_{21} = \sqrt{\frac{C_1}{2}} v_{C_1} - j(R_1 + R_i)\sqrt{\frac{C_1}{L_1}} j\sqrt{\frac{L_1}{2}} i_{L_1} \approx a_1 \overset{d}{=} \sqrt{\frac{C_1}{2}} v_{C_1}$$

$$+ j\sqrt{\frac{L_1}{2}} i_{L_1}$$

$$\omega_2 = \frac{1}{\sqrt{L_2 C_2}} \tag{P1.27}$$

References

[1] Agbinya, J.I. "*Principles of Inductive Near Field Communications for Internet of Things.*" River Publishers: Denmark, ISBN: 978-87-92329-52-3 (2011).

[2] Zhang, F., X. Liu, S.A. Hackworth, R.J. Sclabassi, and M. Sun. "In Vitro and In Vivo Studies on Wireless Powering of Medical Sensors and Implantable Devices." *Proceedings of Life Science Systems and Applications Workshop*, IEEE Xplore, 978-1-4244-4293-5/09/2009, pp. 84–87, April (2009).

[3] Hamam, R.E., A. Karalis, J.D. Joannopoulos, and M. Soljacic. "Coupled-Mode Theory for General Free-Space Resonant Scattering of Waves." *Physical Review A* 75 (2007).

[4] Bhutkar, R., and S. Sapre, "Wireless Energy Transfer using Magnetic Resonance." *Second International Conference on Computer and Electrical Engineering*, December (2009).

[5] Karalis, A., J.D. Joannopoulos, and M. Soljačić. "Efficient Wireless Non-Radiative Mid-Range Energy Transfer." *Annals of Physics* 323 34–48, (2008).

[6] Haus, H.A. *"Microwaves and Fields in Optoelectronics."* Pretince-Hall Inc.: New Jersey, 07632, SSBN: 0-13-946053-5 (1984).

[7] Haus, H.A., and W. Huang. "Coupled-Mode Theory." *Proceedings of the IEEE* 79, no. (10): 1505–1518, October (1991).

[8] Hamam, R.E., A. Karalis, J.D. Joannopoulos, and M. Soljačić. "Coupled-mode theory for general free-space resonant scattering of waves." *Physical Review A* 75, 053801 (2007).

[9] Barybin, A.A., V.A. Dmitriev. *"Modern Electrodynamics and Coupled-Mode Theory: Application to Guided-Wave Optics."* Rinton Press: Princeton, 08540, ISBN: 1-58949-007-X (2002).

[10] Kurs, A., et al. *"Wireless Power Transfer via Strongly Coupled Magnetic Resonances."* *Science Express* 317, no. 5834: 83–86 (2007).

[11] Kurs, A. "Power Transfer Through Strongly Coupled Resonances." *Thesis for Master of Science in Physics under the supervision of M. Soljačić*, September (2007).

[12] Moffatt, R.A. "Wireless Transfer of Electric Power." *Thesis for Bachelor of Science in Physics Under The Supervision of Marin Soljačić*, June (2009).

[13] Imura, T., H. Okabe, and Y. Hori. "Basic Experimental Study on Helical Antennas of Wireless Power Transfer for Electric Vehicles by using Magnetic Resonant Couplings." *Proceedings of Vehicle Power and Propulsion Conference*, IEEE Xplore, 978-1-4244-2601-4/010/2010, pp. 936–940, September (2009).

[14] Jiang, H.C., and Y.E. Wang, "Capacity Performance of an Inductively Coupled Near Field Communication System." *Proceedings of the IEEE International Symposium of Antenna and Propagation Society*, Jul. 5–11, pp. 1–4 (2008).

[15] *ANSOFT Q3D EXTRACTOR*, User Guide, www.ANSOFT.com

[16] Mihai Iordache, Lucia Dumitriu, Daniel Delion, "**SESYMGP**-**S**tate **E**quation **Sym**bolic **G**eneration **P**rogram", User Guide, Library of Electrical Department, Politehnica University of Bucharest, Bucharest (2000).

[17] Rosa, E.B., and F.W. Grover. "Formulas and Tables for the Calculation of Mutual and Self-Inductance." US Government Printing Office Washington (1948).

[18] Kalantarov, P.L., and L.A. Teitlin. *"Inductance computation."* (in romanian), Editura Tehnica Bucuresti (1958).

[19] Timotin, A., V. Hortopan, A. Ifrim, and M. Preda. *Electrical fundamentals* (in romanian), Editura Didactica si Pedagogica, Bucuresti (1970).

[20] Niculae, D., M. Iordache, L. Dumitriu. "Magnetic Coupling Analysis in Wireless Transfer Energy." The 7th International Symposium on Advanced Topics in Electrical Engineering (ATEE), Bucharest, 12–14 May (2011).

[21] Iordache, M., L. Dumitriu, and D. Delion, "Automatic Formulation of Symbolic State Equations for Analog Circuits with Degeneracies." *Proceedings of 6th International Workshop on Symbolic Methods and Applications in Circuit Design*, SMACD 2000, Instituto Superior Técnico, Lisbon, Portugal, pp. 65–72, October 12–13 (2000).

[22] J. i. A Agbinya – Editor, Wireless Power Transfer, River Publishers Series in Communications, 9000 Aalborg Danemark (2012).

[23] www.witricity.com

[24] Zhang, X. Liu, S.A. Hackworth, R.j. Sclabassi, and M. Sun, "In Vitro and In Vivo Studies on Wireless Powering of Medical Sensors and Implantable Devices", Proceedings of Life Science Systems and Applications Workshop, IEEE Xplore, 978-1-4244-4293-5/09/2009, pp. 84–87, April (2009).

[25] A. Karalis, J.D. Joannopoulos, and M. Soljačić, "Efficient wireless non-radiative mid-range energy transfer", Annals of Physics, Vol. 323, pp. 34–48, January (2008).

[26] D. Niculae, M. Iordache, Lucia Dumitriu, "Magnetic coupling analysis in wireless transfer energy", The 7th International Symposium on Advanced Topics in Electrical Engineering (ATEE), 2011, Bucharest, 12–14 May (2011).

[27] N. Tesla "Apparatus for Transmitting Electrical Energy", U.S. patent number 1119732, issued in December (1914).

[28] A. Karalis, J. D. Joannopoulos, M. Soljacic "Efficient Wireless Non-Radiative Mid-Range Energy Transfer", MIT (2006).

[29] A. Ricano, H. Rodriguez, H. Vasquez "Experiment About Wireless Energy Transfer", 1-st International congress on instrumentation and applied sciences, Cancun, Mexico, october (2010).

[30] Ji Wang, "A system of two piezoelectric transducers and a storage circuit for wireless energy transmission through a thin metal wall", IEEE Transactions on Ultrasonics, Ferroelectrics and Frequency Control 55, No. 10 (2008).

[31] E. B. Rosa, F.W. Grover, Formulas and Tables for the Calculation of Mutual and Self-Inductance, US Government Printing Office Washington (1948).

[32] R. A. Moffatt, Wireless Transfer of electric power, Thesis for Bachelor of Science in Physics under the supervision of Marin Soljačić, June (2009).

[33] H. C. Jiang and Yuanxun E. Wang, "Capacity Performance of an Inductively Coupled Near Field Communication System", in Proc. IEEE International Symposium of Antenna and Propagation Society, Jul. 5–11, pp. 1–4 (2008).

[34] D. Niculae, Lucia Dumitriu, M. Iordache, A. Ilie, L. Mandache, "Magnetic Resonant Couplings Used in Wireless Power Transfer to Charge the Electric Vehicle Batteries" – Bulletin AGIR, No. 4, pp. 155–158, ISSN 1224-7928 (2011).

[35] M. Iordache, Lucia Dumitriu, I. Matei, "SYMNAP program – SYmbolic Modified Nodal Analysis Program", User Guide, Department Electrical Library, Politehnica University of Bucharest (2002).

[36] Optimization Toolbox^{TM} User's Guide, MATLAB R2011b, The MathWorks, Inc. (2011).

[37] *ANSOFT Q3D EXTRACTOR*, **User Guide,** www.ANSOFT.com

[38] F. Zhang, X. Liu, S.A. Hackworth, R.J. Sclabassi, and M. Sun, "In Vitro and In Vivo Studies on Wireless Powering of Medical Sensors and Implantable Devices", Proceedings of Life Science Systems and Applications Workshop, IEEE Xplore, 978-1-4244-4293-5/09/2009, pp. 84–87, April (2009).

2

Efficient Wireless Power Transfer based on Strongly Coupled Magnetic Resonance

Fei Zhang and Mingui Sun*

Departments of Neurosurgery and Electrical Engineering,
University of Pittsburgh, 3520 Forbes Ave, Suite 202, Pittsburgh,
PA 15213, USA
*E-mail: drsun@pitt.edu

This chapter describes the principle of wireless power transfer using strongly coupled magnetic resonance, the relay effect for flexible and distant power transmission, and the extension of the relay effect to a network of resonators. Optimization and assessment methods for the resonator network are presented to facilitate the system design and deployment. Theoretical analysis is performed based on a set of coupled-mode equations. Flexible film resonators are designed and prototyped. Experiments are carried out to confirm the theoretical results and demonstrate the effectiveness of the new mechanisms of wireless power transfer.

2.1 Introduction

Transmitting power without wires or cables has attracted considerable interest from both academic and industrial communities since eighteenth centuries [1–4]. Recently, the corresponding utilization standards have been made by Alliance For Wireless Power (A4WP), Wireless Power Consortium (WPC), and Power Matters Alliance (PMA). This technology has many applications, such as wirelessly recharging mobile devices and powering an array of wireless sensors. The currently available wireless power transfer techniques can be roughly classified into the radiative and non-radiative types. The first type includes radio frequency transmission using highly directional antennas and laser, while the second type includes widely utilized inductively coupled transformers. However, high-power radiation poses safety concerns and also

Wireless Power Transfer 2nd Edition, 73–104.

requires a complicated tracking system, while the current non-radiative types only work within a short distance between the energy transmitter and the receiver.

To overcome these problems, a new breakthrough in wireless power transfer technique was reported in 2007 [5]. This technique is often dubbed as "witricity" (abbreviation of wireless electricity) which is also the trademark of Witricity, Inc. This technique is based on non-radiative strongly coupled magnetic resonance and can work over the mid-range distance, defined as several times the resonator size [6–8]. With two identical coils utilized as wireless electric resonators, a 60-watt bulb was fully illuminated wirelessly seven feet away from the power source with a power transfer efficiency of 40% [5]. This pioneer system worked well even when the line of sight between the two resonators was blocked by a non-resonant object, yet was almost transparent to non-resonant objects including the biological tissue. This phenomenon reduces both the safety problem and the inefficiency caused by the energy loss along the path. The wireless electricity has been developed for a variety of practical applications, such as transportation [9], biomedicine [9–15], and consumer electronics [16].

At the present time, numerous wireless electric system designs have been reported, such as powering multiple devices with a single source [17, 18] and transferring power to two devices on both sides of a source with a higher efficiency [19]. Despite these significant developments [6–20], the system configurations are still based on the original scheme of "source–device" or "source–devices." As a result, the effective transmission distance is still limited to the mid-range. In 2011, the report of relay effect extended the original witricity from the source–device(s) scenario to the source–relay–device scenario to enable more flexible and distant wireless power transfer [21]. We will demonstrate by both theoretical treatments and experimental studies that this scheme succeeds beyond the limit of mid-range, allowing much longer and more flexible power transmission without sacrificing efficiency. Additionally, in the latter part of this chapter, an extended concept to a network architecture involving a number of resonators is discussed, and a spectral method to assist in the design and evaluation of wireless power transfer network is described.

2.2 Interaction in Lossless Physical System

The coupled-mode theory based on a set of integrable coupled-mode equations accounts well for the mechanisms of the wireless electric system [22], where power swapping follows a periodic manner and can be expressed analytically.

Considering the modal signals (or state variables in terms of linear dynamic systems) $a_1(t)$ and $a_2(t)$ of two lossless objects with natural frequencies ω_1 and ω_2, we have

$$\frac{da_1(t)}{dt} = i\omega_1 a_1(t) + k_{12}a_2(t) \tag{2.1}$$

$$\frac{da_2(t)}{dt} = i\omega_2 a_2(t) + k_{21}a_1(t) \tag{2.2}$$

where k_{12} and k_{21} are the coupling coefficients between two modes. According to conservation of energy, the time rate of energy change need vanish and then can be expressed as

$$\frac{d}{dt}(|a_1|^2 + |a_2|^2) = a_1(t)\frac{da_1^*(t)}{dt} + a_1^*(t)\frac{da_1(t)}{dt}$$

$$+ a_2(t)\frac{da_2^*(t)}{dt} + a_2^*(t)\frac{da_2(t)}{dt} = 0 \tag{2.3}$$

Substituting Equations (2.1) and (2.2) into Equation (2.3), we have

$$a_1(t)k_{12}^*a_2^*(t) + a_1^*(t)k_{12}a_2(t) + a_2(t)k_{21}^*a_1^*(t) + a_2^*(t)k_{21}a_1(t) = 0 \tag{2.4}$$

Since $a_1(t)$ and $a_2(t)$ can have arbitrary initial amplitudes and phases, the coupling coefficients should be related by

$$k_{12} + k_{21}^* = 0. \tag{2.5}$$

We can obtain two homogeneous equations in $a_1(t)$ and $a_2(t)$ from Equations (2.1) and (2.2). If $a_1(t) = A_1 \cdot e^{j\omega t}$, we will have

$$\omega^2 - (\omega_1 + \omega_2) \cdot \omega + (\omega_1\omega_2 + k_{12}k_{21}) = 0 \tag{2.6}$$

which yields the roots

$$\omega = \frac{\omega_1 + \omega_2}{2} \pm \sqrt{(\frac{\omega_1 - \omega_2}{2})^2 + |k_{12}|^2} \equiv \frac{\omega_1 + \omega_2}{2} \pm \Omega_0 \tag{2.7}$$

Equation (2.7) indicates that the two frequencies of the coupled system are separated by Ω_0. In particular, when $\omega_1 = \omega_2$, the difference between the two natural frequencies of the coupled modes is $2\Omega_0$ (or $2|k_{12}|$). Suppose, initially, that at $t = 0$, $a_1(0)$ and $a_2(0)$ are specified, then the two solutions of Equations (2.1) and (2.2) are expressed by

$$a_1(t) = \left[a_1(0)(\cos\Omega_0 t - j\frac{\omega_2 - \omega_1}{2\Omega_0}\sin\Omega_0 t)\right.$$

$$\left. + \frac{k_{12}}{\Omega_0}a_2(0)\sin\Omega_0 t\right] \cdot e^{j[(\omega_1 + \omega_2)/2]t} \tag{2.8}$$

$$a_2(t) = \left[\frac{k_{21}}{\Omega_0} a_1(0) \sin \Omega_0 t + a_2(0)(\cos \Omega_0 t \right.$$

$$\left. -j \frac{\omega_1 - \omega_2}{2\Omega_0} \sin \Omega_0 t) \right] \cdot e^{j[(\omega_1 + \omega_2)/2]t} \tag{2.9}$$

Let us consider the case where $a_1(0) = 1$, $a_2(0) = 0$, and $\omega_1 = \omega_2 = \omega$, and we have $a_1(t) = \cos \Omega_0 t \cdot e^{j\omega t}$ and $a_2(t) = \sin \Omega_0 t \cdot e^{j\omega t}$. Mode 1 is fully excited at $t = 0$, but at $\Omega_0 t = \pi/2$, all the excitation appears in mode 2. At $\Omega_0 t = \pi$, the excitation returns to mode 1 and mode 2 is unexcited. The process repeats periodically. Hence, the excitation is transferred back and forth with frequency $2\Omega_0$(or $2 |k_{12}|$). Figure 2.1 shows energy exchange between two modes. Here, Figure 2.1a indicates that resonant energy swapping ($k = 500,000$, $f_1 = f_2 = 1$ MHz) can be efficient and complete. However, if $f_1 \neq f_2$ ($k = 500,000$, $f_1 = 1.2$ MHz, $f_2 = 1$ MHz), the energy exchange (Figure 2.1b) is not complete, but partial and inefficient.

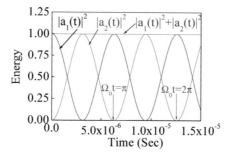

(a) Symmetric resonant case.

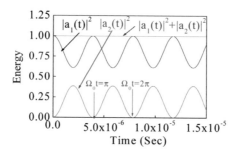

(b) Non-resonant case.

Figure 2.1 Energy exchange in two-resonator lossless system.

2.3 Interaction in Real Two-Resonator Physical System

To easily link coupled-mode theory to the present wireless electric system, the following coupled differential equations are utilized:

$$\frac{da_S(t)}{dt} = (i\omega_S - \Gamma_S)a_S(t) + ik_{SD}a_D(t) \tag{2.10}$$

$$\frac{da_D(t)}{dt} = (i\omega_D - \Gamma_D)a_D(t) + ik_{DS}a_S(t) \tag{2.11}$$

where $a_S(t)$ and $a_D(t)$ denote, respectively, the modal signals at the source and device resonators, $\omega_{S,D} = 2\pi f_{S,D}$ are the individual angular frequencies, $|k_{SD}| = |k_{DS}|$ are the coupling coefficients, and $\Gamma_{S,D}$ are the individual intrinsic decay rates.

Applying the Laplace transform method to Equations (2.10) and (2.11) computing the inverse, we can obtain

$$a_S(t) = e^{\frac{ct}{2}} \left\{ \left[(-(b-a)) \sin\left(\frac{dt}{2}\right) \right] / d + \cos\left(\frac{dt}{2}\right) \right\} \tag{2.12}$$

$$a_D(t) = e^{\frac{ct}{2}} \left\{ \left[2ik \sin\left(\frac{dt}{2}\right) \right] / d \right\} \tag{2.13}$$

where we define $a = i\omega_S - \Gamma_S$, $b = i\omega_D - \Gamma_D$, $c = a + b$, and $d = sqrt(4k^2 - a^2 + 2ab - b^2)$ and assume that the source possesses the full amount of energy (normalized) at $t = 0$.

2.3.1 Fully Resonant Case

If we assume that the source and the device are identical, we have $\omega = \omega_S = \omega_D$, $\Gamma = \Gamma_S = \Gamma_D$, and $k = |k_{SD}| = |k_{DS}|$. Under these conditions, Equations (2.12) and (2.13) yield

$$a_S(t) = e^{(i\omega - \Gamma)t} \cos(kt) \tag{2.14}$$

$$a_D(t) = ie^{(i\omega - \Gamma)t} \sin(kt) \tag{2.15}$$

Then, the total energy of this system, decreasing at a $\exp(-2\Gamma t)$ rate, is expressed by

$$P(t) = |a_S(t)|^2 + |a_D(t)|^2 = e^{-2\Gamma t} \tag{2.16}$$

2.3.1.1 Strong coupling $k/\sqrt{\Gamma_S\Gamma_D} \gg 1$

As a key distance-dependent figure of merit, $k/\sqrt{\Gamma_S\Gamma_D}$ is known as the relative coupling parameter. It represents, intuitively, the ratio of "how fast energy is transferred between the source and the device" to "how fast it is dissipated due to intrinsic losses in these resonators" [6, 18]. Our desired strong coupling regime for wireless electricity is provided by $k/\sqrt{\Gamma_S\Gamma_D} \gg 1$. When this inequality is satisfied, the coupling rate is much higher than the loss rate. However, to realize a wireless electric system, "strong coupling" and "resonance" must be guaranteed simultaneously. The fact is that the coupling, in theory, is inversely proportional to the cube of the separation distance and, in practice, may decay even more steeply, leading to the "mid-range" limitation of the present wireless electric systems.

As shown in Figure 2.2, the current wireless electric system consists of two resonators (source and device), a driving loop, and an output loop. The source resonator is coupled inductively with the driving loop linked to an oscillator to obtain energy for the system. Similarly, the device resonator is coupled inductively with the output loop to supply power to an external load.

The characteristics of the coupled system provide insights into the working of wireless electricity. The energy exchange for $f_S = f_D = 1\,\text{MHz}$, $k = 500,000$, and $\Gamma = 1,000$ is shown in Figure 2.3a. It can be seen that the energies in the source and device resonators are continually and completely exchanged via a strong energy channel. Since the range of the near field surrounding a finite-sized resonant object is proportional to the wavelength, this mid-range non-radiative resonant coupling can only be achieved using subwavelength resonators and thus significantly longer evanescent field tails [6, 18]. This condition is also true for source and device with different decay rates ($\Gamma_S = 100$ and $\Gamma_D = 10,000$) as shown in Figure 2.3b. The total energy decay at a rate of $\exp(-(\Gamma_S + \Gamma_D)t)$ should be compensated by an external source to maintain operation.

Figure 2.2 Basic components in present wireless electric systems.

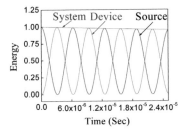

(a) Resonant Strong Coupling with Identical Resonators

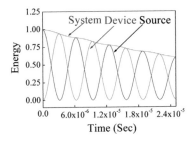

(b) Resonant Strong Coupling for Resonators with Different Decays

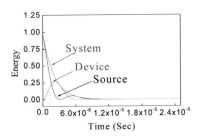

(c) Resonant Weak Coupling.

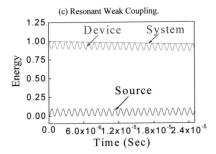

(d) Non-resonant Case.

Figure 2.3 Energy exchanges of a two-resonator lossy system.

2.3.1.2 Weak coupling $k/\sqrt{\Gamma_S\Gamma_D} \approx 1$ or $k/\sqrt{\Gamma_S\Gamma_D} < 1$

If $k/\Gamma \gg 1$ is not satisfied, the system will not resonate since system energy will be lost before an avenue for wireless energy transfer is formed in space. This case is shown in Figure 2.3c with $k = 500,000$ and $\Gamma = 250,000$. It is clear that in order to transmit power at a high efficiency, the coupling rate should be much higher or faster than the decay rate.

2.3.2 General Non-Resonant Case

It is known that two non-resonant objects (e.g., Figure 2.3d with $f_S = 1$ MHz, $f_D = 2$ MHz, $k = 500,000$, and $\Gamma = 1,000$) interact weakly and exchange energy ineffectively. This is also true even under strong coupling, as is the case in Figure 2.3d. It can be observed that the energy absorbed by the device is always very small and the total system energy is also decaying with a rate of $\exp(-(\Gamma_S+\Gamma_D)t)$. Hence, no efficient energy exchange and transfer are achieved in this non-resonant case.

2.4 Relay Effect of Wireless Power Transfer

The previously described results tell us that strongly coupled resonance can increase efficiency by effectively "tunneling" the magnetic field from the source to the device. Specifically, when two resonators are in the mid-range (namely a few times the resonator size), their near fields (evanescent waves) will strongly couple with each other, which will allow the total energy focused in a specific resonant frequency to tunnel/transfer from one resonator to the other resonator within times much shorter than the times of losses in the system.

   However, the current system configuration with the "source–device" scheme must operate in the mid-range, which is often too short in practical applications, especially when the resonator size must be made small. We attack this fundamental limitation by providing a relay solution with additional relay resonator(s) as shown in Figure 2.4.

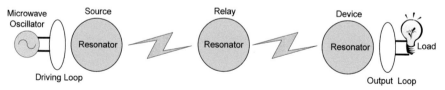

Figure 2.4   Main components of relayed wireless electric system.

For the relayed system, energy transfer operation is modeled in the following matrix form:

$$
\frac{d}{dt}
\begin{bmatrix}
a_1(t) \\
a_2(t) \\
a_3(t) \\
\vdots \\
a_n(t)
\end{bmatrix}
$$

$$
=
\begin{bmatrix}
(i\omega_1 - \Gamma_1) & ik_{12} & ik_{13} & ik_{14} & \cdots & ik_{1n} \\
ik_{21} & (i\omega_2 - \Gamma_2) & ik_{23} & ik_{24} & \cdots & ik_{2n} \\
ik_{31} & ik_{32} & (i\omega_3 - \Gamma_3) & ik_{34} & \cdots & ik_{3n} \\
\vdots & \vdots & \vdots & \vdots & \ddots & \vdots \\
ik_{n1} & ik_{n2} & ik_{n3} & ik_{n4} & \cdots & (i\omega_n - \Gamma_n)
\end{bmatrix}
\cdot
\begin{bmatrix}
a_1(t) \\
a_2(t) \\
a_3(t) \\
\vdots \\
a_n(t)
\end{bmatrix}
$$

$$(2.17)$$

where subscript n represents an n-object physical system. In the above equation, i and j are the decimal numbers ranging from 1 to n (the number of resonators, $n \geq 2$); $a_i(t)$ denotes the signal amplitude of the resonator in the amplitude matrix A; $\omega_i = 2\pi f_i$ is the individual angular frequencies in the frequency matrix; k_{ij} is the coupling coefficient between two resonators in the coupling matrix; and Γ_i is the individual intrinsic decay rate in the loss matrix.

2.4.1 Relay Effect

As shown in Figure 2.4, the relayed wireless electric scheme consists of at least three resonators (source, relay(s), and device), a driving loop, and an output loop. The resonant evanescent strong coupling mechanism between source and device can be mediated by the presence of the relay(s). Through the overlaps of the non-radiative lossless near fields both between source and relay and between relay and device, the distance of efficient energy transfer is extended despite k not being large enough to provide strong coupling at the distance between the source and the device. In this way, "far-range" wireless power transfer can be achieved. In order to clearly understand how this system works, for the ease of analysis, in this section, we will focus on the resonant system having three identical resonators. With the initial conditions $a_s(0) = 1$, $a_R(0) = 0$, and $a_D(0) = 0$, from Equation (2.17), we can obtain

$$a_S = e^{(i\omega - \Gamma)t} \cos(\sqrt{2}kt)/2 + e^{(i\omega - \Gamma)t}/2 \qquad (2.18)$$

$$a_R = ie^{(i\omega - \Gamma)t} \sin(\sqrt{2}kt)/\sqrt{2} \tag{2.19}$$

$$a_D = e^{(i\omega - \Gamma)t} \cos(\sqrt{2}kt)/2 - e^{(i\omega - \Gamma)t}/2 \tag{2.20}$$

2.4.2 Time-Domain Comparison between Relayed and Original Witricity Systems

In Figure 2.5a, the total system energy decays at a rate of $\exp(-3\Gamma t)$ ($f_{S,R,D} = 1\,\mathrm{MHz}$, $k = 500,000$, $\Gamma = 1000$). It can be observed that the energy of the relay resonator is kept at a lower level compared to source and device. It is interesting that the energy leaves the relay (to device) as soon as it reaches relay (from source) and that the frequency of energy oscillation of the relay is twice the device or the source. With the presence of the relay resonator, the device and the source can interact more strongly, exchange energy more quickly, and deliver power more efficiently when compared with the system without the relay resonator (Figure 2.5b). Obviously, if we increase the time

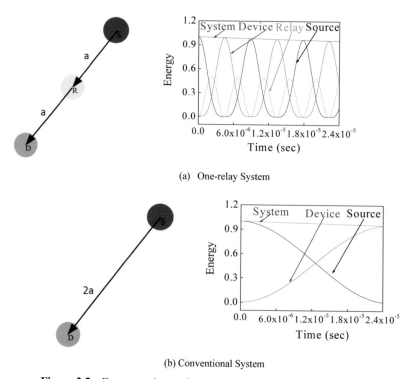

(a) One-relay System

(b) Conventional System

Figure 2.5   Energy exchange for one-relay and conventional systems.

axis in Figure 2.5b, a sinusoidal energy exchange can still take the maximum at a certain time point. However, at this time, the total system energy will be reduced to a much lower level due to the losses of the system. For the clarity, we use red, yellow, and green to represent "source," "relay," and "device" resonators, respectively, in all remaining figures.

Our results show that the efficiency of power transfer can be improved significantly using one or more relay resonators. This approach significantly improves the performance of the present two-resonator system and allows a curved path in space to be defined for wireless power transfer using smaller resonators.

2.5 Wireless Power Transfer with Multiple Resonators

2.5.1 General Solution for Multiple Relays

The general form of the multiresonator wireless power system in Equation (2.17) can be modeled in the following coupled-mode matrix form:

$$\frac{d}{dt}A = UA = KA + (iW - \Gamma)A \qquad (2.21)$$

where

$$A = \begin{bmatrix} a_1(t) \\ a_2(t) \\ a_3(t) \\ \vdots \\ a_n(t) \end{bmatrix}, \quad K = \begin{bmatrix} 0 & ik_{12} & ik_{13} & ik_{14} & \cdots & ik_{1n} \\ ik_{21} & 0 & ik_{23} & ik_{24} & \cdots & ik_{2n} \\ ik_{31} & ik_{32} & 0 & ik_{34} & \cdots & ik_{3n} \\ \vdots & \vdots & \vdots & \vdots & \ddots & \vdots \\ ik_{n1} & ik_{n2} & ik_{n3} & ik_{n4} & \cdots & 0 \end{bmatrix},$$

$$W = \begin{bmatrix} \omega_1 & 0 & 0 & 0 & \cdots & 0 \\ 0 & \omega_2 & 0 & 0 & \cdots & 0 \\ 0 & 0 & \omega_3 & 0 & \cdots & 0 \\ \vdots & \vdots & \vdots & \vdots & \ddots & \vdots \\ 0 & 0 & 0 & 0 & \cdots & \omega_n \end{bmatrix} \quad \text{and}$$

$$\Gamma = \begin{bmatrix} \Gamma_1 & 0 & 0 & 0 & \cdots & 0 \\ 0 & \Gamma_2 & 0 & 0 & \cdots & 0 \\ 0 & 0 & \Gamma_3 & 0 & \cdots & 0 \\ \vdots & \vdots & \vdots & \vdots & \ddots & \vdots \\ 0 & 0 & 0 & 0 & \cdots & \Gamma_n \end{bmatrix}.$$

In the above equation, i and j are the decimal numbers ranging from 1 to n (the number of resonators, $n \geq 2$); $a_i(t)$ denotes the signal amplitude of the resonator in the amplitude matrix A; $\omega_i = 2\pi f_i$ is the individual angular frequencies in the frequency matrix; k_{ij} is the coupling coefficient between two resonators in the coupling matrix; and Γ_i is the individual intrinsic decay rate in the loss matrix.

Assuming that V is the modal matrix whose columns are the eigenvectors of K and that D is the canonical form of K, that is, D is a diagonal matrix with K's eigenvalues on the main diagonal, we have $KV = VD$, or $K = VDV^{-1}$. Using this canonical matrix representation, we note $B = \exp((iKD + iW - \Gamma)t)$. Thus, the analytic solution of Equation (2.17) can be written as

$$A = VBV^{-1}A(t=0) \tag{2.22}$$

where is the initial amplitude matrix. This general solution can work for any number of sources, relays, and devices in 1D, 2D, or 3D scheme to investigate the energy exchange $|A^2|$ between the resonators.

As a key distance-dependent figure of merit, the effective coupling parameter represents, intuitively, the ratio of "how fast energy is transferred between the source and the device" to "how fast it is dissipated due to intrinsic losses in these resonators." Since the intrinsic losses are constant, the speed of energy exchange becomes the evaluation metrics for the wireless power transfer network. The energy exchanges between the resonators can be mediated within a channel or tunnel fully filled with strong resonant evanescent fields. This tunneled energy form will travel between the resonators in an oscillatory fashion, where their loss time between the resonators is required to be much longer than the coupling time to keep the operation of the strong fields. Thus, the Fourier transform could become an effective tool to investigate the oscillatory frequency or the speed of energy exchange in the spectral domain.

2.5.2 Inline Relay(s)

2.5.2.1 One relay

The relay configuration shown in Figure 2.5a can greatly enhance the performance of the wireless power transfer system with a high efficiency, longer distance, and flexible routing. We have analyzed the solutions for one-relay witricity system, where the weaker coupling between the source and device resonators was ignored to observe the difference between original and relayed conditions. Here, we consider all coupling items with a coupling matrix

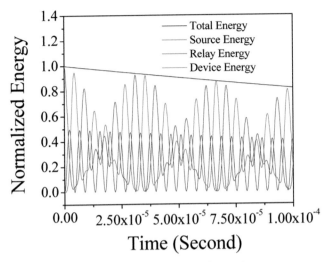

Figure 2.6 Energy exchanges of one-relay case.

[0 1 1/8; 1 0 1; 1/8 1 0] k. After substituting them into Equation (2.22), the energy exchanges among resonators are plotted in Figure 2.6.

2.5.2.2 Two relays

Figure 2.7a gives the locations of two-relay resonators with the corresponding energy exchanges shown in Figure 2.7b. The coupling matrix under this condition is [0 27/8 27/64 1/8; 27/8 0 27/8 27/64; 27/64 27/8 0 27/8; 1/8 27/64 27/8 0] k.

2.5.2.3 Spectral analysis of energy exchanges

Figure 2.8 demonstrates the energy exchange in the spectral domain for the above three inline cases (Figures 2.5b, 2.6a, and 2.7a). Apparently, from the first two waveforms, we can see that the relay for a fixed separation between the source and device resonators accelerates the energy exchanges and improves the efficiency, which conforms to our relay experiments as well. While from the comparisons between the last two waveforms, we can know that additional relays for a certain separation will further speed up the energy exchanges when the source and the device are fixed. This matched with the experimental observation that two relays can improve the effective transmission distance and boost the system efficiency than single relay.

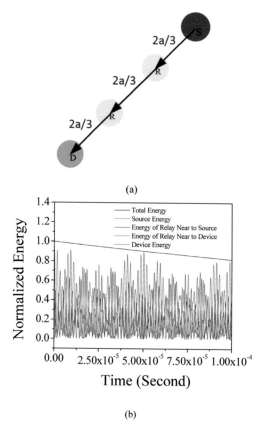

(a)

(b)

Figure 2.7   (a) Resonator position for two relays and (b) energy exchanges.

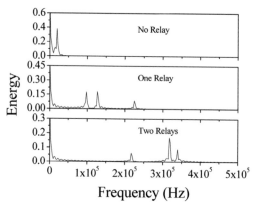

Figure 2.8   Comparisons of energy exchanges for inline cases in the spectral domain.

2.5.3 Optimization of 2D WPTN Scheme

However, it is worth noting that the performance of different conditions even with the same number of relays will also vary. In the next section, we will continue with two conditions with two relays but in 2D scheme to discuss how to obtain the optimum positions to maximize the system efficiency. In order to understand the physical mechanisms of our method more easily, we will focus on three different configurations to show the effectiveness.

2.5.3.1 Case 1 with two relays

In this case, the two relays are placed symmetrically at the two sides of the source and device resonators as shown in Figure 2.9a. The coupling matrix is $[0 \ 1/2/\sqrt{2} \ 1/2/\sqrt{2} \ 1/8; \ 1/2/\sqrt{2} \ 0 \ 1/8 \ 1/2/\sqrt{2}; \ 1/2/\sqrt{2} \ 1/8 \ 0 \ 1/2/\sqrt{2}; \ 1/8 \ 1/2/\sqrt{2} \ 1/2/\sqrt{2} \ 0] \ k$, from which we obtain the energy exchanges in Figure 2.9b.

2.5.3.2 Case 2 with two relays

Figure 2.10a gives another case where one relay is moved to the middle of the source and device resonators. Accordingly, the coupling matrix is changed into $[0 \ 1 \ 1/2/\sqrt{2} \ 1/8; \ 1 \ 0 \ 1 \ 1; \ 1/2/\sqrt{2} \ 1 \ 0 \ 1/2/\sqrt{2}; \ 1/8 \ 1 \ 1/2/\sqrt{2} \ 0] \ k$, and the generated energy exchanges are plotted in Figure 2.10b.

2.5.3.3 Spectral analysis of energy exchanges

According to Figure 2.11, the fact of two similar cases with different energy exchange speeds arrives at the conclusion that the optimization of power network, that is, the optimal location of resonators, significantly affects the performance of the whole energy transfer in case where the number of

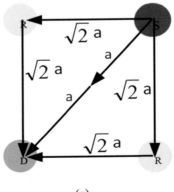

(a)

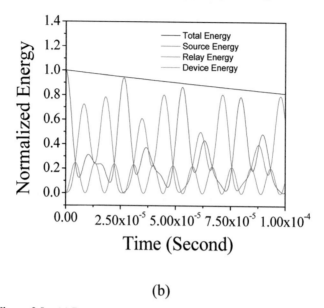

(b)

Figure 2.9   (a) Resonator position for case 1 and (b) energy exchanges.

resonators is the same. The last waveform of Figure 2.8 and the two waveforms in Figure 2.11 together confirm that different configurations of the same number of relays will lead to a different coupling matrix and hence a different energy transfer performance. Apparently, the position strategy of Figure 2.7a

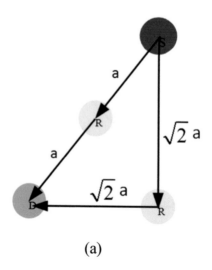

(a)

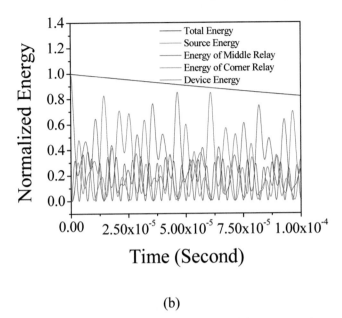

(b)

Figure 2.10   (a) Resonator position for case 2 and (b) energy exchanges.

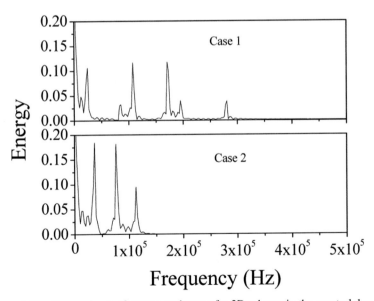

Figure 2.11   Comparisons of energy exchanges for 2D scheme in the spectral domain.

is the best among the three. Inversely, from the other aspect, this fully indicates the necessity of wireless power transfer network (WPTN) optimization and evaluation before real construction and demonstrates the effectiveness of the proposed assessment method by integrating the information from both the time and spectral domains.

2.6 Prototype of Wireless Power Transfer

For practical applications, a flexible thin-film resonator is highly desirable. This type of resonator can be imprinted or integrated with an existing structure, such as a ceiling or a floor, or embedded to the exterior or interior cover of the parent device to which the electric power will be transferred without taking its interior space. This thin-film design not only utilizes the maximum dimensions of the parent device to capture magnetic flux, but also provides the maximum space for the parent device itself. In addition, it facilitates heat dispersion and eases the skin effect of electric current at RF frequencies.

2.6.1 Cylindrical Resonator Design

Our cylindrical resonator design consists of three layers of films as shown in Figure 2.12a: two layers of copper (red) and one layer of insulator (blue). The top and middle panels in Figure 2.12a show, respectively, the top and side views of the resonator. The horizontal narrow copper strips (red) in the middle panel represent a helical inductor. The yellow lines represent the spaces between the copper strips. The bottom panel shows the side view from the interior. Several vertical copper strips (red) are affixed to the insulator film (blue). These vertical strips form physical capacitors with the coil conductors in the exterior layer. Clearly, this thin-film design represents a compact LC tank circuit. In our design, energy transfer is primarily provided by the magnetic field, while the electric field is mostly confined within the physical capacitors. This feature effectively prevents the leakage of the electrical field since the electrically conductive objects interact much more strongly with electric fields than with magnetic fields. Another motivation of our design is to obtain a compact size. By increasing the capacitance using the strips while keeping the same inductance in the LC tank resonator, the operating frequency of the witricity system can be reduced, which is desirable in many practical applications where the size of the parent device is small. In practical applications, the size and shape of the source resonator have fewer restrictions and can be larger than those of the device resonator. A larger source resonator can produce stronger magnetic fields for a longer transmission range. Conversely, the size of the device resonator, though preferred to be as large as possible, is usually limited by the size of the parent device.

2.6.2 Implementation of Cylindrical Resonator

Based on the design in Figure 2.12a, a smaller receiver resonator was made as shown in Figure 2.12b. The exterior copper strip with 0.635 cm width formed a six-turn coil, and six copper strips with 2.54 cm width were affixed to the insulator film on the internal side. The radius and height of the cell were about 8.1 and 5 cm, respectively. Similarly, we also designed much smaller resonators for other body parts such as arms. A seven-turn coil was used as the output coil to connect and power the load. We also set its resonant frequency to be 7 MHz and measured its Q value as 56.38 with the Agilent 8753ES vector

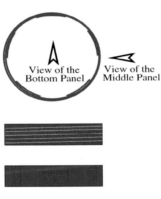

(a) Resonator Design

(b) Real Picture

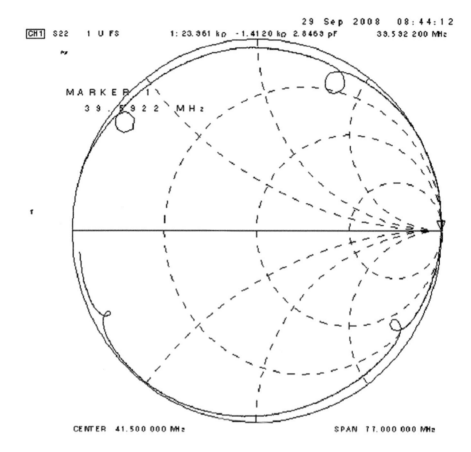

29 Sep 2008   08:44:12
CH1 S22   1 U FS   1: 23.361 kΩ -1.4120 kΩ 2.8463 pF   33.532 200 MHz

MARKER 1
39.3922 MHz

CENTER 41.500 000 MHz   SPAN 77.000 000 MHz

(c) Measured Smith Chart.

Figure 2.12   Construction and measurement of flexible film resonators.

network analyzer as shown in Figure 2.12c. For the purpose of observation, an LED was first used as the load, simulating the load effect of electronic devices and allowing visual examination. This LED was then replaced by a resister from which quantitative measurements was taken. Due to the unique structure of multiple conductor strips, three main resonant frequencies were obtained from the Smith chart at 7.02, 24.64, and 47.85 MHz, which match well with our simulation (6.91, 25.88, and 51.05 MHz, respectively). This enables simultaneous power transfer and data communication through the use of different resonant frequencies.

2.6.3 Evaluation of Cylindrical Resonator

Figure 2.13a shows a functional 7-MHz wireless electric system without relay over a 50-cm separation. Due to the long distance, the LED could not be lit. However, when we placed a relay resonator in the system, the LED was illuminated fully, as shown in Figure 2.13b. Figure 2.13c shows that the distance between the source and the device can be increased and more power can be transferred with one more relay.

After replacing the LED with a resistor, we measured RF power values at both the input terminals of the driving loop and the load terminals of the output loop. The power transfer efficiencies are shown in Figure 2.14a. It can be seen that the conventional wireless electric system can only achieve about 10% efficiency over a 30-cm distance, whereas the efficiency of the relayed system can reach up to 46%. Figure 2.14b shows the power transfer efficiency as the relay resonator moved between the source and device resonators which were separated by a distance of 50 cm. It was observed that, for this particular case, the midpoint between the source and the device is the best position for the relay to achieve the maximum efficiency in the "source–relay–device" scheme.

In a separate experiment, we found that we could utilize the nature of magnetic resonances to guide the direction of magnetic flux and hence control the transmission direction using relays. We believe that this was due to the attraction of the magnetic flux by relay. Thus, with the help of relay(s), a wireless energy connection can turn a corner, just like a wire connection. As shown in Figure 2.15a, if there was no relay, the LED, which was connected to the output coils mounted on the device resonator, did not glow. After we

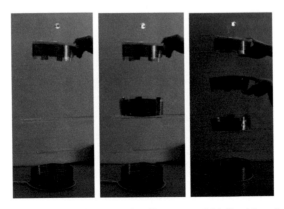

(a) Conventional System, (b) One-relay System, and (c) Double-relay System

Figure 2.13 Conventional and relayed power transfer in the vertical direction.

placed a relay resonator as shown in Figure 2.15b, the LED was illuminated. This indicated that the relay system guided magnetic flux around the 90° corner. After realizing this "L"-shaped wireless connection, in Figure 2.15c, we added two additional relays to form a "Z" shape. Removing any of the two relays turned off the LED, as in Figure 2.15d, where the first relay was removed. These observed phenomena strongly support the promise of future applications of "source–relay(s)–device" scheme.

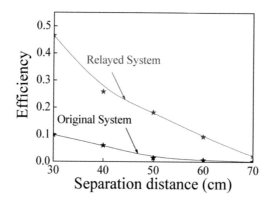

(a) Efficiencies versus Distance.

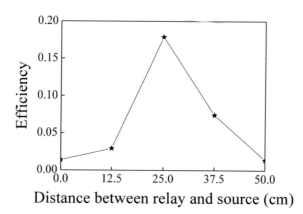

(b) Location of Optimal Relay Position.

Figure 2.14   Measurements of conventional and relayed wireless electric systems.

2.6.4 Application of Cylindrical Resonator

Wireless body sensor networks (wBSN) have emerged in recent years as a key enabling technology to address a number of significant and persistent challenges in health care and medical research, including continuous, noninvasive, and inexpensive monitoring of physiological variables. Typically, a wBSN is composed of a number of sensor nodes dedicated to different forms of measurements, such as the electrocardiogram (ECG), electromyogram (EMG), body temperature, glucose, and blood pressure. Each sensor node, which can be either inside (as an implant) or outside (as a wearable

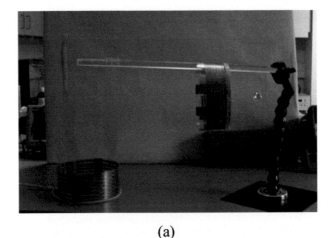

(a)

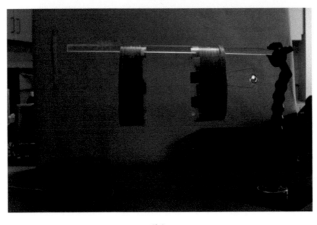

(b)

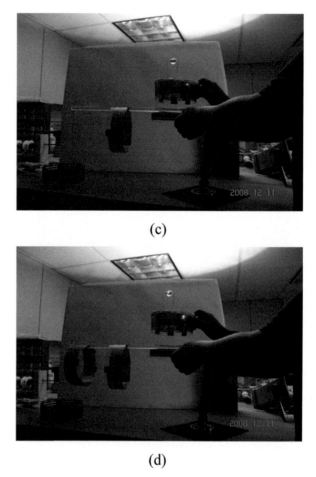

(c)

(d)

Figure 2.15   Directional guidance via the relay resonator.

device) the human body, is usually composed of an analog readout front end, a microprocessor, a radio transmitter/receiver, and a power supply. These sensor nodes, wire-connected to a battery, transmit data continuously through a wireless connection to a central node, typically a PDA or a smart cell phone. The central node collects, visualizes, and analyzes data, and/or wirelessly relays the data or partially processed results to a remote terminal for more advanced off-line processing or evaluation by healthcare professionals. While longer battery capacity, lower power consumption, smaller size of the battery and other circuit components, and higher manufacturing volumes have made wBSN data collection more continuous, noninvasive, and inexpensive, the progress toward *wireless powering* has remained to be a significant, unsolved problem.

In our design, we made a larger source resonator in the form shown in Figure 2.12a. It was a large flexible ring with diameter 350 mm, thickness 0.35 mm, and width 29 mm. The insulator had a dielectric constant of 2.74. The exterior copper tape (width 0.635 cm) formed a four-turn coil. Eight copper strips with 2.54 cm width were affixed to the insulator film on the internal side. The resonator was incorporated into the waist belt, and the one-turn driving coil (a loop) linked to the RF power source can was attached on the belt to make the system wearable. Using vector network analyzer, we set its resonant frequency to be 7 MHz and Q value was measured to be 51.62.

Figure 2.16a shows an experimental 7-MHz witricity system on a work-bench. Wireless energy fully illuminated the low-power LED. When we misaligned the axes of the source and device resonators, the power was still transmitted efficiently as shown in Figure 2.16b. This phenomenon differs from that of the conventional magnetic induction methods. In order to evaluate system performance quantitatively, we replaced the LED with a resistor whose resistance approximates the impedance of the output terminals at resonance. Utilizing a diode-based RF detector, we measured the RF energy at the input terminals of the driving loop and the energy at the load terminals of the output loop. Then, the power transfer efficiencies with and without misalignment are measured and plotted in Figure 2.17. It can be seen that the efficiency without misalignment can reach approximately 80% at a 15-cm separation between the transmitter and the receiver. It can also be seen that certain misalignments (e.g., 5 cm) between the transmitter and the receiver cause only a slight drop in efficiency. This tolerance in misalignment is highly desirable in practical applications since, for example, it allows energy to be transmitted to a moving target or multiple locations as in the case of wBSN.

In our wBSN design, the large cylindrical resonator can be integrated with a waist belt. A separate device containing a battery and electronic circuits is attached to the belt to generate the required RF signal and host electronic circuits. Several smaller cylindrical and planar receiver resonators could be embedded in the clothes or implanted within the body, such as the head, chest, abdomen, arms, and/or legs, after being integrated with the implantable devices as described previously.

In a separate experiment, we utilized a source resonator and several device resonators to simulate the body sensor network. In order to eliminate the health concerns, we use a maximum 500-mW power as the source. Each of these device resonators is magnetically coupled with an energy pickup coil which

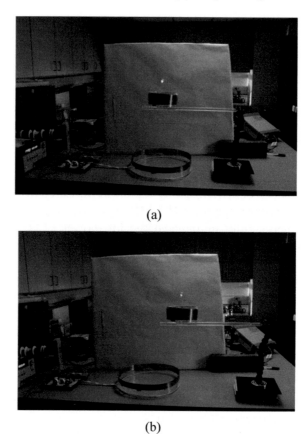

(a)

(b)

Figure 2.16   Witricity system working at 7 MHz: (a) alignment setup and (b) misalignment setup.

is connected to a low-power LED as a load. Figure 2.18 shows that the LEDs, located on the hand, the head (Figure 2.18a), and the limbs (Figure 2.18b), are lit by the energy transmitted from the waist belt transmitter. If each LED is replaced by a rectification and regulation circuit, the power produced can be used to operate a microsensor which is either outside of the body or implanted inside so that we could bypass the batteries needed by these devices to operate. Similarly, if the resonant signal is modulated appropriately, the wireless system can also perform communication tasks. Clearly, this qualitative experiment shows that it is feasible and convenient to use the witricity as a new tool to construct wBSN systems to perform a wide variety of diagnostic or monitoring functions.

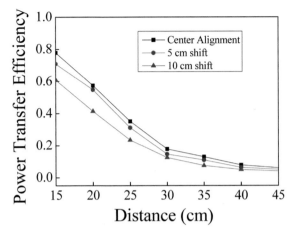

Figure 2.17 Measured transmission efficiency versus distances of separation for cylindrical design.

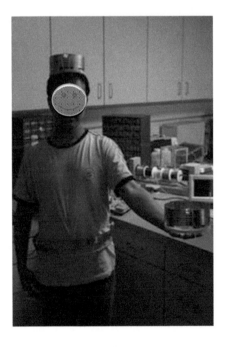

(a)

(b)

Figure 2.18   Wireless power transfer network experiments.

2.7 Discussion

Although, currently, long-life lithium ion batteries and methanol fuel cells have been persuaded as ways to make the electrical components more mobile, consumers' expectations are still far from being met due to the added weight and expenses for battery replacement. The discovery of witricity, as a new option, is revolutionizing the wireless industry and holds great promise to leave the oversized battery as a thing of the past.

With a lock-and-key mechanism, the witricity enables energy to be transferred only to the intended target, in which the source and device resonators are both tuned to the same resonant frequency. The fact that the strongly coupled regime through interaction functions when the two resonators are separated by a long distance and even blocked by an object brings people considerable confidence for the exploration of practical applications. In addition, non-radiative fields ensure that little energy is lost or would adversely affect the environment. The relay [19] and resonance enhancement [21] effects of witricity, with the network optimization concept presented in this chapter, will further enhance the robustness of building a successful WPT network.

As for the practical issues raised in [23], there is still a long way for wireless electricity to be widely and safely adopted in the human society to replace the presently wired systems. In this early stage of development, theoretical advancements are important and necessary [24–26]. This chapter thus focuses on theoretical aspects and proof of concepts rather than aiming at a specific application of wireless electricity. Nevertheless, we envision many potential applications, such as transportation, industry, medicine, and consumer electronics where wirelessly power is very attractive. Its recent success in preliminary applications to consumer electronics and biomedical sensors has received considerable attention in wireless powering and recharging. We believe that the relay effect described in this chapter will alleviate the short-range problem in some of the present designs and accelerate practical adaptations of the wireless electricity.

2.8 Conclusions

Witricity has come to fruition in providing power as Wi-fi and Bluetooth provide wireless data communication. This chapter has presented a general overview of wireless power transfer using strongly coupled magnetic resonance [10–15, 20, 21]. We have analyzed the newly reported relay effect where one or more relay resonators are added to the two-resonator system to extend the WPT distance, increase WPT efficiency, and allow for a curved wireless transmission path in space. This chapter also presents a novel spectral analysis approach to facilitate system optimization and evaluation based on energy exchange in both the time and spectral domains. Theoretical analysis has been presented to enhance the understanding of the oscillatory behaviors when multiple resonators are involved. Our analytical results are in good agreement with the experimental results.

References

[1] Tesla, N. "System of Transmission of Electrical Energy," *U.S. Patent,* 0645576 (1900).

[2] Esser, A., and H.C. Skudelny. "A New Approach to Power Supplies for Robots." *IEEE Transactions on Industry Application* 27: 872–875, September (1991).

[3] Hirai, J., T.W. Kim, and A. Kawamura. "Wireless Transmission of Power and Information for Cableless Linear Motor Drive." *IEEE Transactions on Power Electronics* 15: 21–27, January (2000).

[4] Ka-Lai, L., J.W. Hay, and P.G.W. Beart. "Contact-Less Power Transfer." *U.S. Patent,* 7042196 (2006).

[5] Kurs, A., A. Karalis, R. Moffatt, J. D. Joannopoulos, P. Fisher, and M. Soljacic, "Wireless Power Transfer Via Strongly Coupled Magnetic Resonances." *Science* 317: 83–86, July (2007).

[6] Karalis, A., J.D. Joannopoulos, and M. Soljacic. "Efficient Wireless Non-Radiative Mid-Range Energy Transfer." *Annals of Physics* 323: 34–48, January (2008).

[7] Ho, S.L., J. Wang, W. Fu, and M. Sun. "A Comparative Study Between Novel Witricity and Traditional Inductive Magnetic Coupling in Wireless Charging." *IEEE Transactions on Magnetics:* 1522–1525, May (2011).

[8] Hori, Y. "Motion Control of Electric Vehicles and Prospects Of Supercapacitors." *IEEJ Transactions on Electrical and Electronic Engineering* 4: 231–239, February (2009).

[9] RamRakhyani, A., S. Mirabbasi, and M. Chiao. "Design and Optimization of Resonance-Based Efficient Wireless Power Delivery Systems for Biomedical Implants." *IEEE Transactions on Biomedical Circuits and System* 5: 48–63, February (2009).

[10] Zhang, F., S.A. Hackworth, X. Liu, C. Li, and M. Sun. "Wireless Power Delivery for Wearable Sensors and Implants in Body Sensor Networks." *IEEE EMBC,* Buenos Aires, Argentina, September (2010).

[11] Liu, X., F. Zhang, S. A. Hackworth, R.J. Sclabassi, and M. Sun. "Modeling and Simulation of a Thin Film Power Transfer Cell for Medical Devices and Implants." *IEEE ISCAS*, Taibei, Taiwan, pp. 3086–3089, May 24–27 (2009).

[12] Zhang, F., X. Liu, S.A. Hackworth, R.J. Sclabassi, and M. Sun. "Wireless energy transfer platform for medical sensors and implantable devices." *IEEE EMBS*, Hilton Minneapolis, MA, USA, September 2–6 (2009).

[13] Zhang, F., X. Liu, S.A. Hackworth, R.J. Sclabassi, and M. Sun. "In Vitro and In Vivo Studies on Wireless Powering of Medical Sensors and Implantable Devices." *IEEE LiSSA*, Bethesda, MD, USA, April 9–10 (2009).

[14] Liu, X., F. Zhang, S.A. Hackworth, R.J. Sclabassi, and M. Sun. "Wireless Power Transfer System Design for Implanted and Worn Devices." *35th IEEE Northeast Bioengineering Conference*, Harvard-MIT, Boston, MA, USA, April 3–5 (2009).

[15] Zhang, F., X. Liu, S. A Hackworth, M.H. Mickle, R.J. Sclabassi, and M. Sun. "Wireless Energy Delivery and Data Communication for Biomedical Sensors and Implantable Devices." *35th IEEE Northeast Bioengineering Conference*, Harvard-MIT, Boston, MA, USA, April 3–5 (2009).

[16] http://www.witricity.com/pages/technology.html and http://www.gizmo watch.com/entry/haier-develops-completely-wireless-hdtv-courtesy-witricity/

[17] Cannon, B.L., J.F. Hoburg, D.D. Stancil, and S. C. Goldstein. "Magnetic Resonant Coupling as a Potential Means for Wireless Power Transfer to Multiple Small Receivers." *IEEE Transactions on Power Electronics* 24: 1819–1826, July (2009).

[18] Hamam, R.E., A. Karalis, J.D. Joannopoulos, and M. Soljacic. "Efficient weakly-radiative wireless energy transfer: an EIT-like approach." *Annals of Physics* 324: 1783–1795, August (2009).

[19] Kurs, A., R. Moffatt, and M. Soljacic. "Simultaneous Mid-Range Power Transfer to Multiple Devices." *Applied Physics Letter* 96: 044102, January (2010).

[20] Zhang, F., S.A. Hackworth, W.N. Fu, and M. Sun. "The Relay Effect on Wireless Power Transfer Using Witricity." *The 14th Biennial IEEE Conference on Electromagnetic Field Computation*, Chicago, IL, USA, May 9–12 (2010).

[21] Zhang, F., S.A. Hackworth, W. Fu, C. Li, Z. Mao, and M. Sun, "Relay Effect of Wireless Power Transfer Using Strongly Coupled Magnetic Resonances." *IEEE Transactions on Magnetics* 47: 1478–1481, May (2011).

[22] Haus, H. *Waves and Fields in Optoelectronics*. Prentice-Hall: Englewood Cliffs (1984).

[23] Schneider, D. "Wireless Power at a Distance is Still Far Away." *IEEE Spectrum:* pp. 35–39, May (2010).

[24] Chen, C., T. Chu, C. Lin, and Z. Jou. "A Study of Lossely Coupled Coils for Wireless Power Transfer." *IEEE Transactions on Circuits and Systems Part II* 57, July (2010).

[25] Cheon, S., Y. Kim, S. Kang, M. Lee, J. Lee, and T. Zyung. "Circuit-Model-Based Analysis of a Wireless Energy-Transfer System via Coupled Magnetic Resonances." *IEEE Transactions on Industrial Electronics* 58: 2906–2913, July (2011).

[26] Ho, S., J. Wang, W. Fu, and M. Sun. "A Novel Resonant Inductive Magnetic Coupling Wireless Charger with TiO_2 Compound Interlayer." *Journal of Applied Physics* 109: 07E502, July (2011).

3

Low Power Rectenna Systems for Wireless Energy Transfer

**Vlad Marian, Christian Vollaire, Jacques Verdier
and Bruno Allard**

Ecole Centrale de Lyon, INSA Lyon, Ampère UMR 5005, INL UMR 5270,
36 Av. Guy de Collongues, F-69134, Ecully, France
E-mail: {vlad.marian; christian.vollaire}@ec-lyon.fr;
{jacques.verdier; bruno.allard}@insa-lyon.fr

This chapter gives a general description of RF-DC conversion circuits, particularly those designed for low input power levels. The general concepts and design techniques, as well as the most common circuit topologies, are presented.

3.1 Introduction

The last few years have been characterized by massive development of a wide range of portable electronic devices, not only consumer devices like smartphones but also industrial applications, like wireless sensor networks [1–3]. The trend is toward increased miniaturization of devices in order to facilitate their portability and their integration in the environment. One of the most important problems that need to be solved is ensuring their energy supply. The use of electrical wires is in many cases impossible due to the "wireless" nature of such devices. Most portable electronic devices nowadays rely on batteries for power. For small compact devices, batteries tend to occupy most of the available volume, thus making further miniaturization even more difficult and adding weight and cost.

The autonomy of such devices is limited because the trade-off on batteries regarding size and power density. Batteries need to be periodically replaced or recharged. Used batteries are highly polluting, and battery recycling is a complex and costly process. In the case of rechargeable batteries, the charging process usually relies on a wall plug charger, which somehow limits the portability of a wireless device. Wireless supply systems are supposed to improve the availability, the reliability, and the user-friendliness of portable electronic devices. The natural evolution of power transfer seems to take the same path as information transfer: cutting the cord and stepping in the wireless era.

3.1.1 History of Wireless Power Transfer

Research on wireless power transfer began at the end of the nineteenth century when Hertz and Marconi discovered that energy could be transported from one place to another without the existence of a conductive environment. In the early twentieth century, Nikola Tesla was already working on the Wardenclyffe Tower, a prototype base station serving as an emitter for his "World Wireless System" which would wirelessly supply electrical energy to a distant receiver [4].

Figure 3.1 Nikola Tesla holding a gas-filled phosphor-coated light bulb which was illuminated without wires by an electromagnetic field from the "Tesla coil" [4].

During the first half of the twentieth century, the lack of interest toward wireless power transfer (WPT) can be explained by the lack of technical means, especially high-power and high-frequency generators.

The modern history of WPT began with the Raytheon Airborne Microwave Platform (RAMP) Project initiated by the US Army in the 1950s. Developed in the middle of the cold war, the RAMP Project aimed putting a high-altitude observation platform capable of remaining airborne for long periods of time. The research was led by William C. Brown and led to a demonstration in 1964 of a helicopter platform that flew at an altitude of 18 m while being powered exclusively through a microwave beam from the ground [5].

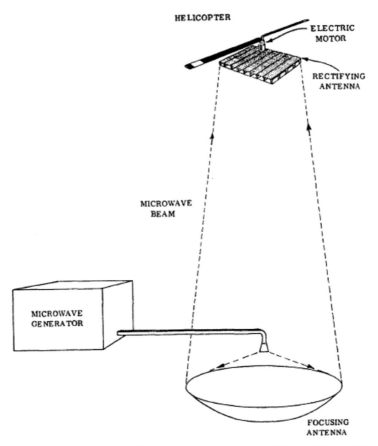

Figure 3.2   General principle of a microwave-powered flying platform [5].

The Solar Power Satellite (SPS) Project is by far the most ambitious and the best-known project dealing with high-power WPT. The general idea was developed by Peter Glaser in 1968. The project consists of placing a satellite in orbit that permanently captures solar energy and sends it toward ground stations through microwave beams. This approach offers the advantage of permanent exposition to a much higher incident power level than on the ground. Although purely theoretical, this project showed that WPT could reach efficiencies close to 100%, provided that the emitter and receiver sizes are large enough and well designed. At the same time, the weight-to-DC power ratio of rectennas was reduced from 5.4 to 1 kg/kW [6]. The SPS Project was never put into practice due to the massive size of the proposed satellite. For 5 GW of transmitted power, the satellite would need to be 5.2 km wide and 10.4 km long, with an emitter antenna of around 1 km in diameter. The cost of such a device is estimated at around 20–40 €/W compared to around 13 €/W for on-ground solar power and around 3.5 €/W for nuclear power [6].

In 1975, the on-ground experiments conducted by the Raytheon Corporation at Goldstone represent a major historic turning point in the domain of WPT. It significantly contributed to the validation of the concept and the credibility of the SPS Project. A total power of 30 kW was transmitted over a distance of 1.6 km with an overall efficiency of 54% [7].

3.1.2 Wireless Power Transfer Techniques

Although the strategies can be very different, wireless power transfer in general is a three-stage process. AC or DC electrical energy is supplied to a high-frequency generator and then fed to the transmitter structure. The electromagnetic wave then travels wirelessly to the receiver structure which feeds a down-converter. The resulting AC or DC energy is thus available for a load situated in a remote or enclosed area.

Several different approaches of wireless power supply can be distinguished. Near-field inductive coupling works on very small distances, typically limited to a few centimeters, but is characterized by very good efficiencies [8, 9]. This is widely used for wireless recharging of the internal battery of consumer items like an electric toothbrush or wireless mouse.

Magnetic resonant coupling between two structures (usually circular coils) allows energy transfer in the near-field area. Operating frequencies are relatively low (in the MHz range), making emitter and receiver quite large [10–12]. The main limitation of this method is that energy can only be transferred over relatively low distances (although higher than the inductive

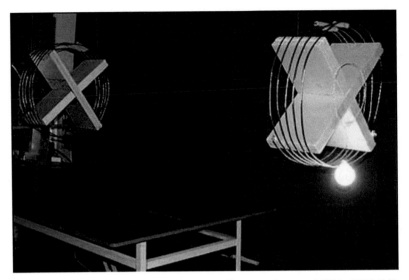

Figure 3.3 WPT by magnetic resonant coupling experiment at MIT in 2008 [10].

coupling zone). Distances are generally in the same order of magnitude as emitter and receiver sizes, and an efficient transfer is only achieved around an optimal operating point [13]. Transmitter-to-receiver efficiencies can reach 70% over distances under one meter, but wall-to-load efficiency is under 20%.

Energy can also be transmitted based on a radiative high-frequency field. It uses high-frequency electromagnetic waves, often above 1 GHz, and energy transfer is done in the far-field region. High power transfer over several kilometers has been achieved with efficiencies sometimes in excess of 70% [6], but the number of viable applications at these power levels tends to be limited due to health and safety regulations and impact of large antenna.

This technique is more often used to supply UHF RFID. Compared to classical 13.56-MHz proximity RFID [14], UHF RFID devices can be supplied at distances in excess of 10 m using high-frequency radio waves [15–17]. The concept of wireless energy transfer can also be applied in order to supply low power electronic devices like industrial sensors or sensor networks. These devices can be supplied exclusively either by the energy from the microwave beam [18] or by batteries that can be remotely recharged [19]. The wall-to-load efficiency is unfortunately very low (1% range).

Low power energy transfer using UHF electromagnetic waves, especially the receiver part, is the main focus of this chapter. UHF WPT can be seen as a three-stage process: the conversion of DC energy into electromagnetic waves, the propagation of the electromagnetic waves from the transmitter to the receiver, and the reception and conversion of the incident electromagnetic waves into DC power. Figure 3.4 gives a schematic view of the whole process.

3.1.2.1 DC-RF conversion

UHF sources are based on various technologies. Semiconductor-based sources are predominant, but suffer from a power limitation at high frequencies. At a frequency of 2.45 GHz, the maximal power delivered by a semiconductor-based source is limited to several hundred watts.

3.1.2.2 Electromagnetic wave propagation

An electromagnetic wave has both electric and magnetic field components, which oscillate in phase perpendicular to each other and perpendicular to the direction of energy propagation. The E and B components obey the following laws:

$$\nabla^2 \vec{E} = \mu_0 \varepsilon_0 \frac{\partial^2 \vec{E}}{\partial t^2} \tag{3.1}$$

$$\nabla^2 \vec{B} = \mu_0 \varepsilon_0 \frac{\partial^2 \vec{B}}{\partial t^2} \tag{3.2}$$

For isotropic transmitter antennas, in a certain point in the space, the power is equally distributed on the surface of a sphere of radius D, D being the distance between the transmitter antenna and the observation point. The power density on the surface of this sphere would then be

$$P_d = \frac{P_t}{4\pi D^2} \tag{3.3}$$

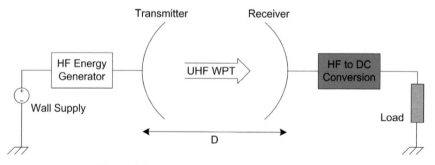

Figure 3.4   Schematic view of the WPT process.

In order to determine the real power at the converter level, this equation must take into account the characteristics of the emitter and receiver antennas. The transmitter-to-receiver power transfer efficiency is evaluated by the Friis equation [20]:

$$P_r = P_t \cdot G_t \cdot G_r \cdot \left(\frac{\lambda}{4\pi D}\right)^2 \qquad (3.4)$$

where P_r, G_r, and P_t, G_t are the power and antenna gain for received and transmitted energy, respectively, λ is the wavelength, and D is the distance between the emitter and the receiver. The form presented is directly derived from Equation 3.3 and is valid for an ideal transmission, without multiple transmission paths. This equation must be completed in order to take into account the losses in the antennas, losses due to multiple paths, and polarization deadaptation:

$$P_r = P_t \cdot G_t \cdot G_r \cdot \left(\frac{\lambda}{4\pi D}\right)^2 \cdot \eta_t \cdot \eta_r \cdot (1 - S_{11}) \cdot (1 - S_{22}) \cdot |\vec{u} \cdot \vec{v}|^2 \cdot \alpha \qquad (3.5)$$

This more realistic form of the equation takes into account the efficiencies of the two antennas (η_t and η_r) as well as the reflection coefficients S_{11} for the emitter antenna and S_{22} for the receiver antenna. The deadaptation due to the polarization is taken into account in the $|\vec{u} \cdot \vec{v}|^2$ term. This term is generally due to two antennas having different polarizations or by two antennas having the same polarization but which are misaligned.

The α term is introduced in order to account for the multiple propagation paths that the transmitted wave can take in order to reach the receiver antenna:

$$\alpha = \left| 1 + \sum_{n=1}^{N} \Gamma_n \cdot \frac{D}{D_n} \cdot e^{-j\frac{2\pi}{\lambda}(D_n - D)} \right|^2 \qquad (3.6)$$

This term takes into account the reflection coefficient Γ_n of each of the obstacles and the length D_n of each propagation path. This implies a perfect knowledge of the environment of the propagating wave.

The power received by the antenna from the propagating wave is an image of the energy contained in the electromagnetic wave. This energy is given by the Poynting vector [20] which is defined as the product of the electric and magnetic field vectors and is having the same direction as the propagation of the electromagnetic wave. The module of the Poynting vector is given by

$$|R| = \frac{1}{\mu_0 c} E^2 \qquad (3.7)$$

3.1.2.3 RF-DC conversion

The RF-DC conversion is the third stage of the WPT process. The basic cell of an RF-DC converter is called a "rectenna" from "rectifying antenna." A rectenna is usually made out of a receiving antenna, an input HF filter, a diode rectifier, and a DC output filter (Figure 3.5). The input filter acts as an impedance match between the antenna and the diode rectifier. The output low-pass filter rejects the harmonics generated by the nonlinear diode behavior. The output load represents any DC load.

The RF-DC conversion efficiency of a rectenna is influenced not only by the amount of power loss in the diodes and by the impedance match between the antenna and the rectifier and between the rectifier and the load, but also by the antenna efficiency. For a rectenna, the RF-DC conversion efficiency is usually defined as the ratio of the total amount of power delivered to the load to the amount of power that the receiving antenna could inject in a perfectly matched circuit:

$$\eta = \frac{P_{\text{DC\_out}}}{P_{\text{RF\_in}}} = \frac{V_{\text{out}}^2}{R_{\text{load}}} \cdot \frac{4\pi \cdot Z_{\text{air}}}{|E|^2 \cdot G \cdot \lambda^2} \tag{3.8}$$

where R_{load} is the load resistance, Z_{air} is the air characteristic impedance, E is the electric field efficient value at receiver position, G is the receiver antenna gain, and λ is the wavelength.

Dipole or printed dipole and linearly polarized patch antennas are most often used in rectenna circuits. Having a high-gain antenna is preferable as more RF power can be collected, and thus, more DC power will be present at load level. The gain of an antenna is, however, proportional to its equivalent

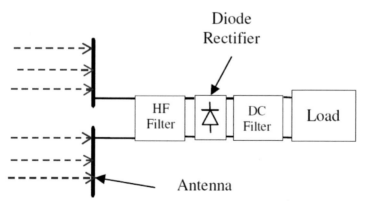

Figure 3.5   Basic schematic of a rectenna circuit.

surface. There is a compromise to make between the gain of an antenna and its surface.

Depending on the application, linear polarization [21] or circular polarization [22] can be used. Circular polarization offers the possibility to keep the DC voltage constant even if there is a rotation of the emitter or the receiver. However, the gain of the receiver antenna usually loses 3 dB compared to a linearly polarized antenna. Double-band [23], triple-band [24], or wideband antennas [25] have been developed for specific applications.

Using UHF electromagnetic waves for power transfer applications is compatible with system miniaturization, but a trade-off is often settled between antenna size and power transfer efficiency. It is easy to see that high-gain antennas improve energy transfer efficiency, but higher gain usually means bigger antenna. In small miniaturized systems, the antenna size is limited, but it is still possible to design rather small antennas with relatively good directivity and gain [26]. A solution in the case of small receiver antenna is to use a high-directivity emitter antenna or higher transmitted power levels in order to ensure a desired energy density (in mW/cm²) at receiver level. The focus is then to optimize the RF-DC energy conversion efficiency at receiver level.

3.2 Low Power Rectenna Topologies

Low power WPT can be defined as the process of sending and receiving energy wirelessly in which the received power is inferior to 1 W. The received power level depends on the application and can be as low as several μW, as in the case of electromagnetic energy harvesting applications. The basic operating principle of a low power electromagnetic energy receiver and converter is shown in Figure 3.6.

The incident low power electromagnetic energy is captured by the receiving antenna and fed to the RF-DC rectifier under the form of a high-frequency sine wave. The rectifier transforms the energy into a DC voltage and current. The DC voltage output level of the rectenna is often too low in order to ensure the direct supply of an electronic circuit or charge a battery, especially when the distance to the power emitter is important. A voltage boost circuit is often used in order to provide the necessary DC voltage level [27, 28], and an maximum power point tracking (MPPT) DC–DC conversion strategy is often used to ensure an optimal power transfer between the rectenna and the load [29, 30]. In addition, following the miniaturization trend of electronic devices, new microscale rechargeable solid-state batteries are being developed

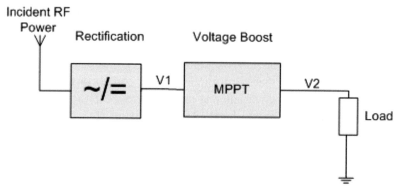

Figure 3.6   Schematic of an EM energy receiver based on a rectenna and an MPPT converter.

and can represent viable energy sources for miniature isolated sensors that can be recharged wirelessly [31, 32].

RF-DC conversion structures are built using diodes or diode-mounted transistors for rectifying the high-frequency sine wave generated by the antenna into DC voltage. Diodes are characterized by a threshold voltage that has to be overcome in order to put them in a conductive state. When important incident power levels are available, diode threshold voltage is not an issue, because incident voltage amplitude is much higher than the threshold voltage [33]. In the case of very low incident power (below 1 mW), loss in the diodes themselves becomes predominant. Most often zero-bias Schottky diodes are used due to their low threshold (around 150 mV) and their low junction capacitance (0.18 pF).

Working frequency is also an important parameter to consider when designing a rectenna. It is often dictated by the desired application. At low frequencies (below 1 GHz), high-gain antennas tend to be quite large. Increasing the frequency thus allows the use of more compact antennas. On the other hand, the amount of available power at a certain distance from the emitter decreases as the frequency increases. Frequencies in the 1–3 GHz range are considered to provide a good compromise between free-space attenuation and antenna dimensions. We chose to design our circuits for a central frequency of 2.45 GHz.

The power transfer between the antenna and the rest of the circuit is optimal when the antennas' characteristic impedance is identical to the conjugate of the impedance of the rest of the circuit:

$$Z_{\text{antenna}} = Z^*_{\text{rectifier}} = Z_0 \tag{3.9}$$

In this case, the power injected by the antenna in the rectifier circuit is

$$PR = \frac{V_{\text{eff}}^2}{4 \cdot Z_0} \tag{3.10}$$

The efficient value of the voltage V_{eff} delivered by the antenna can thus be calculated:

$$V_{\text{eff}} = \sqrt{4 \cdot Z_0 \cdot PR} = \sqrt{4 \cdot Z_0 \cdot \frac{|E| \cdot G \cdot \lambda}{4\pi \cdot Z_{\text{air}}}}$$

$$= |E| \cdot \lambda \cdot \sqrt{\frac{G \cdot Z_0}{\pi \cdot Z_{\text{air}}}} \tag{3.11}$$

Using 3.11, the value of the voltage level of the incident signal can be calculated for different input power levels. Considering an input impedance of 50 Ω, which is the standard, for 1 W of input power at rectifier level, V_{eff} is around 14 V, for 100 mW, it is around 4.5 V, and for 1 mW, it is of only 0.45 V and as low as 140 mV for 100 μW of incident power. As power decreases, the voltage level becomes inferior to the threshold voltage of the diodes used, and as a result, RF-DC conversion efficiency will decrease dramatically.

Diodes are characterized by a threshold voltage, a junction capacitance, and a series resistance. The junction capacitance has an impact on diode switching time; a fast diode should have small junction capacitance. The threshold voltage is a very important factor, especially when low power levels are to be harvested (below 1 mW). In these conditions, an important threshold voltage generates a great amount of loss in the diodes, because the input signal is too weak to overcome the threshold of the diode. When important power levels are available, the threshold voltage is not an issue. In this case, the rectifying efficiency is degraded by resistive losses due to diodes' internal resistance [34].

Microwave rectifiers have different topologies, depending on the position and number of HF diodes.

3.2.1 Circuit Topologies

RF-DC converters are usually diode-based or transistor-based. Schottky diodes are used in the majority of rectenna circuits, but FET-type semiconductors are increasingly used for this purpose. Microwave rectifiers have different topologies, depending on the position and number of diodes or transistors.

3.2.1.1 Series-mounted diode

The simplest and most common configurations are single series-mounted diode topologies, as presented in Figure 3.7.

This structure is based on a single diode mounted in series with the signal path. It is a half-wave rectifier. During the negative half of the input sine wave, diode D_1 is blocked, only the input filter being charged. During the positive half wave, diode D_1 becomes conductive, energy thus flowing from the source and the input filter toward the output filter, which blocks the high-frequency harmonics from reaching the load.

3.2.1.2 Shunt-mounted diode

In the shunt-mounted topology (Figure 3.8), the diode is in parallel between the two HF and LF filters, with the anode or the cathode connected to the electrical ground. The diode is thus directly polarized by the DC voltage it generates. During the negative half wave, diode D_1 is conductive and the input filter is charged. During the positive half wave, D_1 locks and energy flows from the source and the input filter toward the output

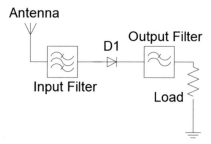

Figure 3.7   Single series-mounted diode rectenna topology.

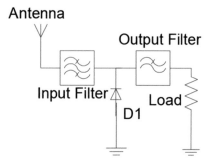

Figure 3.8   Single shunt-mounted diode rectenna topology.

filter, which has the same role as in the case of the series-mounted diode structure.

3.2.1.3 Voltage-doubler topology

The voltage-doubler circuit topology (Figure 3.9) can be seen as a super-position of the series-mounted and shunt-mounted topologies previously described. During the negative half wave, D_1 conducts the current and the input filter gets charged. During the positive half wave, the energy coming from the antenna and the energy stored in the input filter during the negative half wave is transferred through D_2 which becomes conductive, toward the output filter and the load.

The voltage-doubler topology has the advantage of reaching higher DC output voltage levels, almost double of that reachable by single-diode structures. On the other hand, RF-DC conversion efficiency is usually less for single-diode structures. By cascading several times the same topology, voltage multipliers can be obtained, as in Figure 3.10. The final output DC voltage of an N-stage voltage multiplier can be estimated as [35].

$$V_{out} \approx 2 \cdot N \cdot V_{1\_diode} \qquad (3.12)$$

with V_{1_diode} being the output voltage of a single-diode Rectenna for the same input power.

3.2.1.4 Diode bridge topology

The diode bridge topology, widely used in the case of low-frequency rectification, can also be used in RF-DC conversion (Figure 3.11).

The principle of full-wave rectification consists in the restitution of the entire incident wave at load level, compared to the case of half-wave rectifiers in which the load is disconnected from the source for half the time.

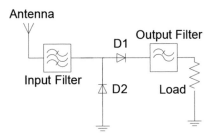

Figure 3.9 Voltage-doubler rectenna topology.

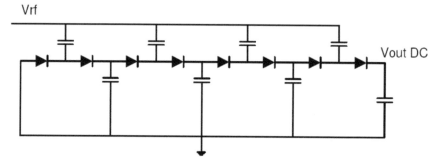

Figure 3.10   Four-stage voltage multiplier topology [35].

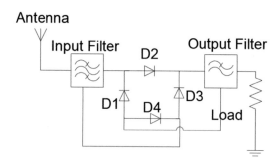

Figure 3.11   Diode bridge rectenna topology.

During the positive wavelength, diodes D_1 and D_4 are conductive, while D_2 and D_3 are blocked. The voltage across the output filter is the same as that of the input. During the negative half wave, diodes D_2 and D_3 are conductive and D_1 and D_4 are blocked. It is then easy to see that the current flow direction is the same as in the previous case. The same can be said about the output voltage of the bridge which remains positive.

The bridge topology offers higher power handling capabilities. As two diodes are conductive simultaneously, the critical power is reduced by the increase in the resistance per diode. On the other hand, as the signal always has to overcome two thresholds, losses increase in the case of low incident power. This is the reason why the bridge rectifier is not adapted to low incident power levels.

3.2.1.5 Transistor-based rectennas
The RF-DC conversion is done using transistors for which the command signal is the rectified signal itself. Several topologies exist.

The circuit in Figure 3.12 is dedicated to supplying a semi-active sensor. The battery present in the system is used to prepolarize the two NMOS transistors in order to increase the sensibility of the structure. The topology has 11% RF-DC conversion efficiency for –6 dBm of input power (250 µW) [36].

Full-wave rectifiers have also been proposed. The circuit in Figure 3.13 works on the same principle as the rectifier bridge with a current return through Schottky diodes. This structure offers good energy conversion efficiency but only for power levels above 1 mW as the sensibility is degraded by the presence of extra losses due to the diodes [37].

Voltage-doubler and voltage multiplier topologies also have their equivalent using MOSFETs. The structure in Figure 3.14 is a multistage voltage multiplier that can be easily adapted to the desired output DC voltage for a specific application. Experimental results have shown excellent detection

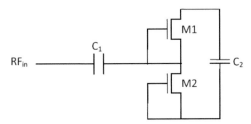

Figure 3.12   NMOS-based RF-DC rectifier presented in [36].

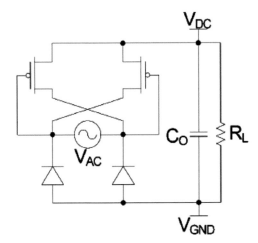

Figure 3.13   Full-wave synchronous rectifier presented in [37].

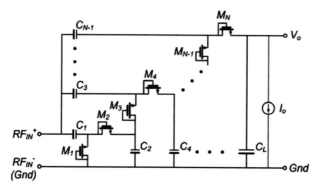

Figure 3.14   Multiple stage MOS-based voltage multiplier [38].

characteristics due to the good output voltage level. The use of this structure is less practical when wanting to provide energy to a circuit due to its low RF-DC conversion efficiency [38].

The majority of MOS-based rectennas are designed for low power applications because MOS transistors generally have lower losses than Schottky diodes when conducting low currents [39]. They have proved a good sensibility for very low incident power levels [40].

For higher power levels, bridge-type rectifiers and rectenna associations have proved to offer better performances, most of all due to their higher power handling capabilities [34].

3.2.2 Rectenna Associations

Supplying sensors or microdevices wirelessly sometimes needs a DC power and/or voltage level that is impossible to achieve with a single-Rectenna element. In order to increase the power and/or voltage level, Rectenna associations are often used in order to convert more of the incident microwave energy. These associations can have different configurations, depending on the interconnection method between the different elements.

Figure 3.15a presents a linear model of a single-Rectenna circuit. Two main types of associations are possible. Figure 3.15b presents a series association, while Figure 3.15c illustrates a parallel Rectenna association.

The electrical parameters of the linear model of the single Rectenna are the following:

$$I_0 = \frac{V_{D0}}{R_{D0} + R_{L0}} \tag{3.13}$$

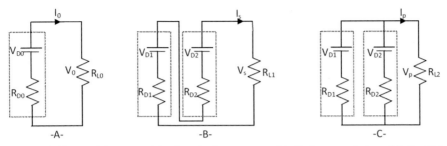

Figure 3.15 (a) Linear model of a single rectenna, (b) Series association, (c) Parallel association.

$$V_0 = \frac{V_{D0} \cdot R_{L0}}{R_{D0} + R_{L0}} \tag{3.14}$$

$$P_0 = \frac{V_{D0}^2 \cdot R_{L0}}{(R_{D0} + R_{L0})^2} \tag{3.15}$$

The theoretical efficiency of one structure is defined as the power P_0 divided by the maximum power that the source can provide.

$$\eta = \frac{4 \cdot R_{L0} \cdot R_{D0}}{(R_{D0} + R_{L0})^2} \tag{3.16}$$

By placing $\sigma = \dfrac{R_{L0}}{R_{D0}}$, the efficiency becomes

$$\eta = \frac{4 \cdot \sigma}{(1 + \sigma)^2} \tag{3.17}$$

For a given input power, the conversion efficiency depends only on the σ factor. A conversion efficiency equal to unity is obtained for $\sigma = 1$, which is equivalent to $R_{L0} = R_{D0}$.

For the series association in Figure 3.15, the electrical parameters are given by the following equations:

$$I_S = \frac{V_{D1} + V_{D2}}{R_{D1} + R_{D2} + R_{L1}} \tag{3.18}$$

$$V_S = \frac{(V_{D1} + V_{D2}) \cdot R_{L1}}{R_{D1} + R_{D2} + R_{L1}} \tag{3.19}$$

$$P_S = \frac{(V_{D1} + V_{D2})^2 \cdot R_{L1}}{(R_{D1} + R_{D2} + R_{L1})^2} \tag{3.20}$$

If the rectennas are identical ($R_{D1} = R_{D2} = R_{D0}$) and the load is optimal ($R_{L1} = R_{D1} + R_{D2} = 2 \cdot R_{L0}$), then the three previous equations become

$$I_S = \frac{V_{D1} + V_{D2}}{2 \cdot (R_{D0} + R_{L0})} = \frac{1}{2}(I_1 + I_2) \tag{3.21}$$

$$V_S = \frac{(V_{D1} + V_{D2}) \cdot R_{L0}}{R_{D0} + R_{L0}} = V_1 + V_2 \tag{3.22}$$

V_1 and V_2 are the voltages delivered by the Rectennas 1 and 2, respectively.

If in addition the two Rectennas receive the same input power, then

$$P_S = I_S \cdot V_S = \frac{1}{2}(I_1 + I_2) \cdot (V_1 + V_2) = \frac{4 \cdot V_{D0}^2 \cdot R_{L0}}{(2 \cdot R_{D0} + R_{L0})^2} \tag{3.23}$$

In this case, $\sigma_{\text{series}} = R_{L0}/(2 \cdot R_{D0})$ and unitary efficiency is obtained for a load $R_{L0} = 2 \cdot R_{D0}$.

In the same way, for the parallel association and using the same hypotheses, the output power is given by

$$P_P = \frac{4 \cdot V_{D0}^2 \cdot R_{L0}}{\left(\frac{R_{D0}}{2} + R_{L0}\right)^2} \tag{3.24}$$

In this case, $\sigma_{\text{parallel}} = 2 \cdot R_{L0}/R_{D0}$ and unitary efficiency is obtained for a load $R_{L0} = R_{D0}/2$.

The equations previously obtained show that by associating several Rectennas, the power delivered to the load increases. A series association does this with an increase in the DC output voltage and a parallel association through the increase in the output current. This purely theoretical and simplified analysis demonstrates the electrical parameter adding capabilities of Rectenna associations. In practice, the results vary slightly due to incertitude concerning the nonlinearity of converter impedances with respect to input power level.

3.2.3 Modeling a Rectenna

HF devices nowadays include increasingly more components and functions. An electronic device can contain power electronic components, radiating components, and transmission lines at the same time. These different components often have different sizes. In numerous cases, separate design and optimization of each subpart prove to be insufficient and important differences are observed between simulated and experimental results. These differences can often be explained by the numerical models that do not always take into account all the complex physical phenomenon involved, for example, couplings and interactions between different parts of the same device.

In order to analyze a HF device, several approaches exist, depending on the situation and the desired modeling accuracy. There is the electromagnetic approach, based on Maxwell's equations which take into account the electromagnetic phenomena in the device. On the other hand, the circuit approach is based on equivalent circuit models. These models, which often suffer simplifications and approximations, are only accurate within a well-defined domain and situation. This underlines the importance of rigorous modeling that takes into account all the physical phenomena in a HF device.

The design and optimization of a Rectenna demands both HF electronics and power electronics know-how. The goal is to maximize either the RF-DC conversion efficiency or the output DC voltage level. It is difficult to consider optimizing each of the elements of a Rectenna individually. In order to account for all the interactions and coupling between the different components, the simulation and optimization must be done on the totality of the circuit and in the same environment. Hybrid cosimulation (electromagnetic–circuit netlist), which combines the laws of electromagnetism and circuit theory, is the most appropriate in this case.

Another source of difficulty comes from the fact that the components in a Rectenna can be very different in size. In the same device, a half-wave radiating structure may be present alongside a localized component (diode, inductor, capacitor) which is sometimes inferior in size to one hundredth of the wavelength. In this case, a mesh based on the smallest device is not practical for obvious reasons due to the necessary amount of computing time and power.

A certain number of electromagnetic simulation tools exist. These tools are usually adapted in the case of simulated devices comparable in size to the wavelength. The main difference between these tools consists in the method used for solving Maxwell's laws, which can be solved either in the time domain or in the frequency domain. Time-domain analysis can do a wideband analysis in a single simulation and can take into account nonlinear elements. The main inconvenient resides in the computing time which can be very long. The most commonly used time-domain methods are finite difference time domain (FDTD) and the transmission line method (TLM).

On the other hand, frequency-domain analysis is generally faster. However, for wideband analysis, the number of simulations necessary is the same as the number of frequency points in the desired frequency range. Nonlinear elements are difficult to model in the frequency domain. Among the most common frequency-domain methods, we can cite the method of moments (MoM) and finite elements method (FEM).

Several circuit modeling approaches have been proposed [41–43]. However, as the RF-DC conversion circuit has a highly nonlinear model behavior and as a consequence, it generates high-order harmonics, it is often difficult to develop an accurate analytical model. One such model has been proposed for the series topology [41].

3.2.4 A Designer's Dilemma

3.2.4.1 Output characteristics

When designing a rectenna, two main parameters are generally evaluated: DC voltage output level and RF-DC energy conversion efficiency. Designers are often confronted with the choice of sacrificing one in order to improve the other. For the purpose of providing an overall comparison tool that includes both of the above-mentioned parameters, a Rectenna figure of Merit (RFoM) can be defined as the product of open-circuit DC output voltage and rectification efficiency when supplying a load of optimal value [44]:

$$RFoM = V_{\text{DC\_open\_circuit}} \cdot \eta_{\text{optimal\_load}} \qquad (3.25)$$

Simulations have allowed us to assess the figure of Merit of four different structures: series- and shunt-mounted single-diode configurations as well as a single-stage and a two-stage voltage-doubler circuit topology. Simulated data at an input frequency of 2.45 GHz is presented in Figure 3.16.

Voltage-doubler topologies have low RFoM for low input power levels, due mainly to their low power conversion efficiency for low power levels. However, the two-stage voltage doubler has high RFoM for high input power levels due to its high DC output voltage. The shunt and series diode topologies are similar for input power levels around 30 µW, but the series-mounted diode topology has the highest RFoM for low input power levels (below 250 µW).

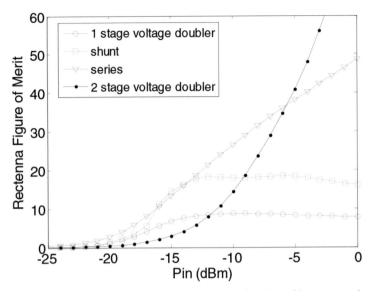

Figure 3.16 Simulated rectenna figure of merit as a function of input power level.

It thus seems to offer the best compromise between DC output level and conversion efficiency at these low incident power levels.

3.2.4.2 Antenna impedance influence

For practical reasons, the majority of antennas and radio frequency circuits have normalized 50 Ω characteristic impedance. Once all the structure is optimized for optimal RF-DC conversion efficiency and high output voltage level, the impact of a non 50 Ω antenna on the overall circuit performances is evaluated. Figure 3.17 traces the evolution of output DC voltage level as a function of antenna internal impedance for a series-mounted diode structure at an incident power level of 30 μW [45].

It appears that for this particular structure, lower antenna impedance is more adequate and allows an increase in output voltage level from around 350 mV with 50 Ω antenna impedance to more than 500 mV with 10 Ω antenna impedance. Antenna characteristic impedance is thus added as a degree of freedom in circuit optimization. This result is very important by considering the fact that with the low input power level, every additional degree of freedom can significantly improve the performances and thus the link budget.

This can be explained by the fact that diode internal junction resistance depends on the current that transits the junction.

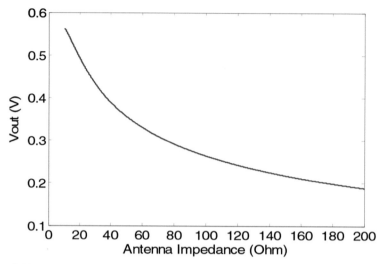

Figure 3.17   Simulated influence of the antenna impedance on rectenna DC output voltage in the case of a single series-mounted diode.

Figure 3.18 presents the equivalent circuit of the packaged Schottky diode. L_p and C_p are parasitic elements introduced by the packaging, R_S is the diode series resistance, R_V is the junction resistance, and C_j is the diode junction capacitance. Diode equivalent resistance is mainly due to the value of the junction resistance. This value depends on the transiting current:

$$R_V = \frac{8.33 \cdot 10^{-5} \cdot n \cdot T}{I_b + I_S} \tag{3.26}$$

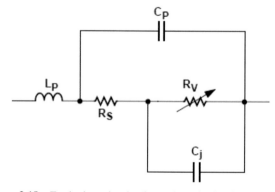

Figure 3.18   Equivalent circuit of a packaged schottky diode [ref].

where I_b is the externally applied current, I_S is the saturation current, T is the temperature, and n is the ideality factor [ref].

The fact of choosing low characteristic impedance for the rectifying circuit increases the amplitude of the current transiting the junction of the diode. An increase in the value of the externally applied current I_b leads to a decrease in the diode resistance and therefore a decrease in the internal diode loss due to the Joule effect. The overall RF-DC conversion efficiency is then improved.

In order to validate the aforementioned results, two different circuits have been designed and manufactured. They are both based on the single series diode structure. To avoid a difficult realization process of the antenna which may lead to inaccurate results, it was decided to represent the antenna behavior by a RF generator connected to a simple L-C matching cell connected to the rectifier. By this way, the effect of the antenna impedance modification is experimentally simulated and allows a direct comparison with simulations.

The first circuit is a reference case because it corresponds to a 50 Ω input impedance design, that is, with non-optimized antenna impedance. The second circuit has been optimized by including the antenna impedance as a degree of freedom. The new characteristic impedance is 10 Ω. The two circuits are presented in Figure 3.19, and performances are shown in Figure 3.20 [45].

The manufactured circuits use printed circuit capacitors. This choice has been retained to minimize the accuracy dispersion of an equivalent lumped component. The results in Figure 3.20 validate the different guidelines applied to the design. Although maximum DC voltage level is different between simulation and measurement (around 100 mV), a relative comparison between the reference circuit and the optimized one leads to a great compliance

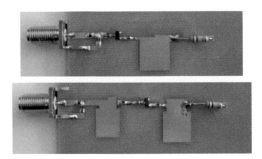

Figure 3.19 Photographs of the fabricated reference circuit with 50 Ω input impedance (top) and fully optimized rectifier (bottom) [45].

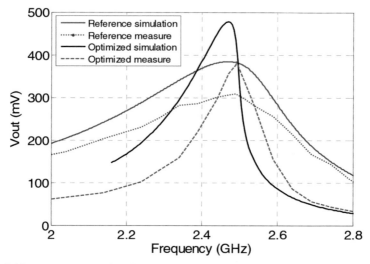

Figure 3.20   Improvement of performances due to the global optimized design compared to the 50-Ω classical design for the single series-mounted diode rectenna for 30 μW of input power [45].

between the predicted result and the measurement. As a whole, an improvement of output voltage of around 100 mV is observed when the antenna impedance is added to the design optimization. In the same time, RF-DC conversion efficiency passes from 20% to more than 30% for 30 μW of incident power. Notice that this last design also leads to a more selective behavior of the rectenna.

By following the same design procedure, the effect of non-50 Ω characteristic impedance has been evaluated on a single-stage voltage-doubler circuit topology (Figure 3.21). As in the case of the single series-mounted diode configuration, a reference 50 Ω and a 10 Ω configuration are compared. Measurement results are shown in Figure 3.22.

The output voltage in the case of the 50 Ω classical design is situated around 300 mV in simulation and 260 mV measured for an input power of 30 μW. Once the antenna impedance is considered as an extra degree of freedom, 10 Ω characteristic impedance offers the best performance. In this case, output voltage level becomes 580 mV in simulation and 520 mV measured for the same input power. This amounts to a 100% improvement for this new design compared to the classical approach. At the same time, RF-DC conversion efficiency for this structure at 30 μW on incident power increases from 15% to around 25%.

Figure 3.21 Photographs of the fabricated reference circuit with 50 Ω input impedance (top) and fully optimized rectifier (bottom).

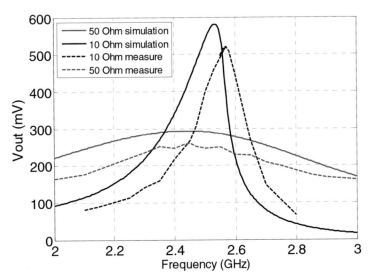

Figure 3.22 Improvement of performances due to the global optimized design compared to the 50-Ω classical design for the single-stage voltage-doubler rectenna for 30 μW of input power.

3.3 Reconfigurable Electromagnetic Energy Receiver

The main limitation of a rectenna circuit is that it is designed for a very well-defined operating point. Good RFDC conversion efficiency is obtained for a given input power level, a central frequency, and a specific load impedance. Outside these defined limits, the energy conversion efficiency decreases dramatically [46]. If load matching is often resolved by the presence of a MPPT DC–DC converter, the power matching is more delicate because each rectenna structure is intrinsically characterized by an optimal input power level at which the conversion efficiency is maximum, but the rectenna quickly becomes inefficient at another power level. This can often be seen as a major limitation in applications in which incident power level can vary considerably, for example, when supplying a moving device or through a changing environment. In another context, the input power level may be very different between an intentional supply of a device and the harvesting of local ambient energy to supply a device.

3.3.1 Typical Application

A typical example of application is described in Figure 3.23. The purpose is to supply energy to a battery-powered sensor placed in an inaccessible area, as in [47, 48].

Batteries can be recharged periodically once their level becomes low. For the same sensor, three possible situations are illustrated. The first situation consists of intentionally sending RF energy from a distant, high-gain emitter antenna, usually of parabolic shape [49, 50]. For an emitted power of 1 W

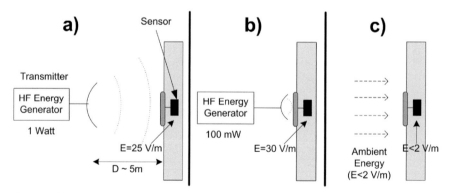

Figure 3.23 Wireless energy transfer scenarios for supplying a remote sensor; (a) distant recharging; (b) proximity recharging; and (c) ambient energy harvesting.

from 5 m, roughly 10 mW can be collected and supplied to the sensor with a compact and high-gain receiving antenna [51, 52] (Figure 3.23a). The second recharge strategy is to provide a proximity wireless energy transfer by placing a much more compact emitter in contact with the sensor area in the direction of the sensor (Figure 3.23b). An estimated 20% of the emitted energy (100 mW) is potentially recoverable at receiver level, much higher than in the previous case. The third scenario takes advantage of the ever-increasing amount of electromagnetic radiation present in our environment, mainly due to the massive development of wireless communications (Figure 3.23c). The most frequently encountered frequency bands are situated around 900/1800 MHz, 2 GHz, and 2.45 GHz, corresponding to standards such as GSM/DCS, UMTS, and WLAN, respectively. Measurement campaigns have shown that typical power levels at 25–100 m distance from a GSM base station reach several $\mu W/cm^2$, especially in the urban areas [53]. About the same power levels have been detected several meters from a WLAN access point. These low energy levels can provide an alternative power source to ubiquitous devices under certain conditions and can be used to continuously harvest energy from ambient environment to boost the life of the device battery [54, 55].

The use of a single-Rectenna device for such a wireless energy transfer system would not be ideal because of the high uncertainty on the incident RF power level. It is highly probable for the rectifier to work outside its optimum power range, and the energy conversion efficiency would be low as a consequence.

In order to overcome these limitations, a reconfigurable electromagnetic harvesting device that is capable of adapting itself to the incident power level was designed, thus ensuring the best possible energy conversion efficiency over a very wide range of input power levels.

The targeted power range is situated between 1 μW (–30 dBm) and 1 W (30 dBm) of input power at receiver level. This power range was divided into four regions: below 1 mW, between 1 and 100 mW, between 100 and 500 mW, and above 500 mW. A different Rectenna structure was designed and optimized for each of the four regions, ensuring that it will have maximum RF-DC conversion efficiency within that region.

3.3.2 Rectenna Circuit Configuration

The four Rectennas are based on topologies previously presented in Section 3.2.1, and their circuit configuration is shown in Figure 3.24. The three structures are tuned for a central frequency of 1.8 GHz, but the central frequency is not a limiting factor here.

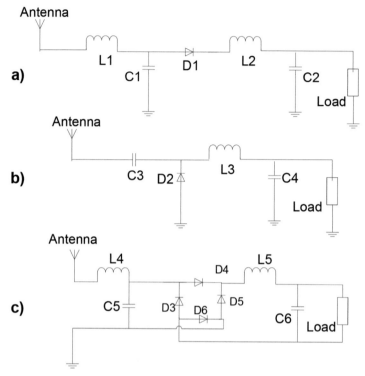

Figure 3.24   Rectenna circuit configuration: (a) Single series-mounted diode; (b) Single shunt-mounted diode; and (c) Diode bridge.

The first Rectenna is based on the single series-mounted diode and is shown in Figure 3.24a. As this structure is dedicated to low power levels, power handling capabilities can be traded for high sensitivity. The choice was made to use zero-bias Schottky diodes that have low power handling capabilities but low threshold voltage (150 mV) and low junction capacitance (0.18 pF).

Figure 3.25 traces the evolution of the RF-DC energy conversion efficiency as a function of the incident RF power level. The rectenna load has been tuned to obtain the maximum power point efficiency for a given input power level. Maximum conversion efficiency of roughly 50% is reached between –5 and 0 dBm (1 mW) of incident power. At lower power levels, the efficiency is lower because of the threshold voltage of the diode which is comparable to the amplitude of the incident signal. For high power levels, internal diode losses become significant due to the diode series resistance. The output DC

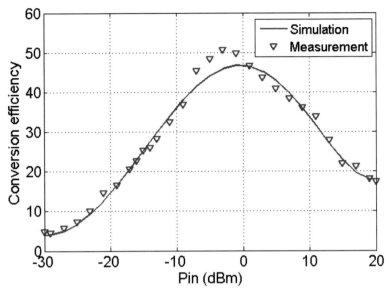

Figure 3.25 Simulated and measured maximum efficiency of the series-mounted diode rectenna (L1 = 5.6 nH, L2 = 6.8 nH, C1 = 1 pF, C2 = 10 pF).

voltage level of the single series-mounted diode rectifier is 400 mV at –15 dBm, 2.1 V at 0 dBm, and 3.75 V at 10 dBm of incident power, respectively.

A second rectenna structure has been designed for the 0 to 20 dBm power input range. The diode is shunt-mounted as shown in Figure 3.24b.

At these power levels, the threshold voltage has less impact on the circuit performances. The main objective is to lower the internal loss inherent to the rectifier diode and to increase the power handling capabilities. The diode used for this structure has a threshold voltage level of 350 mV (HSMS2860 by Agilent). The internal resistance is 6 Ω, and the breakdown voltage is 7 volts. Input and output filters were dimensioned using the same optimization techniques as previously mentioned.

The structure reaches maximum conversion efficiencies of 70% for an input power of +15 dBm, as shown in Figure 3.26. MPPT operating conditions have been considered, and the optimal load is 750 Ω for +15 dBm input power.

The bridge topology (Figure 3.24c) presents the advantage of high power handling capabilities, if diodes with high breakdown voltage are considered (HSMS2820 by Agilent, 15V, 6 Ω series resistance device).

The evolution of RF-DC conversion efficiency of the bridge rectifier is presented in Figure 3.27. MPPT operating conditions have been considered.

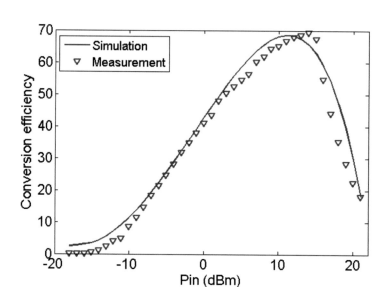

Figure 3.26   Simulated and measured RF-DC energy conversion efficiency of the shunt-mounted diode rectenna (L3 = 15 nH, C3 = 5.6 pF, C4 = 10 pF).

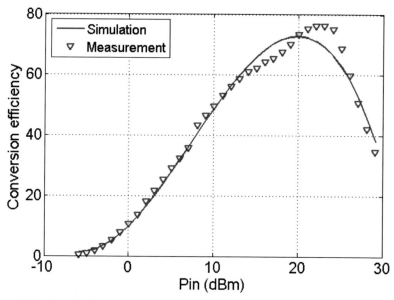

Figure 3.27   Simulated and measured RF-DC energy conversion efficiency of the diode bridge rectenna (L4 = 3.6 nH, L5 = 3.6 nH, C5 = 0.5 pF, C6 = 20 pF).

A peak of 78% is reached at 23 dBm of input power, after which conversion efficiency decreases rapidly. The optimal load is 200 Ω for +23 dBm input power. The 10% efficiency level is reached for 0 dBm. The DC voltage output is 1.1 V @ 10 dBm and 4.2 V @ 20 dBm and reaches 9.3 V @ 30 dBm, respectively.

The different RF-DC rectifier topologies offer good performances over a limited range of power. The use of one or another can be easily decided in the case of a system in which the incident power level is perfectly determined and is not subject to variations. In practice, the available power at the receiver is influenced by many factors. Surrounding environment can change, obstacles can interfere in the path of the propagated power signal, or distance from the emitter can vary. As a result, the rectifier will often work in a region outside its optimum power range, and as a result, the efficiency of the rectification process will be low (less than 20%). A reconfigurable association of rectenna appears as a good solution to overcome this situation. The series-mounted diode rectenna is considered for very low input power level, while the bridge configuration addresses large input power range and the shunt-connected diode rectenna is suitable in the mid-range.

3.3.3 Reconfigurable Architecture

The solution consists in the design of an adaptive rectifier circuit that recon-figures itself depending on the incident RF power level. The general structure of the proposed circuit is illustrated in Figure 3.28.

Simple rectenna as previously presented is connected to a common antenna through an antenna switch, capable of switching between the possible rectenna circuits according to their handling capabilities. At any given moment, the available incident RF power is measured using a passive RF detector, which gives a DC voltage level proportional to the incident power level. A simple three-level logic comparator is used to generate the logic control signals for the switch branches.

3.3.3.1 Antenna switch

An integrated single-pole 4-throw (SP4T) switch structure has been designed and fabricated [56]. The isolation performances are between 42 dB and 53 dB in the 0.8 GHz to 2.5 GHz frequency range, and the insertion loss is kept less than 0.5 dB in each branch. These state-of-the-art figures represent a good trade-off between isolation and insertion loss covering the entire frequency range used in mobile communication devices.

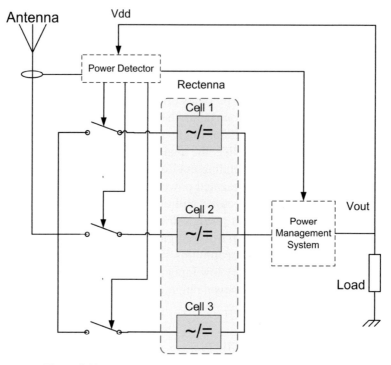

Figure 3.28   Schematics of the reconfigurable rectenna circuit.

A prototype of the SP4T circuit was fabricated using a depletion/accumulation pseudomorphic HEMT process with 0.18 μm gate length. The dimensions of each transistor were individually optimized in order to ensure the best trade-off between isolation and insertion loss. A great advantage is that a switch branch can be turned on or off using simple logic-level signals, with bias current in the sub-μA range.

The circuit layout was realized on a 100-μm GaAs substrate that is characterized by a resistivity exceeding 10^7 Ω.cm. The fabricated chip is shown in Figure 3.29. The total die area is approximately 1.5×2 mm².

3.3.3.2 Global performance

The power detector determines the level of the input RF power available at the antenna. It is obtained using a simple high-impedance diode detector that uses very little of the incident power and provides a DC voltage proportional to the available power. The control circuit is a three-level voltage comparator for which the principle circuit schematic is shown in Figure 3.30. Voltages

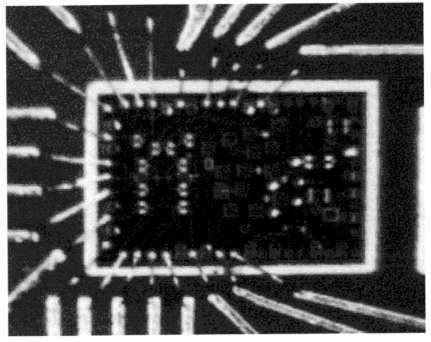

Figure 3.29 Microphotograph of the fabricated integrated switch [56].

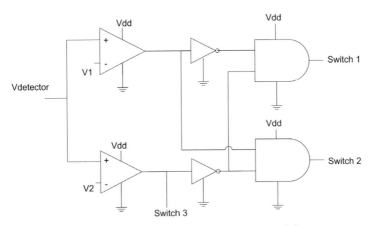

Figure 3.30 Schematic of the incident power level detector.

V1 and V2 represent the RF power threshold level. The power consumption of the circuit is essentially static as the variation of input power level is considered as a slow process. Based on discrete low-power CMOS circuits, power consumption as low as 2.5 μW is experienced from a 2.8 to 3.3 V supply voltage, but this value can be dramatically reduced with an integrated implementation. The power detector circuit only needs a supply voltage if it needs to change its state. By default, when there is very little power available at receiver input, the branch containing the low power rectenna is active. As the power available increases, the voltage supply of the power detector circuit is provided directly from the output of the power management system, and a change of state in the switch is possible.

Thresholds of 1 and 15 dBm are selected to switch between the low-power rectenna (series-mounted single diode), the medium-power rectenna (shunt-mounted single diode), and the high-power rectenna (bridge configuration). The power detector and the logic glue select a branch of the switch. If a 0-V logic level is applied to the gate of a switch transistor, the corresponding switch branch is blocked and no power is supplied toward the corresponding rectenna. A +3-V logic level turns on the switch, and all available incident power is supplied to the corresponding rectenna. At any given moment, only one of the three switch branches is on, and the other two are off. The choice of the rectenna is selected to ensure that the available incident RF power is converted into DC power at the best possible efficiency. The series diode rectenna is used for power levels less than 1 dBm, and the highest conversion efficiency is around 50% for –3 dBm of incident power. For power levels between 1 dBm and 15 dBm, the shunt diode structure is used and a maximum efficiency of 68% is measured at 14 dBm of incident power. For power levels above 15 dBm, the bridge rectifier is turned on with a maximum efficiency point of 78% at 23 dBm of incident power.

The overall efficiency in Figure 3.31 is the maximum value for a given power level. The MPPT operation can be realized in an analog way, and the power consumption is very low, so the impact on efficiency is not significant, except in the –30 dBm range [57, 58].

The overall fabricated structure exhibits measured characteristics that permit to overcome the drawback of classical rectenna which only has good RF-DC conversion efficiencies over a limited power range. The new design is self-reconfigurable and chooses the best-adapted rectenna structure for each power level.

Additional rectenna circuits may be introduced in the lower range and upper range of input power. An integrated bridge rectenna was designed using the same IC technology (Figure 3.32). Though this technology is not dedicated for this kind of application, the efficiency of the integrated bridge circuit reaches values in excess of 75% for very high power levels

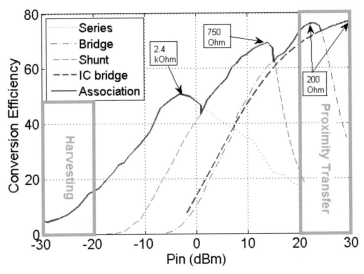

Figure 3.31 Experimental maximum efficiency of the proposed rectenna circuit in Figure 3.28.

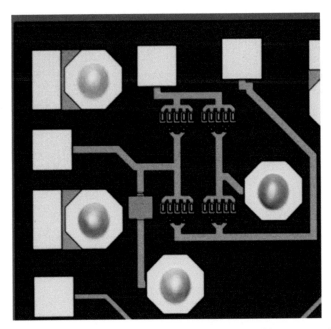

Figure 3.32 Microphotograph of the fabricated diode bridge rectenna IC.

(Figure 3.31). The latter integrated device is one example of improvement to bring to the reconfigurable rectenna for input power levels above 25 dBm.

In order to evaluate the interest of such a reconfigurable rectenna structure, it is interesting to compare the total amount of energy collected using this new structure compared to the energy collected with each of the four previously described structures individually. For this purpose, a random power profile was generated for supplying energy to the system. The evolution of the power level at receiver level is presented in Figure 3.33, along with the switch control signals. A total duration of 10 min is considered, and power level is randomly generated in the [−30 dBm, +30 dBm] range every 5 sec. It is important to have in mind the fact that power levels below −20 dBm (10 μW) are compatible with ambient electromagnetic energy harvesting, while received power levels above 20 dBm (100 mW) are less probable in practice, except for the case of

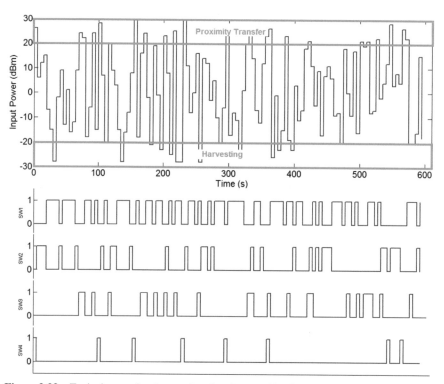

Figure 3.33   Typical power level at receiver level over a 10-minutes period and switch control signals.

proximity high-power wireless energy supply (>1 W of emitted power at a distance below 0.5 m).

For the same power profile, the total amount of DC energy is measured for each of the four rectenna structures taken individually, as well as for the new reconfigurable rectenna structure. Results are presented in Table 3.1. Results are conclusive as the reconfigurable topology collects considerably more energy than any of the other rectennas individually, because the system automatically chooses the structure best suitable for every power level.

The general principle of the reconfigurable microwave power receiver has been demonstrated using a hybrid prototype. For the purpose of accurate individual testing, the different parts of the receiver have been fabricated separately. A low-cost prototype fabricated using standard SMD devices can easily be fitted on the back side of the antenna, behind the antenna ground plane. The footprint of the device is thus defined by the surface of the antenna used. In this case, a standard 5 × 5 cm WLAN patch antenna has been used.

Not only the antenna switch, but also the electronic circuit presented can be integrated, and the chip could be embedded on the antenna backplane. This opens the possibility of miniaturizing the antenna if this would be the need. Furthermore, building the entire circuit on the same chip would improve the overall performance, as there would be less loss in the interconnections and less power consumption in the power detector and power management unit. On the downside, costs would certainly increase, unless fabricated on a large scale.

3.3.3.3 Output load matching

The structure presented so far still needs an output load match device in order to take into account the variations of the load. The most widely used impedance matching technique for energy sources is based on maximum power point tracking (MPPT) technique. In order to determine the specific MPPT algorithm most suitable for the presented concept, the output characteristic of the proposed receiver must be known.

Table 3.1 Total energy collected over a 10-minutes period from the random power profile

Rectenna	Series	Shunt	Bridge (SMD)	Bridge (IC)	Reconfigurable
Energy (J)	5.87	10.05	26.45	14.28	40.65

For different values of input RF power, the evolution of conversion efficiencies as a function of load resistance is studied. Results for the single series-mounted diode rectenna are presented in Figure 3.34.

For input power levels ranging from –20 dBm (10 µW) to 0 dBm (1 mW), a maximum conversion efficiency is observed at the same load resistance value of around 2.4 kΩ. For an input power level of –25 dBm (3.2 µW), maximum conversion efficiency is stable in the 2 to 7 kΩ load resistance range.

According to the optimum power transfer theorem in the case of a voltage source, maximum efficiency is obtained when $R_{\text{source}} = R_{\text{load}}$. This means that the series-mounted diode rectenna behaves as a voltage source with an internal resistance of 2.4 kΩ. Figure 3.35 gives the I(V) representations of the rectenna seen as a voltage source.

These graphs are close to straight lines, confirming the fact that the behavior of a rectenna as a voltage source can be described by

$$U_{\text{load}} = E - R_{\text{source}} \cdot I \tag{3.27}$$

with an internal source resistance of 2.4 kΩ in this case. In addition, the I(V) characteristics are quasi-parallel lines. This means that internal source resistance varies insignificantly with incident RF power level, although the structure is highly nonlinear due to the presence of diodes. Similar results were

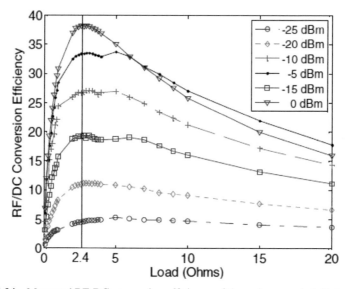

Figure 3.34 Measured RF-DC conversion efficiency of the series-mounted diode structure.

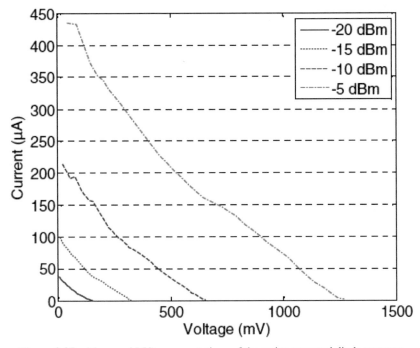

Figure 3.35   Measured I(V) representations of the series-mounted diode rectenna.

obtained in the case of the shunt-mounted diode structure, with an internal source impedance of 750 Ω as well as for the two-diode bridge structures, both diodes with an internal source impedance of 200 Ω. This is an important conclusion, especially in the perspective of designing a power management module based on the maximum power point tracking (MPPT) method. The needed output characteristic of the power supply is formed basically of three impedance steps, 2.4 kΩ for power levels below 0 dBm (1 mW), 750 Ω for power levels between 0 and 15 dBm, and 200 k Ω for power levels above 15 dBm, as shown in Figure 3.36.

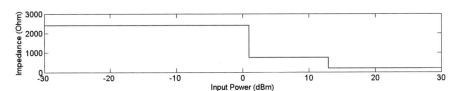

Figure 3.36   Output impedance characteristic needed for MPPT conditions.

Such basic MPPT circuits have already been reported for RF energy harvesting and used for supplying energy to a lithium battery [59]. The circuit can achieve near-constant input resistance using only commercially available discrete circuits, with power consumption of only several µW. Implementing such a device using an appropriate IC process further improves overall performance. making possible effective energy harvesting at input levels below 1 µW, with efficiencies between 40% and 80% and achieve a desired input resistance ranging from several tens of Ω to several tens of kΩ [60].

3.4 Conclusions

This chapter has provided a general overview of low power Rectenna circuits designed for wireless power transfer or electromagnetic ambient energy harvesting. The most commonly used topologies have been presented and compared in terms of RF-DC energy conversion efficiency and output voltage level. Design and optimization techniques were also illustrated. The self-adaptable electromagnetic wave receiver presented opens the way to realistic wireless power supplies that adapt itself to the environment and to the device it needs to power.

References

[1] Al Agha, K., M.-H. Bertin, T. Dang, A. Guitton, P. Minet, T. Val, and J.-B. Viollet. "Which Wireless Technology for Industrial Wireless Sensor Networks? The Development of OCARI Technology." *IEEE Transactions on Industrial Electronics* 56, no. 10: 4266–4278, October (2009).

[2] Gungor, V.C., and G.P. Hancke. "Industrial Wireless Sensor Networks: Challenges, Design Principles, and Technical Approaches." *IEEE Transactions on Industrial Electronics* 56, no. 10: 4258–4265, October (2009).

[3] Tan, Y., and S. Panda. "Optimized Wind Energy Harvesting System Using Resistance Emulator and Active Rectifier for Wireless Sensor Nodes." *IEEE Transactions on Power Electronics* 26, no. 99 (2011): 1.

[4] Cheney, M. *Tesla: Master of Lightning.* MetroBooks (2001).

[5] Maryniak, G.E. "Status of International Experimentation in Wireless Power Transmission." *Sunset Energy Counsel, Solar Energy* 56 (1996).

[6] McSpadden, J., and J. Mankins. "Space Solar Power Programs And Microwave Wireless Power Transmission Technology." *IEEE Microwave Magazine* 3, no. 4: 46–57, December (2002).

[7] Brown, W.C. "The History of Wireless Power Transmission." *Solar Energy* 56, no. 1: 3–21 (1996).

[8] Casanova, J.J., Z.N. Low, and J. Lin. "A Loosely Coupled Planar Wireless Power System for Multiple Receivers." *IEEE Transactions on Industrial Electronics* 56, no. 8: 3060–3068, August (2009).

[9] Low, Z.N., J.J. Casanova, P.H. Maier, J.A. Taylor, R.A. Chinga, and J. Lin. "Method of Load/Fault Detection for Loosely Coupled Planar Wireless Power Transfer System With Power Delivery Tracking." *IEEE Transactions on Industrial Electronics* 57, no. 4: 1478–1486, April (2010).

[10] Karalis, A., J. Joannopoulos, and M. Soljacic. "Efficient Wireless Non-radiative mid-Range Energy Transfer." *Annals of Physics* 323, no. 1: 34–48 (2008).

[11] Cannon, B., J. Hoburg, D. Stancil, and S. Goldstein. "Magnetic resonant coupling as a potential means for wireless power transfer to multiple small receivers." *IEEE Transactions on Power Electronics* 24, no. 7: 1819–1825 (2009).

[12] Valtchev, S., B. Borges, K. Brandisky, and J. Klaassens. "Resonant contactless energy transfer with improved efficiency." *IEEE Transactions on Power Electronics* 24, no. 3: 685–699 (2009).

[13] Sample, A.P., D.a. Meyer, and J.R. Smith. "Analysis, Experimental Results, and Range Adaptation of Magnetically Coupled Resonators for Wireless Power Transfer." *IEEE Transactions on Industrial Electronics* 58, no. 2: 544–554, February (2011).

[14] Hwang, Y.-S., and H.-C. Lin. "A New CMOS Analog Front End for RFID Tags." *IEEE Transactions on Industrial Electronics* 56, no. 7: 2299–2307, July (2009).

[15] Yao, Y., J. Wu, Y. Shi, and F.F. Dai. "A Fully Integrated 900-MHz Passive RFID Transponder Front End With Novel Zero-Threshold RF-DC Rectifier." *IEEE Transactions on Industrial Electronics* 56, no. 7: 2317–2325, July (2009).

[16] Lee, J.-w., and B. Lee. "A Long-Range UHF-Band Passive RFID Tag IC Based on High- Q Design Approach." *IEEE Transactions on Industrial Electronics* 56, no. 7: 2308–2316 (2009).

[17] Mandal, S., and R. Sarpeshkar. "Low-Power CMOS Rectifier Design for RFID Applications." *IEEE Transactions on Circuits and Systems I: Regular Papers* 54, no. 6: 1177–1188, June (2007).

[18] Ashry, A., K. Sharaf, and M. Ibrahim. "A Compact Low-Power UHF RFID Tag." *Microelectronics Journal* 40, no. 11: 1504–1513, November (2009).

[19] Essel, J., D. Brenk, J. Heidrich, H. Reinisch, G. Hofer, G. Holweg, and R. Weigel. "Highly Efficient Multistandard RFIDs Enabling Passive Wireless Sensing." *in 2009 Asia Pacific Microwave Conference. IEEE*, pp. 2228–2231, December (2009).

[20] Balanis, C.A. *Antenna Theory: Analysis and Design*, 3rd ed. Wiley-Interscience: Hoboken (2005).

[21] Douyere, A., J.D. Lan Sun Luk, and F. Alicalapa. "High Efficiency Microwave Rectenna Circuit: Modelling and Design." *Electronics Letters* 44, no. 24, November 20 (2008).

[22] Ren, Y.-J., and K. Chang. "5.8-GHz Circularly Polarized Dual-Diode Rectenna and Rectenna Array for Microwave Power Transmission." *IEEE Transaction on Microwave Theory and Techniques* 54, no. 4: 1495–1502, April (2006).

[23] Ren, Y.-J., M. F. Farooqui, and K. Chang. "A Compact Dual-Frequency Rectifying Antenna With High-Orders Harmonic-Rejection." *IEEE Transaction on Antennas and Propagation* 55, no. 7: 2110–2113, July (2007).

[24] Costanzo, A., F. Donzelli, D. Masotti, and V. Rizzoli. "Rigorous Design of RF Multiresonator Power Harvesters." *European Conference on Antennas and Propagation*, EuCAP 2010, Barcelona, Spain, April 12–16 (2010).

[25] Hagerty, J.A., F.B. Helmbrecht, W.H. McCalpin, R. Zane, and Z. Popovic. "Recycling Ambient Microwave Energy With Broad-Band Rectenna Arrays." *IEEE Transaction On Microwave Theory and Techniques* vol. 52, no. 3: 1014–1024, March (2004).

[26] Lau, P.-Y., K.K.-O. Yung, and E.K.-N. Yung. "A Low-Cost Printed CP Patch Antenna for RFID Smart Bookshelf in Library." *IEEE Transactions on Industrial Electronics* 57, no. 5: 1583–1589, May (2010).

[27] Huang, M., Y. Tsai, and K. Chen. "Sub-1 V Input Single-Inductor Dual-Output (SIDO) DC-DC Converter With Adaptive Load-Tracking Control (ALTC) for Single-Cell-Powered Systems." *IEEE Transactions on Power Electronics* 25, no. 7: 1713–1724 (2010).

[28] Richelli, A., L. Colalongo, S. Tonoli, and Z. Kovacs-Vajna. "A 0.2–1.2 V DC/DC Boost Converter for Power Harvesting Applications." *IEEE Transactions on Power Electronics* vol. 24, no. 5–6: 1541–1546 (2009).

[29] Dolgov, A., R. Zane, and Z. Popovic. "Power Management System for Online Low Power RF Energy Harvesting Optimization." *IEEE Transactions on Circuits and Systems I: Regular Papers* 57, no. 7: 1802–1811, July (2010).

[30] Brunton, S.L., C.W. Rowley, S.R. Kulkarni, and C. Clarkson. "Maximum Power Point Tracking for Photovoltaic Optimization Using Ripple-Based Extremum Seeking Control." *IEEE Transactions on Power Electronics* 25, no. 10: 2531–2540 (2010).

[31] Alahmad, M.a., and H.L. Hess. "*Evaluation and Analysis of a New Solid-State Rechargeable Microscale Lithium Battery.*" *IEEE Transactions on Industrial Electronics* 55, no. 9: 3391–3401, September (2008).

[32] Chen, P.-h., K. Ishida, X. Zhang, Y. Okuma, Y. Ryu, M. Takamiya, and T. Sakurai. "0.18-V Input Charge Pump with Forward Body Biasing in Startup Circuit using 65 nm CMOS." *Custom Integrated Circuits Conference (CICC)*, 19–22 September (2010).

[33] Merabet, B., L. Cirio, H. Takhedmit, F. Costa, C. Vollaire, B. Allard, and O. Picon, "Low-Cost Converter for Harvesting of Microwave Electromagnetic Energy." In *2009 IEEE Energy Conversion Congress and Exposition*, pp. 2592–2599, September (2009).

[34] Merabet, B., F. Costa, H. Takhedmit, C. Vollaire, B. Allard, L. Cirio, and O. Picon. "A 2.45-GHz Localized Elements Rectenna." In *IEEE International Symposium on Microwave, Antenna, Propagation and EMC Technologies for Wireless Communications*, pp. 419–422, October (2009).

[35] Karthaus, U., M. Fischer. "*Fully Integrated Passive UHF RFID Transponder IC With 16.7 μW Minimum RF Input Power.*" *IEEE Journal of Solid-State Circuits* 38, no. 10, October (2003).

[36] Umeda, T., H. Yoshida, S. Sekine, Y. Fujita, T. Suzuki, S. Otaka. "A 950-MHz Rectifier Circuit for Sensor Network Tag with 10-m Distance." *IEEE Journal of Solid-State Circuits* 41, no. 1, January (2006).

[37] Lam, Y.H., W.H. Ki, C.Y. Tsui. "Integrated Low-Loss CMOS Active Rectifier for Wirelessly Powered Device." *IEEE Transaction on Circuit and Systems II: Express Briefs*, 53, no. 12, December (2006).

[38] Yi, J., W.H. Ki, C.Y. Tsui. "Analysis and Design Strategy of UHF Micro-Power CMOS Rectifiers for Micro-Sensor and RFID Applications." *IEEE Transaction on Circuit and Systems* 54, no. 1, January (2007).

[39] Deuty, S. "HDTMOS Power MOSFETs Excel in Synchronous Rectifier Applications." Semiconductor Components Industries, LLC, February (2003).

[40] Yu-Jiun, R., and C. Kai. "5.8-GHz Circularly Polarized Dual-Diode Rectenna and Rectenna Array for Microwave Power Transmission." *IEEE Transactions on Microwave Theory and Techniques* 54, no. 4: 1495–1502, June (2006).

[41] Akkermans, J.A.G., M.C. van Beurden, G.J.N. Doodeman, and H.J. Visser. "Analytical Models for Low-Power Rectenna Design." *IEEE Antennas Wireless Propag. Letters* 4: 187–190 (2005).

[42] Yoo, T.-W. and K. Chang, "Theoretical and Experimental Development of 10 and 35 GHz Rectennas." *IEEE Transaction on Microwave Theory and Techniques* 40, no. 6: 1259–1266, June (1992).

[43] Adachi, S., and Y. Sato. "Microwave-to-dc Conversion Loss of Rectenna." *Space Solar Power Review* 5: 357–363 (1985).

[44] Marian, V., C. Vollaire, B. Allard, and J. Verdier. "Low Power Rectenna Topologies for Medium Range Wireless Energy Transfer." In *Power Electronics and Applications* (EPE 2011), August–September (2011).

[45] Marian, V., C. Menudier, M. Thevenot, C. Vollaire, J. Verdier, and B. Allard. "Efficient Design of Rectifying Antennas for Low Power Detection." In *International Microwave Symposium (IMS),* June (2011).

[46] Takhedmit, H., B. Merabet, L. Cirio, B. Allard, F. Costa, C. Vollaire, and O. Picon. "A 2.45-GHz Dual-Diode RF-to-DC Rectifier for Rectenna Applications." In *2010 European Microwave Conference (EuMC),* pp. 37–40 (2010).

[47] Harms, T., S. Sedigh, and F. Bastianini. "Structural Health Monitoring of Bridges Using Wireless Sensor Networks." *IEEE Instrumentation and Measurement Magazine* 13, no. 6: 14–18, December (2010).

[48] DiStasi, S., C. Townsend, J. Galbreath, and S. Arms. "Scalable, Synchronized, Energy Harvesting Wireless Sensor Networks." In *2010 Prognostics and System Health Management Conference. IEEE,* pp. 1–5 January (2010).

[49] Gardelli, R., G. La Cono, and M. Albani. "A Low-Cost Suspended Patch Antenna for WLAN Access Points and Point-To-Point Links." *IEEE Antennas and Wireless Propagation Letters* 3, no. 1: 90–93, December (2004).

[50] Thongsopa, C., D.-a. Srimoon, and P. Jarataku. "A U-Shaped Cross Sectional Antenna on a U-Shaped Ground Plane with an Offset Parabolic Reflector for WLAN." In *2007 IEEE Antennas and Propagation International Symposium IEEE:* 5159–5162, June (2007).

[51] Kim, S., and W. Yang. "Single Feed Wideband Circular Polarised Patch Antenna." *Electronics Letters* 43, no. 13: 703 (2007).

[52] Shibata, O., H. Koyama, and T. Sawaya. "Small Size, High Gain and High F/B Ratio Patch Antenna Arranging Parasitic Element on the Back." In *2007 European Conference on Wireless Technologies. IEEE,* pp. 264–267, October (2007).

[53] Visser, H.J., A.C.F. Reniers, and J.A.C. Theeuwes. "Ambient RF Energy Scavenging: GSM and WLAN Power Density Measurements." In *2008 38th European Microwave Conference. IEEE,* pp. 721–724, October (2008).

[54] Vullers, R., H. Visser, B. het Veld, and V. Pop. "RF Harvesting Using Antenna Structures on Foil." In *Proceedings of Power MEMS,* pp. 209–212 (2008).

[55] J. Hagerty, F. Helmbrecht, W. McCalpin, R. Zane, and Z. Popovic. "Recycling Ambient Microwave Energy With Broad-Band Rectenna Arrays." *IEEE Transactions on Microwave Theory and Techniques* 52, no. 3: 1014–1024, March (2004).

[56] Marian, V., J. Verdier, B. Allard, and C. Vollaire. "Design of a Wideband Multi-Standard Antenna Switch for Wireless Communication Devices." *Microelectronics Journal* 42, no, 5: 790–797, (May 2011).

[57] Levron, Y., and D. Shmilovitz. "A Power Management Strategy for Minimization of Energy Storage Reservoirs in Wireless Systems With Energy Harvesting." *IEEE Transactions on Circuits and Systems I: Regular Papers:* 1–11 (2010).

[58] Kim, R.-Y., J.-S. Lai, B. York, and A. Koran. "Analysis and Design of Maximum Power Point Tracking Scheme for Thermoelectric Battery Energy Storage System." *IEEE Transactions on Industrial Electronics* 56, no. 9: 3709–3716, September (2009).

[59] Paing, T., J. Shin, R. Zane, and Z. Popovic. "Resistor Emulation Approach to Low-Power RF Energy Harvesting." *IEEE Transactions on Power Electronics* 23, no. 3: 1494–1501, May (2008).

[60] Paing, T., E. Falkenstein, R. Zane, and Z. Popovic. "Custom IC for Ultra-Low Power RF Energy Scavenging." *IEEE Transactions on Power Electronics* 26, no. 6: 1620–1626, Febrauary (2011).

4

Wireless Power Transfer: Generation, Transmission, and Distribution
Circuit Theory of Wireless Power Transfer

4.1 Introduction

Wireless power transfer systems depend strongly on the ability to couple energy from a source to a receiver through the use of a combination of inductors and capacitors. Depending on how the inductors and capacitors are combined, the resulting outcomes may be termed loosely coupled wireless power transfer, strongly coupled WPT, microwave wireless power transfer, or capacitive power coupling. A loosely coupled WPT (LC-WPT) involves the use of only inductors and resistors at the transmitting and receiving sections of the power transfer circuit. A LC-WPT may not need resonance, and therefore, for such cases, optimum power transfer is rarely the objective. Rather, the objective is free placement of coils for energy transfer from one point to another. Usually, for such cases, the coupling coefficients are relatively small. A strongly coupled wireless power transfer (SC-WPT) also known as resonant wireless power transfer requires the use of at least two resonant circuits, one acting as the transmitter and the other as the receiver. Both circuits resonate at the same frequency, and energy is coupled from the transmitter to the receiver inductively. Both LC-WPT and SC-WPT normally operate in the near-field region of an antenna popularly configured as magnetic wire loops.

A microwave wireless power transfer (M-WPT) system is quite different from the above two cases. In M-WPT, an electrical antenna is employed at the transmitter and another electrical antenna is used at the receiver. Both circuits operate at the same frequency and can support long-range wireless power transfer. The mode of power propagation of this type of power transfer system is very similar to conventional radio signal propagation in wireless communication systems.

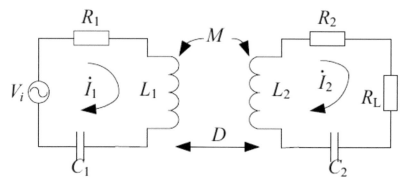

Figure 4.1   Resonant wireless power transfer.

A capacitive wireless power transfer (C-WPT) system employs capacitors to couple energy from source to receiver. C-WPT systems are often used for high-voltage short-range wireless power transfer applications. This chapter is dedicated to only LC-WPT. Other chapters to follow will cover S-WPT, M-WPT, and C-WPT systems.

The foundation for inductive wireless power transfer was laid in Chapter 1. Figure 4.1 cannot be used for optimum power transfer to a receiver load because of inherent problems. First, the apparent power at the load contains both real and reactive powers. Second, both the transmitter and receiver are not synchronized to operate at the same resonant frequency, and hence, optimum power is not transferred to the load. The use of resonating transmitter and receiver leads to elimination of the reactive power inherent in Figure 4.1. Hence, for resonant wireless power transfer, the two circuits are made to resonate at the same frequency. Resonance is established by connecting capacitors in the transmitter and receiver circuits. Although the theory of resonant wireless power transfer is well established, there are fundamental details often clouded by the way and manner authors in current literature have addressed the problem. Two broad approaches have been used to explain resonant wireless power transfer. Both the circuit theory model and coupled-mode theory have been used, although the circuit theory model is used more widely.

4.2 Criteria for Efficient Resonant Wireless Power Transfer

The results in Chapter 1 provide the useful criteria and insight into how to design efficient wireless power transfer systems. These are high power factor, high coupling coefficient, high-Q circuits (low resistive losses), and focused magnetic fields. We examine these criteria in this section.

4.2.1 High Power Factor (cos θ = 1)

Specifically, Equations (14a) to (14d) of Chapter 1 illustrate the importance of high power factor (close to unity) for achieving highly efficient wireless power transfer. Indeed, this requirement is well understood from electric power generation theory. The higher the power factor [(cos θ = 1), the cosine of the angle formed by the real and apparent power delivered to the load], the higher the useful power delivered to a load. This is also true for wireless power transfer. The objective should therefore be to make the power factor to be as close as possible to unity. There is just one and only condition under which the power factor of an RLC circuit is unity. This occurs at resonance. Hence, achieving resonance is essential for achieving optimum wireless power transfer. In situations where circuit imbalances lead to reflected power, matching circuits can be used to achieve better power transfer efficiency to the load.

4.2.2 High Coupling Coefficient ($k \approx 1$)

Coupling coefficient in general decreases with increasing distance between the primary and secondary coils. If we can improve upon the value of the coupling coefficient (k) so that it is very close to unity, high power transfer can be achieved. This condition is also normally achieved when a lossless transformer is used. This is normally achieved by making the air gap between the primary and secondary sections of the transformer as close as possible to zero. Therefore, from now on, we are interested in achieving resonance and also in enhancing the coupling coefficient created by the power transfer system.

4.2.3 High Quality ($Q \gg 1$) Factors

Ultimately, for wireless power transfer, the distance between the "primary" winding (transmitter) and "secondary" winding (receiver) cannot be close to zero as this defeats the purpose of wireless power transfer. Therefore, to help the wireless power transfer circuits, we can avoid having to reduce the gap between the transmitter and receiver to zero by using high Q transmitter (primary) and high Q receiver (secondary) circuits. The quality factors of these two circuits are functions of the self-resistances of the coils used and their inductances. Therefore, the objective is to use wires which have very low self-resistances and to create high-inductance loops. Low self-resistance brings with it the extra bonus of low resistive losses in the circuits.

4.2.4 Matching Circuits

The transmitter circuit is connected to a power supply source, and it has its own resistance. The receiver is also connected to a load. High source/load resistances limit the loaded Q factors of the transmitter and receiver. The primary and secondary coils themselves must therefore have high Qs to compensate for the influences of the source and load. In addition, normal matching circuits may also be designed and used.

4.2.5 Focusing of Magnetic Field

The magnetic field created by a coil spills into a spherical surface. Unfortunately, the receiving coil is rarely spherical. Even if it were, it does not enclose the transmitting coil and hence cannot capture all the flux created by the transmitter. As a first resort, the flux created by a transmitting coil can best be captured by receiving coils when they are axially aligned. Axial misalignment always reduces the amount of power delivered to the load. Better coils such as Helmholtz and Maxwell coils could be used. Helmholtz coils can deliver uniform flux along their axes, and Maxwell coils can produce high-gradient fields in their middle along their axis. Unfortunately, current literature is awash with the use of circular coils.

Over the years, many techniques have been used to improve the efficiency of power delivery to the receiver load and also to enhance the range of power delivery. The techniques include the use of three or more coils in an axial chain network, relaying coils, and flux amplification methods using magnetic or ferrite cores.

4.3 Resonant Wireless Power Transfer

This fundamental model for resonant power transfer can be better understood from the circuit model point of view. A great deal of insight is provided by modeling the system as a pair of resonant coils. The model assumes a resonant transmitter coupled magnetically to a resonant receiver. The receiver is connected to a load R_L as shown in Figure 4.2.

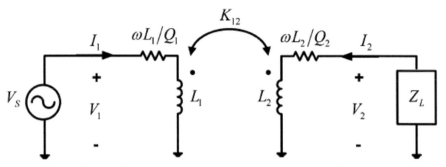

Figure 4.2 Loosely coupled wireless power transfer circuit.

The characteristics of the magnetic transmitter are obtained by applying Kirchhoff's voltage law (KVL) to the first part of the circuit. The impedances of the transmitter and receiver circuits are given by the expressions:

$$Z_1 = R_1 + j\omega L_1 + \frac{1}{j\omega C_1} \qquad (4.1a)$$

and

$$Z_2 = R_2 + R_L + j\omega L_2 + \frac{1}{j\omega C_2} \qquad (4.1b)$$

Hence, the voltage-and-current relationships at the transmitter and receiver circuits are obtained as

$$\begin{bmatrix} V_i \\ 0 \end{bmatrix} = \begin{bmatrix} Z_1 & -j\omega M \\ -j\omega M & Z_2 \end{bmatrix} \begin{bmatrix} I_1 \\ I_2 \end{bmatrix} \qquad (4.2)$$

This equation is identical to the case of non-resonant wireless power transfer systems in Chapter 1. Understandably, the value of the mutual inductance M depends on the inductances L_1 and L_2 of the transmitter/receiver coils and the coupling coefficient k through the expression $M = k\sqrt{L_1 L_2}$. Each circuit excites identical mutual inductance on its neighbor which may be seen as a feedback mechanism between the two circuits. The net effect is that each circuit modulates the impedance of its neighbor through the coupled magnetic flux and the consequent mutual inductance. At resonance, the reactance of each of the impedances Z_1 and Z_2 is identically zero and from which we obtain the resonance equations. For example, for the primary side,

$$\omega L_1 + \frac{1}{j\omega C_1} = 0 \tag{4.3a}$$

or

$$\omega_0 = \frac{1}{\sqrt{L_1 C_1}} \tag{4.3b}$$

At resonance, the two circuits work in near-perfect transfer harmony given by the Equation (4.3) as

$$\omega_0 = \frac{1}{\sqrt{L_1 C_1}} = \frac{1}{\sqrt{L_2 C_2}} \tag{4.4}$$

Equation (4.4) guides the operating frequency of the power transfer system. The inductors and capacitors used do not have to be of the same value. What is essential is that the resonant frequency should be the same for both circuits. The power transfer relationships are the same as in non-resonant two-coil systems given by the Equations (4.5a) and (4.5b):

$$P_{\text{in}} = \frac{V_S^2 Z_2}{Z_1 Z_2 + (\omega M)^2} \tag{4.5a}$$

and

$$P_{\text{out}} = \frac{V_S^2 (\omega M)^2 R_L}{\left[Z_1 Z_2 + (\omega M)^2 \right]^2} \tag{4.5b}$$

At resonance, the power factors in each case become one and the input power and output power become

$$P_{\text{in}} = \frac{V_S^2 R_2}{R_1 R_2 + (\omega M)^2} = \frac{V_S^2 / R_1}{(1 + k^2 Q_1 Q_2)} \tag{4.6a}$$

and

$$P_{\text{out}} = \frac{V_S^2 k^2 Q_1 Q_2 R_L}{R_1 R_2 \left[1 + k^2 Q_1 Q_2 \right]^2} \tag{4.6b}$$

The power transfer efficiency is defined as the ratio of the output power to the input power is

$$\eta = \frac{P_{\text{out}}}{P_{\text{in}}} \times 100\% = \frac{k^2 Q_1 Q_2 R_L}{R_2 \left[1 + k^2 Q_1 Q_2 \right]} \times 100\% \tag{4.7}$$

From Equation (4.7), high coupling coefficients and high quality factors favor high-efficiency power transfer. In the same vein, high load impedance and low self-resistance of the receiver windings are preferred for high-efficiency wireless power transfer systems. At $k = 1$ (ideal transformer) and with very high Q values, the optimum power transfer efficiency is approximately R_L/R_2. The efficiency decreases rapidly with decreasing coupling coefficient. When the coupling coefficient is very small so that $k^2 Q_1 Q_2 << 1$, the power transfer efficiency reduces significantly. These two situations are depicted in Equation (4.8).

$$\eta_{\text{high}} = \frac{R_L}{R_2} \times 100\%; \quad \text{when} \quad k^2 Q_1 Q_2 >> 1$$

$$\eta_{\text{low}} = \frac{k^2 Q_1 Q_2 R_L}{R_2} \times 100\%; \quad \text{when} \quad k^2 Q_1 Q_2 << 1$$

(4.8)

In a nutshell, the system efficiency is a function of the windings, the load, and the architecture (power transfer channel) of the power transfer system.

4.3.1 Higher-Order WPT Systems

The expressions for higher-order wireless power transfer systems involving two, three, four, and indeed N stages can be derived fairly easily by using a recursive method involving the input impedances (using the Equations 4.1a and 4.9a) of the stages as

$$Z_{in,2} = Z_1 + \frac{(\omega M_{12})^2}{Z_2} \tag{4.12a}$$

$$Z_{in,3} = Z_1 + \frac{(\omega M_{12})^2}{Z_2 + \frac{(\omega M_{23})^2}{Z_3}} \tag{4.12b}$$

$$Z_{in,4} = Z_1 + \frac{(\omega M_{12})^2}{Z_2 + \frac{(\omega M_{23})^2}{Z_3 + \frac{(\omega M_{34})^2}{Z_4}}} \tag{4.12c}$$

In general, the input impedance relationship is given by the Equation (4.13)

$$Z_{in,N} = Z_1 + \frac{(\omega M_{12})^2}{Z_2 + \frac{(\omega M_{23})^2}{Z_3 + \frac{(\omega M_{34})^2}{Z_4 + \cdots + \frac{\left(\omega M_{N-2,N-1}\right)^2}{Z_{N-1} + \frac{\left(\omega M_{N-1,N}\right)^2}{Z_N}}}}} \tag{4.13}$$

At resonance, this equation simplifies to

$$Z_{\text{in},N} = R_1 + \cfrac{(\omega M_{12})^2}{R_2 + \cfrac{(\omega M_{23})^2}{R_3 + \cfrac{(\omega M_{34})^2}{R_4 + \cdots + \cfrac{\left(\omega M_{N-2,N-1}\right)^2}{R_{N-1} + \cfrac{\left(\omega M_{N-1,N}\right)^2}{R_N}}}}} \tag{4.13a}$$

We can show that for a four-stage system, the relationship between the currents in each stage, their impedances, and their voltages is given by the expressions:

$$\begin{bmatrix} I_1 \\ I_2 \\ I_3 \\ I_4 \end{bmatrix} = \begin{bmatrix} Z_1 & M_{12} & M_{13} & M_{14} \\ M_{21} & Z_2 & M_{23} & M_{24} \\ M_{31} & M_{32} & Z_3 & M_{34} \\ M_{41} & M_{42} & M_{43} & Z_4 \end{bmatrix}^{-1} \begin{bmatrix} V_S \\ 0 \\ 0 \\ 0 \end{bmatrix} \tag{4.14}$$

Note that in general, for a reciprocal system, the mutual inductances are $M_{ij} = M_{ji}$. Generally, $M_{ij} = j\omega k_{ij}\sqrt{L_i L_j}$. Therefore, we can solve the Equation (4.14). This leads to the following solutions for the input and output currents:

$$I_1 = \frac{\left(1 + k_{23}^2 Q_2 Q_3 + k_{34}^2 Q_3 Q_4\right)}{\left[\left(1 + k_{12}^2 Q_1 Q_2\right)\left(1 + k_{34}^2 Q_3 Q_4\right) + k_{23}^2 Q_2 Q_3\right]} \frac{V_S}{R_S} \tag{4.15a}$$

$$I_4 = \frac{j k_{12} k_{23} k_{34} \sqrt{Q_1 Q_2}\sqrt{Q_2 Q_3}\sqrt{Q_3 Q_4}}{\left[\left(1 + k_{12}^2 Q_1 Q_2\right)\left(1 + k_{34}^2 Q_3 Q_4\right) + k_{23}^2 Q_2 Q_3\right]} \frac{V_S}{\sqrt{R_S R_L}} \tag{4.15b}$$

With these solutions, the expressions for the output power and the system efficiency can be written as follows:

$$\left|\frac{V_L}{V_S}\right| = \left|\frac{I_4 R_L}{I_1 R_S}\right| = \frac{k_{12}^2 k_{23} Q_1 Q_2^2}{\left(1 + k_{12}^2 Q_1 Q_2\right)^2 + k_{23} Q_2^2}\sqrt{\frac{R_L}{R_S}} \tag{4.16}$$

And the efficiency is defined as the power delivered to the load resistance divided by the power in the source and is

$$\eta = \frac{P_{\text{out}}}{P_{\text{in}}} = \frac{V_L^2/R_L}{V_S^2/(4R_S)} \tag{4.17}$$

This results to the equation:

$$\eta = \frac{4 k_{12}^4 k_{23}^2 Q_1^2 Q_2^4}{\left[\left(1 + k_{12}^2 Q_1 Q_2\right)^2 + k_{23}^2 Q_2^2\right]^2} \tag{4.18}$$

The maximum value of the efficiency with respect to the variables in the equation can be determined by setting the differential of Equation (4.18) to zero. We are interested in the maximum efficiency with respect to the quality factor of the first stage because this stage has to transfer maximum power to the subsequent stages. This derivative $(d\eta/dQ_1)$ results to the solution:

$$k_{12}^4 = \frac{\left(1 + k_{23}^2 Q_2^2\right)}{Q_1^2 Q_2^2} \tag{4.19}$$

The parameters of the first two stages determine the maximum efficiency. Specifically, k_{23} must be high and a low value of Q_1 is essential for achieving high coupling of the first stage to the rest of the system. This condition is obtained when the quality factor of the second stage is such that $1 \ll k_{23}^2 Q_2^2$ or at most when $1 \leq k_{23}^2 Q_2^2$. At this point, Q_1 exclusively determines the coupling of the first stage to the rest of the power cascade system.

4.4 Loosely Coupled Wireless Power Transfer System

A loosely coupled wireless power system is shown in Figure 4.2, consisting of a transmitting circuit and a receiving circuit. Both circuits are predominantly series-connected inductor and resistors. The power receiver circuit is connected to a load Z_L.

In Figure 4.2, the resistors are modeled in terms of the quality (Q) factors of the inductors. Since the quality factor of an inductor is given by $Q_i = \omega L_i / R_i (i = 1$ or $2)$, we can show the two resistors in terms of the quality factors of the inductors. These resistors are mostly the self-resistances of the inductors. Hence, the equations supporting loosely coupled wireless power transfer systems are given in Equations (4.20a) and (4.20b):

$$Q_1 = \omega L_1 / R_1; \quad R_1 = \omega L_1 / Q_1 \tag{4.20a}$$

and

$$Q_2 = \omega L_2 / R_2; \quad R_2 = \omega L_2 / Q_2 \tag{4.20b}$$

The coupling coefficient k_{12} is the factor by which the magnetic field in the transmitter is transferred to or induced in the receiver inductor. The alternating current I_1 flowing in the inductor L_1 creates a varying magnetic field Φ_1 surrounding the inductor. The varying magnetic field Φ_1 also links the inductor L_2 and induces an alternating current I_2 in the inductor L_2. The current I_2 in L_2 also creates a varying magnetic field Φ_2. This field also links the inductor L_1. The net effect is mutual inductance which is characterized by the Equation (4.21):

$$M_{12} = k_{12}\sqrt{L_1 L_2} \tag{4.21}$$

This coupling coefficient is usually less than 1, and for loosely coupled power transfer systems, it is very small ($k_{12} \ll 1$), typically less than 0.1. Assume a load of the form in the Equation (4.22):

$$Z_L = R_L + jX_L \tag{4.22}$$

The coupling coefficient is a function of distance and the radii of the coils used at the transmitter and receiver circuits, respectively. At fixed radii of the coils, since the self-inductances of the coils are constants, the coupling coefficient and hence mutual inductance decrease with increasing distance between the transmitter and receiver. In other words, as the receiver recedes from the transmitter, the induced voltage in the receiver load decreases as well. As a result, the loosely coupled wireless power transfer systems cannot provide high-efficiency wireless power transfer. We can demonstrate this quite easily. By using the circuit model of two coupled inductances, we can show that since the induced voltage in the receiver circuit is proportional to $V_2 = \omega^2 k_{12}^2 L_1 L_2$, the transmitter effective impedance can be shown to be given by the Equation (4.23):

$$Z = R_1 + j\omega L_1 + \frac{\omega^2 k_{12}^2 L_1 L_2}{R_2 + j\omega L_2 + Z_L}$$

$$= R_1 + j\omega L_1 + \frac{\omega^2 k_{12}^2 L_1 L_2}{R_2 + R_L + j\left(\omega L_2 + X_L\right)} \tag{4.23}$$

This equation shows complex impedance. It shows that the coupled system is basically an impedance transformer. Hence, both real and reactive powers are induced in the receiver load proportional to the real and imaginary components of the impedance, respectively. More insight may be gained if this equation is written in terms of the quality factors of the coils as in Equation (4.24):

$$Z = \omega L_1 \left(\frac{1 + jQ_1}{Q_1}\right) + \frac{\omega k_{12}^2 Q_2 Q_L L_1 L_2}{Q_L L_2 \left(1 + jQ_2\right) + Q_2 L \left(1 + jQ_L\right)} \tag{4.24}$$

This equation is examined under low Q and high Q values for the transmitting and receiving coils.

4.4.1 Low Q_1 and Q_2

The impedance is approximately real at low Q and is given by Equation (4.25a):

$$Z \cong \frac{\omega L_1}{Q_1} + \frac{\omega k_{12}^2 Q_2 Q_L L_1 L_2}{Q_L L_2 + Q_2 L} \qquad (4.25a)$$

Equation (4.25a) shows that real power can be transferred to the load at low Q.

4.4.2 High Q_1 and Q_2

Again, in this section, we rewrite the expression for the impedance as in Equation (4.25a):

$$Z = \omega L_1 \left(\frac{1 + jQ_1}{Q_1} \right) + \frac{\omega k_{12}^2 Q_2 Q_L L_1 L_2}{Q_L L_2 (1 + jQ_2) + Q_2 L (1 + jQ_L)} \qquad (4.25b)$$

The impedance is approximately complex, which is given by the Equation (4.26) as

$$Z = j \left[R_1 Q_1 - \frac{k_{12}^2 R_1 Q_1 L_2}{L + L_2} \right] \qquad (4.26)$$

Hence, only reactive power is induced in the load, and no real power transfer is possible at high Q.

4.5 Efficiency

From Figure 4.2, we obtain the relationship between the input voltage and induced current in the receiver coil as in Equation (4.27):

$$\begin{bmatrix} V_i \\ 0 \end{bmatrix} = \begin{bmatrix} Z_1 & -j\omega M \\ -j\omega M & Z_2 \end{bmatrix} \begin{bmatrix} I_1 \\ I_2 \end{bmatrix} \qquad (4.27)$$

This leads to the input and output powers at the transmitter and receiver, respectively, as in Equations (4.28a) and (4.28b):

$$P_{\text{in}} = \frac{V_S^2 Z_2}{Z_1 Z_2 + (\omega M)^2} \qquad (4.28a)$$

and

$$P_{\text{out}} = \frac{V_S^2 (\omega M)^2 R_L}{\left[Z_1 Z_2 + (\omega M)^2 \right]^2} \qquad (4.28b)$$

The power transfer efficiency is defined as the ratio of the output power transferred to a load to the input power which is given by the Equation (4.29):

$$\eta = \frac{P_{out}}{P_{in}} = \frac{V_S^2 (\omega M)^2 R_L}{\left[Z_1 Z_2 + (\omega M)^2 \right]^2} \times \frac{Z_1 Z_2 + (\omega M)^2}{V_S^2 Z_2}$$

$$= \frac{(\omega M)^2 R_L}{\left[Z_1 Z_2 + (\omega M)^2 \right] Z_2} \times 100\% \tag{4.29}$$

The denominator has complex variables, meaning that the transferred power contains reactive power. The useful efficiency is determined by the real power delivered to the load. The power transferred to the load depends on the values of Q_1 and Q_2. Three conditions in order for the quality factors to meet are used in the following analysis to estimate the power transfer efficiency. Using the relations in the Equations (4.30a) and (4.30b):

$$Z_1 = R_1 + j\omega L_1 = R_1 (1 + jQ_1) \tag{4.30a}$$

and

$$Z_2 = R_2 + j\omega L_2 = R_2 (1 + jQ_2) \tag{4.30b}$$

we can evaluate the efficiency under three conditions:

a) When $(Q_1, Q_2) << 1$, $Z_1 = R_1$ and $Z_2 = R_2$, and hence, the efficiency of power transfer is

$$\eta = \frac{P_{out}}{P_{in}} = \frac{(\omega M)^2 R_L}{\left[R_1 R_2 + (\omega M)^2 \right] R_2} \times 100\% \tag{4.31a}$$

The power transferred to the load that is given by the Equation (4.31a) is purely real and limited by the self-resistances of the inductors and the load resistance. Low self-resistances and high load resistance are required for high efficiency.

b) When $Q_1 = Q_2 = 1$

$$Z_1 = \sqrt{2} R_1 \angle 45° \text{ and } Z_2 = \sqrt{2} R_2 \angle 45°$$

The system efficiency is given by the Equation (4.31b):

$$\eta = \frac{P_{out}}{P_{in}} = \frac{(\omega M)^2 R_L}{\left[2R_1 R_2 + (\omega M)^2 \right] \sqrt{2} R_2 e^{\frac{j\pi}{4}}} \times 100\% \tag{4.31b}$$

The efficiency is a complex number. The real system efficiency is given by the Equation (4.31c):

$$\eta = \frac{P_{out}}{P_{in}} = \frac{(\omega M)^2 R_L}{\left[2R_1 R_2 + (\omega M)^2\right] R_2} \times 100\% \qquad (4.31c)$$

This efficiency is smaller than the case when $Q_1 = Q_2 = 1$.

c) When $(Q_1, Q_2) >> 1$

$Z_1 = jR_1 Q_1$ and $Z_2 = jR_2 Q_2$, the efficiency is

$$\eta = \frac{P_{out}}{P_{in}} = \frac{-j\,(\omega M)^2 R_L}{\left[-R_1 R_2 Q_1 Q_2 + (\omega M)^2\right] R_2 Q_2} \times 100\% \qquad (4.31d)$$

The efficiency is imaginary (reactive efficiency) as no real power is transferred to the load. Thus, under high Q, the system power transfer is purely reactive.

In Chapter 2, we will show that when resonance is used, real and optimum power transfer to the load is achieved. This is because at resonance, all the reactive power is converted to real power delivery.

4.6 Summary

This chapter has revealed that power can be transferred wirelessly using an inductive transmitter and an inductive receiver without resonance. Lack of resonance causes creation of both real and reactive powers, and the amount of real power created depends on the coupling coefficients of the transmitter and receiver coils. At very low Q, the transferred power is purely real. When the quality factors of the receiver and transmitter coils are equal to unity, a mixture of real and reactive powers is developed across the receiver load. When the quality factors are very high, the power delivered to the load is purely reactive. High Q wireless power transfer as we have shown is only desirable under resonance conditions.

5

Inductive Wireless Power Transfer Using Circuit Theory

Kyriaki Fotopoulou and Brian Flynn

Institute for Integrated Micro and Nano Systems, The University of Edinburgh, Edinburgh, UK
E-mails: k.fotopoulou@ieee.org; Brian.Flynn@ed.ac.uk

This chapter discusses the field of inductive coupling for wireless power transfer. In recent years, wireless power transfer has witnessed an increased interest from academia and industry alike due to its obvious advantages over conventional wired power transfer techniques. In fields such as low-power implantable biomedical electronics and passive radio frequency identification (RFID) systems, a wireless powering scheme is very desirable. Although the principles of inductive power transfer are not new, the task of achieving a reliable yet efficient inductive link requires a detailed knowledge of the specific requirements for the application field and a deep understanding of the underlying physics. Therefore, the scope of this chapter is to shed light on the mechanisms of inductive coupling without the use of convoluted mathematical discussions. The content presented in this chapter provides a concise overview of the field, and it is most suited as a first introduction to the topic at an undergraduate level. A detailed list of references is at the disposal of the advanced researcher aiming to venture further into the topic.

5.1 Introduction

Wireless power transfer is a fascinating field of research which enjoys a long- and well-established history. The origins of power transmission via radio waves date back to the early work of Heinrich Hertz [36]. In addition to Hertz's experiments, Nikola Tesla, an acknowledged genius in the area of low-frequency electrical power generation and transmission, conducted pioneering work into this topic at the turn of the nineteenth century. Tesla became interested in the broad concept of resonance and sought to apply the principle to the transmission of electrical power from one point to another without the aid of wires. By means of alternating surges of current running up and down a mast, he strived to set up oscillations of electrical energy over large areas of the surface of the Earth. In this manner, Tesla hoped to set up standing waves into which receiving antennas could be immersed at the optimum points. The first attempts of this type were carried out at Colorado Springs, Colorado, in 1899. The famous Tesla coils were resonated at a frequency of 150 kHz with an input power of 300 kW obtained from the Colorado Springs Electric Company [11]. Unfortunately, there is no clear evidence as to the specific amount of power that could be transferred to a distant point. Tesla's work on high-power transmitters continued after the Colorado Springs in another large installation situated in Long Island 60 miles from New York. However, the unorthodox experimental methods employed by Nikola Tesla and the scale of his ambitious plans resulted in the termination of the project primarily due to the lack of financial resources. Despite the fact that this work was prematurely halted, Telsa managed to produce an impressive number of patents on transmission of electrical energy [89, 90].

In the past few decades, a considerable amount of work has been done in the area of wireless powering. There are two distinct scenarios for wireless powering, namely inductive powering for short ranges in the LF, MF, and HF region, and high power density directive radiated powering in the microwave region. The difference between the two wireless powering methods is reflected in the preliminary work of Hertz and Tesla. In the former case of Hertz, radio wave propagation was employed to transmit power from one point in space to another by propagating electromagnetic waves between antennae. In the latter case, which is the focus of this chapter, Tesla transferred power via inductive coupling between two resonating coils [89, 90]. With the advantage of historical perspective, one can realize that Tesla's attempts at efficient wireless power transfer were decades ahead of the available technology. It was not until the dawn of the biomedical implants era that the true potential of wireless energy transfer was realized.

The last four decades have witnessed an increasing interest in the area of biomedical engineering and RFID. The development of the artificial heart in the mid-1950s led the way in the evolution of biomedical implants. More recently, advances in microelectronics and system-on-chip architectures (SoC) directed the interest of the research community to the implementation of low-power biomedical systems for *in vivo* diagnostic devices [35, 51, 95]. Some common applications of implanted devices include biomedical sensors for biometric data measuring, biotelemetry implants, drug delivery systems, transducers, prostheses, artificial organs, and neurostimulators to mention a few [10, 33, 53, 74, 95]. Reliability, safety, and size are of prime importance in the design of such systems. Consequently, in designing implantable or autonomous microsystems, the development of a contactless, wireless powering scheme that is capable of powering a system is a major challenge for engineers [102]. Wireless data and power transfer is a very attractive option for such systems as it frees them from wire tethering. The implementation of miniaturized embedded systems is not limited to biomedical applications. Sensors and actuating microsystems are widely used for environmental monitoring and other industrial applications. Other fields that employ inductive coupling are RFID, contactless smart cards, and wireless microelectronic mechanical systems (MEMS).

An inductive link between two magnetically coupled coils is now one of the most common methods for contactless power transfer from the external world to implantable biomedical devices and sensor systems. A number of factors, such as size constraints, cost, battery lifetime issues, and reliability, forbid the use of an integrated power source. Real-time powering using inductive energy transfer is a favored alternative for such systems.

5.2 Advantages of Inductive Coupling for Energy Transfer

Recent work in low-power wireless communication systems and distributed sensing work has explored two main power supply options, these being batteries and ambient power scavenging. Battery technology is mature, completely self-contained, and extensively commercialized. Improvements in the miniaturization and the energy density have been accomplished, with the new generation of primary lithium cells being a good example. A promising trend in energy-containing devices is in thin-film super capacitors.

Despite the advancements in this area, using a battery to power an implant has a clear disadvantage; once the source is exhausted, it needs to be replaced using a surgical procedure. Even for relatively large batteries and conservative communication schedules, an optimistic estimate for the mean time to replacement is approximately two years. The problem is significantly aggravated for systems with batteries of more inconspicuous form factors. For example, in RFID scenarios, which involve many sensors interacting, integrated batteries will need a replacement every few months. This rate is clearly unsustainable for many applications particularly so for implanted devices.

The alternative is to deliver wireless energy in order to (a) provide online power directly to the system or (b) recharge the implanted batteries. This can be achieved by using electromagnetic powering of the system (EMP). EMP includes two main options: (1) inductive RF coupling of energy using a carrier frequency in the range of a few kHz to a few MHz, depending on the application, and (2) infrared powering [39]. Infrared powering that appears attractive at first, due to their use of photodiodes as receivers rather than coils, results in smaller implants [29, 58]. However, the process is inefficient, hence non-practical for many implant applications.

Other power sources that have been explored include ambient power scavenging from sources like thermal and kinetic energy from the body. Moreover, energy harvesting from the external environment as a source of energy is another possibility. However, challenges remain, particularly with ambient power constraints since this technology is still in its infancy. Furthermore, although ambient power sources can be used for wireless communication devices, they are unsuitable for implanted systems.

Clearly, the most attractive option is near-field coupling which can provide sensor operation in the range of a meter without line of sight. Inductive RF links are currently at the center of attention for short-range power transfer and bidirectional data communication. Communication between the implanted device and the outside world can be established through the link by modulating the RF signal used to power the implant [72]. The use of inductive coupling can reduce the dimensions of such systems as well as extend their expected operational life. As such, implanted devices can be cost-effective to manufacture and more reliable, a fact that renders inductive RF links very appealing for biomedical and RFID applications. Finally, this type of inductive coupling is advantageous because it avoids the undesirable surgical replacement of implanted power sources. Hence, the possibility of infection where wires would pierce the skin is diminished, and patient discomfort is minimized.

5.3 Applications of Inductive Power Transfer

Having established the main advantages of inductive coupling, the main applications of inductive coupling will be now discussed in more depth. The main focus of this chapter is on applications involving inductively coupled RFID and biomedical embedded sensor systems. However, the inductive coupling model suggested in this chapter can be applied in any loosely coupled system where a magnetic link is adopted for wireless power transfer, so long as the operation range of the device remains well within the near-field.

Radio frequency identification is a rapidly developing technology with a wide range of applications in various areas. RFID is an automatic identification method, based on remotely retrieving information via radio waves from miniature electronic circuits called RFID tags. An early work and possibly the first which explored RFID technology was presented by Harry Stockman in a landmark paper given in [84]. By the early 1950s, several technologies related to RFID were being pursued. Among the most famous applications of RFID from this era is the long-range transponder system of *Identification Friend or Foe* (IFF) for aircraft. Although the fundamental principle of RFID was established in the late 1940s, it is only recently that the technology has taken off due to a decrease in the cost and increasing capability [68].

An RFID system has two main components, a reader and a tag. RFID tags are categorized as either passive or active tags, depending on their source of power [14]. A diagram depicting the common ISM frequency bands available to RFID systems is shown in Figure 5.1. Active RFID tags are autonomous and carry their own power source, usually in the form of an on-board battery. Active tags transmit a stronger signal and can achieve a larger read range (20–100 m) and higher data rates. Active tags operate at higher frequencies—commonly 455 MHz, 2.45 GHz, or 5.8 GHz, depending on the application and memory requirements [96]. However, their high cost and considerable size are important disadvantages.

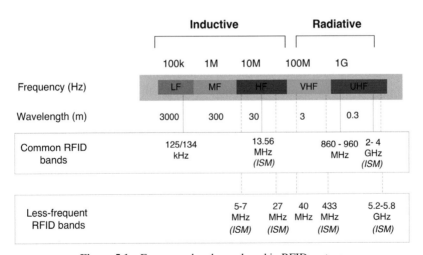

Figure 5.1 Frequency bands employed in RFID systems.

When a smaller read range is required and data rate is not critical, passive tags offer a desirable alternative [65]. This type of tag is read by intercepting the magnetic field of the reader. Usually, passive tags are low-power CMOS devices, very compact and inexpensive. The process by which a passive tag is powered is inductive near-field coupling, and typical read ranges are about 1 m. Passive RFID systems can be subdivided into low-frequency (LF) and high-frequency (HF) devices.

In addition, implantable RFID transponders have been used for livestock tracking for many years and they are bridging the gap between the RFID and embedded sensor domain. A recent but increasing development of this technology which evokes enormous interest is the use of RFID in humans. Until a few years ago, human RFID implants were limited to the domain of cybernetics provocateurs like Kevin Warwick or hobbyists like Amal Graafstra [30]. However, in 2004, the US Food and Drug Administration approved the first RFID tag for human implantation as a method of accessing and tracking medical records in hospitalized or incapacitated patients [23]. For example, sensitive information about the identity, physiological characteristics, health, and nationality can be easily embedded in a miniaturized but robust transponder about the size of the grain of rice [23]. Such a system can be lifesaving for people with chronic conditions and in emergency situations. Despite the advantages of this technology, there are serious ethical issues that should be properly addressed such as personal data privacy and security. Nevertheless, with the proliferation of radio frequency technology, the use of embedded RFID devices for human implantation is expected to slowly but steadily increase in the near future.

Additional applications of inductive coupling are emerging from the increasing interest in developing new standards for near-field communications. In recent years, a new principle of near-field communication (NFC) has originated from the evolution of more traditional RFID schemes. This technology is targeting simplified, standardized, short-range communication modules similar to the Contactless Smart Card protocol. The operation of the NFC protocol is based on inductive coupling. This technology opens new applications for the RFID technology, such as automatic payment using mobile phones in close proximity communication as a transaction vehicle. The forecasts for the future development of this technology are promising. Market researchers anticipate that by the year 2012, twenty percent of the worldwide mobile phone sales will be NFC-enabled [101].

In the biomedical sensors' arena, inductive coupling is also a major technology. The rapid scaling of CMOS technologies to smaller dimensions has resulted in very high integration densities. This enables circuitry of ever-increasing complexity to be implemented on very small chip areas, which is very well suited for use in implanted systems. Implanted electronics are used in medical devices for diagnosis as well as treatment. Among the first applications of such systems were pacemakers for cardiac arrhythmia and cochlear implants for partially restoring the hearing of deaf people. Since the early 1980s, the use of such systems has been extensive.

According to the US Food and Drug Administration 2002 data, it is estimated that approximately 59,000 people worldwide have received cochlear implants [56]. More recently, new applications like intracranial or intraocular systems have become possible. Such devices include deep-brain stimulators for Parkinson's disease, spinal cord stimulators for control of pain, and brain–machine interfaces for paralysis prosthetics [5]. Ophthalmic applications of embedded devices are also extensive. For instance, devices for monitoring intraocular pressure in glaucoma patients are common. Other retinal prostheses include visual cortex stimulators where the idea is to restore vision in blind patients by coupling electrodes directly to the visual cortex [56]. This system is designed with an integrated planar *RX* microcoil in order to minimize the size of the overall system. Figure 5.2 illustrates an implanted drug delivery system, developed at the University of Edinburgh, which can be employed in a wide range of treatments such as chemotherapy, and *in vivo* drug release for conditions like glaucoma or diabetes.

It follows from the previous discussion that there is a considerable variety of biomedical implanted devices. Although the applications of such systems are diverse, their common denominator is that embedded devices need to

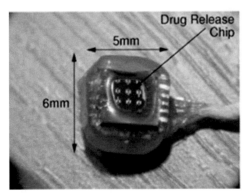

Figure 5.2 Photograph of a drug delivery microsystem developed at the University of Edinburgh [80].

have low power consumption, be compact, reliable, and fully autonomous. Therefore, wireless operation of implantable systems is key to their successful deployment in clinical applications. Applications involve biomedical embedded or RFID systems where power is supplied by means of a loosely coupled magnetic power link as discussed earlier.

5.4 Fundamentals of Inductive Coupling

Wireless inductive power transfer systems are defined as systems where energy is transferred from an external primary *TX* coil to a secondary *RX* coil using an alternating magnetic field. Essentially, this principle is very similar to a transformer action. A typical low-power inductive link is illustrated in Figure 5.3. Although improved designs have emerged in the industry, the electromagnetic modeling of the link efficiency and the optimization of the coil design have received less attention by researchers. Such systems can be divided into two categories, namely closely coupled and loosely coupled systems.

The efficiency of an inductive link depends on the magnetic coefficient of coupling, κ, which is a function of the geometrical parameters of the link, such as the coil size and shape and the coil separation distance. The magnetic links of micromodule systems considered in this chapter are loosely coupled systems and are characterized by extremely low coefficients of coupling. The coupling factor, which can be as low as 1%, presents a formidable problem for powering weakly coupled RFID and implanted micromodules.

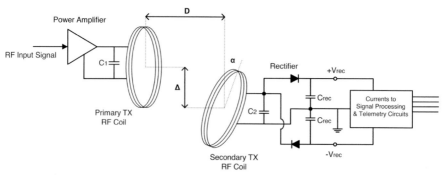

Figure 5.3 Example of low-power inductive link for implanted sensor system illustrating *RX* coil displacement. *D* is the coil separation distance, Δ represents the lateral misalignment, and α is the angular misalignment angle.

An added complication in the design of an efficient inductive link is the variation of the relative position of the *TX* and *RX* coils. The primary or *TX* coil is located outside the body and is driven by an external transmitter circuit. The secondary or *RX* coil is implanted with the device and connected to the receiver electronics. A comprehensive study on the topic of injectable electronic identification, monitoring, and stimulation systems has been carried out by Troyk, who produced a review paper presented in [91]. Referring to this study, in embedded devices such as cochlear or visual prostheses, the coils are separated by a layer of skin and tissues in the region of 2–6 cm. Also, a typical implanted micromodule has a diameter of less than 3 mm, whereas an external transmitter coil has a diameter of at least 9 cm. Essentially, the external coil dimensions are only restricted by issues related to patient comfort and aesthetic considerations. In fact, external coils can usually be as large as necessary. In an ideal situation, the *TX* and *RX* coils are coaxially orientated such that maximum coupling results. However, in the biomedical domain, misalignment of the coils can easily occur due to anatomical requirements such as skin mobility and variations in the thickness of subcutaneous fatty tissue.

In addition, coil displacement is common in conventional HF passive RFID devices. In most passive RFID systems, the reader (*TX*) and tag (*RX*) coils are separated by a distance D, typically in the range of a few centimeters, depending on the application and frequency. Often the tag coil is not placed directly on top of the reader coil. This can be easily demonstrated in spatially selective antennas for very close proximity HF RFID applications. Classic examples of this type of RFID systems include contactless smart cards for access control, e-ticketing, and label item tracking. For these devices also referred to as *dynamic objects*, the mutual reader–tag alignment can vary drastically. Consequently, for issues like anti-collision, safety, and reliability for the user, it is critical to be able to predict the misalignment tolerance of such systems and specify geometric boundaries of operation.

There have been several approaches to the analysis and design of inductively coupled transcutaneous links, targeting optimal efficiency in the majority of the cases. However, previous work mainly concentrated on steady-state circuit analysis for coupling optimization and validation through experiment. Transcutaneous links have been analyzed by Donaldson et al. [17], Galbraith et al. [28], Heetderks et al. [35], and Ko et al. [93]. Although more limited, finite element analysis has been also used for transcutaneous link modeling [73]. However, the definitive work on the coupling of air-cored coils was done by Grover [32] and Terman [88]. It can be demonstrated that the coupling coefficient is related to the mutual inductance M. Earlier research by Soma et al. [83] and Hochmair [37] attempted to present complete analytical solutions for the calculation of the mutual inductance M in order to determine the effect of misalignment on the coupling factor. Unfortunately, most of the developed solutions so far are semi-analytical and mathematically complex. Furthermore, existing studies usually target a specific design and cannot be universally applied for different coil geometries and orientations. To date, the only attempt to present a simplified analytical method for the calculation of the power transfer efficiency under different coil orientations and characteristics was given by Fotopoulou et al. in [25]. Based on the observation that for loosely coupled inductive links, the mutual inductance is very low, the problem can be significantly simplified. This can be achieved by studying the magnetic field of the *TX* coil and avoiding the involved mathematical treatment of the mutual inductance. Coil dimensions, shape, and orientation substantially affect the magnitude of the magnetic field in the near-field. Since the magnitude of the magnetic field is closely related to the efficiency of the inductive link, it is critical to identify a coil structure that maximizes the coupling between the *TX* and *RX*. However, it is beyond the scope of this chapter to go into the detailed analysis of misalignment effects on the inductive link efficiency for loosely coupled systems. Hence, for the researcher interested in this topic, a comprehensive study on coil misalignment is presented in [24].

5.4.1 Inductive Coupling and Transformer Action

An inductive link consists of two weakly coupled resonant circuits with the *TX* and *RX* coils forming the primary and secondary windings of a loosely coupled air-cored transformer, illustrated in Figure 5.4. The inductive term suggests that power transmission takes place in the near-field and transformer action theory is adequate to describe the coupling of energy from the *TX* to the *RX* load.

A time-varying current, $I = I_o \sin(\omega t)$, circulating in the transmitter coil generates a magnetic field flux which, in turn produces a time-varying voltage across the inductance of the receiving coil. when detected by the receiver a time-varying voltage appears across the inductance of the receiving coil. The magnitude of the e.m.f. induced in the receiver circuit is expressed by Faraday's law as follows:

$$V = M\frac{di}{dt} = \omega_o M\, I_o \sin\left(\omega_o t + \frac{\pi}{2}\right) \qquad (5.1)$$

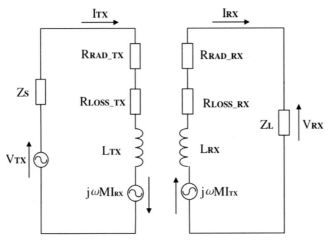

Figure 5.4 Equivalent circuit of near-field transmission model using the principle of the transformer action [21].

A physical representation of an inductive link is shown in Figure 5.5 where the air-cored primary coil is wired to the transmitter and the pickup coil is incorporated in the receiver unit (RFID tag or biomedical sensor). An AC power source excites the primary coil creating the magnetic field that surrounds the antenna coil. The secondary coil intersects the magnetic field lines created by the transmitter coil, and a current is induced in it. This process is prone to power losses with three main categories involved, these being radiation, ohmic, and absorption losses.

The associated *TX* and *RX* coil resistance introduces some ohmic losses, denoted as $R_{LOSS_{TX}}$ and $R_{LOSS_{RX}}$, respectively. The ohmic losses comprise the most important loss component and are dissipated as heat in the coil. The

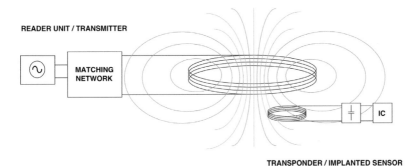

Figure 5.5 Physical representation of inductive coupled coils.

radiation losses are represented by an additional series resistance, $R_{RAD_{TX}}$ and $R_{RAD_{RX}}$, as shown in Figure 5.4. It should be noted that the radiation resistance for the applications considered in this work is negligible and can be omitted. In biomedical applications, there are some additional losses present due to energy absorption in the tissue. This is mainly caused by the current circulating in the tissue due to the magnetic field. The generated current increases the temperature in the tissue which can be potentially dangerous. Higher frequencies have a smaller skin depth, and therefore, more radiation is absorbed. Therefore, in order to be able to eliminate this type of loss, it is important to work with lower frequencies. The presence of a ferrite-cored receiver antenna permits the use of lower frequencies without compromising for range and power transfer efficiency.

Using the equivalent circuit of a transformer as shown in Figure 5.4, a near-field power transmission model can be developed. The general procedure adopted follows the approach introduced by Yates [103] and Earnshaw [19].

By employing the nodal analysis, the current and voltage gains can be easily evaluated. Consequently, the power transferred from the transmitter antenna to the receiver can be determined. For simplicity in the following mathematics, the ohmic resistance and radiation losses can be combined in as follows:

$$R_{TX} = R_{LOSSTX} + R_{RADTX}$$
$$R_{RX} = R_{LOSSRX} + R_{RADRX}$$

For the purpose of simplicity, any absorption losses will not be included in the circuital modeling at this stage. It follows that the nodal equations can be determined from the equivalent circuit in Figure 5.4 usingk Kirchhoff's laws:

$$V_{TX} = (Z_s + R_{TX})I_{TX} + j\omega L_{TX} I_{TX} - j\omega M I_{RX} \qquad (5.2)$$
$$V_{RX} = (Z_L + I_{RX}) = j\omega M I_{TX} - j\omega L_{RX} I_{RX} - R_{RX} I_{RX} \qquad (5.3)$$

Thus, the current gain can be expressed as follows:

$$\frac{I_{RX}}{I_{TX}} = \frac{j\omega M}{j\omega L_{RX} + R_{RX} + Z_L} \qquad (5.4)$$

The relationship between V_{TX} and V_{RX} can be obtained by algebraic manipulation of the previous expressions. Solving for I_{TX} in Equation (5.3) and substituting in Equation (5.4), it yields

$$I_{TX} = \frac{j\omega M - \omega^2 M^2 I_{TX}}{(j\omega L_{RX} + R_{RX} + Z_L)(Z_s + R_{TX} + j\omega L_{RX})} \qquad (5.5)$$

Referring back to Equation (5.3), it follows that by multiplying Equation (5.5) by Z_L and rearranging, the voltage transfer function can be determined

$$\frac{V_{RX}}{V_{TX}} = \frac{j\omega M Z_L}{(j\omega L_{RX} + R_{RX} + Z_L)(Z_s + R_{TX} + j\omega L_{RX}) + \omega^2 M^2} \quad (5.6)$$

5.4.2 Resonant Circuit Topologies

In order to increase the efficiency of power transmission by inductive coupling, resonant coupled coils are usually utilized in RFID and embedded devices [8, 17, 28, 40, 92]. The incentive behind a resonant implementation lies in the following parameters:

- In a tuned *TX* circuit configuration, the current circulating in the antenna coil is maximized. As demonstrated by Galbraith in [28], there are two methods for driving the *TX*. For a current-driven *TX*, a capacitor needs to be connected in parallel in order for the impedance of the circuit to be maximized and the voltage across the inductor coil to resonate. For the voltage-driven transmitter, the situation is reversed and series resonance is required to achieve minimum impedance. Both of these techniques optimize the current flow, which in return maximizes the magnetic field generated by the *TX* antenna coil [28].
- The second reason for adopting a tuned configuration is justified for impedance matching purposes. Resonance is considered as a means of matching the *RX* antenna coil to the load introduced by the electronics in the embedded device. The decision to use series or parallel resonance depends on the load resistance as indicated by Vandevoorde in [92]. For low-power inductive links studied in this work, the choice between a series or parallel connection is determined by practical considerations regarding the secondary coil inductance and the impedance of the resonant capacitor. It can be argued that the series scenario requires values for the *RX* coil which are impossible to attain practically.

This is especially true for an implanted scheme due to the size restrictions imposed on the embedded coil. It is evident from Figure 5.6 that both the series and parallel topologies achieve an optimal efficiency but for very different resonant capacitor values. The high values for the *RX* inductor coil can be explained by the high voltage drop across the series capacitor. Therefore, for low-power links, a series secondary L-C tank behaving as a voltage source is more suitable.

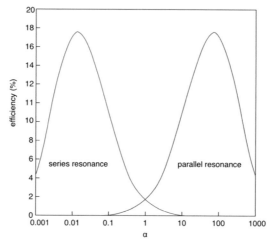

Figure 5.6   Efficiency versus, $a = \omega\, C_{RX}\, R_L$.

The value of the tuning capacitor in the series configuration can be computed by the following formula:

$$C = \frac{1}{\omega^2 L} \tag{5.7}$$

However, if parallel resonance is used, Expression (5.7) is only valid for an unloaded Q greater than 5. In inductive coupled links, there are four possible combinations of series or parallel tuned resonant coils in the transmitter and receiver circuits, as follows:

- Series resonant *TX* : parallel resonant *RX*
- Series resonant *TX* : series resonant *RX*
- Parallel resonant *TX* : parallel resonant *RX*
- Parallel resonant *TX* : series resonant *RX*.

The equivalent circuit of Figure 5.4 can equally well represent any one of the four tuned resonant coil topologies. The mutual inductance of the coils can be expressed as follows:

$$M = \kappa \sqrt{L_{TX} L_{RX}} \tag{5.8}$$

and the unloaded quality factors Q as follows:

$$Q = \frac{\omega L_{TX}}{R_{TX}} \tag{5.9}$$

$$Q = \frac{\omega L_{RX}}{R_{RX}} \tag{5.10}$$

where κ, $0 \le \kappa \le 1$, is the coefficient of coupling [19]. The quality factor, Q, is defined as the ratio of capacitive or inductive reactance to the resistance. Any distributed *mutual capacitance* denoted as C_m, in Figure 5.7, is associated with the transmitter and receiver coil coupling. However, for the link analysis described, any mutual capacitance will be omitted, as it is small enough to be neglected.

The equations expressing voltage and current transfer functions for all four topologies depicted in Figures 5.8a, b and 5.9a, b can be derived using the same method as the one described previously. However, analyzing these topologies using the standard nodal equations is tedious and does not facilitate intuitive insight into the problem. Although it could be useful to know the voltage and current transfer ratios, the main aim of this work is to suggest a power transfer function, which allows a direct comparison between different antenna coils for several possible geometries and orientations.

5.4.3 Power Transfer across a Poorly Coupled Link

The explosion in the development and popularity or RFID technologies and embedded devices has directed the interest of designers for wireless systems to issues such as power consumption and harvesting. Until recently, these parameters have been almost last on the list of specifications with implementations focusing more on reliability and data rate. The importance of consistent power supply becomes even more pronounced in implanted devices where coupling variations are common and the coils are weakly coupled which is the case for most of the applications discussed in this chapter so far. A circuital representation of a loosely coupled link is given in Figure 5.10.

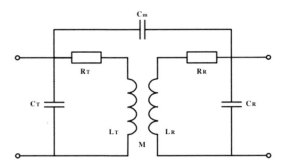

Figure 5.7 Equivalent circuit of an imperfect inductive link [19].

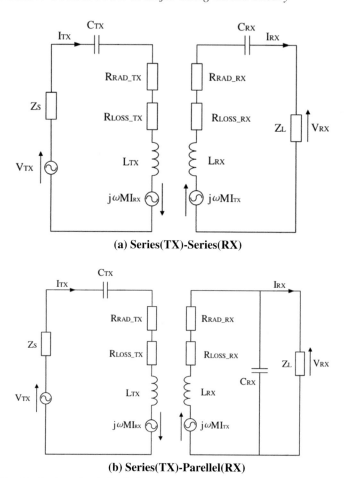

(a) Series(TX)-Series(RX)

(b) Series(TX)-Parellel(RX)

Figure 5.8   Equivalent circuit for Series resonant *TX* topology and series or parallel resonant *RX* configuration.

In the development of a power transfer function, it is critical to define the physical meaning of real power transfer. Real power transfer can be identified as the ratio which determines the amount of power dissipated in the *TX* in order to transfer a specific amount of power across the link to an implanted load, as given below:

$$\frac{P_{RX}}{P_{TX}} = \frac{V_L I_L \cos \phi_L}{V_{TX} I_{TX} \cos \phi_{TX}} \tag{5.11}$$

The real power dissipated in the load is expressed by P_{TX}, and P_{RX} represents the power provided by the source driving the *TX* coil. V_{TX},

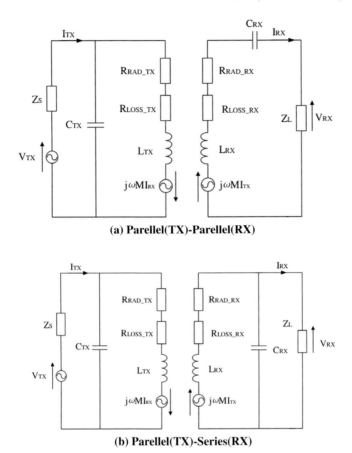

(a) Parellel(TX)-Parellel(RX)

(b) Parellel(TX)-Series(RX)

Figure 5.9   Equivalent circuit for Parallel resonant *TX* topology and series or parallel resonant *RX* configuration.

I_{TX}, V_L, and I_L define the voltage and current amplitude at the resonant *TX* and *RX* tuned circuits, respectively. The phase differences between the voltage and current signals in the primary and secondary tuned circuits are given by φ_{TX} and φ_L. It should be noted that the phase difference does not affect the power transfer function, and therefore, only the modulus of Equation (5.11) and not the associated argument of the load impedance is of interest. Finally, the efficiency of the link can be defined as follows:

$$\eta = \frac{V_L I_L}{V_{TX} I_{TX}} \qquad (5.12)$$

An equivalent circuit model for a poorly coupled system is depicted in Figure 5.10 [103]. Using this equivalent circuit, a formula expressing the power transfer from transmitter to receiver can be derived, following the method suggested by Yates et al. [103]. On the transmitter side, the impedance Z_s is assumed to resonate with the transmitter coil inductance, L_{TX}, enabling the maximum currentflow. It follows that

$$Z_s = \frac{1}{j\omega C}, \quad \text{where} \quad C = \frac{1}{\omega^2 L_{TX}} \tag{5.13}$$

The *TX* coil is exited by a sinusoidal current $I_{TX} = I_o e^{j\omega t}$. The real input power under these conditions is given by

$$P_{TX} = I_{TX\_rms}^2 \cdot R_{TX} \tag{5.14}$$

On the receiver side, the load impedance denoted Z_L in Figure 5.10 should be conjugate-matched to the impedance of the *RX* coil to achieve maximum power transfer.

$$Z_{LOAD} = R_{RX} - j\omega L_{RX} \tag{5.15}$$

In this case, the available real power from the *TX* that is delivered to the load is given as follows:

$$P_{RX} = \frac{V_L^2}{\mathbb{R}(Z_{LOAD})} \tag{5.16}$$

However, to optimize the power transfer in the radio frequency region, it is necessary to use Jacobi's theorem. (For a fixed source impedance, maximum

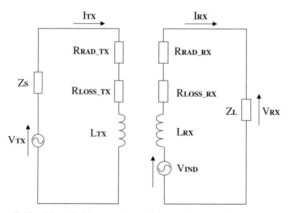

Figure 5.10   Near-field power transfer model assuming poor coupling.

power is always transferred into a conjugate-matched load.) Consequently, the *RX* circuit is transformed to a potential divider where

$$V_L = \left(\frac{Z_L}{Z_L + R_{RX}} \right) \cdot V_{IND} \tag{5.17}$$

Following from Equations (5.16) and (5.17), the received power is expressed in terms of the induced voltage across the *RX* coil:

$$P_{RX} = \frac{V_{IND}^2}{4R_{RX}} \tag{5.18}$$

The resulting magnetic field vector at the receiver coil can be obtained by integrating Biot–Savart law around the *TX* loop [67, 81]. Thus, for a short solenoid coil with *N* turns, the magnetic field at the center of the *RX* coil becomes

$$\mathbf{H} = \frac{I \cdot N_{TX}}{4\pi} \oint \frac{\mathrm{dl} \times \mathbf{r}}{r^2} \tag{5.19}$$

By Faraday's law, the induced voltage at the receiver antenna is expressed by the rate of change of flux linkage as follows [82]:

$$V_{IND} = \mu_o N_{TX} A_{RX} j\omega H \tag{5.20}$$

where N_{RX} is the number of turns of the receiver coil, A_{RX} is the loop area ($A_{RX} = \pi b^2$), *b* is the radius of the *RX* loop, and μ_o is the permeability of the free space.

Equation (5.20) is valid only for poorly coupled systems. In order for this condition to be true, the distance of separation (d) between the *TX* and *RX* needs to be much larger than the dimensions of the *RX* antenna. This is indeed true for biomedical and short-range RFID applications where the dimensions of the link are between 1 and 5 cm and the *RX* coil is usually millimeter- and submillimeter-sized [66]. This condition ensures a uniform magnetic field around the *RX* such that Faraday's law is applicable. Combining Equations (5.19) and (5.20), the induced voltage at the *RX* can be expressed in terms of the transmitter current:

$$V_{IND} = \frac{\mu_o \cdot N_{TX} \cdot N_{RX} \cdot A_{RX} \cdot j\omega I_{TX}}{16\pi^2 \cdot R_{TX} \cdot R_{RX}} \cdot H_{INT} \tag{5.21}$$

Substituting for V_{IND} and $I_{TX\mathrm{rms}}$ in Equation (5.18) and rearranging yields

$$\frac{P_{RX}}{P_{TX}} = \frac{\mu_o^2 \cdot N_{TX}^2 \cdot N_{RX}^2 \cdot A_{RX}^2 \cdot \omega^2}{16\pi^2 R_{TX} R_{RX}} \cdot H_{INT}^2 \tag{5.22}$$

where

$$\mathbf{H_{INT}} = \int\limits_0^\pi \frac{\mathbf{dl_{TX}} \times \mathbf{r}}{r^2} H_{INT} = \int\limits_0^\pi \frac{\mathbf{dl_{TX}} \times \mathbf{r}}{r^2} \qquad (5.23)$$

The function derived by Yates et al. is based on an analysis of two identical solenoid coils assumed aligned on a common central axis as depicted in Figure 5.11. Clearly, this scenario represents the ideal case which is adequate as an introduction to the topic.

5.4.4 Near-and Far-Field Regions

The space surrounding an antenna is usually subdivided into three regions: the reactive near-field region, the radiating near-field referred to as Fresnel region, and the far-field or Fraunhofer region. These regions are designed to identify the changes in the field structure as the observation point is crossing between boundaries. For an electrically small antenna, there are only two separate regions, the reactive near-field and the radiating field. Energy is stored in the former, while energy propagates as electromagnetic waves in the latter. The boundary between the two regions is generally accepted to be at a distance, r, from the antenna:

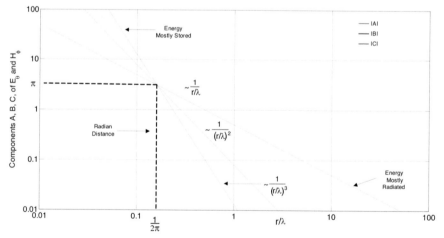

Figure 5.11   Variation of the magnitudes of the components of E and H of a short electric dipole as a function of distance (rl). The magnitudes of all components equal at the radian distance $1/(2)$. At larger distance (far-field) energy is mostly radiated, at smaller distances (near-field) mostly stored [46].

λ being the wavelength [21].

$$r = \frac{\lambda}{2\pi} \tag{5.24}$$

In the case of an electrically large antenna, a further distinction becomes significant where the radiating field is split into the radiating near-field and the radiating far-field regions. There are distinct differences between regions posing a great difficulty in a universal field representation for any point in space. However, approximations can be made to simplify the formulation of the field and yield a closed-form solution. At this stage, it is very important to introduce the field regions properly and discuss their influence on the field solution. Over the years, various criteria have been established and are commonly used to identify the different field regions. The following definition and quotations are presented by the IEEE in [1].

The reactive near-field is defined as "that region of the field immediately surrounding the antenna wherein the reactive field predominates." For most antennas, the outer boundary of this region is commonly taken to exist at a distance: $R < 0.62\sqrt{D^3/\lambda}$ from the antenna, where λ is the wavelength and D is the largest dimension of the antenna.

The radiating near-field or Fresnel region is defined as "that region of the field of an antenna between the reactive near-field region and the far-field region wherein radiation predominates and the angular field distribution is dependent upon the distance from the antenna." The radial distance R over which this region is defined is $0.62\sqrt{D^3/\lambda} \leq R \leq 2D^2/\lambda$. In this region, the field pattern is, in general, a function of the radial distance and the radial field component may be appreciable.

The far-field (Fraunhofer) region is defined as "that region of the field of an antenna where the angular field distribution is essentially independent of the distance from the antenna." In this region, the real part of the power density is dominant. The radial distance R over which this region exists is $R \geq 2D^2/2$. The outer boundary is situated ideally at infinity. In this region, the field components are essentially transverse to the radial distance, and the angular distribution is independent of the radial distance. In order to illustrate the difference between the three field regions, as described above, the magnetic field of a small loop antenna is discussed. At a radial distance r from a small circular loop carrying a sinusoidal uniform current $I = I_o e^{j\omega t}$, two components of magnetic field exist, these being H_θ and H_r [46]. Due to the circular symmetry of the antenna, there is no variation with respect to ϕ.

Hence, H_θ and H_r are expressed as follows:

$$H_\theta = \frac{m_o \sin\theta e^{j(\omega t - \beta r)}}{4\pi}\left[-\frac{\beta_o^2}{r} + j\frac{\beta_o}{r^2} + \frac{1}{r^3}\right] \tag{5.25}$$

$$H_\theta = \frac{m_o \cos\theta e^{j(\omega t - \beta r)}}{4\pi}\left[j\frac{\beta_o}{r^2} + \frac{1}{r^3}\right] \tag{5.26}$$

where m_o is the magnetic moment, equal to $I_o \pi a^2$, and β_o is defined as the phase constant of free space and is equal to $2\pi/\lambda_o$. Conventionally, the terms in the magnetic field intensity expressions H_θ and H_r are described as follows:

- $1/r$, radiation component in the far-field
- $1/r^2$, induction component in the radiating near-field
- $1/r^3$, magnetostatic component in the reactive near-field

Solving Maxwell's equations for the fields of a localized oscillating source like an antenna, surrounded by a homogeneous and isotropic material, the field decays proportionally to the terms listed above. For instance, the fields of a source in a homogeneous isotropic medium can be written as multipole expansion. The terms in this expansion are spherical harmonics, which provide the angular dependence, multiplied by spherical Bessel functions, which support the radial dependence. For large r, the spherical Bessel functions decay as $1/r$ in the radiating far-field. For a distance closer to the source, the reactive near-field dominates and other powers of r become significant. In this case, the field strength is proportional to the term $1/r^2$. This term is sometimes referred to as the induction term of the radiating near-field. It can be thought of as the energy stored in the field and returned to the antenna in every half-cycle. For even smaller r, terms proportional to $1/r^3$ become significant; this is sometimes called the magnetostatic field term at regions very close to a loop antenna where the magnetic field dominates. The difference between the field regions can be effectively illustrated using the diagram of Figure 5.12.

5.4.5 The Importance of the Loop Antenna

The loop antenna is a fundamental and simple antenna and it has received tremendous attention for short-range wireless power transfer and communications. Loop antennas have been used continually since the early days of radio and they have become particularly widespread for applications in the HF band (3–30 MHz), due to their small size. Loop antennas are very

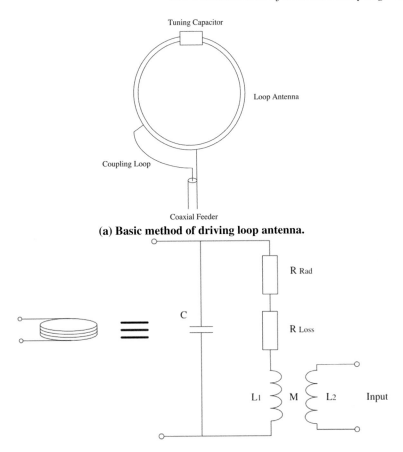

(a) Basic method of driving loop antenna.

(b) Equivalent circuit of multiturn loop antenna using lumped parameter.

Figure 5.12 L_1 main loop, L_2 coupling loop, M mutual inductance, C resonating capacitance, R_{LOSS} loss resistance, R_{RAD} radiation resistance, C resonating capacitor.

adaptable structures, and except from the widely known circular geometry, they can take many different forms such as rectangle, square, triangle, and ellipse. The loop antenna has been identified as being particularly suited to applications where wireless power delivery is required such as the powering of biomedical implants [28, 86, 94] and passive RFID tags [21]. The ability to receive energy wirelessly is essential to these devices since true autonomy is a key factor. Inductive power and telemetry are implemented in a variety of applications where the size of the RFID tag or implant is limited and batteries are not allowed. Currently, inductive energy scavenging techniques for ubiquitous computing systems, RFID devices, and sensor networks are receiving extensive attention [15, 62, 65].

The use of a loop antenna in near-field systems yields several advantages, among these being:

- The relatively non-directional nature of loop antennas can improve the operating range of the device in the near field.
- In the HF band, the antenna may be reasonably described as electrically small allowing the assumption of uniform current distribution which significantly simplifies the analysis.
- The effect of any conductive media, such as biological tissue, is decreased using an implantable loop antenna due to the fact that the dominant magnetic near-field suffers less attenuation compared to an electric field.
- A small loop is primarily inductive. In most magnetically coupled systems, resonance is employed using the inductance of the loop antenna in the oscillator resonant tank. In this manner, the loop can operate as a transmitting and receiving antenna, at the specified resonant frequency. This can be demonstrated in simple low-power short-range transmitters such as those presented in [3, 103].

The loop antenna is a very versatile structure classified into two categories: (a) electrically small and (b) electrically large. A loop antenna is considered to be electrically small if the radius is very small compared to the wavelength $(a \ll \lambda)$ or the overall length (l) is less than one-tenth of the wavelength $(1 < \lambda/10)$. On the contrary, electrically large loops are those whose circumference (C) is comparable to the wavelength of the operating frequency $(C \approx \lambda)$. Most of the formulae presented in this chapter are valid for electrically small coils. As discussed earlier, a coil is generally considered to be small if the total conductor length is less than a tenth of the wavelength [6]. Consequently, the range of frequencies over which the analysis presented in this chapter is restricted.

5.4.6 Small Loop of Constant Current

A small loop antenna (circular or square) is equivalent to an infinitesimal magnetic dipole whose axis is perpendicular to the plane of the loop. As a result, the fields radiated by an electrically small circular or square loop antenna are of the same mathematical form as those radiated by an infinitesimal dipole. A schematic antenna configuration is depicted in Figure 5.12a representing the loop with its self-inductance, resistance, and tuning capacitor in series. In practice, an unbalanced feeder is connected via a coupling loop, the magnetic flux of which links the main loop. This principle is represented in Figure 5.12b.

5.4.7 The Loop in Transmitting Mode

A small magnetic loop antenna is primarily inductive and can be delineated by a lumped element equivalent circuit as shown in Figure 5.12b. The circuital representation of its input impedance when it is employed as a transmitting antenna is shown in Figure 5.13, which is similar with Figure 5.12b. Consequently, the input impedance Z_{in} is represented as follows:

$$Z_{in} = R_{in} + jX_{in} = (R_{RAD} + R_{LOSS}) + j(X_A + X_i) \qquad (5.27)$$

where

- R_{RAD} is the radiation resistance
- R_{LOSS} is the loss resistance of loop conductor
- X_A is the inductive reactance of magnetic loop antenna, $X_A = \omega L_A$
- X_i is the reactance of the loop conductor, $X_i = \omega L_i$

In Figure 5.13, the capacitor C_r is used to resonate the antenna. In addition, any distributed or interwinding parasitic capacitance present at the loop antenna can be incorporated in the capacitor C_r. The resonant magnetic loop can be employed as a transmitter to transfer energy by means of inductive coupling. In this manner, power can be transferred using transformer action from a primary transmitting coil to the secondary receiving coil, where both coils are tuned resonant magnetic loops at a common frequency. At resonance, the susceptance B_r of the capacitor C_r is chosen in order to cancel the imaginary

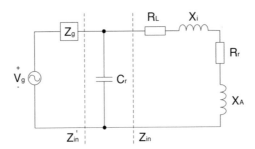

Figure 5.13 Equivalent circuit of loop antenna in transmitting mode.

part of Bi. Hence, at resonance, the capacitor value is determined by the following expression [6]:

$$C_r = \frac{B_r}{2\pi f} = -\frac{B_{in}}{2\pi f} = \frac{X_{in}}{2\pi f(R_{in}^2 + X_{in}^2)} \tag{5.28}$$

The equivalent admittance Y_{in} of the capacitor C_r is given as follows:

$$Y_{in} = G_{in} + jB_{in} = \frac{1}{X_{in}} = \frac{1}{R_{in} + jX_{in})} \tag{5.29}$$

where

$$G_{in} = \frac{R_{in}}{R_{in}^2 + X_{in}^2} \tag{5.30}$$

$$B_{in} = -\frac{X_{in}}{R_{in}^2 + X_{in}^2} \tag{5.31}$$

Hence, the equivalent input impedance, Z'_{in}, of resonant magnetic loop antenna is derived as follows:

$$Z'_{in} = R'_{in} = \frac{1}{G_{in}} = \frac{R_{in}^2 + X_{in}^2}{R_{in}} = R_{in} + \frac{X_{in}^2}{R_{in}} \tag{5.32}$$

The losses in the tuned transmitter coil are expressed by R_{RAD} and R_{LOSS}. In electrically small loop antennas, the radiation losses are negligible, as shown in Figure 5.14. The radiation resistance of loop antennas with uniform current and dimensions small compared to the wavelength can be easily evaluated using the following expression [6, 69]:

$$R_{\mathrm{RAD}} = 20\pi^2 N^2 \left(\frac{l}{\lambda}\right)^4 \Omega \tag{5.33}$$

The proof for the radiation resistance Expression (5.33), for an electrically small loop antenna, is given by [6]. For an N-turn loop, the magnetic field passes through all the loops and the radiation resistance is increased by the term N^2 in Equation (5.33), where l is the circumference of the loop and λ is the wavelength.

The radiation and ohmic losses are important parameters in the design of an antenna as they determine the radiation efficiency. In general, for the electrically small loop antennas being investigated in this work, the loss resistance is generally much larger than the radiation resistance.

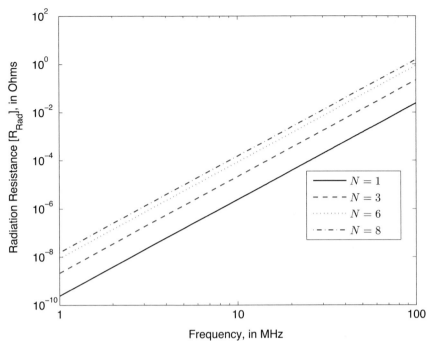

Figure 5.14 Radiation resistance for a constant current single-loop and multiloop antenna with respect to frequency.

Therefore, the corresponding radiation efficiency is very low. As a result, small loop antennas are very poor radiators and they are seldom employed in conventional far-field transmission. The radiation efficiency can be computed based on the two experimental techniques referred to as the *Wheeler method* and the *Q method* [60]. It should be noted that due to the complexity of the current distribution between the windings of a multiturn antenna, great confidence has not yet been placed in an analytical solution to the radiation efficiency riddle. However, until greater understanding of the theoretical approach to this problem has been established, greater confidence should be placed in the experimental techniques mentioned previously. A detailed discussion of the *Wheeler* Equation (5.34) and the *Q* Equation (5.35) methods is given in [60, 77, 76]. The radiation efficiency as defined by the *Wheeler* method is given by the following expression:

$$\eta = \frac{R_{\text{RAD}}}{R_{\text{RAD}} + R_{\text{LOSS}}} \tag{5.34}$$

The radiation efficiency as defined by the Q method is given below:

$$\eta = \frac{Q_{RL}}{Q_R} = \frac{\text{Power Radiated}}{\text{Power Radiated} + \text{Power Received}} \qquad (5.35)$$

where Q_{RL} and Q_R are the quality factors of practical and ideal antennae, respectively. Consequently, the loop antenna is a poor radiator since $R_{\text{LOSS}} \gg R_{\text{RAD}}$. Therefore, when employing loop antennas, magnetic coupling is a more efficient method to transfer energy in the near-field.

5.4.8 The Loop in the Receiving Mode

A magnetic loop is often used as a receiving antenna or as a probe to measure magnetic flux density. The Thevenin equivalent circuit of a loop antenna in the receiving mode is depicted in Figure 5.15. An electrically small loop enclosing an effective area A and placed in a uniform alternating magnetic field can be now considered. Assuming that the axis of the winding is parallel to the field strength vector H and the incident field is uniform over the plane of the loop, the open-circuit voltage for a multiturn loop antenna can be written as, [6, 82]:

$$V_{\text{IND}} = j\,\omega\mu_\text{o}\,N\,H\,A \qquad (5.36)$$

Close investigation of the previous expression shows that Equation (5.37) is an alternative representation of Faraday's law, where the *emf* across the terminals of the receiving antenna is equal to the time rate of change of magnetic flux through the area of the loop. For the static antenna, [67]:

$$E \bullet dl = \frac{\partial \psi}{\partial t} m = -\frac{\partial S}{\partial t} B \bullet dS \qquad (5.37)$$

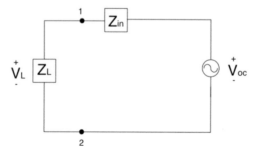

Figure 5.15   Equivalent circuit of loop antenna in transmitting mode.

where the flux ψ_m is found by evaluating the normal component of flux density B over the surface of the loop as shown in Figure 5.16.

By revisiting the Thevenin equivalent circuit of Figure 5.15 it is evident that when the load impedance Z_L is connected to the output terminals of the loop, the voltage *VL* across the load impedance Z_L is related to the output impedance and the induced open-circuit voltage of Equation (5.37) as follows:

$$V_L = V_{\text{IND}} \frac{Z_L}{Z'_{IN} + Z_L} \tag{5.38}$$

It should be noted that the open-circuit voltage of Equation (5.37) is also related to the vector effective length of the loop antenna expressed as follows:

$$l_e\,(\theta, \phi) = \widehat{a_\theta}\, l_\theta\,(\theta, \phi) + \widehat{a_\phi} l_\phi\,(\theta, \phi) \tag{5.39}$$

However, since the quantity in Equation (5.39), alternatively referred to as the effective height, is a far-field quantity it can be omitted from the near-field induced voltage calculation in Equation (5.37). In a far-field investigation where the radiation field contributes to the open-circuit voltage, the following representation is more appropriate:

$$V_{\text{IND}} = jk_o A \cos\psi_1 \sin\theta_1 \tag{5.40}$$

where the factor $\cos\psi_1\sin\theta_1$ is introduced as the magnetic flux density component which is normal to the plane and $k_o = 2\pi/\lambda$ is the free space propagation constant. Since we are focused on the magnetostatic case in the near-field, Faraday's law can be utilized in the form expressed by Equation (5.40), ignoring any far-field contributions.

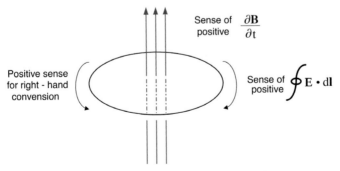

Figure 5.16 Sense relations for Faraday's law.

Generally, it is common to incorporate ferrite material at the center of a receiver antenna in order to improve the signal reception. The advantages of ferrite-loaded loop antennas of small volume at HF have been known since the 1940s. Advances in ferrite technologies have made their use even more attractive.

In applications where it is important to minimize the volume occupied by the device, as in remotely controlled biomedical implants, the use of ferrite antennas is desirable. The ferromagnetic material concentrates the magnetic flux lines toward the receiver antenna. Thus, the flux density at the center of the loop is increased by a factor μ_r. The voltage induced across the terminals of a ferrite antenna can be expressed by [64, 82]:

$$V_{\text{IND}} = j\omega\mu_o\mu_r NHA \tag{5.41}$$

where N is the number of turns and μ_r is the magnetic permeability of the ferromagnetic material.

5.5 Mutual Inductance of Coupled Coils

Following from the circuital analysis of the resonant *TX*, *RX* topologies discussed earlier on, the concept of mutual induction and its implications to the modeling of near-field power transfer will be presented here. When a variable magnetic field links one part of a circuit to another part, an induced voltage is generated according to Faraday's law. This coupling of energy is represented in the circuit by means of a mutual inductor M, as shown in Figure 5.17. The value of M is defined as the magnetic flux ψ_{12} linking path 1 in Figure 5.17, divided by the current I_2, as shown in [67]:

$$V_{21} = \frac{\Psi_{12}}{I_2} \tag{5.42}$$

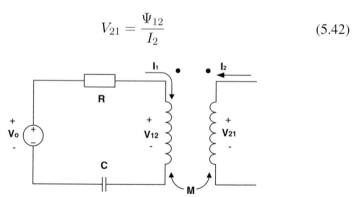

Figure 5.17 Designation of mutual coupling using the mutual inductance principle.

The voltage induced in the first path, depicted on the diagram of Figure 5.17, is given as follows:

$$V_{21} = \frac{d\Psi_{12}}{dt} = M\frac{I_2}{dt} \tag{5.43}$$

In the same manner, the time-varying current circulating in circuit 1, shown in Figure 5.17, induces a voltage in circuit 2 given as follows:

$$V_{21} = \frac{d\Psi_{21}}{dt} = M\frac{I_1}{dt} \tag{5.44}$$

At this point, it should be noted that for isotropic materials, there is a reciprocal relation showing that the same M generates a voltage in circuit 2.

The mutual inductance M varies according to the resonant frequency and the inductance of the coils under consideration. However, M is greatly affected by geometrical parameters such as the *TX*, *RX* coil proximity, shape, and orientation. The mutual inductance as defined previously arises from the induced voltage in one circuit due to current circulating in another circuit. Several approaches to its calculation can be now discussed, some of which will be used to justify the decision of adopting a loosely coupled system.

Flux linkages. The most straightforward approach follows from Faraday's law, finding the magnetic flux linking one circuit related to the current flowing in the other circuit, as in Equation (5.49). Therefore, for two circuits denoted 1 and 2, the mutual inductance can be written as follows:

$$M12 = \frac{\int \mathbf{B2} \cdot \mathbf{dS1}}{I2} \tag{5.45}$$

where **B2** is the magnetic flux due to current I_2 and integration is over the surface of circuit 1. By reciprocity, $M_{21} = M_{12}$, which is true for isotropic materials, and as a result, the calculation can be done using the inducing current at each circuit. Two parallel coaxial conducting loops can be now considered as pictured in Figure 5.18. Using Biot–Savart law, the magnetic field generated by the current circulating in one loop can be calculated at a point on the axis as given in [21, 47, 67]:

$$B_z(0, d) = \frac{\mu I 2 a^2}{2(a^2 + b^2)^{3/2}} \tag{5.46}$$

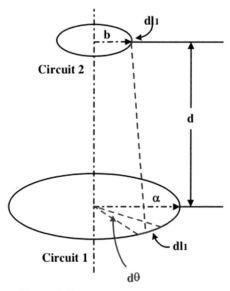

Figure 5.18   Two parallel coaxial loops.

For the applications envisaged in this work, such as in the biomedical and RFID domains, loop 2 which acts as the *RX* is usually small enough compared with the loop spacing *d*.

Thus, the magnetic flux cutting through loop 2 is considered to be relatively constant and uniform over the second loop. In retrospect, the relation for the mutual inductance becomes

$$M = \frac{\mu \pi a^2 b^2}{2(a^2 + b^2)^{3/2}} \qquad (5.47)$$

Magnetic Vector Potential Approach. The application of Stoke's theorem to Equation (5.47) yields an equivalent expression in terms of the magnetic vector potential:

$$M = \frac{\int (\nabla \times A2) \cdot dS1}{I2} \qquad (5.48)$$

In cases where the magnetic field is difficult to calculate directly, compared to the vector potential, this form becomes useful. In particular, this is true for problems where the circuit has straight-line segments or can be approximated in such a manner.

Neumann's Form Another standard technique used in the calculation of mutual inductance of two filamentary conductors is given by Neumann's formula given below. This approach originates from calculating the magnetic vector potential A arising from circuit 2, assuming the current to be in line filaments and neglecting retardation.

$$A2 = \oint \frac{\mu I2 \cdot dI}{4\pi R} \tag{5.49}$$

Substitution in Equation (5.49) yields Neumann's formula:

$$M = \frac{1}{I2} \oiint \frac{\mu I2 \cdot dI1 \cdot dI2}{4\pi R} = \frac{\mu}{4\pi} \oiint \frac{dl1 \cdot dl2}{R} \tag{5.50}$$

where R is the distance between current element $dl2$ and the point at which the magnetic fieldshould be computed.

This standard form is named after Neumann which is used in the derivation of the mutual inductance calculation for a system of parallel coaxial loops as illustrated in Figure 5.18.

Having presented the most popular methods for the derivation of the mutual inductance, it is now possible to use these to estimate the coupling between the *TX* and *RX* coils using the flux linkage method for a number of practical antenna coils. An estimate of the mutual coupling will justify the loosely coupled idea adopted in order to simplify the problem of modeling near-field inductive coupling.

Referring back to Equation (5.47), the mutual inductance of two coaxial coils of radii a and b with N_{TX} and N_{RX} number of turns at a distance d is given by

$$M = \frac{\mu \pi N_{TX} N_{RX} a^2 b^2}{2(a^2 + b^2)^{3/2}} \tag{5.51}$$

Figure 5.19 depicts the mutual inductance for a number of *TX* and *RX* coil dimensions, representative in biomedical and RFID scenarios. Figures 5.19 and 5.20 show that the mutual inductance between coils with radii within practical bounds for the application areas studied is very low even for larger coils.

Once the mutual inductance is known, the concept of coupling coefficient, k, can be introduced, as it provides a qualitative prediction about the coupling of the conductor loops independent of their geometric dimensions. The following relation applies as referred to in numerous textbooks [21, 19]:

$$\kappa = \frac{M}{\sqrt{L1 \cdot L2}} \tag{5.52}$$

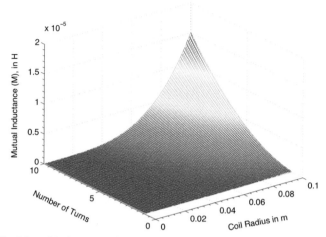

Figure 5.19   Mutual inductance of *TX* and *RX* coils separated by a distance of 0.5 m for varying number of turns and coil radius.

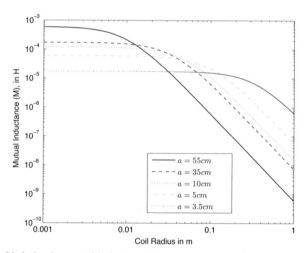

Figure 5.20   Variation in mutual inductance between the *TX* and *RX* antenna coils, for a *RX* coil radius of 3.5 cm, as the distance between the coil increases.

where $L1$ and $L2$ represent the inductance of the *TX* and *RX* coils, respectively, and the coupling coefficient k always varies between two extreme cases as $0 \leq k \leq 1$.

An analytic calculation for complex antenna structures can be very difficult to achieve. However, an approximation for the coupling coefficient for the simple case of two parallel and coaxial circular loops is given in Equation (5.53), [70, 71]. This expression can be used as a first step to approximate the magnitude of the coupling factor:

$$\kappa = \frac{a^2 b^2}{\sqrt{ab} \cdot \sqrt{d^2 + a^2}^3} \tag{5.53}$$

In implantable devices, the external primary or transmitter coil and an implanted secondary coil are separated by a layer of skin and tissues usually not exceeding 1–3 cm in thickness. The magnetic link allows the transfer of energy and information, through the biological tissue medium. However, often misalignment of the coils is possible due to skin mobility and variations in the thickness of subcutaneous fatty tissue. Since some degree of misalignment is inevitable between *TX* and *RX* coils in transcutaneous RF links, its effect on the mutual inductance needs to be addressed. To the best of the author's knowledge, the practical issues of coil misalignment and orientation and their implications on transmission characteristics of RF links have been overlooked by researchers with very few exceptions. Papers presented by Flack et al. [22] and Hochmair [37] consider only the effects of lateral displacement of coils on the mutual inductance. The study by Flack et al. is based on experimental data, whereas Hochmair reduces the double integral to a single and implements numerical integration to solve it. Neither of these works takes into account the geometric characteristics of the coils and their shape focusing only on parallel circular loops of zero thickness, these being essentially flat filaments. Almost all designs still depend on an uneasy alliance between the experimental work by Terman [88], empirical data or the extensive data for the self- and mutual inductances of coils presented by Grover [32] based on numerical methods. Although these techniques are acceptable, they are not applicable unless the actual coil size and shape are specified, which renders them unsuitable for optimization purposes with respect to displacement and angular tolerance of the system.

Any theoretical investigation of the mutual inductance in arbitrary coil configurations is extremely complex due to the lack of symmetry and the tedious work required to solve the double integral in Neumann's formula. The only semi-analytical solution for the mutual inductance for flat loop coils in lateral and angular misalignment is introduced by Soma et al. [83]. In the case of two coaxial loops, the computation of the mutual inductance has been treated in detail by numerous textbooks and is given by

$$M_i = \mu_o \sqrt{ab}\left[\left(\frac{2}{k} - k\right)K(k) - \frac{2}{k}E(k)\right] \qquad (5.54)$$

where $K(k)$ and $E(k)$ are the complete elliptic integrals of the first and second kind, respectively, and k is the modulus of $K(k)$ and $E(k)$.

Based on the mutual inductance for the ideal coaxial scenario, Soma derived upper and lower bounds for the mutual inductance in lateral and angular misalignment and an arithmetic average of these bounds. In the lateral misalignment configuration, the lower and upper bounds of the mutual inductance are derived by the following expressions [83]:

$$M_L\,(\text{min}) = \frac{\mu_o ab}{\sqrt{a\,(b+\Delta)}}G(r_{\text{min}}) \qquad (5.55)$$

$$M_L\,(\text{max}) = \frac{\mu_o ab}{\sqrt{a\,(b-\Delta)}}G(r_{\text{max}}) \qquad (5.56)$$

$$r\,(\text{min}) \equiv \left(\frac{4a\,(b-\Delta)}{(a+b-\Delta)^2 + d^2}\right)^{1/2} \qquad (5.57)$$

$$r\,(\text{max}) \equiv \left(\frac{4a\,(b+\Delta)}{(a+b+\Delta)^2 + d^2}\right)^{1/2} \qquad (5.58)$$

and $G(r)$ is the bracketed expression in Equation (5.59), denoted as follows:

$$G(r) = \left(\frac{2}{k} - k\right)K(k) - \frac{2}{k}E(k) \qquad (5.59)$$

A closer approximation of the mutual inductance can be evaluated by substituting for the maximum value of $G(r\text{max})$ in the lower bound of the mutual inductance as given by Equation (5.60), which yields

$$M_{L1} = \frac{\mu_o ab}{\sqrt{a\,(b+\Delta)}}G(r_{\text{max}}) \qquad (5.60)$$

In addition, an arithmetic mean of the upper and lower bounds is also derived in [83]:

$$M_{L2} = \frac{M_L\,(\text{min}) + M_L(\text{max})}{2} \qquad (5.61)$$

In the same manner, approximate formulas for the mutual inductance under angular misalignment, expressed by a tilt angle α, are presented as follows:

$$M_{A1} = \frac{\mu_o \sqrt{ab}}{\sqrt{\cos a}} \tag{5.62}$$

$$M_{A2} = \frac{M_i}{\sqrt{\cos a}} \tag{5.63}$$

where

$$r \equiv \frac{r_{\max} + r_{\min}}{2} \tag{5.64}$$

$$r_{\max} = \left(\frac{4ab \cos a}{a^2 + b^2 + d^2 - 2bd \sin a + 2ab \cos a} \right) \tag{5.65}$$

$$r_{\min} = \left(\frac{4ab \cos a}{a^2 + b^2 + d^2 + 2bd \sin a + 2ab \cos a} \right) \tag{5.66}$$

Expressions (5.59), (5.60), (5.61), and (5.62) were implemented in MATLAB and plotted for two equal coils with their centers displaced by a distance Δ that does not exceed the radius of the *RX* coil b and an angular misalignment that does not exceeds $25°$, respectively. Expressions (5.59)–(5.62) are illustrated in Figures 5.21a and b which were produced to test the accuracy of the MATLAB implementation of Soma's model and are identical to Figures 3 and 5 in Soma's paper [83]. Again, according to Soma in the angular misalignment case, Expression (5.61) tends to underestimate the value of the mutual inductance with an error in the order of 20% compared to the numerical integration results. On the contrary, the approximation estimating the mutual inductance given in Equation (5.62) is believed to be correct within 3% of the numerical integration values showing a significantly improved performance with respect to Grover's data even at large misalignment angles.

Consequently, as mentioned above, it is critical to note that the analysis presented by Soma in [83] has some limitations as it is based on simplifying the solution of Newman's formula by evaluating the upper and lower bounds of the double integral. Although the upper and lower bounds for the lateral displacement can be rather conservative since the integrand in Neumann's formula is not treated as a whole, Soma's method shows good agreement with Grover's data. By utilizing this procedure, formulae for the upper and lower bounds of the mutual inductance in lateral and angular misalignment are given without the need of resorting to numerical integration. However, this method provides only a semi-analytical solution, meaning that the expressions

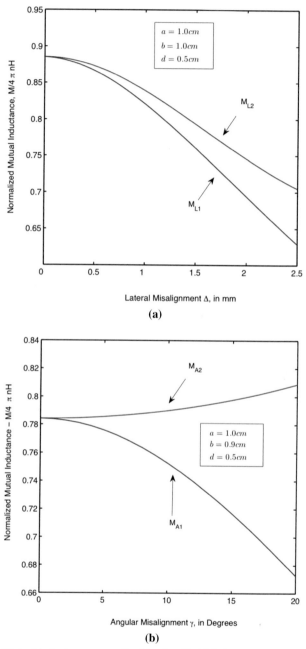

Figure 5.21 Mutual inductance normalized by $4\pi nH$: (a) in lateral misalignment configuration and (b) in angular misalignment configuration.

evaluating the bounds for the mutual inductance can be valid only for a certain range of misalignment values. In addition, this method overlooks the geometrical parameters of practical coils such as number of turns and shape.

Contrary to these limitations, Soma's method has some significant advantages such as it provides simple analytical, computationally efficient, and accurate expressions for the mutual inductance under misalignment.

Soma's cohesive analytical derivation of the mutual inductance for coils under lateral and angular misalignment is utilized here in order to compute the mutual inductance of misaligned coils. Figures 5.21 and 5.22 depict the variation of the mutual inductance of two coils in the lateral and angular orientations using the analytic forms of the mutual inductance introduced previously. In the design of transdermal RF magnetic coupling of power for an implant, there are a number of different factors that need to be deliberated. Among the most crucial parameters in the design of an RF link are the size and shape of the coils, the location of the implant, and the displacement tolerance of the system. Each application has its own unique requirements that influence the priority of these factors. Specifically, in the biomedical scheme, the implantable unit should be minimized in order to eliminate patient discomfort and physiological problems. This is in fact reflected in Figures 5.22 and 5.23 where a realistic value, in the order of 10 ($a/b = 10$), was adopted for the ratio of the *TX* to the *RX* coil radii.

In spite of the fact that Figures 5.22 and 5.23 represent the mutual inductance between single-turn coils based on Soma's method, it is evident that the value of the mutual inductance is very low for realistic coil dimensions and separation distance. Hence, referring to these plots, the assumption of weakly coupled coils becomes apparent in the misaligned case.

Referring to Figure 5.23, an unexpected increasing trend of the mutual inductance with the misalignment angle is shown. This behavior might seem surprising at first, but it can be easily interpreted as pointed out by Hochmair in [37]: *The tilting of the receiver coil brings half of the coil closer to the perimeter of the transmitting coil, and since the magnetic field is maximized at the coil perimeter, we expect increasing coupling, which overcompensates some losses due to the larger distance between the transmitting coil and the other half of the receiving coil.*

Based on this principle, Soma suggests that this behavior is the principal reason behind the error in Grover's data.

The results presented in this section support the hypothesis of a weakly coupled system. Accordingly, for the weakly coupled case, the mutual inductance is not critical and can be overlooked as a simplification in the analysis

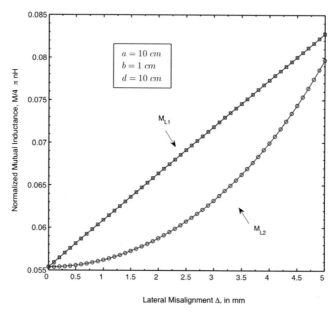

Figure 5.22   Mutual inductance versus lateral displacement of the coil centers for *TX* and *RX* coils with radii of 10 and 1 cm, respectively.

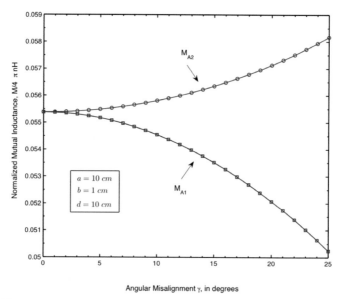

Figure 5.23   Mutual inductance versus angular misalignment for *TX* and *RX* coils with radii of 10 and 1 cm, respectively.

of the power transfer under coil misalignment conditions being discussed in detail in the following chapter. Nevertheless, mutual inductance is a key factor in closely coupled systems and for telemetry and communication purposes, especially when LSK[1] is employed using the reflected impedance technique of an inductive coupled transformer [50, 87].

5.6 The Loosely Coupled Approximation

The mutual inductance and hence the coupling factor are always small in the application areas considered in this chapter. This fact simplifies the modeling of the energy coupled across the inductive link. As a result, a simple equation can be derived following the approach suggested by Yates et al. [103], which will enable us to observe the relation of the power transfer on various application parameters and most importantly the impact of misalignment on the link effency as demonstrated by Fotopoulou et al. in [24, 25].

The validity of the poor coupling scenario can be tested against the results presented by Vandevoorde et al. [92] representing the closely coupled situation. According to Vandevoorde et al. the total link efficiency for parallel, ηp, and series resonant, ηs, secondary coils can be written as follows:

$$\eta_p = \frac{\kappa^2 Q_{TX} Q_{RX}}{\left(1 + \frac{Q_{RX}}{a} + \kappa^2 Q_{TX} Q_{RX}\right) + \left(a + \frac{1}{Q_{RX}}\right)} \tag{5.67a}$$

$$\eta_s = \frac{\kappa^2 Q_{TX} a}{\left(1 + \frac{1}{Q_{RX}} + \kappa^2 Q_{TX}\right) + \left(a + \frac{1}{Q_{RX}}\right)} \tag{5.67b}$$

where α is a unitless constant defined as follows:

$$a = \omega C_{RX} R_L \tag{5.68}$$

In both Equations (5.67a) and (5.67b), Q_{TX} and Q_{RX} express the quality factor of the primary and secondary coils, respectively. An analytic calculation of the coupling factor is only possible for very simple antenna configurations. For two parallel conductor loops centered on a single z-axis, as indicated in Figure 5.11, the coupling coefficient, k, can be approximated by 5.60 according to [70].

In order to achieve maximum power transfer across the link, both the primary and secondary circuits are tuned to the same resonant frequency. In addition, in low-power inductive links, the combination of the receiver coil and capacitor impedance should be matched to the impedance of the implanted device. Hence, Expressions (5.67a) and (5.67b) can be combined to represent the maximum efficiency in both the series and parallel configurations as derived in [92]:

$$\eta_{opt} = \frac{\kappa^2 Q_{TX} Q_{RX}}{1 + (\sqrt{1 + \kappa^2 Q_{TX} Q_{RX}})^2} \tag{5.69}$$

and the new optimal value for α, in parallel and series configurations respectively, is given by the following expressions:

$$a_{opt} = \frac{Q_2}{\sqrt{1 + \kappa^2 Q_{TX} Q_{RX}}} \tag{5.70}$$

$$a_{opt} = \frac{\sqrt{1 + \kappa^2 Q_{TX} Q_{RX}}}{Q_{RX}} \tag{5.71}$$

In poorly coupled systems, the term $k^2 Q_{TX} Q_{RX}$, in the above equations, is usually much smaller than 1 even for coils with high quality factors in the order of 100. It follows that by employing $k^2 Q_{TX} Q_{RX} + 1$, Expressions (5.67) and (5.68) can be extended to account for poorly coupled systems. Consequently, the maximum efficiency and the constant α in loosely coupled inductive links can be represented using the following formulae:

$$\eta_{opt} = \frac{\kappa^2 Q_{TX} Q_{RX}}{4}$$

$$a_{opt} = Q_{RX} \text{ for Parallel Resonance.} \tag{5.72}$$

$$a_{opt} = \frac{1}{Q_{RX}} \text{ for Series Resonance.}$$

The validity of the poor coupling case can be tested against the modeling of closely coupled coils presented in [92]. This comparison has been performed for maximum link efficiency where expressions 5.106 and 5.108 are plotted against the receiver coil radius. The results are depicted in Figure 5.23 for a parallel tuned receiver configuration as this is the most common topology currently employed in the applications considered. Referring back to Figure 5.23, the full lines represent the power transfer taking into account the energy coupled back to the transmitter as calculated by (5.106), whereas the dashed lines show the power transfer as evaluated by (5.108) using the poor coupling theory. It is evident from Figure 5.23 that the poor coupling approximation indicates a level of deviation at large coil radii in the order of 2 dB. However, for a small receiver coil radii, the discrepancy is minimal and considered negligible. It could be argued that for a minimum separation distance between the coils of twice the transmitter diameter, the poor coupling approximation is valid up to a receiver radius half that of the transmitter coil. Since in implanted devices, the receiver coil is usually much smaller than the transmitter, it can be concluded that the range of values over which this methodology is valid is first constrained by other factors in the analysis other than the poor coupling concept.

5.7 Summary

The scope of this chapter was to introduce the reader to the background theory of near-field inductive coupling for loosely coupled inductive links. During the course of this chapter, the figures of merit for efficient inductive coupling were discussed. The resonant circuit configurations classically employed in inductive coupled systems were introduced. The fundamental theory behind the loop and dipole antennas, which are critical for the modeling of inductive resonant *RX-TX* systems, was explained. Another important factor, that is the mutual inductance between two inductively coupled coils situated in the ideal and non-ideal orientations, was derived using Soma's analysis. The results yield that for the applications where the mutual coupling is very low, the decision to model the inductive coupling action based on the magnetostatic scenario is well justified. Hence, this technique can be a powerful tool for the calculation of the link efficiency where the loosely coupled approximation is valid. In addition, the source of losses in the power transfer process between two inductively coupled coils was identified. The principles of inductive power transfer were presented both from a circuital and from an electromagnetic perspective. Last but not least, the concept of a loosely coupled inductive link, in conjunction with the loop antenna theory and its suitability for near-field operation, was discussed. Finally, the limitations of the existing research were addressed, and a detailed literature review is available for the reader.

References

[1] IEEE standard definitions of terms for antennas, IEEE Std. 145-1993, March (1993).

[2] SL1ICS3101 I-CODE1 Label IC. Data Sheet, Product Specification Rev. 1.3, Philips Semiconductors, January (2005).

[3] Ahmadian, M., B.W. Flynn, A.F. Murray, and D.R.S. Cumming. "Data Transmission for Implantable Microsystems Using Magnetic Coupling." *IEEE Proceedings Communications*, 152, no. 2: 247–250 (2005).

[4] Akin, T., K. Najafi, and R.M. Bradley. "A Wireless Implantable Multi-channel Digital Neural Recording System for a Micromachined Sieve Electrode." *IEEE Journal of Solid-State Circuits*, 33, no. 1: 109–118 (1998).

[5] Baker, M.W., and R. Sarpeshkar. "Feedback Analysis and Design of RF Power Links for Low-Power Bionic Systems." *IEEE Transactions on Biomedical Circuits and Systems* 1, no. 1: 28–38 (2007).

[6] Balanis. C. *Antenna Theory: Analysis and Design*, 2nd edition. New York: Wiley (1997).

[7] Balanis, C.A. "Antenna Theory: A Review." *Proceedings of the IEEE* 80, no. 1: 7–23 (1992).

[8] Basset, P., A. Kaiser, B. Legrand, D. Collard, and L. Buchaillot. "Complete System for Wireless Powering and Remote Control of Electrostatic Actuators by Inductive Coupling." *IEEE Transactions on Mechatronics* 12, no. 1: 23–31 (2007).

[9] Bottomley, P.A., and E.R. Andrew. "RF Magnetic Field Penetration, Phase Shift and Power Dissipation in Biological Tissue: Implications for NMR Imaging." *Physics in Medicine and Biology* 23, no. 4: 630–643 (1978).

[10] Bronzino. J.D. *The Biomedical Engineering Handbook*. IEEE Press: New York (1995).

[11] Brown. W.C. "The History of Power Transmission by Radio Waves." *IEEE Transactions on Microwave Theory and Techniques* 32, no. 9: 1230–1242 (1984).

[12] Butterworth, S. "Note on the Alternating Current Resistance of Single Layer Coils." *Physical Review* 23: 752–755 (1925).

[13] Butterworth. S. "On the Alternating Current Resistance of Solenoidal Coils." *Proceedings of the Royal Society of London. Series A, Containing Papers of a Mathematical and Physical* 107, no. 744: 693–715 (1925).

[14] Chen, S.C.Q., and V. Thomas. "Optimization of Inductive RFID Technology." pp. 82–87, May (2001).

[15] Chevalerias, O., T. O'Donnell, D. Power, N. O'Donnovan, G. Duffy, G. Grant, and S.C.O'Mathuna. "Inductive Telemetry of Multiple Sensor Modules." *IEEE Pervasive Computing* 4, no. 1: 46–52 (2005).

[16] Daniel, J.P. "Mutual Coupling Between Antennas for Emission of Reception—Application to Passive and Active Dipoles." *IEEE Transactions on Antennas and Propagation* 22, no. 2: 347–349 (1975).

[17] Donaldson, N.N., and T.A. Perkins. "Analysis of Resonant Coupled Coils in the Design of Radio Frequency Transcutaneous Links." *Medical & Biological Engineering & Computing*, 21, no. 5: 612–627 (1983).

[18] Dunbar. R.M. "The Performance of a Magnetic Loop Transmitter-Receiver System Submerged in the Sea." *The Radio and Electronic Engineer* 42, no. 10: 457–463 (1972).

[19] Earnshaw. J.B. *An Introduction to AC Circuit Theory*. New York: Macmillan & Co. Ltd. (1960).

[20] Fenwick, R.C., and W.L. Weeks. "Submerged Antenna Characteristics." *IEEE Transactions on Antennas and Propagation* 11, no. 3: 296–305 (1963).

[21] Finkenzeller. K. *RFID Handbook-Radio-Frequency Identification Fundamentals and Applications*. New York: John Wiley & Sons Ltd. (1999).

[22] Flack, F.C., E.D. James, and D.M. Schlapp. "Mutual Inductance of Air-Cored Coils: Effect on Design of Radio-Frequency Coupled Implants." *Medical & Biological Engineering & Computing* 9: 79–85 (1971).

[23] Foster, K.R., and J. Jaeger. "RFID Inside—the Murky Ethics of Implanted Chips." *IEEE Spectrum* 44, no. 3: 8–29 (2007).

[24] Fotopoulou. K. *Inductive Wireless Power Transfer for RFID and Embedded Devices: Coils Misalignment Analysis and Design*. PhD thesis, University of Edinburgh, Edinburgh (2008).

[25] Fotopoulou, K. and B. Flynn. "Wireless Power Transfer in Loosely Coupled Links: Coil Misalignment Model." *IEEE Transactions on Magnetics* 47, no. 2: 416–430 (2011).

[26] Gabriel, C., S. Gabriel, and E. Corthout. "The Dielectric Properties of Biological Tissues:I. Literature Survey." *Physics in Medicine and Biology* 41, no. 11: 2231–2249 (1996).

[27] Gabriel, S., R. Lau, and C. Gabriel. "The Dielectric Properties Of Biological Tissues: III Parametric Models of the Dielectric Spectrum

of Tissues." *Physics in Medicine and Biology* 41, no. 11: 2271–2293 (1996).

[28] Galbraith, D.C., M. Soma, and R.L. White. "A Wide-Band Efficient Inductive Transdermal Power and Data Link with Coupling Insensitive Gain." *IEEE Transactions on Biomedical Engineering* 34, no. 4: 265–275 (1987).

[29] Goto, K., T. Nakagama, O. Nakamura, and S. Kawata. "An Implantable Power Supply with An Optically Rechargable Lithium Battery." *IEEE Transactions on Biomedical Engineering* 48, no. 7: 830–833 (2001).

[30] Graafstra, A. Hands on - How Radio-Frequency identification and I got personal. *IEEE Spectrum*, 44, no. 3: 18–23 (2007).

[31] Greene. F.M. "The Near-Zone Magnetic Field of a Small Circular-Loop Antenna." *Journal of Research of the National Bureau of Standards-C* 71C, no. 4: 319–325 (1967).

[32] Grover, F.W. *Inductance Calculations, Working Formulas and Tables*, 2nd edition. New York: D. Van Nostrand Company, Inc. (1946).

[33] Hamici, Z., R. Itti, and J. Champier. "A High-Efficiency Power and Data Transmission System for Biomedical Implanted Electronic Devices." *Measurement Science and Technology* 7: 192–201 (1996).

[34] Hansen. R.C. "Radiation and Reception with Buried and Submerged Antennas." *IEEE Transactions on Antennas and Propagation* 11, no. 3: 207–216 (1963).

[35] Heetderks. W.J. "Rf Powering of Millimeter and Submillimeter Sized Neural Prosthetic Implants." *IEEE Transactions on Biomedical Engineering* 35, no. 5: 323–327 (1988).

[36] Hertz. H. Ueber sehr schnelle electrische schwingungen. *Annalen der Physik und Chemie* 267, no. 7: 421–448 (1887).

[37] Hochmair, E.S. "System Optimization for Improved Accuracy in Transcutaneous Signal and Power Transmission." *IEEE Transactions on Biomedical Engineering* BME-31, no. 2: 177–186 (1984).

[38] Huang, Q., and M. Oberle. "A 0.5-mW Passive Telemetry IC for Biomedical Applications." *IEEE Journal of Solid-State Circuits* 33, no. 7: 937–946 (1998).

[39] Jeutter, D.C. "Overview of Biomedical Telemetry Techniques." *Engineering in Medicine and Biology Magazine* 2: 17–24 (1983).

[40] Jiang, B., J. Smith, M. Philipose, S. Roy, K. Sundara-Rajan, and A. Mamishev. "Energy Scavenging for Inductively Coupled Passive RFID

Systems." *IEEE Transactions on Instrumentation and Measurement* 56, no. 1: 118–125 (2007).

[41] Johnson, C.C., and A.W. Guy. Nonionizing electromagnetic wave effects in biological materials and systems. *Proceedings of the IEEE* 60, no. 6: 692–718 (1972).

[42] Kaiser, U., and W. Steinhagen. "A Low-Power Transponder IC for High-Performance Identification Systems." *IEEE Journal of Solid-State Circuits* 30: 306–310 (1995).

[43] Kennedy, P.A. "Loop Antenna Measurements." *IRE Transactions on Antennas and Propagation* 4, no. 4: 610–618 (1956).

[44] King, R. "The Rectangular Loop Antenna as a Dipole." *IEEE Transactions on Antennas and Propagation* 7, no. 1: 53–61 (1959).

[45] Ko, W.H., R. Plonsey, and S.R. Kang. "The Radiation from an Electrically Small Circular Wire Loop Implanted in a Dissipative Homogeneous Spherical Medium." *Annals of Biomedical Engineering* 1: 135–145 (1972).

[46] Kraus, J.D. *Antennas*, 2nd edition. McGraw-Hill, Inc. (1988).

[47] Kraus, J.D. and D.A. Fleisch. *Electromagnetics with Applications*, 5th edition. New York: McGraw Hill (1999).

[48] Levin, B. "Field of a Rectangular Loop." *IEEE Transactions on Antennas and Propagation* 52, no. 4: 948–952 (2004).

[49] Li, L., Mook-Seng Leong, P. Kooi, and T. Yeo. "Exact Solutions of Electromagnetic Fields in Both Near and Far Zones Radiated by Thin Circular-Loop Antennas: A General Representation." *IEEE Transactions on Antennas and Propagation* 45, no. 12: 1741–1748 (1997).

[50] Liang, C.K., J.J. Chen, C.L. Chung, C.L. Cheng, and C.C. Wang. "An Implantable Bidirectional Wireless Transmission System for Transcutaneous Biological Signal Recording." *Physiological Measurement* 26, no. 1: 83–97 (2005).

[51] Liu, W., K. Vichienchom, M. Clements, S.C. DeMarco, C. Hughes, E. McGucken, M.S. Humayun, E de Juan, J.D. Weiland, and R. Greenbery. "A Neuro-Stimulus Chip with Telemetry Unit for Retinal Prosthetic Device." *IEEE Journal of Solid-State Circuits* 35, no. 10: 1487–1497 (2000).

[52] Lou, E., N. Durdle, V. Raso, and D. Hill. "Measurement of the Magnetic Field in the Near-Field Region and Self-Inductance in Free Space Due to a Multiturn Square Loop." *IEEE Proceedings Science Measurement and Technology* 144, no. 6: 252–256 (1997).

[53] McGray, J.E. "Theoretical Foundation for Real-Time Prostate Localization Using an Inductively Coupled Transmitter and a Superconducting Quantum Interference Device (SQUID) Magnetometer System." *Journal of Applied Clinical Medical Physics* 5, no. 4: 29–45 (2004).

[54] Medhurst, R.G., "H.F. Resistance and Self-Capacitance of Single Layer Solenoids—Part I." *Wireless Engineer* 24: 35–43, February (1947).

[55] Medhurst, R.G., "H.F. Resistance and Self-Capacitance of Single Layer Solenoids—Part II." *Wireless Engineer* 24: 80–92, March (1947).

[56] Mokwa, W. "Medical Implants Based on Microsystems." *Measurement Science and Technology* 18, no. 5: 47–57 (2007).

[57] Moore, R.K. "Effects of Surrounding Conducting Medium on Antenna Analysis." *IEEE Transactions on Antennas and Propagation* 11, no. 2: 216–225 (1963).

[58] Murakawa, K., M. Kobayashi, O. Nakamura, and S. Kawata. "A Wireless Near-Infrared Energy System for Medical Implants." *IEEE Engineering in Medicine and Biology Magazine* 18, no. 6: 70–72 (1999).

[59] Neihart, N. and R.R. Harrison. "Micropower Circuit for Bidirectional Wireless Telemetry in Neural Recording Applications." *IEEE Transactions on Biomedical Engineering* 52, no. 11: 1950–1959 (2005).

[60] Newman, E.H., P. Bohley, and C.H. Walter. "Two Methods for the Measurement of Antenna Efficiency." *IEEE Transactions on Antennas and Propagation* 23, no. 4: 457–461 (1975).

[61] Overfelt, P. L. "Near Fields of the Constant Current thin Circular Loop Antenna of Arbitrary Radius." *IEEE Transactions on Antennas and Propagation* 44, no. 2: 166–171 (1996).

[62] Paradiso, J.A., and T. Starner. "Energy Scavenging for Mobile and Wireless Electronics." *IEEE Pervasive Computing* 4, no. 1: 18–27 (2005).

[63] Parramon, J., P. Doguet, D. Marin, M. Verleyssen, R. Munoz, L. Leija, and E. Valderrama. "Asic-Based Battery Less Implantable Telemetry Microsystem for Recording Purposes." In *Proceedings of the IEEE 19th Annual International Conference of Engineering in Medicine and Biology Society* 5, pp. 2225–2228 (1997).

[64] Pettengill, R., H. Garland, and J. Meindl. "Receiving Antenna Design for Miniature Receivers." *IEEE Transactions on Antennas and Propagation* 25, no. 4: 528–530 (1977).

[65] Philipose, M., J. Smith, B. Jiang, A. Mamishev, S. Roy, and K. Sundara-Rajan. "Battery-Free Wireless Identification and Sensing." *IEEE Pervasive Computing* 4, no. 1: 37–45 (2005).

[66] Pichorim, S.F., and P.J. Abatti. "Design of Coils for Millimeter- and Submillimeter-Sized Biotelemetry." *IEEE Transactions on Biomedical Engineering* 51, no. 8: 1487–1489 (2004).

[67] Ramo, S., J.R. Whinnery, and T. VanDuzer. *Fields and Waves in Communication Electronics,* 3rd edition. New York: John Wiley & Sons, Inc. (1994).

[68] Seshagiri Rao, K.V., P.V. Nikitin, and S.F. Lam. "Antenna Design for uhf rfid Tags: A Review and a Practical Application." *IEEE Transactions on Antennas and Propagation* 53, no. 12: 3870–3876 (2005).

[69] Richtscheid, A. "Calculation of the Radiation Resistance of Loop Antennas with Sinusoidal Current Distribution." *IEEE Transactions on Antennas and Propagation* 24, no. 6: 889–891 (1976).

[70] Roz, T., and V. Fuentes. "Using Low Power Transponders and Tags for RFID Applications." pp. 1–8.

[71] Sauer, C., M. Stanaćević, G. Cauwenberghs, and N. Thakor. "Power Harvesting and Telemetry in CMOS for Implanted Devices." *IEEE Transactions on Circuits and Systems-I: Regular Papers* 52, no. 12: 2605–2613 (2005).

[72] Sawan, M., Y. Hu, and J. Coulombe. "Wireless Smart Implants Dedicated to Multichannel Monitoring and Microstimulation." *Proceedings of the IEEE International Conference on Pervasive Computing* 5, no. 1: 21–39 (2005).

[73] Schmidt, S., and G. Lazzi. "Use of FDTD Thin-Strut Formalism for Biomedical Telemetry Coil Designs." *IEEE Transactions on Microwave Theory and Techniques* 52, no. 8: 1952–1956 (2004).

[74] Scuder, J. "Powering an Artificial Heart: Birth of the Inductively Coupled-Radio Frequency System in 1960." *Artificial Organs* 26, no. 11: 909–915 (2002).

[75] Smith, G.S. "Proximity Effect in Systems of Parallel Conductors." *Journal of Applied Physics* 43, no. 5: 2196–2203 (1972).

[76] Smith, G.S. "Radiation Efficiency of Electrically Small Multiturn Loop Antennas." *IEEE Transactions on Antennas and Propagation* 20, no. 5: 656–65 (1972).

[77] Smith, G.S. "Efficiency of Electrically Small Antennas Combined with Matching Networks." *IEEE Transactions on Antennas and Propagation* 25, no. 3: 369–373 (1977).

[78] Smith, G.S. *Loop Antennas,* in antennas engineering handbook edition. New York: McGraw-Hill (1984).

[79] Smith, G.S. *An Introduction to Classical Electromagnetic Radiation.* Cambridge University Press, New York (1997).

[80] Smith, S., T.B. Tang, J.G. Terry, J.T.M Stevenson, B.W. Flynn, H.M. Reekie, A.F. Murray, A.M. Gundlach, D. Renshaw, B. Dhillon, A. Ohtori, Y. Inone, and A.J. Walton. Development of a Miniaturised Drug Delivery System with Wireless Power Transfer and Communication. *IET Nanobiotechnology* 1, no. 5: 80–86 (2007).

[81] Smythe, W.R. *Static and Dynamic Electricity,* 2nd edition. New York: McGraw-Hill (1950).

[82] Snelling, E.C. *Soft Ferrites: Properties and Applications*, 2nd edition. Butterworths (1988).

[83] Soma, M., C.D. Galbraith, and R. White. "Radio-frequency coils in implantable devices: Misalignment Analysis and Design Procedure." *IEEE Transactions on Biomedical Engineering* 34, no. 4: 276–282 (1987).

[84] Stockman, H. "Communication by Means of Reflected Power." *Proceedings of the IRE*, 36, no. 10: 1196–1204 (1948).

[85] Stuchly, M.A., and T.W. Dawson. "Interaction of Low-Frequency Electric and Magnetic Fields with the Human Body." *Proceedings of the IEEE* 88, no. 5: 643–664 (2000).

[86] Suster, M., D.J. Young, and W.H. Ko. "Micro-Power Wireless Transmitter for High Temperature MEMS Sensing and Communication Applications." In *Proceedings of theIEEE International Conference on Microelectromechanical Systems*, pp. 641–644 (2002).

[87] Tang, Z., B. Smith, J. Schild, and P. Peckham. "Data Transmission from an Implantable Biotelemeter by Load-Shift Keying Using Circuit Configuration Modulator." *IEEE Transactions on Biomedical Engineering* 42, no. 5: 524–528 (1995).

[88] Terman. F. *Radio Engineers Handbook.* New York: McGraw-Hill (1943).

[89] Tesla, N. *Apparatus for Transmission of Electrical Energy*, U.S. Patent 649621, May (1900).

[90] Tesla. N. *Method of Utilizing Effects Transmitted Through Natural Media.* U.S. Patent 685954, November (1901).

[91] Troyk, P.R. Injectable Electronic Identification, Monitoring, and Stimulation Systems. *Annual Review of Biomedical Engineering* 1: 177–209 (1999).

[92] Vandevoorde, G., and R. Puers. "Wireless Energy Transfer for Stand-Alone Systems: a Comparison Between Low and High Power

Applicability." *Sensors and Actuators A: Physical* 92, no. 1: 305–311 (2001).

[93] Liang, S., W. Ko and C. Fung. "Design of Radio-Frequency Powered Coils for Implant Instruments." *Medical and Biological Engineering and Computing* 15, no. 6: 634–640 (1977).

[94] Wang, G., W. Liu, M. Sivaprakasam, and G.A. Kendir. "Design and Analysis of an Adaptive Transcutaneous Power Telemetry for Biomedical Implants." *IEEE Transactions on Circuits and Systems-I: Regular Papers* 52, no. 10: 2109–2117 (2005).

[95] Wang, L., Tong Boon Tang, E. Johannessen, A. Astaras, M. Ahmadian, A. Murray, J. Cooper, S. Beaumont, B. Flynn, and D. Cumming. "Integrated Micro-Instrumentation for Dynamic Monitoring of the Gastro-Intestinal Tract." In *Proceedings of the IEEE-EMBS Annual International Special Topic Conference on Microtechnologies in Medicine & Biology*, pp. 219–222 (2002).

[96] Weinstein, R. "RFID: A Technical Overview and its Application to the Enterprise." *IT Professional* 7, no. 3: 27–33 (2005).

[97] Werner, D.H. "An Exact Formulation for the Vector Potential of a Cylindrical Antenna with Uniformly Distributed Current and Arbitrary Radius." *IEEE Transactions on Antennas and Propagation* 41, no. 8: 1009–1018 (1993).

[98] Werner, D.H. "An Exact Integration Procedure for Vector Potentials of Thin Circular Loop Antennas." *IEEE Transactions on Antennas and Propagation* 44, no. 2: 157–165 (1996).

[99] Wheeler, H.A. "Fundamental Limitations of Small Antennas." *Proceedings of the IRE*, 35, no. 12: 1479–1484 (1947).

[100] Wheeler, H.A. "Small Antennas." *IEEE Transactions on Antennas and Propagation* 23, no. 4: 462–469 (1975).

[101] Wiechert, T.J.P., F. Thiesse, F. Michahelles, P. Schmitt, and E. Fleisch. "Connecting Mobile Phones to the Internet of Things—A Discussion of Compatibility Issues Between EPC Technology and NFC technology. Technical report." Auto-ID Lab Switzerland (2008).

[102] Wu, J., V. Quinn, and G.H. Bernstein. "A Simple, Wireless Powering Scheme for Mems Devices." *Proceedings of SPIE* 4559: 43–52 (2001).

[103] Yates, D.C., A.S. Holmes, and A.J. Burdett. "Optimal Transmission Frequency for Ultralow-Power Short Range Radio Links." *IEEE Transactions on Circuits and Systems I: Regular Papers* 51, no. 7: 1405–1413 (2004).

[104] Zborowski, M., B. Kligman, R. Midura, A. Wolfman, T. Patterson, M. Ibiwoye, and M. Grabiner. "Decibel Attenuation of Pulsed Electromagnetic Field (PEMF) in Blood and Cortical Bone Determined Experimentally and from the Theory of Ohmic Losses." *Annals of Biomedical Engineering* 34, no. 6: 1030–1041 (2006).

6

Recent Advances on Magnetic Resonant Wireless Power Transfer

Marco Dionigi[1], Alessandra Costanzo[2], Franco Mastri[2], Mauro Mongiardo[1] and Giuseppina Monti[3],*

[1]DI, University of Perugia, Italy
[2]DEI, University of Bologna, Italy
[3]DII, University of Salento, Italy
*E-mail: giuseppina.monti@unisalento.it

This chapter provides a general overview of magnetic resonant wireless power transfer systems based on network models. The power transferred to a single and multiple receivers loads at resonance is derived and explained. It is also shown the importance of using appropriate matching networks and how to design the oscillator and the load rectifier.

6.1 Introduction

Efficient energy transfer is nowadays becoming an essential topic: it allows saving of economical resources, quality-of-life improvement, and pollution reduction, to name just a few issues. In particular, energy transfer is often realized by using electromagnetic power, which can be transmitted either along a guiding medium (transmission lines), or without a supporting medium. Waveguides with bounded cross sections possess a discrete spectrum, for example metallic waveguides [1]. Open waveguides, that is, with unbounded cross section, exhibit a continuous spectrum and, possibly, a few guided modes [2]. Naturally, guided waves require the presence of a medium to support wave propagation; in addition, energy decays along the transmission lines in an exponential manner.

A different mechanism for transmitting electromagnetic energy is by using radiative fields and antennas [3]. However, radiated energy is spread over

Wireless Power Transfer 2nd Edition, 217–270.

radiation angles, and its attenuation goes with an inverse quadratic law with respect to distance, even in the case of vacuum. Therefore, in order to realize an efficient power transfer, we need good focusing properties for the transmitting antenna and a large receiving antenna for avoiding energy spillover.

Another possibility for transmitting electromagnetic energy is by using the reactive fields present nearby open resonators, as recently suggested in [4–6]. While purely inductive coupling has several applications, it is only suitable for very short range and necessitates a quite precise positioning. However, if we add to inductive coupling a resonant circuit, we can realize energy exchange at significantly larger distances (see e.g., [7, 8]). Thus, two, or more, synchronous (i.e., operating at the same resonant frequency) open resonators may exchange energy via their reactive fields. We call this process resonant wireless power transfer (WPT).

This type of phenomenology has been referred to in the literature with several different names: wireless resonant energy links (WREL), wireless power transfer (WPT), WITRICITY (WIreless elecTRICITY), wireless energy transfer (WET), etc.

As a matter of fact, wireless power transfer plays an important role in many different contests: in bioengineering for the use of implanted and worn devices [9, 10], in sensors' networks for avoiding the use of batteries [11, 12], in robotics, for 3D integrated circuits [13], for charging electrical vehicles (EV) [14–16], and, last but not least, to seamlessly recharge mobile devices [17, 18]. In [8, 12, 19], wireless energy transfer to multiple devices has been considered. An excellent review of the possibilities, characteristics, and challenges for WPT systems has been presented in [20]. It is remarkable that WPT shares significant similarities with near-field magnetic communications [21] and with RFID [22].

In resonant WPT, the coupling should mainly take place via the magnetic field; in fact, magnetic field coupling has the advantage, with respect to electric field coupling, to be rather insensitive to the presence of different dielectrics.

The resonators should be coupled via their reactive fields, possibly avoiding radiation, which poses problems of electromagnetic compatibility. To this end, it is preferable that the electric field storage takes place in a confined region of space, which may be physically realized either by a lumped capacitance or by other suitable structures. Note that we are considering open *resonators* and not antennas (which mainly serve to produce a radiated field).

It suffices a very low coupling for the energy transfer to take place, and in the presence of "perfect" infinite Q resonators, even a very weak coupling will provide the possibility of energy transmission to the load (which is the only place where energy can be dissipated since, by definition, the resonators are lossless).

As we will see in this chapter, resonant WPT can be investigated in an accurate and rigorous manner by using network theory. In particular, we will see that by using mutual inductors and resonant impedances, several important results can be derived.

In Figure 6.1, we have reported the sketch of a typical system for WPT. Apart from AC/DC converters, we note the presence of a RF oscillator and a transmitting (TX) resonator, a receiving (RX) resonator, and matching networks to couple the energy from the RF oscillator to the TX resonator and from the RX resonator to the AC/DC converter. Note that the subnetwork between the output of the RF oscillator and the input of the AC/DC converter is a linear system; this part can be described as a two-port network containing only linear, passive components. An example of a practical implementation

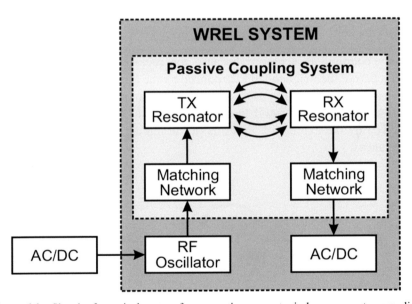

Figure 6.1 Sketch of a typical system for magnetic resonant wireless resonant energy link (WREL). Apart for AC/DC converters, we note the presence of a RF oscillator and a transmitting (TX) resonator, a receiving (RX) resonator, and matching networks to couple the energy from the RF oscillator to the TX resonator and from the RX resonator to the AC/DC converter. Note that the part between the output of the RF oscillator and the input of the AC/DC converter, denoted as passive coupling system in the figure, is a two-port network containing only linear, passive components.

of this part is reported in Figure 6.12. Naturally, when more than one transmitter/receiver is present, a suitable description can be done in terms of a linear N-port network. In the next sections, we will show why the type of structure illustrated in Figure 6.1 is advantageous for realizing efficient WPT.

6.2 Coupled Inductors

6.2.1 Coupled Inductors

Consider two inductors of values L_1 and L_2 coupled via their mutual inductance M as shown in Figure 6.2. The first inductor, which is excited by an external source, will be referred as the transmitter coil; the second inductor, which receives energy from the transmitter, will be referred as the receiver coil. In the frequency domain, the two-port network corresponding to coupled inductances has an impedance representation of the type:

$$V_1 = j\omega L_1 I_1 + j\omega M I_2$$
$$V_2 = j\omega M I_1 + j\omega L_2 I_2$$

(6.1)

The energy coupling is achieved because the receiver inductor actually intercepts a part of the magnetic field produced by the transmitter. The coupling coefficient k is typically used to represent the efficiency of energy transfer from the transmitter coil to the receiver coil; this coupling coefficient is given by the expression in terms of the mutual inductance and the self-inductances:

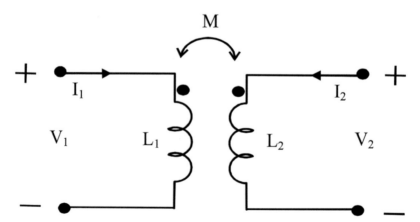

Figure 6.2 Coupled inductors may be used to represent magnetic coupling.

$$k = \frac{M}{\sqrt{L_1 L_2}} \tag{6.2}$$

The power coupling is seen in the transmitter as a product of the mutual inductance M and the current I_2 flowing in the receiver; at the receiver, power coupling is due to the current I_1 flowing in the transmitter being coupled through the mutual inductance as well.

The actual values of the network elements can be derived as described in many references, for example, Section 4.1 of [23] or [24, 25] for inductance calculation. As an example, the coil inductance expressed in μH of a coil of total length L_e, diameter D (expressed in meters), and N turns is given by

$$L = \frac{(ND)^2}{\sqrt{L_e + 0.45D}} \tag{6.3}$$

By denoting with R_i the radius of the ith coil and by r the distance between the centers of the resonators, we can compute the magnetic coupling factor as follows:

$$k = 1.4 \frac{(R_1 R_2)^2}{\sqrt{\left(R_1^2 + r^2\right)^3} \sqrt{R_1 R_2}} \tag{6.4}$$

The mutual inductance is finally obtained from Equation (6.2). As an example in Figure 6.3, we have plotted the coupling coefficient for the case specified in the caption. The dependence with r in the above equation is very important. In fact, as we will demonstrate later, power coupling depends on k^2, thus going with r^{-6}. This fact shows why we need to be very close in order to use non-resonant magnetic coupling. On the other hand, as we will see in the next sections, by using resonance, we can greatly enhance the coupling.

A useful network representation for coupled inductors is illustrated in Figure 6.4.

It is convenient to introduce an ABCD network representation for the coupled inductors. We recall that the ABCD definition is

$$\begin{pmatrix} V_1 \\ I_1 \end{pmatrix} = \begin{pmatrix} A & B \\ C & D \end{pmatrix} \begin{pmatrix} V_2 \\ I_2 \end{pmatrix} \tag{6.5}$$

In the above equation, V_1, V_2 are the voltages at ports 1 and 2, respectively; I_1 is the current entering port 1, and I_2 is the current *going out* from port 2. The great advantage of the ABCD representation is that cascading several different elements simply corresponds to multiplication of the relative ABCD matrices.

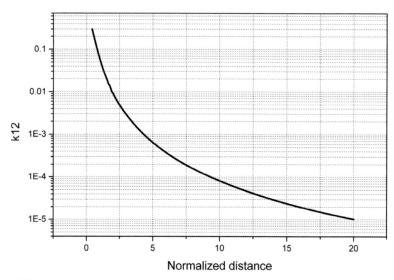

Figure 6.3   Coupling coefficient of a 2-turn, 60-mm diameter coil versus center coil distance normalized to diameter.

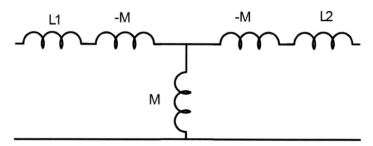

Figure 6.4   Coupled inductors' network representation. Note in the central part, a T section with–M series inductors and M shunt inductor. This part is equivalent to an impedance inverter.

The series impedance Z has an ABCD representation given by:

$$\begin{pmatrix} A & B \\ C & D \end{pmatrix} = \begin{pmatrix} 1 & Z \\ 0 & 1 \end{pmatrix}$$

(6.6)

A shunt impedance Y has the following ABCD representation:

$$\begin{pmatrix} A & B \\ C & D \end{pmatrix} = \begin{pmatrix} 1 & 0 \\ Y & 1 \end{pmatrix}$$

(6.7)

It is also convenient to introduce an element, the *impedance inverter* that has the correspondent ABCD matrix defined as follows:

$$\begin{pmatrix} A & B \\ C & D \end{pmatrix} = \begin{pmatrix} 0 & -jK \\ -j/K & 0 \end{pmatrix} \tag{6.8}$$

If we consider an ABCD matrix placed before a load impedance, Z_L, the input impedance seen from port 1, Z_{in}, when port 2 is closed on the load impedance Z_L, is given, in general, by

$$Z_{in} = \frac{V_1}{I_1} = \frac{AV_2 + BI_2}{CV_2 + DI_2} = \frac{AZ_L + B}{CZ_L + D} \tag{6.9}$$

For the impedance inverter case, $A = D = 0$; $B = -jK$; $C = -\frac{1}{jK}$ yields:

$$Z_{in} = \frac{K^2}{Z_L} \tag{6.10}$$

This explains the name of the impedance inverter.

With reference to Figure 6.4, we note in the central part a T section with—M series inductors and M shunt inductor; let us introduce the series impedance Z_M, the shunt admittance Y_M, and the impedance inverter value K, defined as follows:

$$Z_M = -j\omega M = -1/Y_M = -jK \tag{6.11}$$

It is apparent that this T section corresponds to an impedance inverter.

$$\begin{pmatrix} 1 & Z_M \\ 0 & 1 \end{pmatrix} \begin{pmatrix} 1 & 0 \\ Y_M & 1 \end{pmatrix} \begin{pmatrix} 1 & Z_M \\ 0 & 1 \end{pmatrix} = \begin{pmatrix} 0 & -jK \\ -j/K & 0 \end{pmatrix} \tag{6.12}$$

Note that being the impedance inverter frequency independent, the above representation is valid only at a single frequency. By using impedance inverters, the central part of the coupled inductors may be represented simply as an impedance inverter. A series inductance, an impedance inverter and another series inductance on the other hand, is equivalent to the coupled inductors.

For future use, it is also convenient to introduce the ABCD representation for a 1:n transformer, see Figure 6.5, which takes the following form:

$$\begin{pmatrix} A & B \\ C & D \end{pmatrix} = \begin{pmatrix} 1/n & 0 \\ 0 & n \end{pmatrix} \tag{6.13}$$

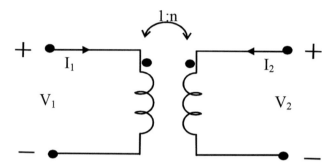

Figure 6.5   A 1:n transformer; the relative ABCD description is given by Equation (6.13).

It is possible to establish a relationship between the coupled inductors and a transformer, as shown in Figure 6.6, with the following equivalences:

$$L_a = (1 - k^2)L_1; \; L_b = k^2 L_1; \; n = \frac{1}{k}\sqrt{\frac{L_2}{L_1}} \qquad (6.14)$$

It is also advantageous to find out the relationship between impedance inverters and a transformer. By cascading two impedance inverters, we note the following relationship:

$$\begin{pmatrix} 0 & -jK_1 \\ -j/K_1 & 0 \end{pmatrix}\begin{pmatrix} 0 & -jK_2 \\ -j/K_2 & 0 \end{pmatrix} = \begin{pmatrix} -K_1/K_2 & 0 \\ 0 & -K_2/K_1 \end{pmatrix}$$

$$= \begin{pmatrix} 1/n & 0 \\ 0 & n \end{pmatrix} \qquad (6.15)$$

Therefore, [17] a transformer can be represented as two impedance inverters. Finally, we analyze what happens when cascading three impedance inverters:

$$\begin{pmatrix} 0 & -jK_1 \\ -j/K_1 & 0 \end{pmatrix}\begin{pmatrix} 0 & -jK_2 \\ -j/K_2 & 0 \end{pmatrix}\begin{pmatrix} 0 & -jK_3 \\ -j/K_3 & 0 \end{pmatrix}$$

$$= \begin{pmatrix} 0 & jK_1K_3/K_2 \\ \dfrac{jK_2}{K_1K_3} & 0 \end{pmatrix} \qquad (6.16)$$

It is apparent that by *cascading three impedance inverters, we get another impedance inverter.* Equivalently, by cascading a transformer and an impedance inverter, we get another impedance inverter.

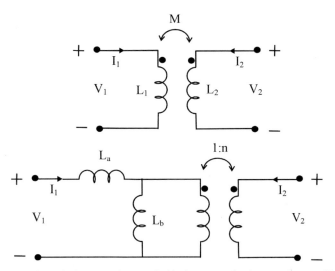

Figure 6.6 Relationship between the coupled inductors and a 1:n transformer. Note that the series inductance L_a, with expression given in Equation (6.14), represents the part of magnetic field that is not coupled, while the shunt inductance L_b is the part of the magnetic field that is coupled.

The above relationships will become extremely useful when we will consider resonators coupled by mutual inductance. To this end, let us now briefly recall the network properties of a simple resonant circuit.

6.2.2 The Series Resonant Circuit

An open resonator may be described, to a first approximation, by a lumped network either of parallel or of series type. In the following, we consider only the series representation, shown in Figure 6.7. The impedance of a series resonant circuit is given by

$$Z\left(j\omega\right) = R + j\left(X_L - X_C\right)$$
$$X_L = \omega L = 2\pi f L \tag{6.17}$$
$$and\ X_C = {}^{1}\!/_{\omega C} = {}^{1}\!/_{2\pi f C}$$

The terms X_L and X_C are the inductive and capacitive reactance of the inductor and capacitor, respectively, and f (Hertz) is the frequency. The various types of losses, such as ohmic resistance, radiation, and parasitic resistances, are accounted for in the resistance R. Let us first consider the lossless ideal case. In this instance, when $X_L = X_C$, that is, when $\omega_0 = 1 \big/ \sqrt{LC}$,

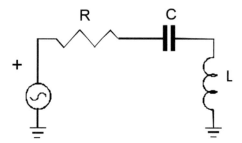

Figure 6.7   A series resonant circuit.

we have the resonant condition of the circuit and the impedance becomes zero. When we take into account the losses as well, we have that, at resonance, the total impedance is equal to R.

Another representation of the series resonant circuit may be given by introducing the slope parameter and the quality factor; the slope parameter is defined as follows:

$$ \chi = \sqrt{\frac{L}{C}} \tag{6.18} $$

The Q factor is defined by the ratio of the energy stored in the circuit to the energy dissipated by the circuit:

$$ Q = 2\pi \frac{Energy\ stored\ in\ the\ circuit\ per\ cycle}{Energy\ dissipated\ by\ the\ circuit\ per\ cycle}. \tag{6.19} $$

Since the quality factor is a ratio of two similar quantities, it has no dimensions. The quality factor refers therefore to the qualities of the inductors and capacitors in the circuit and pertains to their ability to transfer energy with respect to the losses occurring in the series resistor. Hence,

$$ Q = \frac{\omega_0 L}{R} = \frac{1}{R\omega_0 C} \tag{6.20} $$

The series resonant circuit may be therefore represented in terms of R, L, and C or in terms of Q, χ, and ω_0. In the latter case, the impedance is found as follows:

$$ Z = \chi \left[j \left(\frac{\omega}{\omega_0} - \frac{\omega_0}{\omega} \right) + \frac{1}{Q} \right] \tag{6.21} $$

The relevance of a resonant circuit is its ability to generate relatively high voltages and currents with respect to those provided by the source. Let us consider the voltage measured along the inductor; this is obtained using a voltage divider principle as follows:

$$V_0 = \frac{jX_L}{R + j(X_L - X_C)} V_{in} \tag{6.22}$$

If we now consider this voltage along the inductor at resonance, we obtain

$$V_0 = \frac{jX_L}{R} V_{in} = jQV_{in} \tag{6.23}$$

The above equation shows that the voltage along the inductor is Q times that provided by the generator. For WPT applications, typically resonators with high Q, of the order of one thousand, are often considered. This will therefore provide considerable strength to the reactive magnetic field generated by the coils.

6.2.3 Adding Resonators to the Coupled Inductors

Let us now add, on both sides of the coupled inductors, a series resonant circuit, thus obtaining the network shown in Figure 6.8. While the element values may be different, we consider the case of **synchronous** resonators, that is, with same resonant frequency. The Kirchhoff voltage equations are as follows:

$$
\begin{aligned}
V_1 &= [R_1 + j\omega L_1 + (1/j\omega C_1)] I_1 + j\omega M I_2 \\
0 &= j\omega M I_1 + [R_2 + j\omega L_2 + (1/j\omega C_2)] I_2
\end{aligned} \tag{6.24}
$$

Let us now assume to be at resonance (i.e., $\omega = \omega_0$); in this case, the impedances are purely resistive and Equation (6.24) simplifies as follows:

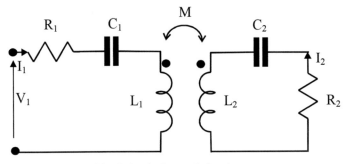

Figure 6.8 Inductively coupled series resonators.

$$V_1 = R_1 I_1 + j\omega_0 M I_2$$
$$0 = j\omega_0 M I_1 + R_2 I_2 \tag{6.25}$$

From Equation (6.25), the following expression can be obtained for the currents:

$$\begin{pmatrix} I_1 \\ I_2 \end{pmatrix} = \begin{pmatrix} \dfrac{R_2 V_1}{\omega_0^2 M^2 + R_1 R_2} \\[2ex] \dfrac{j\omega_0 M V_1}{\omega_0^2 M^2 + R_1 R_2} \end{pmatrix} \tag{6.26}$$

Therefore, the output power P_2, that is, the power dissipated on the resistance R_2, is given by:

$$P_2 = \frac{1}{2} R_2 |I_2|^2 = \frac{\omega_0^2 M^2 R_2 |V_1|^2}{2 \left(R_1 R_2 + \omega_0^2 M^2 \right)^2} \tag{6.27}$$

while the input power P_1 is:

$$P_1 = \frac{1}{2} V I^* = \frac{R_2 |V_1|^2}{2 \left(R_1 R_2 + \omega_0^2 M^2 \right)} \tag{6.28}$$

By considering the power ratio (η), and by recalling Equations (6.2) and (6.20), we obtain the following equation:

$$\eta = \frac{P_2}{P_1} = \frac{\omega_0^2 M^2}{\omega_0^2 M^2 + R_1 R_2} = \frac{k^2 Q_1 Q_2}{k^2 Q_1 Q_2 + 1} \tag{6.29}$$

The above equation shows that, for small coupling, the power transfer depends on the square of the coupling coefficient multiplied by the quality factors of the resonators. We have previously seen that the inductive coupling coefficient decays as r^{-6}; this rather fast decay is mitigated by using resonators which increase by Q times the field, hence providing a means for mid-range power transfer.

Let us now consider a more realistic case, that is, when adding also the source resistance R_S and the load resistance R_L, as shown in Figure 6.9, where R_1 and R_2 are the ohmic resistances of the transmitter and receiver coils, respectively. We assume that the resonators, as before, are synchronous and $\omega_0 = \dfrac{1}{\sqrt{L_1 C_1}} = \dfrac{1}{\sqrt{L_2 C_2}}$.

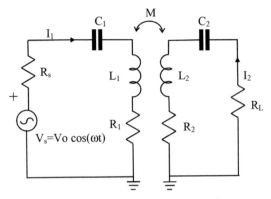

Figure 6.9 Mutually coupled resonators taking into account the source and load resistance.

Referring to Figure 6.9, we have:

$$V_1 = (R_1 + R_S) I_1 + j\omega_0 M I_2$$
$$0 = j\omega_0 M I_1 + (R_2 + R_L) I_2$$

(6.30)

According to Equation (6.30), the following expression can be derived for the currents:

$$\begin{pmatrix} I_1 \\ I_2 \end{pmatrix} = \begin{pmatrix} \dfrac{(R_2 + R_L) V_1}{\omega_0^2 M^2 + (R_1 + R_S)(R_2 + R_L)} \\[4mm] \dfrac{j\omega_0 M V_1}{\omega_0^2 M^2 + (R_1 + R_S)(R_2 + R_L)} \end{pmatrix}$$

(6.31)

Using Equation (6.31), for P_1 and P_2 we get:

$$P_1 = \frac{1}{2}VI^* = \frac{(R_2 + R_L)|V_1|^2}{2\left[(R_1 + R_S)(R_2 + R_L) + \omega_0^2 M^2\right]}$$

(6.32)

$$P_2 = \frac{1}{2}R_L|I|^2 = \frac{\omega_0^2 M^2 R_L |V_1|^2}{2\left[(R_1 + R_S)(R_2 + R_L) + \omega_0^2 M^2\right]^2}$$

(6.33)

Let us introduce the following quantities:

$$Q_1^L = \frac{\omega_0 L_1}{R_1 + R_S}; \quad Q_2^L = \frac{\omega_0 L_2}{R_2 + R_L}; \quad Q_2^e = \frac{\omega_0 L_2}{R_L}$$

(6.34)

By using Equations (6.2) and (6.34), for the power ratio (η) we have:

$$
\eta = \frac{P_2}{P_1} = \left[\frac{\omega_0^2 M^2}{\omega_0^2 M^2 + (R_S + R_1)(R_L + R_2)} \right] \frac{R_L}{(R_L + R_2)}
$$

$$
= \left[\frac{k^2 Q_1^L Q_2^L}{k^2 Q_1^L Q_2^L + 1} \right] \frac{Q_2^L}{Q_2^e} \tag{6.35}
$$

In order to put in evidence that the values of the unloaded Qs have changed significantly, in Equation (6.35) we have introduced the external and loaded Q (i.e., Q_2^e, Q_1^L, Q_2^L) defined as in Equation (6.34). This causes a significant degradation of the performances for a WPT system. As an example, if we consider two resonators with $Q = 1000$, the loaded quality factors Q_1^L, Q_2^L, with respect to a source/load impedance of 50 Ω, become 20. This means that we are able to receive just 4×10^{-4}, the power received in the unloaded case.

From Equation (6.35), it is evident the dependence of η from the load and source resistance, so that we can pose the question if, by changing their values, we can increase the power exchange.

According to this observation, in the following part of this section, closed form matching impedances for maximum efficiency, maximum power on the load, or conjugate matching are reported and discussed.

6.2.4 Maximum Efficiency, Maximum Power on the load, and Conjugate Matching: Two-Port Case

Let us consider a two-port network (see Figure 6.10) implementing a WPT link and represented by its impedance matrix:

$$
\overline{\overline{Z}} = \begin{pmatrix} z_{11} & z_{12} \\ z_{21} & z_{22} \end{pmatrix}, z_{ij} = r_{ij} + jx_{ij}. \tag{6.36}
$$

We assume that the network is reciprocal (i.e., $z_{12} = z_{21}$), so that we have:

$$
V_1 = z_{11}I_1 + z_{12}I_2 \tag{6.37}
$$

$$
V_2 = z_{12}I_1 + z_{22}I_2 \tag{6.38}
$$

Referring to Figure 6.10 and to power transfer applications, we look for the optimum value of $Z_L = R_L + jX_L$. As shown in [26], three different approaches can be adopted:

- *maximize the efficiency* (defined as the ratio between the active power delivered to the load and the active power provided by the generator),

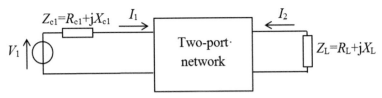

Figure 6.10 Schematic representation of a two-port network with a voltage generator (V_1) on port 1 and a load impedance (Z_L) on port 2.

- *maximize the power* delivered to the load R_L,
- *realize conjugate matching* (i.e., power matching).

According to the network theory reported in [26], it can be demonstrated that the above reported approaches require very different values of Z_L. This is highlighted in Table 6.1 where the optimum values of R_L and X_L corresponding to efficiency maximization, power maximization, and conjugate matching are reported.

The parameters ξ, χ, θ_r, and θ_x that appear in Table 6.3 are defined as follows:

$$\xi = r_{12}/\sqrt{r_{11}r_{22}}, \chi = x_{12}/\sqrt{r_{11}r_{22}}, \theta_r = \sqrt{1+\chi^2}\sqrt{1-\xi^2}, \theta_x = \chi\xi. \tag{6.39}$$

Values reported in Table 6.2 refer to the case where the generator impedance is not present (i.e., $R_S = 0$); however, if the generator impedance is present, we can simply include it into z_{11}. In order to illustrate the application of data

Table 6.1 Optimum value of the load impedance $Z_L = R_L + jX_L$ for efficiency maximization. The following definitions are used: $\xi = r_{12}/\sqrt{r_{11}r_{22}}$, $\chi = x_{12}/\sqrt{r_{11}r_{22}}$, $\theta_r = \sqrt{1+\chi^2}\sqrt{1-\xi^2}$, $\theta_x = \chi\xi$, $\eta_e = (\xi^2+\chi^2)/[(1+\theta_r)^2+\theta_x^2]$.

Parameter	Maximum efficiency
R_L	$r_{22}\theta_r$
X_L	$r_{22}\theta_x - x_{22}$
R_{c1}	0
X_{c1}	$x_{12}r_{12}/r_{22} - x_{11}$
R_{in}	$r_{11}\theta_r$
X_{in}	0
P_{in}	$4/\theta_r$
P_L	$4\eta_e/\theta_r$
η	η_e

Table 6.2   Optimum value of the load impedance $Z_L = R_L + jX_L$ for power maximization. The following definitions are used: $\xi = r_{12}/\sqrt{r_{11}r_{22}}$, $\chi = x_{12}/\sqrt{r_{11}r_{22}}$, $\theta_r = \sqrt{1+\chi^2}\sqrt{1-\xi^2}$, $\theta_x = \chi\xi$.

Parameter	Maximum power
R_L	$r_{22}\theta_r^2/(\theta_x^2+1)$
X_L	$r_{22}\theta_x + (r_{22}\theta_x\theta_r^2)/(\theta_x^2+1) - x_{22}$
R_{c1}	0
X_{c1}	$x_{12}r_{12}/r_{22} - x_{11}$
R_{in}	$2r_{11}\theta_r^2/(1+\theta_r^2+\theta_x^2)$
X_{in}	0
P_{in}	$2(1+\theta_r^2+\theta_x^2)/\theta_r^2$
P_L	$(\xi^2+\chi^2)/\theta_r^2$
η	$(\xi^2+\chi^2)/[2(1+\theta_r^2+\theta_x^2)]$

Table 6.3   Optimum value of the load impedance $Z_L = R_L + jX_L$ for conjugate matching. The following definitions are used: $\xi = r_{12}/\sqrt{r_{11}r_{22}}$, $\chi = x_{12}/\sqrt{r_{11}r_{22}}$, $\theta_r = \sqrt{1+\chi^2}\sqrt{1-\xi^2}$, $\theta_x = \chi\xi$, $\eta_e = (\xi^2+\chi^2)/[(1+\theta_r)^2+\theta_x^2]$.

Parameter	Conjugate matching
R_L	$r_{22}\theta_r$
X_L	$r_{22}\theta_x - x_{22}$
R_{c1}	$r_{11}\theta_r$
X_{c1}	$x_{12}r_{12}/r_{22} - x_{11}$
R_{in}	$2r_{11}\theta_r$
X_{in}	0
P_{in}	$2/\theta_r$
P_L	$2\eta_e/\theta_r$
η	$\eta_e/2$

reported in Table 6.3, let us consider two coupled inductors described by the following impedance matrix:

$$\overline{\overline{Z}} = \begin{pmatrix} r_1 + j\omega L_1 & j\omega M \\ j\omega M & r_2 + j\omega L_2 \end{pmatrix}. \tag{6.40}$$

In this case, from Table 1-III we have:

$$X_{c1} = -x_{11} = -j\omega L_1. \tag{6.41}$$

By comparing results summarized in Table 1.3, some important remarks are:

- Conjugate matching and efficiency maximization require the same value of R_L, while a different value is required to maximize power.
- Conjugate matching and efficiency maximization require the same value of X_L, while a different value is required to maximize power. To this regard, it should be noticed that all the three values coincide when $r_{12} = 0$, and, as a consequence, we have $\theta_x = 0$. Only this case corresponds to a resonant circuit at port 2.
- The required value for R_{c1} is equal to zero for efficiency and power maximization, while it is different from zero for conjugate matching. It should be noticed that a value of R_{c1} different from zero is useful to find the reference impedance, however, in a WPT system it realizes an unnecessary waste of energy.
- The same value of X_{c1} is necessary for all the three approaches.
- The value R_{in} required for efficiency and power maximization may be significantly different. In the case of power maximization it remains quite low even for large values of θ_r.
- The value required for the input reactance (i.e., X_{in}) is zero in all the three approaches.

6.2.5 Maximum Efficiency: N-port Case

Results reported in the previous section refer to a two-port network modeling a WPT system consisting of a single transmitter and a single receiver. However, in some practical applications, solving the problem in the case of multiple receivers is of interest; to this regard, useful results are reported in [27].

With reference to Figure 6.11, in [27] the problem of an N-port having a transmitter on port 1 and (N-1) receivers on the remaining ports is analyzed by determining the values of the load impedances which maximize efficiency.

By assuming that the (N-1) receivers have a negligible mutual coupling and that the N−port network is described by the following $\overline{\overline{Z}}$ matrix:

$$\overline{\overline{Z}} = \begin{pmatrix} r_{11} & jx_{12} & \cdots & jx_{1N} \\ jx_{21} & r_{22} & \cdots & jx_{2N} \\ \vdots & \vdots & \ddots & \vdots \\ jx_{N1} & jx_{N2} & \cdots & r_{NN} \end{pmatrix} \tag{6.42}$$

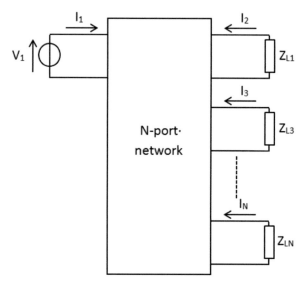

Figure 6.11   N-port network modeling a WPT system having a transmitter on port 1 and (N-1) receivers on the remaining ports.

it can be derived that the values of the load impedances (i.e., $Z_{Li} = R_i$) which maximize efficiency are:

$$R_i = r_{ii} A_N.$$ (6.43)

Being A_N defined as:

$$A_N^2 = 1 + \sum_{i=2}^{N} \chi_{1i}^2, \chi_{1i}^2 = \frac{x_{ij}^2}{r_{ii} r_{jj}}.$$ (6.44)

When the N-port is terminated on the load impedances expressed by Equation (6.43), the efficiency for each load is:

$$\eta_i = \frac{P_i}{P_{in}} = \frac{\frac{1}{2} R_{pi} I_i^2}{\frac{1}{2} \Re\{Z_{in}\} I_1^2} = \frac{\chi_{1i}^2}{(A_N + 1)^2}.$$ (6.45)

Where Z_{in} is the input impedance seen by the generator at port 1 when the remaining ports are closed on the optimal loads given in Equation (6.43). If we denote as total efficiency (η_{TOT}) of the N-port network the ratio between the sum of the power delivered to the N-1 loads and the power delivered by the generator, by using Equation (6.45), we can derive:

$$\eta_{TOT} = \frac{\sum\limits_{i=2}^{N} P_i}{P_{in}} = \frac{\sum\limits_{i=2}^{N} \frac{1}{2} R_{pi} I_i^2}{\frac{1}{2} \Re\left\{Z_{in}\right\} I_1^2} = \frac{(A_N - 1)}{(A_N + 1)}. \tag{6.46}$$

Results reported in this section solve the problem of determining the optimum value of the load impedances of a multiple receivers WPT system with reference to efficiency maximization. However, as highlighted in the previous section, the values necessary for efficiency maximization could be quite different from those required by power maximization, which remains a problem of interest and that has not yet been solved in the literature.

In addition to this, it is worth underlining that results summarized in Equation (6.43) does not take into account a possible mutual coupling between receivers. As a consequence, taking into account that depending on the receiver spatial configuration their mutual coupling could be negligible or not, the solution of a WPT system with mutually coupled receivers is also a case that deserves to be solved.

6.2.6 Scattering Matrix Representation of a Wireless Power Transfer Network

In order to achieve a better understanding of the resonant WPT, it may be convenient to refer to the scattering matrix representation of a two-port network. A matched WPT link corresponds to a scattering matrix with $S_{11} = S_{22} = 0$. This *matching condition* is necessary in order to transfer all the incident power to the other port. Normally, scattering matrices are closed on the reference impedance on both ports; unfortunately, closing the scattering matrix on relatively large impedances will destroy resonance by severely lowering the Qs. On the other hand, in order to preserve the resonant condition, it is appropriate to close the scattering matrix ports either on short or on open circuits, which cause total reflection.

We have previously introduced the ABCD representation of two-port networks. For a symmetrical network, the scattering parameters can be obtained from the following relationships:

$$S_{11} = \frac{AR_L + B - CR_S R_L - DR_S}{AR_L + B + CR_S R_L + DR_S}, S_{22} = \frac{AR_S + B - CR_S R_L - DR_L}{AR_S + B + CR_S R_L + DR_L}$$

$$\tag{6.47}$$

$$S_{12} = S_{21} = \frac{2\sqrt{R_S R_L}}{AR_L + B + CR_S R_L + DR_S} \tag{6.48}$$

Where R_s and R_L are the impedances assumed for normalization at port 1 and at port 2, respectively.

Two mutually coupled lossless resonators, at resonance, become simply an impedance inverter. This gives for the scattering parameters the following values:

$$S_{11} = S_{22} = \frac{2K^2}{K^2 + R_S R_L} - 1 \tag{6.49}$$

$$S_{12} = S_{21} = \frac{2jK\sqrt{R_S R_L}}{K^2 + R_S R_L} \tag{6.50}$$

The above equations, for small values of K (corresponding to small coupling), give, respectively, $S_{11} \approx -1$ and $S_{12} \approx 0$. From Equations (6.49) and (6.50), it can be also derived that, for a WPT link consisting of two resonators, the matching condition with respect to the normalization impedances R_s and R_L can be satisfied by imposing:

$$K = \sqrt{R_S R_L} \tag{6.51}$$

As evident from the above relation, in order to satisfy the matching condition for a given value of the generator impedance (R_s) and of the load impedance (R_L), K is uniquely determined by Equation (6.51). As illustrated in the next section, additional degrees of freedom in satisfying the matching condition can be obtained by adding one or more resonators to the WPT link.

6.3 Four Coupled Resonators

In order to extend the distance covered by a WPT link using magnetically coupled resonators, a configuration of interest is based on the use of relay resonators (i.e., resonators coupled to the transmitting and the receiving one but not directly excited or connected to the load). The properties of a such WPT link will be illustrated in this section with reference to a WPT link composed by four resonators (i.e., a transmitting resonator, a receiving resonator and two relay elements) and thus corresponding to three impedance inverters. The structure is shown in Figure 6.12, with its schematic representation shown in Figure 6.13. The equivalent network of this arrangement is composed by

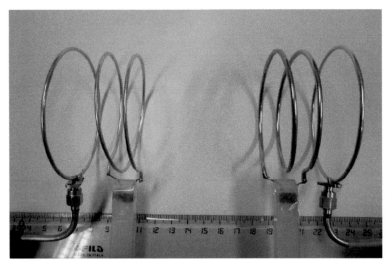

Figure 6.12 Photograph of four coupled resonator system connected to a network analyzer for measurement purposes.

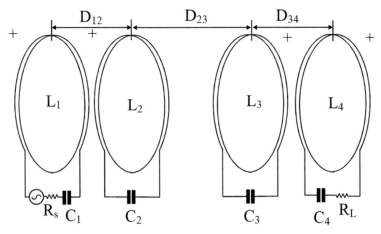

Figure 6.13 Schematic of the structure shown in the previous figure.

four resonators, which magnetically couple with each other. Naturally, when considering an inline arrangement, it is sufficient to consider just the coupling between adjacent resonators, as illustrated in Figure 6.14. By considering the network representation of the coupled inductances in terms of impedance inverters, we can derive the narrow-band network shown in Figure 6.15.

A further simplification arises when considering the network operating at the resonant frequency, since the impedances become purely resistive, as illustrated in Figure 6.16. Finally, the simplest case occurs when considering lossless resonators, giving rise to network composed, apart for the source and load impedance exclusively of impedance inverters (Figure 6.17). Note that in this last case, the coils widely spaced (i.e., the coils where the power

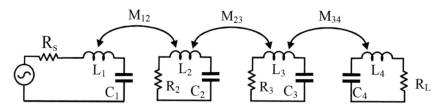

Figure 6.14   Wideband equivalent network of the four coupled resonators shown in the previous two figures. Note that we have assumed that each resonator only couples with adjacent ones. Naturally, this hypothesis can be easily removed if necessary (see, e.g., [28, 29]).

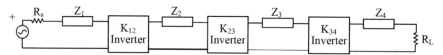

Figure 6.15   Narrow-band representation of the four coupled resonator system with impedance inverters.

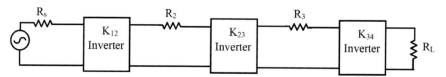

Figure 6.16   As in the previous figure but considering the system operating at the resonant frequency; therefore, the impedances become purely resistive. Resistances of the first and last resonators have been absorbed into the source and load resistances, respectively.

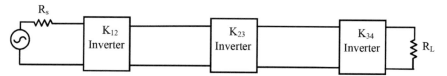

Figure 6.17   As in the previous figure but considering the lossless resonator. In practice, this case is quite useful for deriving a starting point approximation.

transfer takes place) are coil 2 and coil 3. This means that K_{23} is given once the transmitting and receiving positions are fixed. On the other hand, the couplings K_{12} and K_{34} can be used in order to match the structure to the source and to the load resistance.

In the lossless case, at resonance, the impedances are equal to zero and we have just the cascade of three impedance inverters. With reference to Equation (6.16), assuming a symmetrical structure with $K_{12} = K_{34} = K'$, we get

$$\begin{pmatrix} 0 & -jK' \\ -j/K' & 0 \end{pmatrix} \begin{pmatrix} 0 & -jK_{23} \\ -j/K_{23} & 0 \end{pmatrix} \begin{pmatrix} 0 & -jK' \\ -j/K' & 0 \end{pmatrix}$$
$$= \begin{pmatrix} 0 & jK'^2/K_{23} \\ \frac{jK_{23}}{K'^2} & 0 \end{pmatrix} \tag{6.52}$$

By using Equations (6.47) and (6.48), the following expressions can be obtained for the scattering parameters:

$$S_{11} = S_{22} = \frac{2K'^4}{K'^4 + R_S R_L K_{23}^2} - 1 \tag{6.53}$$

$$S_{12} = S_{21} = \frac{2jK'^2 K_{23}\sqrt{R_S R_L}}{K'^4 + R_S R_L K_{23}^2} \tag{6.54}$$

The matching condition at port 1 and 2 (i.e., $S_{11} = S_{22} = 0$) can be satisfied by choosing K' and K_{23} according to the following relation:

$$\frac{K'^2}{K_{23}} = \sqrt{R_S R_L} \tag{6.55}$$

By comparing Equations (6.55) and (6.51), it can be seen that, with respect to the two resonators case, in the case of a symmetrical WPT link using 4 resonators, one more degree of freedom is available for matching. In fact from Equations (6.55), it is evident that in this case we can use both K' and K_{23} to achieve a perfectly matched network.

However, in order to realize also the resonant condition, we have to close this network on appropriate impedances. If the source and load impedances are fixed from other considerations, we have to interpose appropriate transformers that reduce the source and load impedances to suitable values. Naturally, as we have noted before, the combination of impedance inverter and a transformer is equivalent to another impedance inverter. Therefore, we can conclude that a network with three impedance inverters and four resonators is appropriate for realizing efficient resonant magnetic wireless power transfer.

By denoting with Z_i the impedance of the ith resonator and with $K_{i,i+1}$ the impedance inverter relating the ith and ($i + 1$)th resonators, have

$$T_{ABCD} = \begin{bmatrix} A & B \\ C & D \end{bmatrix} = \begin{bmatrix} 1 & Z_1 \\ 0 & 1 \end{bmatrix} \cdot \begin{bmatrix} 0 & -jK_{12} \\ \frac{-j}{K_{12}} & 0 \end{bmatrix} \cdot \begin{bmatrix} 1 & Z_2 \\ 0 & 1 \end{bmatrix} \cdot$$

$$\begin{bmatrix} 0 & -jK_{23} \\ \frac{-j}{K_{23}} & 0 \end{bmatrix} \cdot \begin{bmatrix} 1 & Z_3 \\ 0 & 1 \end{bmatrix} \cdot \begin{bmatrix} 0 & -jK_{34} \\ \frac{-j}{K_{34}} & 0 \end{bmatrix} \cdot \begin{bmatrix} 1 & Z_4 \\ 0 & 1 \end{bmatrix} \quad (6.56)$$

which gives

$$A = \frac{j\left[Z_3 K_{12}^2 + Z_1 K_{23}^2 + Z_1 Z_2 Z_3\right]}{K_{12} K_{23} K_{34}}$$

$$B = \frac{j\left[K_{12}^2 K_{34}^2 + Z_3 Z_4 K_{12}^2 + Z_1 Z_4 K_{23}^2 + Z_1 Z_2 K_{34}^2 + Z_1 Z_2 Z_3 Z_4\right]}{K_{12} K_{23} K_{34}}$$

$$C = \frac{j\left[K_{23}^2 + Z_2 Z_3\right]}{K_{12} K_{23} K_{34}}$$

$$D = \frac{j\left[Z_2 K_{34}^2 + Z_4 K_{23}^2 + Z_2 Z_3 Z_4\right]}{K_{12} K_{23} K_{34}}$$

$$(6.57)$$

By transforming the ABCD matrix to the scattering matrix, and by considering $|S_{21}|$, we obtain

$$|S_{21}|^2 = \left| \frac{2\sqrt{R_s R_L}}{A \cdot R_s + B + C \cdot R_s R_L + D \cdot R_s} \right|^2 \quad (6.58)$$

In the above equation, we have denoted with R_S the source and with R_L the load impedance. Equation (6.58) expresses the efficiency of the network. When considering a symmetrical system, we have

$$Z_1 = Z_4 \qquad Z_2 = Z_3 = Z \qquad R_S = R_L = R$$

$$K_{12} = K_{34} = K$$

$$(6.59)$$

and Equation (6.57) can be simplified as follows:

$$A = \frac{j\left[ZK^2 + Z_1 K_{23}^2 + Z^2 Z_1\right]}{K^2 K_{23}}$$

$$B = \frac{j\left[K^4 + ZZ_1 K^2 + Z_1^2 K_{23}^2 + Z^2 Z_1^2\right]}{K^2 K_{23}}$$

$$(6.60)$$

$$C = \frac{j\left[K_{23}^2 + Z^2\right]}{K^2 K_{23}}$$

$$D = \frac{j\left[ZK^2 + Z_1 K_{12}^2 + Z^2 Z_1\right]}{K^2 K_{23}}$$

providing for the scattering parameter:

$$|S_{21}|^2 = \left|\frac{2R}{A \cdot R + B + C \cdot R^2 + D \cdot R}\right|^2 \qquad (6.61)$$

We note that once the resonators have been selected, the values of the relative impedances are given. In addition, once the distance between resonators 2 and 3 has been selected, the relative coupling is also fixed. Therefore, we can optimize our system by choosing only R and K in the symmetrical case. Nonetheless, it is possible to find out a procedure that allows us to obtain an optimal design as discussed in [30, 31].

For a simplified design, a good starting point can be obtained by considering the lossless case; in this instance, we have

$$A = D = 0 \qquad B = \frac{jK^2}{K_{23}} \qquad C = \frac{jK_{23}}{K^2} \qquad (6.62)$$

and, as already seen in Equation (6.55), by using $K^2/K_{23} = \sqrt{R_S R_L}$, we can recover the matched condition case.

6.4 Travelling Waves, Power Waves and Conjugate Image Impedances

The scattering matrix representation as described in section 6.2.6 is related to the selection of a reference impedance at each port; the corresponding formalism is denoted by traveling waves [32] when real values are selected as reference impedances. In the following part of this section, it will be shown that the use of real reference impedances has some important drawbacks when used to represent a WPT link, since it does not provide a realistic representation of the reflection of power in the case of complex impedances. To this regard, the power waves formalism will be introduced.

6.4.1 Travelling Waves and Power Waves

Let us consider a transmission line with propagation constant γ and characteristic impedance $Z_c = Z_{cr} + jZ_{ci}$, the voltage and current along the line are:

$$V(z) = V_0^+ e^{-\gamma z} + V_0^- e^{+\gamma z}$$

$$I(z) = \frac{V_0^+}{Z_c} e^{-\gamma z} - \frac{V_0^-}{Z_c} e^{+\gamma z}$$

(6.63)

Where V_0^+ and V_0^- are the amplitude of the forward and reflected traveling waves. At $z = 0$, Equation (6.63) becomes:

$$V(z = 0) = V = V_0^+ + V_0^-$$

$$I(z = 0) = I = \frac{V_0^+}{Z_c} - \frac{V_0^-}{Z_c}$$

(6.64)

The normalized voltage wave amplitudes are defined as:

$$\bar{a} = \frac{V + Z_c I}{2\sqrt{Z_{cr}}}$$

$$\bar{b} = \frac{V - Z_c I}{2\sqrt{Z_{cr}}}$$

(6.65)

According to this definition, the voltage and current can be expressed as:

$$V = \left(\bar{a} + \bar{b}\right)\sqrt{Z_{cr}}$$

$$I = \left(\bar{a} - \bar{b}\right)\frac{\sqrt{Z_{cr}}}{Z_c}$$

(6.66)

From Equations (6.65)–(6.66), the following expression can be derived for the average power (here and in the following the asterisk denotes the complex conjugate):

$$P = \frac{1}{2}\Re e\,(VI^*) = \frac{1}{2}\frac{Z_{cr}^2}{Z_{cr}^2 + Z_{ci}^2}\left(|\bar{a}|^2 - |\bar{b}|^2\right) + \frac{1}{2}Z_{cr}\Re e\left\{\frac{2j\Im m\left(\bar{a}^*\bar{b}\right)}{Z_{cr}^*}\right\}$$

(6.67)

Equation (6.67) highlights that the total power P is the difference between the powers of the incident and reflected waves only for real values of Z_c, i.e. for the traveling waves formalism; in this case, as observed in [33], Equation (6.67) also satisfies superposition of power: a principle that is generally not valid in electrical engineering.

Let us now consider the circuit illustrated in Figure 6.18 consisting of a generator V_R with a series impedance Z_R, that can be seen as the Thevenin representation of a generic single port network, terminated on an load impedance Z_L. The reflection coefficient for traveling waves is:

$$\bar{\Gamma} = \frac{\bar{b}}{\bar{a}} = \frac{Z_L - Z_R}{Z_L + Z_R} \tag{6.68}$$

In the case of complex Z_R and Z_L it is well known that the maximum power transfer (MPT) is obtained when $Z_L = Z_R^*$, however from Equation (6.68), for $Z_L = Z_R^*$ we get:

$$\bar{\Gamma} = \frac{Z_R^* - Z_R}{Z_R^* + Z_R} = -j\frac{\Im m\,(Z_R)}{\Re e\,(Z_R)} \tag{6.69}$$

This result highlights that the reflection coefficient $\bar{\Gamma}$ as defined in Equation (6.68) and relating traveling waves is not suitable for power transfer applications, in fact, it does not represent the reflection of power in the case of complex impedances.

As deeply discussed in the literature [34–41], when the focus is on power reflections, a more suitable definition of the reflection coefficient can be obtained referring to power waves (PW) instead of traveling waves. Let us introduce the definition of PW:

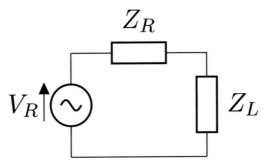

Figure 6.18 Thevenin equivalent circuit of a generator V_R, generator impedance Z_R and a load impedance Z_L.

$$a = \frac{V + Z_R I}{2\sqrt{\Re e\,(Z_R)}}$$

$$b = \frac{V - Z_R^* I}{2\sqrt{\Re e\,(Z_R)}} \tag{6.70}$$

By comparing Equations (6.70) and (6.65), it can be seen that the difference between the definition of traveling waves and PW consists in the use of the complex conjugate in defining the normalized reflected wave. From Equation (6.70), the following expression can be obtained for the voltage and the current:

$$V = \frac{Z_R^* a + Z_R b}{\sqrt{\Re e\,(Z_R)}}$$

$$I = \frac{a - b}{\sqrt{\Re e\,(Z_R)}} \tag{6.71}$$

As a consequence, the power can expressed as follows:

$$P = \frac{1}{2}\Re e\,(V I^*) = \frac{1}{2}\left(|a|^2 - |b|^2\right)$$

$$+ \frac{1}{2\Re e\,(Z_R)}\Re e\,\{2j\Im m\,(a^* b Z_R)\}$$

$$= \frac{1}{2}\left(|a|^2 - |b|^2\right) \tag{6.72}$$

As for the reflection coefficient, with reference to PW, we have [34]:

$$\Gamma = \frac{b}{a} = \frac{Z_L - Z_R^*}{Z_L + Z_R} \tag{6.73}$$

It should be noticed that also PW are not fully satisfactory. As an example, if we use Equation (6.73) to compute the reflection coefficient of a short circuit, we obtain a value different from -1. Furthermore, all measurements at microwave frequency use real reference impedances. According to the analysis developed in this section we can conclude that both formalisms (i.e., traveling and power waves) should be used keeping in mind their respective limits of validity. To this regard, the conjugate impedance approach [34, 41] which allows to use both the traveling and power waves taking advantage of their relative merits will be illustrated in the next section.

6.4.2 Conjugate Image Impedances

The conjugate impedance approach is an elegant procedure for determining the load impedances of a two-port power link which maximize the efficiency [41].

Let us consider the two-port network illustrated in Figure 6.19, Z_{c1}^* is the input impedance at port 1 when port 2 is terminated on Z_{c2}, while Z_{c2}^* is the input impedance at port 2 when port 2 is terminated on Z_{c1}. We assume that the impedance matrix of the 2-port network is:

$$\overline{\overline{Z}} = \begin{pmatrix} z_{11} & z_{12} \\ z_{21} & z_{22} \end{pmatrix}, z_{ij} = r_{ij} + jx_{ij} \tag{6.74}$$

The conjugate image impedances are [34, 41]:

$$z_{c1} = r_{11}\left(\theta_r + j\theta_x\right) - jx_{11}$$
$$z_{c2} = r_{22}\left(\theta_r + j\theta_x\right) - jx_{22} \tag{6.75}$$

Where the parameters θ_r and θ_x are defined as follows:

$$\theta_r = \sqrt{\left(1 - \frac{r_{12}^2}{r_{11}r_{22}}\right)\left(1 + \frac{x_{12}^2}{r_{11}r_{22}}\right)}, \theta_x = \frac{r_{12}x_{12}}{r_{11}r_{22}} \tag{6.76}$$

The efficiency of the two-port network, defined as the ratio between the power delivered to the load (P_L) and the power available from the generator (P_{in}), is maximized when the network is terminated with the conjugate impedances expressed in Equation (6.75), i.e. the generator impedance is Z_{c1}, while port 2 is terminated on a load with impedance Z_{c2}. In particular, it can be demonstrated that for a reciprocal network the maximum of the efficiency is:

$$\eta = \frac{P_L}{P_{in}} = \left|\left(\frac{1 - \theta_r + j\theta_x}{1 + \theta_r + j\theta_x}\right)\frac{z_{21}}{z_{12}}\right| = \left|\frac{1 - \theta_r + j\theta_x}{1 + \theta_r + j\theta_x}\right| \tag{6.77}$$

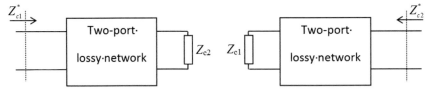

Figure 6.19 Definition of conjugate image impedances.

According to the above reported results, for a generic reciprocal two-port network the efficiency can be maximized at the operating frequency by adding to the input and output port appropriate matching networks that transform the source impedance into Z_{c1} and the load impedance into Z_{c2}. Furthermore, any problem related to the selection of traveling or power waves can be eliminated by adding the imaginary parts of the conjugate image impedances to the two-port network, so that the real part of the conjugate image impedances can be used as reference impedances.

In order to illustrate this procedure, let us consider the WPT link reported in Figure 6.20 which consists of four magnetically coupled coils. The second and the third coils are resonant at the operating frequency, while the first and fourth coils can be resonant or not.

Figure 6.21 illustrates the corresponding measured scattering parameters. Measurements were performed by using a Vector Network Analyzer (VNA) with 50 Ω reference impedances.

In particular, at 68 MHz the measured scattering matrix is:

$$\overline{\overline{S}} = \begin{pmatrix} 16.46 + j9.8 & 12.4 - j79.16 \\ 12.4 - j79.16 & 13.22 + j32.31 \end{pmatrix} \quad (6.78)$$

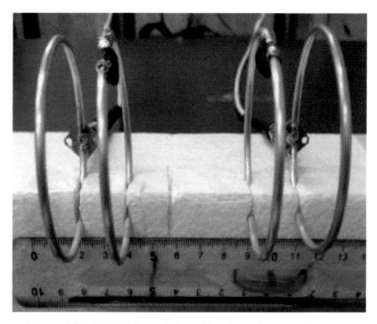

Figure 6.20   WPT link consisting of four inductively coupled coils.

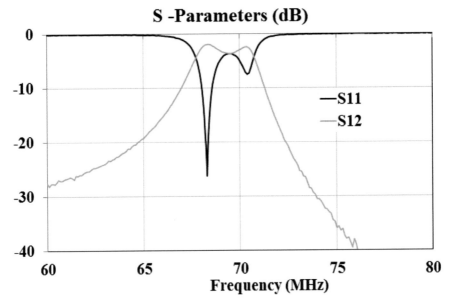

Figure 6.21   Scattering parameters of the WPT link illustrated in Figure 6.20.

Thus resulting in the conjugate image parameters reported in Table 6.4.

According to data reported in Table 6.4, for conjugate matching we need to:

1. add a series capacitor of 27.6 pF and one of 25.3 pF at port 1 and 2 respectively;
2. use for the two-port network including the series capacitors added at point 1) a reference impedance of 47.5 Ω at port 1, and a reference impedance of 38.1 Ω at port 2.

The improvement of the performance corresponding to this approach is highlighted in Figure 6.22 where the scattering parameters measured when the two-port network is terminated on the conjugate image parameters are reported.

Table 6.4   Conjugate image impedances and image impedances of the WPT link of Figure 6.20.

	Port 1	Port 2
Conj. image impedances	47.5414 – j84.799	38.1850 – j92.5459

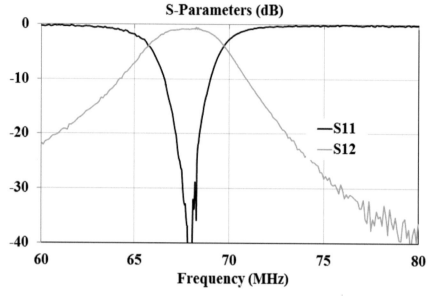

Figure 6.22   Scattering parameters of the WPT link illustrated in Figure 6.20 when port 1 and port 2 are terminated on the conjugate image parameters given in Table 6.4.

6.5 Measurement of the Resonator Quality Factor

The measurement of the resonant frequency and Q factor of resonators is based principally on two techniques: the reflection method and the transmission method [42]. It is interesting to note that it is not possible to measure directly the unloaded Q of a resonator but only its loaded Q. This is due to the necessity to couple the resonator to an external circuitry in order to pick the measurement signal. However, a simple modeling of the test structure allows one to de-embed the unloaded Q from the measured loaded Q.

In Figure 6.23, an equivalent circuit of a resonator, composed by L_2, C_2, and R_2, coupled to a measuring probe, is described. The measuring probe construction depends on the characteristics of the resonator. In order to measure the Q of a resonator for wireless power transfer, a simple inductive loop is sufficient. By introducing $\omega_0 = \frac{1}{\sqrt{L_2 C_2}}$ and $Q_0 = \frac{\omega_0 L_2}{R_2}$, the input impedance Z_i at the probe port is given by

$$Z_i = j\omega L_1 + \cfrac{\dfrac{(\omega M)^2}{R_2}}{1 + jQ_0 \left(\dfrac{\omega}{\omega_0} - \dfrac{\omega_0}{\omega}\right)} \tag{6.79}$$

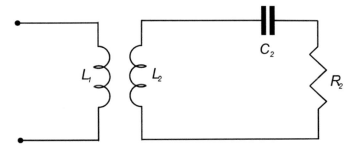

Figure 6.23 Equivalent circuit of a resonator coupled to a measuring probe.

By choosing a small reactance value ωL_1, we can neglect it, and at the resonant frequency, the input impedance is given by

$$Z_i = R_i = \frac{(\omega M)^2}{R_2} \tag{6.80}$$

It is convenient to further simplify the frequency dependence as follows:

$$\frac{\omega}{\omega_0} - \frac{\omega_0}{\omega} \approx 2\frac{\omega - \omega_0}{\omega_0} \tag{6.81}$$

accordingly, Equation (6.79) can be rewritten in the form:

$$Z_i = \frac{R_i}{1 + jQ_0\left(2\frac{\omega - \omega_0}{\omega_0}\right)} \tag{6.82}$$

with the corresponding input reflection coefficient expressed as follows:

$$\Gamma_i = \frac{Z_i - R_c}{Z_i + R_c} \tag{6.83}$$

In the above equation, R_c is the reference impedance of the probe port. When we consider a resonant frequency $\omega \to \infty$, we detune the resonator and the input impedance becomes $Z_i = 0$; the coefficient of reflection Γ_d for the detuned resonator becomes $\Gamma_d = -1$.

With reference to Figure 6.24, the complex number $\Gamma_i - \Gamma_d$ can be expressed by the following equation:

$$\Gamma_i - \Gamma_d = \frac{2\dfrac{R_i}{R_c}}{1 + \dfrac{R_i}{R_c} + jQ_0 2\dfrac{\omega - \omega_0}{\omega_0}} \tag{6.84}$$

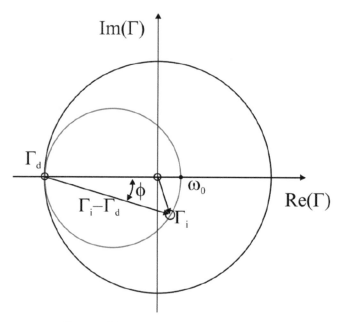

Figure 6.24   Reflection coefficient versus ω.

The parameter $\kappa = R_i/R_c$ represents the ratio of the power dissipated in the internal resistance to that transmitted to the output port; by denoting with Γ_{ir} the reflection coefficient at resonance, we have

$$\Gamma_i - \Gamma_d = \frac{2\kappa}{1+\kappa} \tag{6.85}$$

which is a real number. A graphical representation of $\Gamma_i - \Gamma_d$ is illustrated in Figure 6.24.

As shown in [9, 11, 15, 42, 43], the following relation holds for loaded and unloaded Q_0:

$$Q_L = \frac{Q_0}{1+\kappa} \tag{6.86}$$

by introducing the above expression into Equation (6.84), we get

$$\Gamma_i - \Gamma_d = \frac{2}{\left(1+\dfrac{1}{\kappa}\right)\left[1+jQ_L 2\dfrac{\omega-\omega_0}{\omega_0}\right]} \tag{6.87}$$

It is advantageous to introduce the auxiliary variable

$$\alpha = Q_L 2 \frac{\omega - \omega_0}{\omega_0} \tag{6.88}$$

and to rewrite Equation (6.87) as follows:

$$\Gamma_i - \Gamma_d = \frac{2}{\left(1 + \dfrac{1}{\kappa}\right)[1 + j\alpha]} = \frac{2(1 - j\alpha)}{\left(1 + \dfrac{1}{\kappa}\right)[1 + \alpha^2]}$$

$$= x + jy = \rho e^{j\phi} = \sqrt{x^2 + y^2}\, e^{jarctg\frac{y}{x}}. \tag{6.89}$$

From the above equation, it can be observed that

$$\varphi = arctg(-\alpha) \tag{6.90}$$

and therefore,

$$\tan(\phi) = -Q_L 2 \frac{\omega - \omega_0}{\omega_0} \tag{6.91}$$

In order to measure Q_L, one may select two frequencies, denoted by f_3 and f_4, where $\varphi = -45°$ and $\varphi = 45°$, respectively, thus obtaining

$$Q_L = \frac{f_0}{f_3 - f_4} \tag{6.92}$$

Finally, once κ is computed from Equation (6.85), we have the value of the unloaded Q:

$$Q_0 = Q_L(1 + \kappa) \tag{6.93}$$

Let us now refer to the two-port response of a resonator, as shown in Figure 6.25. By assuming negligible coupling between the input and output probes, the loaded quality factor may be obtained from the following equation:

$$Q_L = \frac{f_0}{\Delta f} \tag{6.94}$$

with the meaning of the symbols defined in Figure 6.25. By measuring the insertion loss at the resonant frequency $|S_{21}|$, it is possible to obtain the value of the unloaded quality factor by the well-known formula:

$$Q_0 = \frac{Q_L}{1 - |S_{21}|^2} \tag{6.95}$$

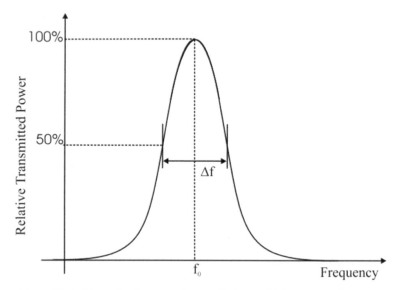

Figure 6.25 Example of measured transmission coefficient versus frequency.

6.6 Examples of Coupled Resonators for WPT

Following the previous introduced theory, we have designed and tested several WPT systems. As a first example, let us consider the four-coil WPT system shown in Figure 6.12, with the two inner loops resonating, by using a lumped capacitor, at 14.56 MHz. The central inductors consist of a 2-turn coil fabricated by using a 2 mm silver-plated copper wire with a diameter of 60 mm. The calculated input loop inductance is 130 nH, and the resonators' Q factor is about 300. The measured and simulated efficiencies are given in Figure 6.26.

Another system has been designed, composed by a pair of resonators made with a single-loop inductor of diameter 20 cm and a couple of capacitors; the latter were made by two rectangular pieces of high-quality Teflon substrate, metalized on both sides. In this way, a high-quality capacitor is obtained. The Teflon substrate is a Taconic TLY5 with the thickness of 0.51 mm. A capacitance of about 72 pF is easily obtained, and by trimming the edge of the substrate, an accurate tuning of the resonators is obtained. The system has been tuned and measured with the arrangement shown in Figure 6.27. In Figures 6.28–6.30, we have reported the efficiencies, defined as the square of the modulus of the S_{21} multiplied by 100, of the linear part, as obtained by using a vector network analyzer.

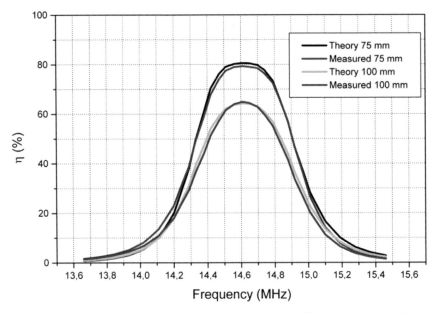

Figure 6.26   Simulated and measured efficiencies at 75 and 100 mm resonator coil center distance.

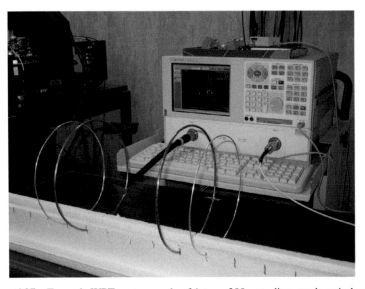

Figure 6.27   Example WPT system made of 1-turn, 200-mm-diameter loop inductors.

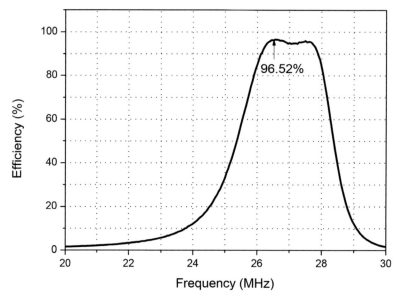

Figure 6.28   Measured efficiency at 100 mm resonator coil center distance.

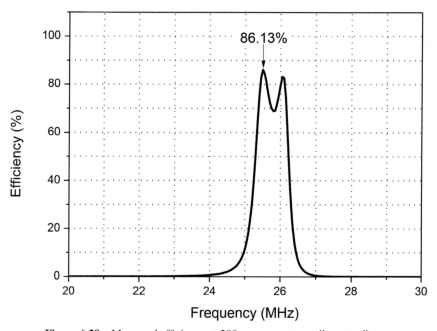

Figure 6.29   Measured efficiency at 200 mm resonator coil center distance.

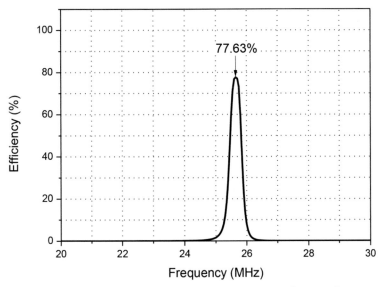

Figure 6.30   Measured efficiency at 300 mm resonator coil center distance.

Due to the resonators' quality factor of about 350, a high efficiency is obtained at a distance of 10 cm. In this configuration, the linear part of the WPT system delivers the power to the load almost like a direct connection. As the distance between the two resonators increases, a drop of the efficiency is measured, although at the distance of 30 cm, an efficiency of about 77% is reached.

6.7 Design of the Oscillator Powering the Resonant Link

The power oscillator is a key part of any WPT system as it is responsible for the efficiency of the entire link. When resonant magnetic links are concerned, the oscillator cannot be designed as a stand-alone subsystem loaded by a standard 50 Ω termination. On the contrary, the steady-state oscillatory regime needs to be designed concurrently with the magnetic resonant circuit, which at the same time settles on the oscillator frequency and represents its actual load. This allows us to simultaneously ensure the oscillation condition and the highest conversion efficiency for any possible working state. Indeed, as discussed in the previous sections of this chapter, the magnetic link resonant frequency varies with the distance between the transmitter and the receiver and this requires an oscillator with a sufficient band to ensure a stable oscillatory regime and flat system efficiency for any possible working condition.

The primary requirement is to use oscillators operating in high-conversion-efficiency conditions. This may be obtained by ideal switching-mode operation of the transistors where the switch current flows with zero voltage across the switch, when the switch is on, and the voltage builds up with zero current, when the switch is off. The ideal behavior of the transistor output waveforms is plotted in Figure 6.31 Virtually, 100% DC-to-RF conversion efficiency may be obtained if the internally dissipated power is effectively cancelled. The drawback of this choice is the need for oscillator embedding network able to effectively reject the higher harmonics associated with such strongly nonlinear regimes. Resonant links characterized by high Q factors at the oscillation frequency intrinsically provide a strong attenuation of higher harmonic components.

The topology to be chosen is mainly dependent on the link operating frequency and on the power levels involved. For ultra-low-power and UHF operating bands, typical of wireless sensor network applications, power oscillator approach, and injection-locking scheme have been reported with a conventional class-E PA to improve the overall transmitter efficiency. Excellent results for medium power link operating in the 400 MHz band are demonstrated by the design of a class-E oscillator making use of harmonic balance techniques [44, 45]. The optimum operating point, for the highest conversion efficiency, is obtained after the design of the reactive network. For this reason, these solutions are certainly appropriate when the oscillator load is likely known and is not expected to change significantly. When the resonator distances are not fixed, the coupling factor influences their resonant frequency, and thus, the oscillator is required to be tuned accordingly. A solution to these aspects can be found in the Royer-type topology schematically reported in Figure 6.32.

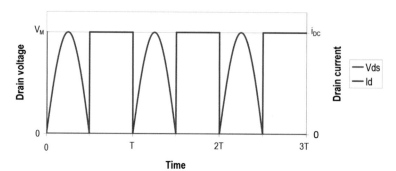

Figure 6.31   Ideal drain voltage and current waveforms.

The oscillator consists of two cross-coupled MOSFETs connected to a resonant circuit (*transmitter resonator*) formed by the center-tapped inductor L_{1a}–L_{1b} and the capacitor C_1. This resonator acts as the primary-side link coil of the power system and is coupled to a second identical resonant circuit (*receiver resonator*), formed by L_2 and C_2, which is connected to a rectifier and a DC load.

In order to describe the principle of operation of the oscillator, we refer to the simplified scheme depicted in Figure 6.33, where the MOSFETs

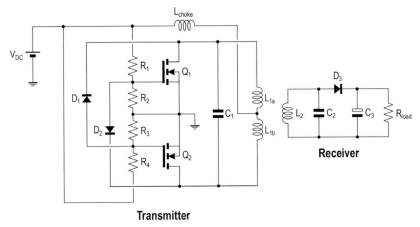

Figure 6.32 Circuit schematic of a WPT link driven by a Royer-type oscillator.

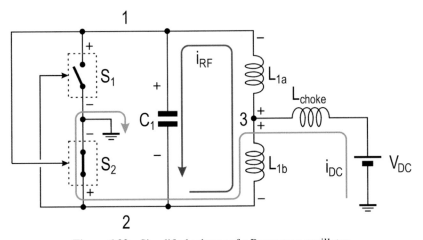

Figure 6.33 Simplified scheme of a Royer-type oscillator.

are replaced by voltage-controlled ideal switches. Owing to the connection structure, when either one of the switches is in the "on" state, it forces to zero the control voltage of the other one, which is thus set into the "off" state.

If the transmitter resonator is weakly coupled to the receiver (i.e., the distance between the coils is very large), its angular resonant frequency is given by

$$\omega_0 = \frac{1}{\sqrt{2(1 + k_{ab})LC_1}} \tag{6.96}$$

where L is the common value of the inductance of the two sections of L_1 and k_{ab} their coupling factor. If the loaded Q factor of this resonant circuit is high enough (in practice, it could be shown that a value of about 10 is sufficient), we can assume that the voltage across C_1 is sinusoidal

$$v_{C1}(t) = V_M \sin(\omega_0 t) \tag{6.97}$$

Let us also assume that at $t = 0$, S_1 is "off" and S_2 is "on," as it is shown in Figure 6.33. This implies that during the first half period of v_{C1}, the voltage at node 1 is equal to v_{C1} and is positive, so that S_2 is held in the "on" state. As a consequence, the voltage across S_2 remains equal to zero and S_1 is kept "off." At $t = \pi/\omega_0$, v_{C1} becomes equal to zero and S_2 turns "off"; thus, the voltage at node 2 starts increasing. Hence, in the subsequent half period, S_1 is held "on" and the voltage at node 1 is kept to zero. This situation holds until v_{C1} crosses zero again at $t = 2\pi/\omega_0$ and the cycle restarts.

The inductance L_{choke} is chosen in such a way that its impedance is very large (ideally infinite) at ω_0, so that the capacitor current (i_{RF}) can circulate only in the loop formed by C_1 and L_1. The DC voltage source and L_{choke} thus behave as a DC current source. The current i_{DC}, injected in the central tap of L_1, can only flow through the switch that is in the "on" state. This also shows that the current in the switches is constant during the "on" state. As a result, the ideal drain voltage and current waveforms are of the kind represented in Figure 6.31. Each device switches when v_{ds} becomes equal to zero, and i_{d} flows through the device only when v_{ds} is zero. Hence, in the ideal case, the power dissipated by the devices is zero.

In order to compute the amplitude of the resonator voltage V_M, which also represents the peak value of v_{ds}, we notice that the voltage across the two sections of L_1 (L_{1a}, L_{1b}) is

$$v_{L1b}(t) = -v_{L1a}(t) = \frac{V_M}{2} \sin(\omega_0 t) \tag{6.98}$$

As a consequence, the voltage of the central tap of L_1 (node 3) is a rectified sinusoid of amplitude $V_M/2$, whose average value is given by

$$V_{3Av} = \frac{\omega_0}{\pi} \int_0^{\frac{\pi}{\omega_0}} \frac{V_M}{2} \sin(\omega_0 t) dt = \frac{V_M}{\pi} \tag{6.99}$$

Since the DC component of the voltage across L_{choke} is zero, this value must be equal to the DC source voltage. Hence, the resonator voltage amplitude and the DC source voltage are related by

$$V_M = \pi V_{DC} \tag{6.100}$$

The behavior of the actual oscillator of Figure 6.32 can be analyzed by time-domain simulation or by means of a harmonic balance technique, specialized for autonomous circuits [45]. The latter approach provides higher computational performance when the steady-state regime is addressed, and thus, it is more efficient if the oscillator regime has to be analyzed several times, as it happens when the effects of circuit parameter variations have to be investigated or when a circuit optimization is performed.

Figures 6.34–6.36 report some results obtained by a harmonic balance simulation of Royer-type oscillator employing two IRF740 power MOSFETs. The resonant frequency of the transmitter resonator is 232.6 kHz, and its Q factor is 78. The DC supply voltage is 12 V.

In Figure 6.34, the actual waveforms of the resonator voltages are plotted. In this case, the amplitude of the voltage across C_1 is slightly lower than πV_{DC} due to the voltage drops introduced by the nonzero resistances of the devices in the "on" state. This voltage drop is evident in Figure 6.35, reporting the drain voltage and current waveforms. In the actual oscillator, when current flows through a device, its drain voltage is not zero; moreover, due to device capacitances, also the switching times are not zero, and hence, voltage starts rising before current has reached zero. As a consequence, power dissipation occurs. However, since voltage and current never assume large values simultaneously, the power absorbed by the devices is usually small, and practical conversion efficiencies of the order of 80% can be easily obtained. Finally, Figure 6.36 shows the gate voltage and current waveforms, resulting from the charging and discharging of the gate capacitance.

A further advantage of the Royer-type oscillator is represented by the fact that its oscillation frequency coincides with the resonant frequency of the

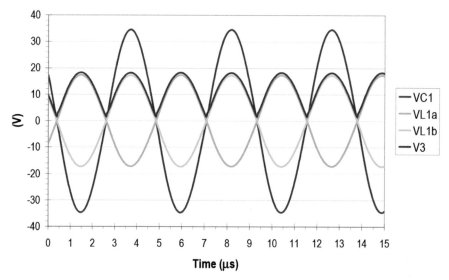

Figure 6.34   Voltages across the components of the transmitter resonator.

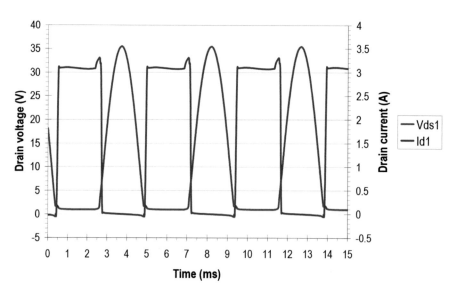

Figure 6.35   Actual drain voltage and current waveforms of a Royer-type oscillator.

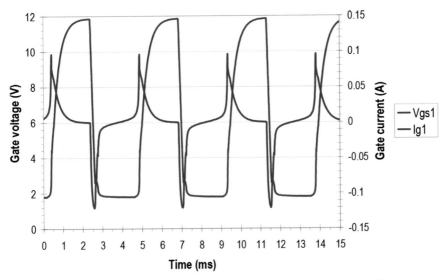

Figure 6.36 Gate voltage and current waveforms of a Royer-type oscillator.

reactive part of its load, which also coincides with the resonant frequency of the power link. Hence, when the distance between the coils is varied, the oscillation is automatically tuned accordingly and the system, for any coupling level, operates at the frequency providing the maximum power transfer to the load. In order to illustrate this property, we can investigate the dependence of the load power on frequency by replacing the MOSFETs with a sinusoidal current source, connected in parallel to C_1. As a matter of fact, the MOSFETs act as two square wave current sources in phase opposition. The DC component of their current can only flow through L_{1a}–L_{1b} and L_{choke}, while the RF ones flow through the resonant circuit. If the Q factor is high enough, only the fundamental component gives rise to a significant voltage across the resonator, while the higher harmonics are practically short-circuited by C_1 and thus give negligible contribution to the output power.

Figure 6.37 shows the results obtained with the same power link considered in the previous example, for several values of the coupling factor k between the transmitter and the receiver resonator. For low values of k, the output power has a maximum for $f \approx \omega_0/(2\pi)$, while for high values of k, the curves exhibit two resonant peaks, approximately located at $f = \omega_0/(2\pi\sqrt{1+k})$ and $f = \omega_0/(2\pi\sqrt{1-k})$. The frequencies corresponding to the absolute maxima of the curves of Figure 6.37 are also reportedin

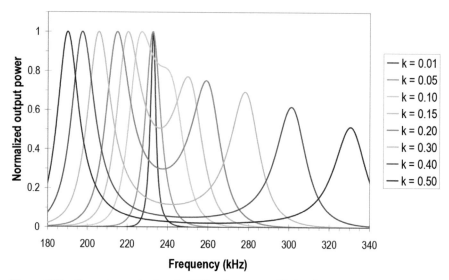

Figure 6.37 Power transmitted to the DC load versus oscillation frequency for several values of the coupling factor.

Figure 6.38 (indicated by the blue dots) where the oscillator frequency is plotted against the coupling factor. This plot clearly shows that the oscillator frequency is always very close to the optimal one from power transfer point of view.

Finally, some representative results of the entire system performance, obtained by a harmonic balance simulation, are summarized in Figures 6.39 and 6.40. In Figure 6.39, the output DC power is compared with the input DC power and with the oscillator RF power for a wide range of possible coupling factor and thus of possible transmitter and receiver distances. It can be observed that the design of the system as whole allows us to guarantee a fairly flat behavior of the available output power for a wide range of coupling distances. A suitable behavior of the system is also demonstrated in Figure 6.40 where the different contributions to the system efficiency versus the coupling factor are plotted. The transmitter (TX) efficiency, that is, the oscillator conversion efficiency, is comprised between 74 and 80% all over its operating band. Due to the rectifier losses, the receiver exhibits a slightly lower efficiency, between 74 and 76%. As a consequence, also the resulting total efficiency is not critically dependent on the coil distance and is better than 55% all over even coupling factor range.

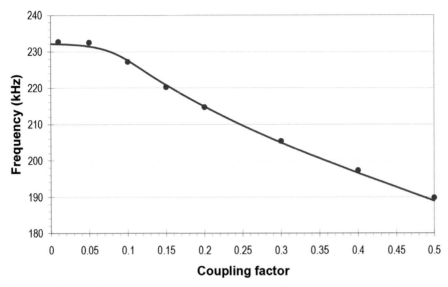

Figure 6.38 Tuning characteristic of a Royer-type oscillator versus the coupling factor. The blue dots indicate the frequencies that correspond to the maximum power transfer to the load.

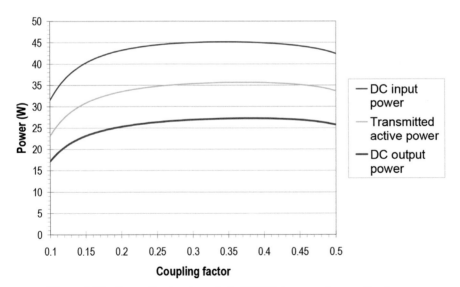

Figure 6.39 Power flowing through the WPT link versus the coupling factor.

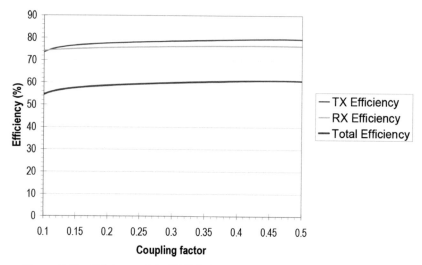

Figure 6.40   Efficiency of the WPT link components versus the coupling factor.

6.8 Conclusions

This chapter has provided a general overview of resonant magnetic wireless power transfer. It has been shown that a basic system consists of resonators coupled via their magnetic fields. Using impedance inverter representation facilitates the analysis and design of such systems. It has also been shown that in order to maintain a high quality factor, the source and load should not be directly attached to the resonators, thus making necessary to introduce appropriate matching networks. Finally, the power oscillator challenges, for the best conversion efficiency, have been discussed for WPT systems with varying resonator distances. These also represent resonant varying loads for the oscillator and need to be accounted for to properly design its operating band. A systematic procedure has been discussed to compute the overall system performances with respect to the resonator behavior and coupling factors.

6.9 Exercises

1. With reference to Figure 6.8, assuming zero resistances and $L_1 = L_2$; $C_1 = C_2$, find the resonant frequencies.
2. Show that a symmetrical network with three equal, lossless resonators behaves, at the resonant frequency, as an ideal transformer with $n = -1$. Let us assume that the resonators are placed along a line and each resonator couples only with the adjacent ones.

3. Show that a symmetrical network with five equal, lossless resonators behaves, at the resonant frequency, as an ideal transformer with $n = 1$. As before, let us assume that the resonators are placed along a line and each resonator couples only with the adjacent ones.

4. Inductance computation of a single-loop inductor (use the MATLAB code given on next pages). With reference to Figure 6.13, compute the inductance of a single-loop inductor of radius $A = 0.1$ m composed of a conducting wire of radius $r = 0.0015$ m.

5. Inductance computation of mutual coupling between single-loop inductors (use the MATLAB code given on next pages). With reference to Figure 6.15, compute the mutual inductance of a couple of loop inductors with equal radius $A1 = A2 = 0.1$ m with their centers at a distance $d = 0.2$ m.

6. Computation of mutual inductance of tightly wrapped loop inductance. Consider now the case study of inductors made of infinitely thin conductors wrapped on the same circumference. Calculate the mutual inductance between the two coils of radius $A1 = 0.1$ m and $A2 = 0.05$ m at the distance $d = 0.1$ m composed, respectively, of 5 and 3 turns.

6.9.1 MATLAB function for single-loop inductance computation

```
function [L] = LoopL(A,r);
%%%%%%%%%%%%%%%%%%%%%%%%%%%%%%%%%%%%%%%
% computation of inductance of a conducting loop
% dimensions meters
% A radius of the loop (m)
% r radius of the conducting wire (m)
% L inductance (H)
mu0 = 4*pi*10^(-7);
k=(4*A*(A-r)/(2*A-r)^2);
[K,E] = ellipke(k);
L=mu0*(2*A-r)*((1-k/2)*K-E);
Return
```

6.9.2 MATLAB function for two coaxial conducting loops mutual inductance computation

```
function [Lm] = LoopLM(A1,A2,d);
%%%%%%%%%%%%%%%%%%%%%%%%%%%%%%%%%%%%%%%
% computation of mutual inductance between two coaxial conducting % loops
```

```
% dimensions meters
% A1 radius of loop 1 (m)
% A2 radius of loop 2 (m)
% d distance between the loops (m)
% Lm mutual inductance (H)
mu0 = 4*pi*10^(-7);
a=(A1^2+A2^2+d^2)/(A1^2*A2^2);
b=2/A1/A2;
k=(2*b/(a+b));
[K,E] = ellipke(k);
Lm=2*mu0*sqrt(a+b)/b*((1-k/2)*K-E);
Return
```

References

[1] Collin, R.E. *Field Theory of Guided Waves.* New York: Mc-Graw-Hill Book Co. (1960).

[2] Mongiardo, M., T. Rozzi. *Open Electromagnetic Waveguides.* London: IEE (1997).

[3] Felsen, L.B., M. Mongiardo, and P. Russer. *Electromagnetic Field Computation by Network Methods.* Berlin: Springer (2009).

[4] Karalis, A., R. Moffatt, J. D. Joannopoulos, P. Fisher, and M. Soljacic A. Kurs. "Wireless Power Transfer via Strongly Coupled Magnetic Resonances." *Science* 317: 83–86 (2007).

[5] Joannopoulos, J.D., M. Soljacic, and A. Karalis. "Efficient Wireless Non-Radiative Mid-range Energy Transfer." *Annals of Physics* 323: 24–48 (2008).

[6] Tomassoni, C., P. Russer, R. Sorrentino, and M. Mongiardo. "Rigorous Computer-Aided Design of Spherical Dielectric Resonators for Wireless Non-Radiative Energy Transfer." In *MTT-S International Microwave Symposium*, Boston, pp. 1–4 (2009).

[7] Fanti, J.O., Giulia Feng, Yifei Omanakuttan, K. Ongie, R. Setjoadi, A. Sharpe, and N. Mur-Miranda. "Wireless Power Transfer Using Weakly Coupled Magnetostatic Resonators." *Energy Conversion Congress and Exposition (ECCE)*, pp. 4179–4186 (2010).

[8] Hoburg, J.F., D.D. Stancil, S.C. Goldstein, and B.L. Cannon. "Magnetic Resonant Coupling as a Potential Means for Wireless Power Transfer to Multiple Small Receivers." *Transactions on Power Electronics* 24, no. 7: 1819–1825 (2009).

[9] Ping, Hu, A.P., S. Malpas, D. Budgett Si. "A frequency control method for regulating wireless power to implantable devices." *Transactions on Biomedical Circuits and Systems* 2, no. 1: 22–29 (2008).

[10] Xuelin Liu, Hao Li, G. Shao, Qi Li, Hongyi Fang. "Wireless Power Transfer System for Capsule Endoscopy Based on Strongly Coupled Magnetic Resonance Theory." in *International Conference on Mechatronics and Automation*, pp. 232–236 (2011).

[11] Zhi, Ian F. Akyildiz Sun. "Magnetic Induction Communications for Wireless Underground Sensor Networks." *Antennas and Propagation* 58, no. 7: 2426–2435 (2010).

[12] Mahanfar, A., B. Kaminska and S.J. Mazlouman. "Mid-Range Wireless Energy Transfer Using Inductive Resonance for Wireless Sensors." in *International Conference on Computer Design*, pp. 517–522 (2009).

[13] Wentzloff Han Sangwook, D.D. "Wireless Power Transfer Using Resonant Inductive Coupling for 3D Integrated ICs." in *3D Systems Integration Conference (3DIC), 2010 IEEE International* (2010).

[14] Okabe, H., Y. Hori T. Imura. "Basic Experimental Study on Helical Antennas of Wireless Power Transfer for Electric Vehicles by using Magnetic Resonant Couplings." in *Vehicle Power and Propulsion Conference*, pp. 936–940 (2009).

[15] Wenzhen Zhang, Bo Qiu, Dongyuan Fu. "Analysis of Transmission Mechanism and Efficiency of Resonance Coupling Wireless Energy Transfer System." *Journal of Electrical Machines:* 2163–2168 (2008).

[16] Hori, Yoichi. "Future Vehicle Society Based on Electric Motor, Capacitor and Wireless Power Supply." in *Power Electronics Conference (IPEC), 2010 International*, pp. 2930–2934 (2010).

[17] Meyer, D.A., J.R. Smith, A.P. Sample. "Analysis, Experimental Results, and Range Adaptation of Magnetically Coupled Resonators for Wireless Power Transfer." *Transactions on Industrial Electronics:* 544–554 (2011).

[18] Sanghoon Kim, Yong-hae Kang, Seung-youl Lee, M. Lee, Jong-moo Zyung, Taehyoung Cheon. "Circuit Model Based Analysis of a Wireless Energy Transfer System via Coupled Magnetic Resonances." *Transactions on Industrial Electronics* 58, no. 99 (2011).

[19] Chang Son, H., D. Hyun Kim, K. Ho Kim, Y. Jin Park, and J. Wook Kim. "Analysis of Wireless Energy Transfer to Multiple Devices using CMT." in *Proceedings of Asia-Pacific Microwave Conference:* 2149–2152 (2010).

[20] Shoki, H. "Issues and Initiatives for Practical use of Wireless Power Transmission Technologies in Japan." in *Microwave Workshop Series on Innovative Wireless Power Transmission: Technologies, Systems, and Applications (IMWS)*, pp. 87–90 (2011).

[21] Agbinya, J.I. *Principles of Inductive Near Field Communications for Internet of Things.* Aalborg, Denmark: River Publishers, ISBN: 978-87-92329-52-3 (2011).

[22] Mongiardo, M., and M. Dionigi. "CAD of Wireless Resonant Energy Links (WREL) Realized by Coils." in *MTT-S International Microwave Symposium,* Anaheim, CA, USA, 1760–1763 (2010).

[23] Finkenzeller, Klaus. *RFID Handbook: Fundamentals and Applications in Contactless Smart Cards, Radio Frequency Identification and Near-Field Communication.* New York: Wiley (1998).

[24] Grover, W. *Inductance Calculations.* New York: Dover (1946).

[25] Okuyama, Yuki, Nobuyoshi Kikuma, Kunio Sakakibara and Hiroshi Hirayama. "A Consideration of Equivalent Circuit of Magnetic-Resonant Wireless Power Transfer." in *Antennas and Propagation (EUCAP), Proceedings of the 5th European Conference on,* pp. 900–903 (2011).

[26] Marco Dionigi, Mauro Mongiardo, Renzo Perfetti. "Rigorous Network and Full-Wave Electromagnetic Modeling of Wireless Power Transfer Links." *IEEE Transaction on Microwave Theory and Techniques* 63, no. 1: 65–75 (2015).

[27] Minfan Fu, Tong Zhang, Chengbin Ma, and Xinen Zhu. "Efficiency and Optimal Loads Analysis for Multiple-Receiver Wireless Power Transfer Systems." *IEEE Transaction on Microwave Theory and Techniques* 63, no. 3: 801–812 (2015).

[28] Mongiardo, M., R. Sorrentino, C. Tomassoni, and M. Dionigi. "Networks Methods for Wireless Resonant Energy Links (WREL) Computations." in *ICEAA,* Turin, Italy (2009).

[29] Mezzanotte, P., M. Mongiardo, and M. Dionigi. "Computational Modeling of RF Wireless Resonant Energy Links (WREL) Coils-based Systems." *ACES Conference*, Tampere, Finland (2010).

[30] Mongiardo, M., M. Dionigi. "CAD of Efficient Wireless Power Transmission Systems." in *MTT-S International Microwave Symposium*, Baltimore, MD, USA (2011).

[31] Mongiardo, M., and M. Dionigi. "Efficiency Investigations for Wireless Resonant Energy Links Realized with Resonant Inductive Coils." in *GEMIC, German Microwave Conference.*, Darmstadt (2011).

[32] R. B. Marks and D. F. Williams. "A general waveguide circuit theory." *Journ. of Research-National Institute of Standards and Technology* 97: 533–533 (1992).

[33] S. Llorente-Romano, A. Garca-Lamperez, T. K. Sarkar, and M. Salazar-Palma. "An Exposition on the Choice of the Proper S Parameters in Characterizing Devices Including Transmission Lines with Complex Reference Impedances and a General Methodology for Computing Them." *IEEE Antennas and Propagation Magazine* 55, no. 4: 94–112 (2013).

[34] S. Roberts, "Conjugate-Image Impedances," in Proceedings of the IRE, pp. 198–204 (1946).

[35] K. Kurokawa. "Power Waves and the Scattering Matrix." *IEEE Transactions on Microwave Theory and Techniques* 13, no. 2: 194–202 (1965).

[36] P. Penfield. "Noise in Negative-Resistance Amplifiers." *Circuit Theory, IRE Transactions on* 7, no. 2: 166–170 (1960).

[37] J. Rahola. "Power Waves and Conjugate Matching." *IEEE Trans. Circuits Syst. II* 55, no. 1: 92–96 (2008).

[38] N. Inagaki. "Theory of Image Impedance Matching for Inductively Coupled Power Transfer Systems." *IEEE Transactions on Microwave Theory and Techniques* 62, no. 4: 901–908 (2014).

[39] D. A. Frickey. "Conversions between S, Z, Y, H, ABCD, and T parameters which are valid for complex source and load impedances." IEEE Trans. on Microwave Theory and Techniques 42, no. 2: 205–211 (1994).

[40] D. Williams. "Traveling Waves and Power Waves: Building a Solid Foundation for Microwave Circuit Theory." *IEEE Microwave Magazine* 14, no. 7: 38–45 (2013).

[41] Alessandra Costanzo, Marco Dionigi, Diego Masotti, Mauro Mongiardo, Giuseppina Monti, Luciano Tarricone, and Roberto Sorrentino. "Electromagnetic Energy Harvesting and Wireless Power Transmission: A Unified Approach." *Proceedings of the IEEE* 102, no. 11: 1692–1711 (2014).

[42] Kajfez, D., and P. Guillon. *Dielectric Resonators*. Atlanta: Noble Publishing Corporation (1998).

[43] Ikuo Awai. "Design Theory of Wireless Power Transfer System Based on Magnetically Coupled Resonators." in *International Conference on Wireless Information Technology and Systems (ICWITS)* (2010).

[44] Suarez, A., D. B. Rutledge S. Jeon. "Nonlinear Design Technique for High-Power Switching-Mode Oscillators." *Transaction on Microwave Theory and Techniques* 54, no. 10: 888–899 October (2006).

[45] Costanzo, A., F. Mastri and C. Cecchetti V. Rizzoli. "Harmonic-Balance Optimization of Microwave Oscillators for Electrical Performance, Steady-State Stability, and Near-Carrier Phase Noise." in *MTT-S International Microwave Symposium Digest,* San Diego, pp. 1401–1404 (1994).

7

Techniques for Optimal Wireless Power Transfer Systems

This chapter is an ensemble of innovative techniques in wireless power transfer. In it, we present new methods of delivering flux efficiently from an inductive transmitter to an inductive receiver by using either flux concentrator or separator. A concentrator increases the flux coupling coefficient and hence leads to increased flux delivered to a receiver by a large order of magnitude, while the separator reduces crosstalk between two identical types of nodes and also leads to significant increase in power delivery. Using a separator coil system without increasing the number of coils increases the efficiency of a four-coil wireless power transfer system from 34.62% to 72.74%. The effects of eliminating a coil by doubling the radii of the primary and secondary coils without using a concentrator directly are also studied. This provides an elegant approach for efficient wireless power delivery systems.

After these analyses, we present a method of how to quickly derive the power transfer function equation for an inductive array of N loops including split ring arrays. Nearest neighbor interaction concept is employed, and magneto-inductive wave transmission is assumed. The easy-to-use algorithm for the transfer function equations assumes also low power coupling approximation which applies to many current applications of inductive methods. The assumptions lead to the power transfer equation for any N coils. Correction terms are suggested for larger coupling coefficients and quality factors. Interpretation of the overall system of loops based on the approximation is suggested and shown to be a very reasonable approach for explaining what takes place in such systems from the electronic communication point of view.

The chapter concludes with proposals and analysis of a new framework for wireless feedback control systems. The feedback control problem is modeled as inductive flux link from a coil to its neighbors. This provides a foundation for system control of wireless Internet of things and an easier path to the design, analysis, control, and performance analysis of such systems.

7.1 Introduction

Over the past few years, the efficiencies of wireless power transfer systems using two, three, and four coils [1–4, 8] have been investigated. It became very clear from them that crosstalk and weak coupling between inductive antennas reduce the efficiency of wireless power transfer systems. Loose coupling and crosstalk result in poor wireless power delivery systems, for example, in biomedical implants and inductive communication systems, and in many cases, the efficiencies are less than 40% as shown in [1–4]. In practice, to increase the power delivery efficiency, the number and sizes of coils in the system are increased. Recently, the use of four-coil systems consisting of a driver, primary, secondary, and load coils became popular. Peng in 2011 [1], Kiani in 2011 [2], and Ramrakhyani in 2011 [3] presented such systems recently. The driver and primary coil, and the secondary and load coil pairs are tightly coupled, leading to improved power transfer.

Generally, coils located within the near-field region of other coils also receive inductive crosstalk from them. Near-field edge is defined by $(d_0 = \lambda/2\pi)$. At 10 MHz, the near-field edge is about 4.78 m [4] where λ is the resonant wavelength of the power transfer system. To reduce the crosstalk and enhance the power transfer, a modification of the four-coil system was studied by Ng and Bai in [5]. The primary and secondary coils were connected directly and separated by a long distance. This eliminates a source of crosstalk one of the coupling coefficients in the system.

Other investigators have also recently highlighted the effects of non-adjacent inductive resonators which affect not only the power delivery efficiency but also the optimal frequency of wireless power transfer [6]. One of such effects is a shift in the location of maximum efficiency away from the resonant frequency of the resonator. This in effect implies non-optimal performance due to the shift away from resonance. This effect is due to cross-coupling of energy back into the system through multiple reflective impedances [4].

In a four-coil system, the most critical distance of interest is the separation d_{23} between primary and secondary coils because it is the longest and also causes weak coupling between the primary and secondary coils. Duong and Lee [7] suggested a method for mitigating the effects of this separation by using impedance matching of the input impedance of the system which should be matched to the source resistance. They suggested either reducing the quality factor Q_1 of stage 1 or reducing the coupling coefficient k_{12} between the driver and primary coils. Changing the quality factor means changing either the operating resonant frequency or the resistance of the loop, which practically means using a better-quality wire for the coils. This adds to the cost of the wireless power transfer system. Changing the resonant frequency invariably affects the rest of the stages (primary, secondary, and load coils), which is counter-productive. Similarly, changing the driver wires means redesigning the system entirely with new coils, which is not always the best thing to do in an operational system. Therefore, in [7], the authors increased the separation between the driver and primary coils and thus reduced the coupling coefficient k_{12} (this means loose coupling) as opposed to the preferred tight coupling between the driver and primary coils. Hence, in this paper, focus is placed on how to reduce the effects of this distance, not by reducing k_{12}, but rather using two innovative methods for maximum power transfer. This is achieved by introducing flux concentrators and separators for maximum power transfer. Since the induced power in the load is proportional to the sixth power of distance between it and the secondary coil and also since there is cross-coupling between pairs of coils, a clever design of the coupling coefficients is essential. These methods of maximum power delivery to the load are discussed, modeled, and analyzed. The first two methods use flux concentrator coils to boost the required flux for delivering the power efficiency, while the third method uses flux separator coils to reduce crosstalk in the system (recover lost flux) to enable the required flux to be delivered as well.

7.2 Flux Conentrators

7.2.1 Splitting of Coupling Coefficients

The amount of power transferred to a load is proportional to the flux density (flux per unit area) in the load coil. To increase the flux intercepted by the load coil, a flux concentrator is used. This is a coil which is deployed between two neighboring coils so as to split one of the coupling coefficients into two larger coupling coefficients. This results in $(N + 1)$ coupling coefficients. To illustrate this, we start from a four-coil system and add a fifth coil. The fifth coil is called the flux concentrator. Its function is to strengthen the magnetic field delivered to the secondary coil. It is a well-established fact that the amount of magnetic field coupled to another coil is larger if the distance between the coils is smaller, and it is therefore reasonable to conceive continuously, strengthening the flux by using concentrators located at shorter distances to bridge the distance between the primary and secondary coils without losing range. In the following analysis, without the loss of generality, let the concentrator be positioned midway between the primary and secondary coils but maintain the distances between the driver and primary coils and between the secondary and load coils as in a four-coil system. Figure 7.1 shows the orientations of the coils for maximum flux coupling such that the distance between the circumferences of the load and primary coils (or radius r) is $d = \sqrt{r^2 + x^2}$ for vertical separation of x units. By placing the concentrator in between the primary and secondary coils, the original coupling coefficient k_{ps} between them is split into two larger coupling coefficients $(k_{pc}, k_{cs}) > k_{ps}$; the coupling coefficient between the primary and concentrator and between the concentrator and secondary can be shown to be given by the equation:

$$k_{ps}^2 = \frac{r_p^3 r_s^3}{\left(r_p^2 + (x_{ps}/2)^2\right)^3} = \frac{64 r_p^3 r_s^3}{\left(4 r_p^2 + x_{pc}^2\right)^3} \tag{7.1}$$

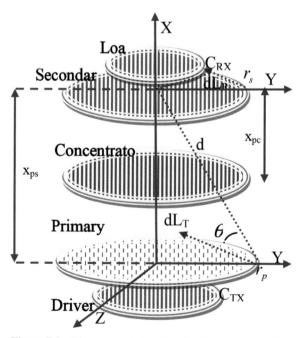

Figure 7.1   Flux concentrators in inductive power transfer.

In Equation (7.1), x_{ps} is the distance between the primary and secondary coils in the four-coil system and $x_{pc} = x_{cs} = x_{ps}/2$ is the distance of the concentrator to the primary or secondary coil, r_p is the radius of the primary coil, and r_s is the radius of the secondary. This equation shows that the effect of placing the concentrator between these two coils is to enhance the flux coupling by a factor of 64. Similar results could be obtained by increasing the radii of the coils as shown in the next section.

7.2.2 Doubling of Coil Radius

Equation (7.1) suggests a redesign of the coils which could achieve the same objective for which the concentrator is used. This could be achieved by doubling the radii of the primary and secondary coils.

To demonstrate this, we first absorb the factor 2 in the denominator of Equation (7.1) (also reflected in the numerator) into the radius of the primary coil, implying that its effective radius is $r'_p = 2r_p$ and creates the coupling coefficient:

$$k'^2_{ps} = \frac{8r'^3_p r^3_s}{\left(r'^2_p + x^2_{pc}\right)^3} \tag{7.2}$$

Next, absorb again the factor 8 in the numerator as well into the radius of the secondary coil by using the relation $r'_s = 2r_s$ and write the coupling coefficient equation as

$$k'^2_{ps} = \frac{r'^3_p r'^3_s}{\left(r'^2_p + x^2_{pc}\right)^3} \tag{7.3}$$

k'_{ps} in Equation (7.3) is 64 times k_{ps} in the original system of four coils! Hence, a five-coil system in which a concentrator is placed midway between the primary and secondary coils is equivalent to a three-coil system if we double the radii of the primary and secondary coils of a four-coil system. Therefore, the designer has a choice of either inserting a fifth coil of a given radius in between the primary and secondary coils or doubling the radii of the primary and secondary coils in a four-coil system. In fact, by recursively using this design process, a four-coil system can be shown to be equivalent to a three-coil system if one of the four coils is a concentrator placed midway between the second and third coils of a three-coil system. The distance between the driver and secondary coils is kept to what it was in the three-coil system. For the three-coil system to be equivalent to the four-coil system, double the radii of the second and third coils while keeping all the other system parameters the same as before. Therefore, in the limit, a two-coil system with twice the radii is equivalent to a three-coil system when the distance between the second and third coils is halved.

7.3 Separators

In the third method, the objective is to eliminate a source of cross-coupling which often results in reduced power delivery efficiency. This is achieved by using a separator which eliminates a coupling coefficient. This reduces a source of crosstalk and the number of coupling coefficients from N to $(N - 1)$. Thus, it harvests what should have been crosstalk as useful flux which is delivered to the load. The separator method consists of alternate transmitter and receiver deployment system in which two transmitters are separated by a receiver (vice versa). The separation is either hard-wired or by unconnected transmitter and receiver. The first transmitter is always directly excited. The second is inductively excited using mutual inductance. In the hard-wired case, the first receiver and second transmitter are connected as in Figure 7.2. The second receiver is mutual inductance-coupled. To prove that in fact, this method concentrates flux by reducing crosstalk, we use the theory of magnetic waveguides. We formulate the system equations for the set up.

Generally, the node equations are given by the expressions ($1 \leq n \leq 4$):

$$Z_n I_n + j\omega \left(M_{n-1,n} I_{n-1} + M_{n+1,n} I_{n+1} \right) = 0$$
$$M_{01} = M_{10} = 0 \qquad (7.4)$$
$$\text{and} \quad I_0 = 0$$

Z_n are impedances, I_n are currents in the coils, $M_{n-1,n}$ are mutual inductances, L_n are self-inductances, and Z_L is the load impedance. To simplify the analysis, the fact that transmitter TX2 and receiver RX1 are connected directly is initially ignored and circuit is treated as if there are two

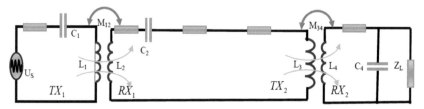

Figure 7.2   Flux concentration using separators.

non-directly connected nodes (as in the waveguide system of Figure 7.3). We will account for their direct connection during the solution of the system Equations (7.5) (where U_S is the transmitter voltage and V_L is the load voltage):

$$Z_1 I_1 + j\omega M_{21} I_2 = U_S$$
$$Z_2 I_2 + j\omega(M_{12} I_1 + M_{32} I_3) = 0$$
$$Z_3 I_3 + j\omega(M_{23} I_2 + M_{43} I_4) = 0 \qquad (7.5)$$
$$Z_4 I_4 + j\omega M_{34} I_3 = 0$$
$$I_4 Z_L = V_L$$

In Figure 7.3, r is the radius of the individual coils and d is the length of the waveguide which is divided into sections of length x. The mutual inductances between coils i and j are given by M_{ij}, L_i are self-inductances, Z_i are impedances of the coils, and ω is the radian frequency shown to be identical in all the coils. Let the currents in each loop be given by I_i.

In a four-node system, Equations (7.5) need to be solved. In the real sense, the same current flows in the first receiver and second transmitter. We will account for this in the equations. Since the first receiver and the second transmitter are connected, we make the following assumptions before

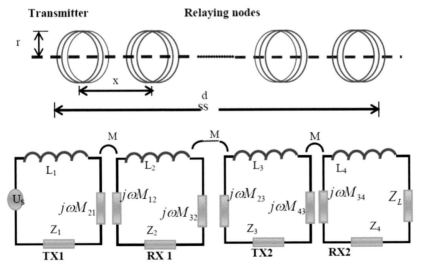

Figure 7.3 Flux separator system.

solving the equations. Since the current flowing in the connected receiver and transmitter is the same, we set $I'_2 = I_2 = I_3$ and $I'_3 = I_4$. This modifies the simultaneous equations when we set $I_2 = I'_2$ and $I_3 = I'_3$ to

$$Z_1 I_1 + j\omega M_{21} I_2 = U_S$$
$$(Z_2 + Z_3)I_2 + j\omega(M_{12}I_1 + M_{32}I_2) + j\omega(M_{23}I_2 + M_{43}I_3) = 0 \quad (7.6)$$
$$Z_4 I_3 + j\omega M_{34} I_2 = 0$$
$$I_3 Z_L = V_L$$

We therefore have three equations to solve three unknowns. This leads to the solution

$$\frac{V_L}{U_S} = \frac{-\omega^2 M_{12} M_{34} Z_L}{\left[\begin{array}{c} Z_1 (Z_2 + Z_3) Z_4 + \omega^2 M_{12} M_{21} Z_4 \\ +\omega^2 M_{34} M_{43} + j\omega (M_{23} + M_{32}) Z_4 \end{array}\right]} \quad (7.7)$$

Apart from the imaginary component in the denominator of Equation (7.7), the solution is similar to the traditional solution for the four-coil systems (Equation (7.10)). By connecting receiver 1 to transmitter 2, the same current is flowing in the two coils. Separating them far from each other means that there is approximately no flux coupling and no crosstalk between them, and hence, $M_{23} = M_{32} = 0$. Their impedances are, however, not zero. Therefore, the system transfer function is purely real at resonance and can be simplified to

$$\frac{V_L}{U_S} = \frac{-\omega^2 M_{12} M_{34} Z_L}{Z_1 (Z_2 + Z_3) Z_4 + \omega^2 M_{12} M_{21} Z_4 + \omega^2 M_{34} M_{43}} \quad (7.8)$$

At resonance, $M_{ij} = M_{ji}$ and $k_{ij} = k_{ji}$. Let Q_n be the quality factors of the coils and R_L and R_n the load and coil resistances, respectively. Define $\delta = (1/R_2 + 1/R_3)$, and then Equation (7.8) becomes

$$\frac{V_L}{U_S} = \frac{-R_L k_{12} k_{34} \left(\sqrt{Q_1 Q_2 Q_3 Q_4}\right)}{\delta \sqrt{R_1 R_2 R_3 R_4} \left[1 + \dfrac{k_{12}^2 Q_1 Q_2}{R_3 \delta} + \dfrac{k_{34}^2 Q_3 Q_4}{R_1 R_2 \delta}\right]} \quad (7.9a)$$

$$\eta_c = \frac{P_L}{P_T} = \frac{R_L k_{12}^2 k_{34}^2}{\delta^2 R_2{}^* R_3{}^* R_4} \frac{Q_1 Q_2 Q_3 Q_4}{\left[1 + \dfrac{k_{12}^2 Q_1 Q_2}{R_3 \delta} + \dfrac{k_{34}^2 Q_3 Q_4}{R_1 R_2 \delta}\right]^2} x100\% \quad (7.9b)$$

The transmitted power is $P_T = U_S^2/2R_1$, and the power delivered to the load is $P_L = V_L^2/2R_L$. From Equations (7.9a) and (7.10b), direct connection of TX2 and RX1 has decoupled coil 2 from coil 3 and eliminated the coupling coefficient k_{23} in the numerator of the equation as well as reduced the effect of loops two and three to the square roots of their quality factors. It has also eliminated two terms as shown in the denominator of Equations (7.10a and 7.10b) (the traditional four-resonant-coil wireless power transfer system), the first related to k_{23} and the second related to the product of the quality factors of the four coils.

$$\left| \frac{V_L}{U_S} \right| = \frac{R_L Q_2 Q_3 k_{12} k_{23} k_{34} \sqrt{Q_1 Q_4}}{\sqrt{R_1} \left[(1 + k_{12}^2 Q_1 Q_2)(1 + k_{34}^2 Q_3 Q_4) + k_{23}^2 Q_2 Q_3 \right]} \tag{7.10a}$$

$$P_L = \frac{k_{12}^2 k_{23}^2 k_{34}^2}{R_4} \frac{Q_2^2 Q_3^2 Q_1 Q_4 P_T}{\left[(1 + k_{12}^2 Q_1 Q_2)(1 + k_{34}^2 Q_3 Q_4) + k_{23}^2 Q_2 Q_3 \right]^2} \tag{7.10b}$$

$$\eta = \frac{P_L}{P_T} x 100\% \tag{7.10c}$$

The power efficiency in Equation (7.9b) is larger than the value in Equation (7.10c). This increase is due to reduced crosstalk between the primary and secondary coils and also a reduction in second-order crosstalk between the coils in the unconnected system. Clearly, the constant $\delta = (1R_2 + R_3)$ in Equations (7.9a and 7.9b) indicates that the impedances of the receiver 1 and transmitter 2 appearin parallel to the flux induced in the pair.

7.3.1 Simulations

The effects of the separator (Figure 7.2) were studied using MATLAB simulation. Choosing $r_1 = r_4 = 5$ cm, $r_2 = r_3 = 30$ cm, $R_1 = R_4 = 0.25\,\Omega$, $R_2 = R_3 = 1.0\,\Omega$, and resonant frequency of 10 MHz, when $d_{12} = d_{34} = 40cm$ and $d_{23} = 69cm$, the power efficiency given by Equations (7.9b) and (7.10) is $\eta_c = 44.8\%$ and $\eta = 4.3\%$. Keeping $d_{12} = d_{34} = 30$ cm and varying the distance d_{23}, a constant efficiency was obtained for the directly connected coils of value 23.72%. This is because the distance d_{23} was replaced by a direct connection of the coils. The effect of the direct connection is btter appreciated by varying the distances d_{12} and d_{24} between the driver and primary and between the secondary and load coils while keeping d_{23} constant. We chose to keep $d_{23} = 74$ cm, the distance where the normal coupling architecture

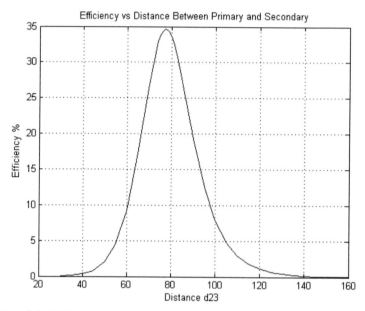

Figure 7.4   Efficiency of 4-coil wireless power transfer system without separator.

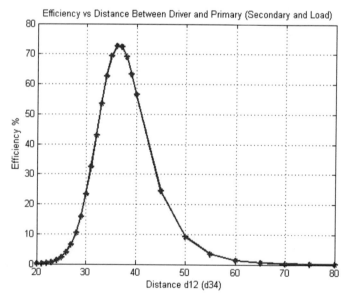

Figure 7.5   Efficiency of 4-coil wireless power transfer system using separators.

gives the highest efficiency of 34.6%. Figure 7.5 shows the effect of using separator coils, leading to a nominal efficiency of about 73%. This more than doubles the efficiency of the system compared with when there were no separator nodes in use. From Figure 7.5, the maximum efficiency occurs at about $d_{12} = d_{23}/2$. Thus, maximum system efficiency is obtained when the load position is at $2d_{23}$.

The ratio of Equations (7.10b) to (7.9b) was simulated using the values in Table 7.1. The ratio of Equation (7.10b) when no separator coil is used to Equation (7.9b) with separated coils 2 and 3 is defined as gain G given as follows:

$$G = \frac{k_{23}^2 Q_2 Q_3 \left[\delta + \dfrac{k_{12}^2 Q_1 Q_2}{R_3} + \dfrac{k_{34}^2 Q_3 Q_4}{R_1 R_2}\right]^2}{[(1 + k_{12}^2 Q_1 Q_2)(1 + k_{34}^2 Q_3 Q_4) + k_{23}^2 Q_2 Q_3]^2} \qquad (7.11)$$

With the coil radii as shown in Table 7. 1, when the distance between the coils 2 and 3 is $d_{23} = 62$ ($d_{12} = 40$; $d_{34} = 40$) cm, G = 1.0256. When $d_{23} > 62$ cm, G < 1, the separator begins to reduce crosstalk (showing the effectiveness of the architecture in Figure 7.2). Below 62 cm, it is better to inductively separate coils 2 and 3. The effect of doubling the radii of coils 2 and 3 was also studied. By setting both radii at 30 cm, coils 2 and 3 when $d_{23} = 69$, G = 0.9738.

At this distance, the performances of the two architectures are about the same. The effect of concentrator coil was also studied. The receiver power equation for the five-coil system is given by Equation (7.12):

$$P_L = \frac{R_L k_{12}^2 k_{23}^2 k_{34}^2 k_{45}^2 Q_2^2 Q_3^2 Q_4^2 Q_1 Q_5 P_T}{R_5 \left[\begin{array}{l} 1 + \left(k_{23}^2 Q_2 Q_3 + k_{34}^2 Q_3 Q_4 + k_{45}^2 Q_4 Q_5\right) \\ + Q_2 Q_4 \left(\begin{array}{l} k_{12}^2 k_{34}^2 Q_1 Q_3 \\ + k_{23}^2 k_{45}^2 Q_3 Q_5 + k_{12}^2 k_{45}^2 Q_1 Q_5 \end{array}\right) \end{array}\right]^2} \qquad (7.12)$$

Table 7.1   Parameters for wireless power transfer system

Description	Coil 1	Coil 2	Coil 3	Coil 4
Radius of coil	$r_1 = 5cm$	$r_2 = 15cm$	$r_3 = 15cm$	$r_4 = 5cm$
Resistance of coil	$R_1 = 0.25\Omega$	$R_2 = 1.0\Omega$	$R_3 = 1.0\Omega$	$R_1 = 0.25\Omega$
Inductance of coil	$L_1 = 1.06\mu H$	$L_2 = 20.0\mu H$	$L_3 = 20.0\mu H$	$L_4 = 1.06\mu H$
Capacitance of coil	$C_1 = 235_p F$	$C_2 = 12.6_p F$	$C_3 = 12.6_p F$	$C_4 = 235_p F$
Resonant frequency	10 MHz	10 MHz	10 MHz	10 MHz
Quality factors	266	1257	1257	266

The ratio of the power received in the five-coil to the four-coil systems is defined as the gain:

$$G = \frac{R_4 k_{23}^2 k_{34}^2 Q_C^2 Q_5 \left[\left(1 + k_{12}^2 Q_1 Q_2\right) \left(1 + k_{34}^2 Q_3 Q_4\right) + k_{23}^2 Q_2 Q_3 \right]^2}{R_5 k_{23}^2 Q_{4(4)} \left[\begin{array}{c} 1 + \left(k_{23}^2 Q_2 Q_C + k_{34}^2 Q_C Q_4 + k_{45}^2 Q_4 Q_5\right) \\ + Q_2 Q_4 \left(\begin{array}{c} k_{12}^2 k_{34}^2 Q_1 Q_C + \\ k_{23}^2 k_{45}^2 Q_C Q_5 \\ + k_{12}^2 k_{45}^2 Q_1 Q_5 \end{array} \right) \end{array} \right]^2}$$

(7.13)

where $G = P_L^{(5)}/P_L^{(4)}$, $Q_1^{(4)} = Q_1^{(5)}$, $Q_2^{(4)} = Q_2^{(5)}$, $Q_3^{(4)} = Q_4^{(5)}$, $k_{12}^{(4)} = k_{12}^{(5)}$, $k_{34}^{(4)} = k_{45}^{(5)}$, $P_T^{(4)} = P_5^{(5)}$, and the same load is used in both cases. In the simulation, we set $d_{12} = d_{45} = 10$ cm and $d_{23} = d_{34} = 40$ cm, meaning that for the equivalent four-coil system, $d_{23} = 80$ cm. This gives the maximum gain $G = 16.7519$. Values of d_{23} more than or less than 80 cm result in less gain. Figure 7.4 shows that the gain with concentrator reaches a maximum at a location and asymptotes with nothing more to be gained by moving it closer to the receiver.

In Figure 7.6, the position of the concentrator is varied from 1 cm from coil 2 until 100 cm from it, and the ratio of the power expected in a five-coil system with concentrator and four-coil system without concentrator is computed using Equation (7.13). The four blue curves from left to right are for when the radius of the concentrator is varied from 10 to 30 cm. The red curve is for when the radius is 30 cm. The gain asymptotes after a certain distance from the primary coil and remains the same after that.

7.3.2 Effect of Concentrator Quality Factor

The quality factor Q of a concentrator may be varied by changing either the resonant frequency, the value of the inductor, or the resistance of the wire used. It is, however, preferable to keep the resonant frequency unchanged as well as the inductor but to use wires that could sustain different ohmic losses. The effect of varying concentrator quality factor was therefore studied by changing the resistance of the concentrator wire from 1.0Ω to 4.5Ω. Figure 7.7 shows the effects of varying Q where the lowest curve is for low $R_3 = R_c = 4.0$ Ω (smaller Q) and the uppermost curve is for $R_c = 1.0$ Ω (larger Q). Lower gains result when wires with large resistive losses are used.

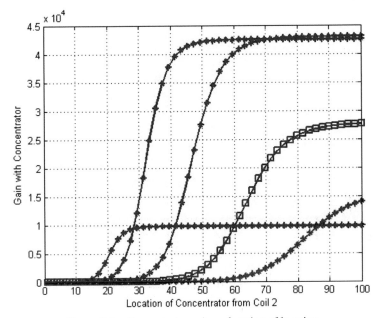

Figure 7.6 Concentrator gain as function of location.

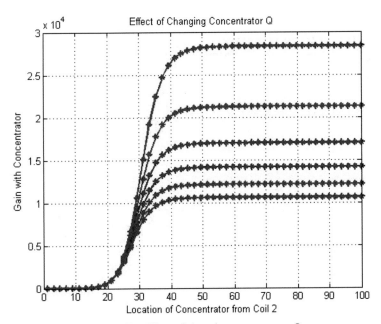

Figure 7.7 Effect of changing concentrator Q.

7.3.3 Effect of Concentrator Radius

The effect of changing concentrator radius was studied. The radius was varied from 10 cm (red curve in Figure 7.8) to 30 cm (rightmost blue curve). The gain asymptotes to about 10, 800 and maintains this value for all values of the radius. Smaller gains are, however, recorded at low distances from the primary coil.

In summary, flux concentrators deliver maximum power to receiver loads. The concentrator is a coil placed between the primary and secondary coils to increase the coupling coefficients to orders of magnitude more than 16 for modest coil radii. It is shown that this can also be achieved by increasing the radii of the primary and secondary coils. A crosstalk separator may also be used. The separator is a transmitter which is deployed between two receivers. In doing so, the distance between the receivers is doubled, and the presence of the separator prevents any meaningful cross-coupling of fluxes between the two receivers thus separated.

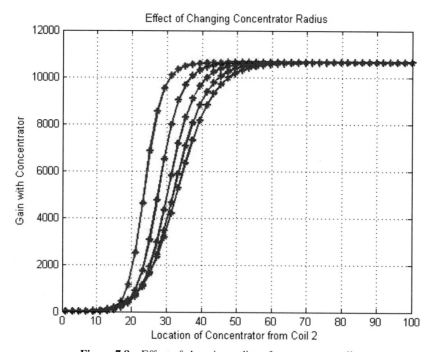

Figure 7.8 Effect of changing radius of concentrator coil.

7.4 Approximate Magneto-Inductive Array Coupling Functions

Interest in magnetic induction (MI) systems operating as power communication transceivers has increased a lot within the last few years because they do not interfere with the existing traditional electromagnetic wave radiators in most bands. A typical magnetic induction device draws as low current as 7 mA to transfer voice or data over a couple of meters. To their advantage, magnetic induction communication systems are not generally affected by the environment. In fact, the only parameter in the MI power equation and link budget that has to do with the environment is the permeability of the materials in the link and source (sink) which acts to enhance the received signal. Hence, the permeability of the medium can be used to advantage to amplify the signal power in the link. Therefore, issues such as fading, multipath propagation, interference, and noise which plague electromagnetic (EM) systems are not problems in MI communication systems. Rather, of interest is how to improve upon the rapid decline in MI power due to the inverse sixth path loss with range.

Applications of inductive methods have become more and more widespread in transcutaneous systems [9], near-field voice communications [10], wireless power transfer [3], data transmission systems [11], underground communications [12], and links and communication channels inside integrated circuits [13]. Many of these applications use several coils either to extend the range of the application or to deliver power more efficiently. The more the coils used, the more the number of equations that must be solved to determine the transfer function of the system of coils. One of such applications where many coils are used is in magneto-inductive waveguides. Magneto-inductive waveguides have recently emerged as a method of extending the range of MI communication systems. The pioneering works of Syms and Solymar [14–19], Shamonina [19], and Kalinin [21] have established some of the theories for the MI waveguides. Recently, the authors also demonstrated relaying in MI systems [20]. These systems involve arrays of coils arranged as resonant chain networks or in multiple paths to create multipath relay nodes. The solutions to their lumped circuit models normally involve solving systems of simultaneous equations which are prone to mistakes because of the number of variables and equations involved. A system of N resonating nodes requires $(N + 1)$ simultaneous equations to be solved and becomes very difficult when N is large. The objective of this paper is to propose a fast approximation method in other to simplify this rigor.

7.4.1 System Specifications

In its simplest form, a one section peer-to-peer flux coupling system is used for communication in the traditional MI system with no options of range extension (Figure 7.9). This model has been well analyzed and discussed by Agbinya et al [11]. The lumped circuit model of the system is also shown in Figure 7.9. An N-section waveguide extends this simple formulation by using N resonating coils or split-ring resonators. Figure 7.9 shows the multiple coil array version of the system. We assume that each coil is loaded with a capacitor to resonate and the receiver has a load Z_L. The current in each loop n is therefore I_n. We assume only nearest neighbor interaction in which only currents in nodes $n - 1$ and $n + 1$ affect node n. We also assume that the N nodes are resonating at frequency ω_0. Each node n has

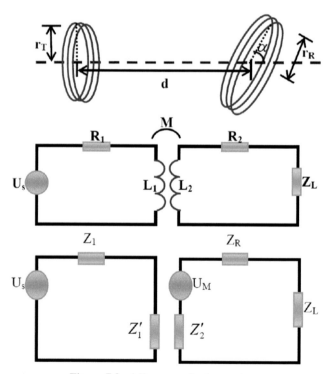

Figure 7.9   MI communication system.

impedance Z_n and current I_n. Node $n = 1$ is excited with input voltage and the rest couple magnetic fields from one to the other until the receiver node is reached. The intermediate nodes are passive, but the receiver has a load impedance Z_L.

7.4.2 Power Relations in Inductive Systems

The purpose of this section is to establish a system equation which is repeatable and easily usable in a form similar to the propagation equation in basic electromagnetic communication systems. To do this, let us consider a peer-to-peer inductive communication consisting of two loops, a transmitter loop and a receiver loop. The governing equations for the system (Figure 7.9) are as follows:

$$Z_1 I_1 + j\omega M_{12} I_2 = U_s \tag{7.14}$$

$$Z_2 I_2 + j\omega M_{12} I_1 = 0$$
$$M_{ij} = k_{ij} \sqrt{L_i L_j} \tag{7.15}$$
$$0 \leq k_{ij} \leq 1$$

where the transmitter and receiver loop impedances are Z_1 and Z_2 with currents flowing through them as I_1 and I_2, respectively.

The mutual inductance and the coupling coefficient between them written in general terms are M_{ij} and k_{ij}, respectively ($i = 1$, $j = 2$). Let the transmitter parameters have index "1" and index "2" be reserved for the receiver variables. The impedances Z_2' refer to the influences of the transmitter on the receiver and of the receiver on the transmitter Z_1'. The voltage developed across the receiver load due to inductive action is given by the expression:

$$V_L = I_2 Z_L \tag{7.16}$$

Where in general for a system consisting of multiple intermediate (or relay) nodes, each loop impedance is

$$Z_n = R_n + j\omega L_n + \frac{1}{j\omega C_n} \tag{7.17}$$

The impedance of the receiver contains the load impedance and is

$$Z_r = R_r + j\omega L_r + \frac{1}{j\omega C_r} + Z_L \tag{7.18}$$

By solving Equations (7.14) to (7.16) simultaneously, the ratio of the load voltage to that of the source is given by the expression:

$$G_v = \frac{V_L}{U_S} = \frac{-j\omega M_{12} Z_L}{\omega^2 M_{12}^2 + Z_1 Z_2}$$

(7.19)

The transfer function G_v relates the input voltage (U_S) to the voltage V_L developed across the receiver load impedance Z_L. This may also be written in terms of the current induced in the receiver coil as follows:

$$G_v(\omega) = \frac{I_L}{U_S} = \frac{-j\omega M_{12}}{\omega^2 M_{12}^2 + Z_1 Z_2}$$

(7.20)

where $I_L = I_2 = V_L/Z_L = G_v U_S$.

7.4.3 Algorithm for Approximate Transfer Function

Equation (7.19) is a simple case because it involves solving only three equations, but once the number of loops increases and for N loops (N + 1) equations are required, solving the equations to obtain $G_v(\omega)$ becomes a lot more difficult and time-consuming. Hence, the objective of this paper is to provide a fast algorithm for the transfer function for multiple loops which relates the input transmitter voltage to the inductive current flowing through the load impedance. Let the quality factors of the transmitter and receiver be $Q_1 = \omega L_1/R_1$ and $Q_2 = \omega L_2/R_2$.

By substituting Equation (7.15) in the transfer function Equation (7.20), the expression for the inductive system at resonance becomes

$$\frac{I_L}{U_S} = \frac{-jk_{12}\sqrt{Q_1 Q_2}}{\sqrt{R_1 R_2}\left(k_{12}^2 Q_1 Q_2 + 1\right)}$$

(7.21)

At low coupling when also the quality factors (Q) are small, the inequality $k_{12}^2 Q_1 Q_2 \ll 1$ holds or $k_{12}^2 \ll 1/Q_1 Q_2$. The inductive transfer function at resonance reduces to

$$\left|\frac{I_L}{U_S}\right| = \frac{k_{12}\sqrt{Q_1 Q_2}}{\sqrt{R_1 R_2}} = \frac{k_{ij}\sqrt{Q_i Q_j}}{\sqrt{R_i R_j}}$$

(7.22)

Figure 7.10a and b shows the low coupling approximation when the two coils have identical $Q = 2$ and $Q = 10$, respectively, and $k = 0.1$. The approximation is very accurate for low k and low Q but starts to deviate for higher Q. Figure 7.10c shows the variation of the transfer function with Q

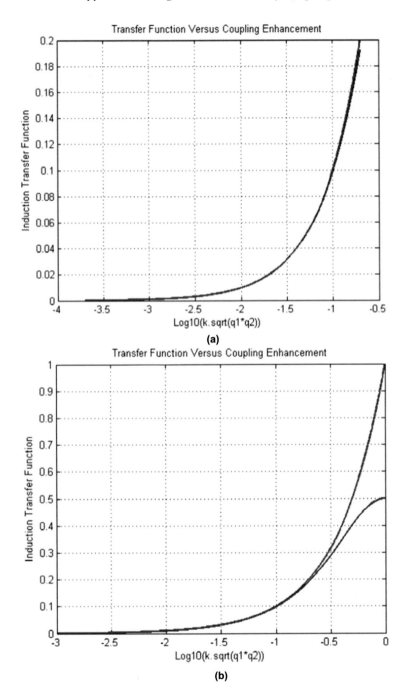

(a)

(b)

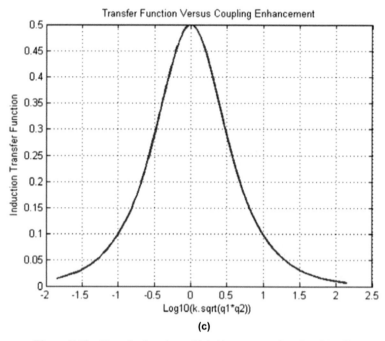

Figure 7.10   Transfer function with/without approximation ($N = 2$).

and plotted as a function of the logarithm of the magnitude of the numerator of Equation (7.21).

We then insert a new resonating node between the transmitter and receiver as a means of extending the communication range. This sets up a three-node system with the following Kirchhoff voltage law (KVL) equations. The equations for the three-section waveguide are as follows:

$$
\begin{aligned}
Z_1 I_1 + j\omega M_{12} I_2 &= U_s \\
Z_2 I_2 + j\omega M_{12} I_1 + j\omega M_{23} I_3 &= 0 \\
Z_3 I_3 + j\omega M_{23} I_2 &= 0 \\
V_L &= I_3 Z_L
\end{aligned}
\tag{7.23}
$$

Solving the simultaneous Equation (7.23) results in the expression:

$$
\frac{V_L}{U_S} = \frac{\omega^2 M_{12} M_{23} Z_L}{\omega^2 \left(M_{12}^2 Z_3 + M_{23}^2 Z_1 \right) + Z_1 Z_2 Z_3}
\tag{7.24}
$$

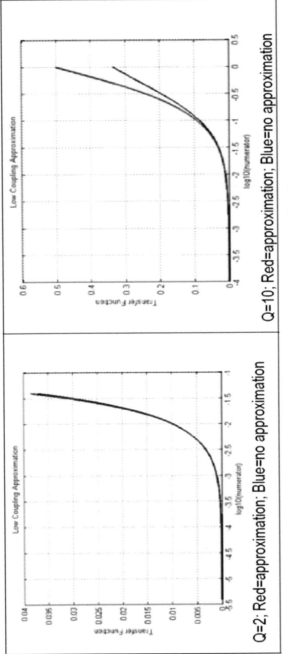

Figure 7.11   Transfer functions with approximation ($N = 3$).

At resonance, this equation can be simplified and becomes

$$\frac{I_L}{U_S} = \frac{k_{12}k_{23}Q_2\sqrt{Q_1Q_3}}{\sqrt{R_1R_3}\left[1 + k_{12}^2Q_1Q_2 + k_{23}^2Q_2Q_3\right]} \tag{7.25}$$

As in Equation (7.21), we consider low coupling when $k_{12}^2Q_1Q_2 + k_{23}^2Q_2Q_3 \ll 1$ or $Q_2 \ll 1/(k_{12}^2Q_1 + k_{23}^2Q_3)$, and writing $k_{ij} = k_{i,j}$, the low coupling transfer function becomes

$$\left|\frac{I_L}{U_S}\right| = \frac{k_{12}k_{23}Q_2\sqrt{Q_1Q_3}}{\sqrt{R_1R_3}} = \frac{\sqrt{Q_1Q_N}\prod_{i=1}^{N-1}k_{i,i+1}\prod_{i=1}^{N-2}Q_{i+1}}{\sqrt{R_1R_N}} \tag{7.26}$$

The approximation for $k = 0.1$ and $Q = 2$ is held at $N = 3$ with slight deviations only for higher $Q = 10$. We extend the system one more time to demonstrate the concept further for $N = 4$. In this case, the system of equations is

$$\begin{aligned}
Z_1I_1 + j\omega M_{12}I_2 &= U_s \\
Z_2I_2 + j\omega M_{12}I_1 + j\omega M_{23}I_3 &= 0 \\
Z_3I_3 + j\omega M_{23}I_2 + j\omega M_{34}I_4 &= 0 \\
Z_4I_4 + j\omega M_{34}I_3 &= 0 \\
V_L &= I_4Z_L
\end{aligned} \tag{7.27}$$

This has the solution

$$\frac{V_L}{U_S} = \frac{j\omega^3 M_{12}M_{23}M_{34}Z_L}{\begin{aligned}Z_1Z_2Z_3Z_4 &+ \omega^4 M_{12}^2 M_{34}^2 \\ +\omega^2 &\begin{pmatrix} M_{12}^2Z_3Z_4 + M_{23}^2Z_1Z_4 \\ +M_{34}^2Z_1Z_2 \end{pmatrix}\end{aligned}} \tag{7.28}$$

At resonance, it reduces to

$$\left|\frac{I_L}{U_S}\right| = \frac{Q_2Q_3k_{12}k_{23}k_{34}\sqrt{Q_1Q_4}}{\sqrt{R_1R_4}\left[1 + \begin{pmatrix} k_{12}^2Q_1Q_2 \\ +k_{23}^2Q_2Q_3 + k_{34}^2Q_3Q_4 \\ +k_{12}^2k_{34}^2Q_1Q_2Q_3Q_4 \end{pmatrix}\right]} \tag{7.29}$$

The low coupling approximation of the transfer function for $N = 4$ with $\left(k_{12}^2Q_1Q_2 + k_{23}^2Q_2Q_3 + k_{34}^2Q_3Q_4\right) + k_{12}^2k_{34}^2Q_1Q_2Q_3Q_4 \ll 1$ is

$$\left|\frac{I_L}{U_S}\right| = \frac{k_{12}k_{23}k_{34}Q_2Q_3\sqrt{Q_1Q_4}}{\sqrt{R_1R_4}}$$

$$= \frac{\sqrt{Q_1Q_4}\displaystyle\prod_{i=1}^{3}k_{i,i+1}\prod_{i=1}^{2}Q_{i+1}}{\sqrt{R_1R_4}} \tag{7.30}$$

Normally, even when the loops are arranged equidistant from each other, the strongest coupling is between the first and second loops with coefficient k_{12} and the rest of the coefficients become smaller and smaller toward the receiver.

In general, for N magneto-inductive loops (Figure 7.12), the low coupling transfer function with only nearest neighbor interaction becomes

$$\left|\frac{I_L}{U_S}\right| = \frac{\sqrt{Q_1Q_N}\displaystyle\prod_{i=1}^{N-1}k_{i,i+1}\prod_{i=1}^{N-2}Q_{i+1}}{\sqrt{R_1R_N}} \tag{7.31}$$

This is a general expression which holds for all N, provided that the approximations are made in the denominators for the solutions to the simultaneous KVL equations. Equation (7.31) represents a general simplification of the transfer function when multiple loops are involved. To use the equation, it is required that the number of loops N be selected including the electrical dimensions of

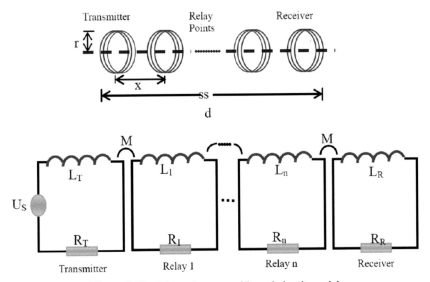

Figure 7.12 Magnetic waveguide and circuit model.

each loop. The electrical dimensions include the radius r_i, the number of turns N, the resistance of each loop R_i, and the distance between the loops l, which are then used to compute the quality factors of the loops. Select the excitation voltage U_S and the load impedance. The load could be a sensor or a device being driven by the array of inductive loops.

7.4.4 Interpretation of Algorithm

It is essential to overview what the above equation represents in terms of communication using inductive nodes operating either as a normal wireless power transfer system or as a normal inductive communication system. By defining Q as the gain of a node, the general equation has the following interpretation as in traditional RF systems. The product $\sqrt{Q_1 Q_N}$ is the gain of the transmitter and receiver stages. This is similar to the product of the transmitter and receiver antenna gains in RF systems. The quantity $\prod_{i=1}^{N-2} Q_{i+1}$ is the product of the gains of the intermediate or relay stages. This too is similar to the product of the gains of the intermediate stages in a transponder system. The term $\prod_{i=1}^{N-1} k_{i,i+1}$ is the path loss of the general magnetic channel. Thus, we can model the overall system of N resonating inductive loops as in Figure 7.13.

The resistors R_1 and R_N are the inherent ohmic losses of the transmitter and receiver stages. The inductive system has a channel with path loss (gain) equal to $G_c = \prod\limits_{i=1}^{N-1} k_{i,i+1} \prod\limits_{i=1}^{N-2} Q_{i+1}$. If as in transcutaneous system applications or in embedded biomedical systems when biological tissues are part and parcel of the channel between the transmitter and receiver, and if the biological channel has impedance Z_b, the channel gain is modified to $G_c = Z_b \prod_{i=1}^{N-1} k_{i,i+1} \prod_{i=1}^{N-2} Q_{i+1}$ and the channel becomes a function of frequency and leads to further coupling losses between the transmitter and receiver.

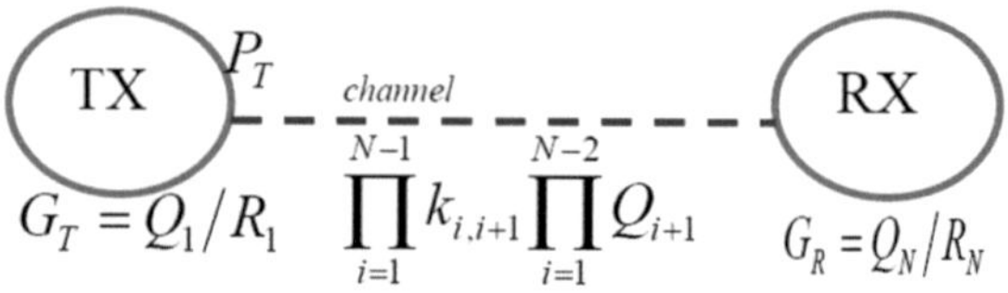

Figure 7.13 Peer-to-peer inductive communication system.

7.4.5 Correction Terms

Progressive correction terms may be added to the denominator of the transfer function equation. These corrections apply for when only the first loop is excited with an input voltage at frequency ω. For example, when $N = 2$, the correction term is $k_{12}^2 Q_1 Q_2$. When $N = 3$, there are two terms $k_{12}^2 Q_1 Q_2 + k_{23}^2 Q_2 Q_3$ that could be used for corrections and are functions of k_{12}^2 and k_{23}^2. If the nodes are distributed evenly in a chain network so that the distance between nodes is constant x, then $k_{23}^2 < k_{12}^2$, and to improve upon the approximation, the term $k_{12}^2 Q_1 Q_2$ may be used for correction. For $N > 3$, progressive corrections may be made for higher $k\,(Q)$ based on adding terms in k_{ij}^2 where i and j are larger integers compared with the previous correction terms.

The algorithm presented enables fast derivation of inductive loop system transfer functions used in embedded biomedical data and wireless power transfer, personal area networks, and communications underground. The algorithm applies for all N and eliminates the need to solve a large system of equation, provided that the coupling coefficients and the quality factors of the loops are small. Progressive corrections to account for larger k and Q can be made for large k and Q applications.

7.5 Wireless Feedback Modelling

Classical feedback control systems feed the output of the system back into the input as a means of controlling the behavior of the overall system. Normally, this is achieved with direct wire connections. Realistically, the emergence of wireless systems means wireless feedback control is essential as a means of remote control of the performance of the systems. That, however, requires wireless interconnections of the input and outputs of the systems. In wireless communication systems where permanent directly connected feedback is not available, pilot signals are sent when needed to provide information about the state of the transmitter, channel, or receiver either to the transmitter or to the receiver or both. While this does not provide a realistic feedback system, it does help to provide occasional information. A realistic wireless connection of the output of the system to its input will usher in a new era of remote linear control system applications and sensing. While direct transmissions of radio signals could be used, this requires two transceivers operating in a duplex mode using either Bluetooth or some other radiations. A better but permanent feedback can be achieved using mutual induction. Feedback is an essential tool for the control and optimum performance of many electronic systems. Feedback control systems are usually hardwired into the design with components that are directly linked with wires. The purpose of this paper is to demonstrate wireless feedback or feedback without wires. Magnetic coupling can result in wireless feedback as shown below. This lays a new foundation for automatic control of short-range and near-field communication systems wirelessly.

To date, how to limit crosstalk in inductive communication systems and wireless power transfer systems is still an unsolved problem. Crosstalk in inductive systems is, however, due to feedback of flux through mutual inductance between coils and hence current from neighboring coils interfering with the desired signals. From the feedback point of view, crosstalk appears as undesirable poles in the system transfer function. Such poles could therefore be identified clearly and be nullified with zeros inserted into the circuits.

Figure 7.14 shows three coils that are linked in pairs with magnetic fields. The first coil creates the field due to varying current and this field is sympathetically picked up by the second and third coils, and the second coil also induces flux on the first and third coils. The third coil induces flux on both. Since the magnetic lines of force are circular in nature, they pass through both the coil creating them and the one receiving them. Figure 7.14 represents a wireless feedback system. The next sections provide proofs. To do so, we first

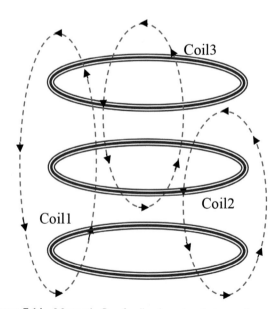

Figure 7.14   Magnetic flux feedback system between three coils.

illustrate the traditional two-loop wireless power transfer function. To show that a system of resonating loops indeed forms a feedback control system, we consider a peer-to-peer inductive communication consisting of two loops, a transmitter loop and a receiver loop (Figure 7.15). The equivalent circuit for the two-loop systems is given in Figure 7.16. The governing equations for the system from Figure 7.16 are as follows:

$$Z_1 I_1 + j\omega M_{12} I_2 = U_s \qquad (7.32)$$

$$Z_2 I_2 + j\omega M_{12} I_1 = 0$$
$$M_{ij} = k_{ij} \sqrt{L_i L_j} \qquad (7.33)$$
$$0 \le k_{ij} \le 1$$

where the transmitter and receiver loop impedances are Z_1 and Z_2 with currents flowing through them as I_1 and I_2, respectively. The mutual inductance and the coupling coefficient between them written in general terms are M_{ij} and k_{ij}, respectively ($i = 1, j = 2$). Let the transmitter parameters have index "1" and index "2" be reserved for the receiver variables. The impedances Z_2' refer to the influences of the transmitter on the receiver and of the receiver

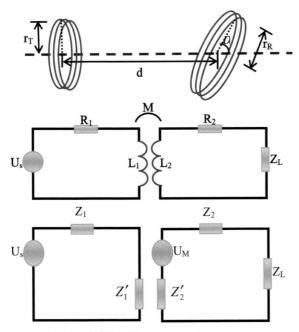

Figure 7.15 MI communication system.

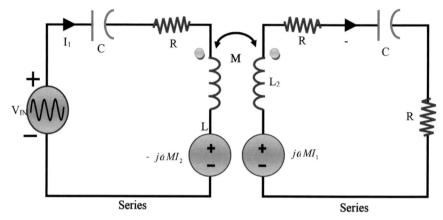

Figure 7.16   System model of inductive transmitter and receiver.

on the transmitter Z_1'. The voltage developed across the receiver load due to inductive action is given by the expression:

$$V_L = I_2 Z_L \tag{7.34}$$

Where in general for a system consisting of multiple intermediate (or relay) nodes, each loop impedance is

$$Z_n = R_n \mid j\omega L_n + \frac{1}{j\omega C_n} \tag{7.35}$$

When the last stage is the receiver, the impedance of the receiver load should be added and is

$$Z_2 = R_2 + j\omega L_2 + \frac{1}{j\omega C_2} + Z_L \tag{7.36}$$

By solving Equations (7.32) to (7.34) simultaneously, the ratio of the load voltage to that of the source is given by the expression:

$$G_v = \frac{V_L}{U_S} = \frac{-j\omega M_{12} Z_L}{\omega^2 M_{12}^2 + Z_1 Z_2} \tag{7.37}$$

The transfer function G_v relates the input voltage (U_S) to the voltage V_L across the receiver load impedance Z_L.

7.5.1 Wireless Feedback

Coupling of magnetic flux between neighboring coils produces wireless feedback due to mutual inductance and can be used to control the communication system. In the feedforward mode, flux linkage between a transmitting source is picked up by a receiver, and in the feedback mode, the receiver induces also flux on the transmitter. Assuming this is correct, we should be able to show through the theory of inductive couplers that the transfer function of coil systems is indeed feedback control systems whether the system is for power transfer or for communication of data, transcutaneous systems or in magneto-inductive waveguides. Wireless feedback means that a receiver circuit can be used to control the transmitter remotely. The idea is indeed seen in the fact that we have reflected impedance in inductive loop systems, which is a reflection of the receiver impedance on the transmitter circuit.

A clear relationship between the effects of the receiver on the transmitter as a feedback control system needs to be defined. This paper demonstrates this missing link clearly. The basic feedback control system consists of an open-loop gain A (amplifier) and a feedback gain β (attenuator). Figure 7.17 illustrates the two-coil situation in which the output of the open-loop gain amplifier is the induced current. The overall closed-loop transfer function is given as in Equation (7.32). Since

$$I_2 = (U_S - \beta I_2).A \qquad (7.38)$$

$$\frac{I_2}{U_S} = G = \frac{A}{1 + A\beta} \qquad (7.39)$$

The two-coil system in Figure 7.15 forms a wireless closed-loop system in which the mutual inductance and the coupling coefficients play the roles of

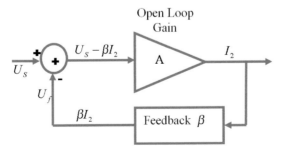

Figure 7.17 Closed-loop feedback control system.

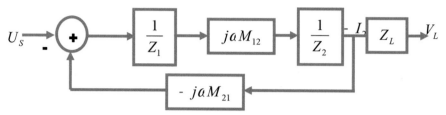

Figure 7.18   Wireless feedback using magnetic induction.

forward gain and negative feedback factors. This is illustrated as shown in Figure 7.18. Comparing Figure 7.18 with Figure 7.15, we can derive the following system equations by defining

$$I_0 = I_2; \quad A = \frac{j\omega M_{12}}{Z_1 Z_2}; \quad V_f = -j\omega M_{21} I_2 = \frac{\omega^2 M_{12} M_{21}}{Z_1 Z_2}$$

$$I_2 = A\left(U_s + j\omega M_{21} I_2\right) = \frac{j\omega M_{12}}{Z_1 Z_2}\left(U_s + j\omega M_{21} I_2\right)$$

$$\left(1 + \frac{\omega^2 M_{12} M_{21}}{Z_1 Z_2}\right).I_2 = \frac{j\omega M_{12} U_s}{Z_1 Z_2}$$

$$\frac{I_2}{U_s} = \frac{j\omega M_{12}}{Z_1 Z_2 \left(1 + \dfrac{\omega^2 M_{12} M_{21}}{Z_1 Z_2}\right)} \tag{7.40}$$

Thus, the induced current in the receiver is proportional to the mutual inductance and attenuated by the feedback coupling from the receiver into the transmitter circuit. This reduction is further exacerbated by the impedances of the transmitter and receiver circuits. The voltage gain of the overall system therefore is

$$\frac{V_L}{U_s} = \frac{-I_2.Z_L}{U_s} = \frac{-j\omega M_{12} Z_L}{Z_1 Z_2 \left(1 + \dfrac{\omega^2 M_{12} M_{21}}{Z_1 Z_2}\right)} \tag{7.41}$$

Equations (7.39) and (7.41) are identical when $M_{12} = M_{21}$. Equation (7.41) proves that an inductive wireless power transfer system implements a wireless feedback system whose forward and feedback gains are functions of the mutual inductance between them. Indeed, any two coils with mutual inductance between them is a wireless feedback system. Therefore, design, stability, and control of wireless power transfer systems can be facilitated using our framework. The closed-loop system has a zero at zero frequency $\omega = 0$ and two poles at $\omega_{1,2} = \pm j(\sqrt{Z_1 Z_2}/M_{12})$. De Moivre's theorem is used to find the location of the poles. Since

$$\sqrt{Z_1 Z_2} = [(R_1 R_2 - X_1 X_2) + j(R_1 X_2 + R_2 X_1)]^{\frac{1}{2}}$$

Using $[a + jb]^n = r^n(\cos n\theta + j \sin n\theta)$ and since $n = 1/2$, $a = (R_1 R_2 - X_1 X_2); b = (R_1 X_2 + R_2 X_1)$

$$r^{1/2} = \sqrt{(R_1 R_2 - X_1 X_2)^2 + (R_1 X_2 + R_2 X_1)^2}$$

$$\cos\left(\frac{\theta}{2}\right) = \frac{R_1 R_2 - X_1 X_2}{(R_1 R_2 - X_1 X_2)^2 + (R_1 X_2 + R_2 X_1)^2}$$

$$\sin\left(\frac{\theta}{2}\right) = \frac{R_1 X_2 + R_2 X_1}{(R_1 R_2 - X_1 X_2)^2 + (R_1 X_2 + R_2 X_1)^2}$$

$$\text{Let } p = \frac{R_1 X_2 + R_2 X_1}{\sqrt{(R_1 R_2 - X_1 X_2)^2 + (R_1 X_2 + R_2 X_1)^2}} \quad \text{and}$$

$$q = \frac{R_1 R_2 - X_1 X_2}{\sqrt{(R_1 R_2 - X_1 X_2)^2 + (R_1 X_2 + R_2 X_1)^2}}.$$

Then, the poles of the transfer function occur at $\omega_1 = (-p + jq)/M_{12}$ and $\omega_2 = (p - jq)/M_{12}$. Thus, with strong coupling, the poles are close to the origin of the axes. In weak coupling, the poles are a lot removed from the origin. At resonance, the poles are purely imaginary and equal to $\omega_{1,2} = \pm j(\sqrt{R_1 R_2}/M_{12})$. Thus, the locations of the poles can be varied either by changing the impedances of the receiver and transmitter or by varying the mutual inductance. The mutual inductance is a function of inductances and the coupling coefficient which may also be varied at will.

7.5.2 Q-Based Explanation of Wireless Closed-Loop Transfer Function

The receiver has influence on the transmitter and is given as a reflected impedance in Figure 7.15 and expressed as

$$Z_{r1} = \frac{-j\omega M I_2}{I_1} = \frac{\omega^2 M^2}{Z_2} \tag{7.42}$$

The transmitter also has similar influence on the receiver. To illustrate this effect on the transmitter clearly, let the quality factors of the transmitter and receiver be $Q_1 = \omega L_1/R_1$ and $Q_2 = \omega L_2/(R_2 + R_L)$, respectively. Hence, at resonance frequency, $\omega_0 = 1/\sqrt{L_1 C_1} = 1/\sqrt{L_2 C_2}$, the reflected impedance of the receiver on the transmitter becomes

$$Z_{r1} = k_{21}^2 Q_1 Q_2 R_1 \tag{7.43a}$$

$$Z_{r2} = k_{12}^2 Q_1 Q_2 (R_2 + R_L) \tag{7.43b}$$

The reflected impedance of the transmitter on the receiver is also given as in Equation (7.42). Hence, the effect of the receiver on the transmitter is to modulate the resistance of the transmitter (Equation (7.41)). Given that the resonant frequency is fixed, the depth of modulation can be varied either by changing the coupling coefficient or by varying the quality factors of the coils. The wireless feedback behavior is explained further below.

Next, it can be shown that the transfer function of a power transfer system is indeed a wireless closed-loop control system with negative feedback. By substituting Equation (7.33) in the transfer function Equation (7.41), the expression for the inductive system at resonance becomes

$$G_v = \frac{V_L}{U_S} = \frac{-jZ_L}{\sqrt{R_1 R_2}} \frac{k_{12}\sqrt{Q_1 Q_2}}{\left(1 + k_{12}^2 Q_1 Q_2\right)} \tag{7.44}$$

When Equation (7.44) is factored in terms of the quality factors, the open-loop and feedback gains are given by the relations:

$$\begin{aligned} A &= k_{12}\sqrt{Q_1 Q_2} \\ A\beta &= k_{12}^2 Q_1 Q_2 = k_{12}k_{21}Q_1 Q_2 \\ \beta &= k_{21}\sqrt{Q_1 Q_2} \end{aligned} \tag{7.45}$$

Notice that we have written $k_{12}^2 = k_{12}k_{21}$ in its proper form as the product of the forward coupling between inductor 1 and inductor 2 and the reverse coupling between inductor 2 and inductor 1 which is really what it is in Equation (7.39) but often appears as a square k_{12}^2 because $k_{12} = k_{21}$ in practice and k_{21} is due to crosstalk. Therefore, the feedback loop gain is due to unwanted flux from the receiver into the transmitter. Finally, we can write the expression for the closed-loop gain as follows:

$$G_v = \frac{V_L}{U_S} = \frac{-jZ_L}{\sqrt{R_1 R_2}} \frac{A}{(1 + A\beta)} \tag{7.46}$$

The term $-jZ_L/\sqrt{R_1 R_2}$ is a scaling factor on the input voltage and the induced voltage in the receiver load. The form of Equation (7.46) provides the basis for further investigation of wireless power transfer systems as control systems that could lead to how best to control them. Also, by this formulation, it becomes easier to null the effects of crosstalk on the overall performance of the system. Hence, from Equation (7.46), it is observed that Equations (7.43a) and (7.44b) represent a wireless feedback from the receiver to the transmitter and vice versa so that

$$Z_{r1} = A\beta.R_1 \tag{7.47a}$$

$$Z_{r2} = A\beta.(R_2 + R_L) \tag{7.47b}$$

7.6 Conclusions

This chapter provides innovative techniques for enhancing the performance of wireless power transfer through the use of concentrators and separators. It also demonstrates how to derive the power transfer relations using circuit theory techniques with approximations given. The comparison between communication channels and antenna theory reduces the problem of using gain equations. It concludes with demonstrating that wireless power transfer systems actually also provided wireless feedback.

References

[1] Hou, P., M.-J. Jia, L. Feng, Y. Mao, and Y.-H. Cheng "An Analysis of Wireless Power Transmission Based on Magnetic Resonance for Endoscopic Devices" in *Proceedings 5th International Conference on Bioinformatics and Biomedical Engineering, (ICBBE)*, pp. 1–3 (2011).
[2] Kiani, H., U.-M. Jow, and M. Ghovanloo. "Design and Optimization of a 3-Coil Inductive Link for Efficient Wireless Power Transmission" *IEEE Transactions On Biomedical Circuits and Systems:* doi:10.1109/TBCAS.2011.2158431 (2011).

[3] Ram Rakhyani, A.K., S. Mirabbasi, and M. Chiao. "Design and Optimization of Resonance-Based Efficient Wireless Power Delivery Systems for Biomedical Implants." *IEEE Transactions on Biomedical Circuits and Systems* 5, no. 2: 48–63 (2011).

[4] Agbinya, J.I. *Principles of Inductive Near Field Communications for Internet of Things.* ISBN: 978-87-92329-52-3; River Publishers Denmark, Aalborg (2011).

[5] Ng, D.C., G. Felic, E. Skafidas, and S. Bai. "Closed-Loop Inductive Link for Wireless Powering of a High Density Electrode Array Retinal Prosthesis." 92–97 (2009).

[6] Lee, C.K., W.X. Zhong, and S.Y.R. Hui. "Effects of Magnetic Coupling of Non-adjacent Resonators on Wireless Power Domino-Resonator Systems." *IEEE Transactions on Power Electronics* (2011).

[7] Duong, T.P., and J.-W. Lee. "Experimental Results of High-Efficiency Resonant Coupling Wireless Power Transfer Using a Variable Coupling Method." *IEEE Microwave and Wireless Components Letters* 21, no. 8: 442–444 (2011).

[8] Cannon, B.L., J.F. Hoburg, D.D. Stancil, and S.C. Goldstein. "Magnetic Resonant Coupling as a Potential Means for Wireless Power Transfer to Multiple Small Receivers." *IEEE Transactions on Power Electronics* 24, no. 7: 1819–1825 July (2009).

[9] Chen, Q., S.C. Wong, C.K. Tse, and X. Ruan. "Analysis, Design, and Control of a Transcutaneous Power Regulator for Artificial Hearts." *IEEE Transactions on Biomedical Circuits & Systems* 3, no. 1: 23–31 (2009).

[10] Freelinc "FreeLinc Near-Field Magnetic Induction Technology" pp. 1–5.

[11] Agbinya, J.I., N. Selvaraj, A. Ollett, S. Ibos, Y. Ooi-Sanchez, M. Brennan, and Z. Chaczko. "Size and Characteristics of the 'Cone of Silence' in Near Field Magnetic Induction Communications." *Journal of Battlefield Technology* March (2010).

[12] Sun, Z., and I.F. Akyildiz. "Underground Wireless Communication Using Magnetic Induction." in: *Proceedings of the IEEE ICC 2009,* Dresden, Germany, June (2009).

[13] Akyildiz, I.F., Z. Sun, and M. C. Vura. "Signal Propagation Techniques for Wireless Underground Communication Networks." *Physical Communication* 2: 167–183 (2009).

[14] Syms, R.R.A., E. Shamonina, and L. Solymar. "Magneto-Inductive Waveguide Devices." in *Proceedings of IEE Microwaves, Antenna and Propagation* 153, no. 2: 111–121 (2006).

[15] Syms, R.R.A., and L. Solymar. "Bends in Magneto-Inductive Waveguides." *Metamaterials* (2010).

[16] Syms, R.R.A., L. Solymar, IR Young, and T Floume. "Thin-Film Magneto-Inductive Cables." *Journal of Physics D* 43 (2010).

[17] Syms, R.R.A., L. Solymar, and I.R. Young. "Three-Frequency Parametric Amplification in Magneto-Inductive Ring Resonators." *Metamaterials* 2: 122–134 (2008).

[18] Syms, R.R.A., I.R. Young, and L. Solymar. "Low-Loss Magneto-Inductive Waveguides." *Journal of Physics D: Applied Physics* 39: 3945–3951 (2006).

[19] Syms, R.R.A., O. Sydoruk, E. Shamonina, and L. Solymar. "Higher Order Interactions in Magneto-Inductive Waveguides." *Metamaterials* 1: 44–51 (2007).

[20] Masihpour, M., and J.I. Agbinya. "Cooperative Relay in Near Field Magnetic Induction: A New Technology for Embedded Medical Communication Systems." in Proceedings of IB2Com, Malaga, Spain, December 15–18 (2010).

[21] Kalinin, V.A., K.H. Ringhofer, and L. Solymar. "Magneto-Inductive Waves in One, Two and Three Dimensions." *Journal of Applied Physics* 92, no. 10: 6252–6261 (2002).

8

Directional Tuning/Detuning Control of Wireless Power Pickups

This chapter presents an algorithm named directional tuning/detuning control (DTDC) for power flow regulation of wireless power transfer (WPT) systems. The controller regulates the power being delivered to the pickup by deliberately tuning/detuning the center frequency of the pickup tuning circuit with the operating frequency according to the load demand and circuit parameter variations. This is essentially achieved by controlling the duty cycle of a switch-mode tuning capacitor in the pickup resonant tank, which in turn gives a desired equivalent tuning capacitance. The controller allows the wireless pickup to operate with full-range tuning and eliminates the tedious fine-tuning process associated with traditional fixed tuning methods. This therefore eases the components selection of WPT system design and allows the system to have higher tolerance in circuit parameter variations.

8.1 Introduction

Wireless power transfer (WPT) technology has successfully gained a wide range of applications owing to its unique capability of contactless power transfer [1–6]. However, one of the major technical problems that has limited its further development is the lack of efficient, accurate, and fast power-flow control (PFC) of the system [7, 8]. Circuit parameter variations such as load change, magnetic coupling variations between the primary and secondary coils, and the system operating frequency drift can all cause the output voltage of wireless power pickups to deviate from the desired operating condition and fail to meet the load demand where a constant and stable output voltage is required [9–13]. The PFC of WPT system has therefore been a major focus of research and development in this field.

Generally, the PFC of a WPT system can be applied in either a primary power supply or secondary power pickups. For the primary control, it is commonly achieved by regulating either the input voltage or the operating frequency of the primary resonant converter to vary the pickup output voltage [14–16]. Despite the fact that controlling the primary side can minimize the physical size and heat generation of the pickup which is desirable in applications such as biomedical implantable devices, it is only suited to single-pickup situations. This chapter is about directional tuning/detuning control on individual power pickups, so it can be used for multiple-load wireless power transfer applications.

8.1.1 Shorting Control

Shorting control is a most commonly used PFC method due to its simplicity [17–20]. Figure 8.1 shows a general structure of a pickup with shorting control.

From Figure 8.1, it can be seen that the pickup is magnetically coupled to the primary track that carries a high frequency current of I_P, with a mutual inductance of M. An open-circuit voltage V_{OC} of the pickup coil (with a self-inductance of L_S) is induced under such a condition and can be expressed by:

$$V_{OC} = j\omega_0 M I_P \tag{8.1}$$

where ω_0 is the nominal operating frequency. A tuning capacitance C_S is generally required to be fine-tuned to the nominal operating frequency so that the power transfer capacity of the WPT system can be maximized. The dc inductor L_{DC} and the dc capacitor C_{DC} are used to filter off the high-frequency components of the rectified voltage for providing a smooth dc output voltage to the load. A controller, a shorting switch S_{SC}, and a freewheeling diode D_{DC} form the control circuitry for regulating the output

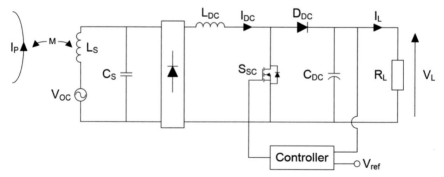

Figure 8.1   General structure of an LC power pickup with shorting control.

voltage V_L by switching on/off S_{SC} to obtain a desired average load current. In a PWM-controlled pickup, if I_{DC} is continuous, the load current I_L is expressed as:

$$I_L = D_S I_{DC} = \frac{D_S I_{SC}}{\sqrt{2}} \qquad (8.2)$$

where D_S is the duty cycle of the PWM signal for the shorting switch S_{SC}, and I_{SC} is the short-circuit current of the pickup coil which can be expressed by:

$$I_{SC} = \frac{V_{OC}}{j\omega_0 L_S} \qquad (8.3)$$

Considering $V_L = I_L R_L$, the output power of the parallel-tuned pickup can be determined by:

$$P_L = V_L I_L = (D_S I_{DC})^2 R_L \qquad (8.4)$$

Despite the fact that the shorting control is easy to implement and can effectively control the power flow of the pickups, it however suffers from inefficient operations when R_L is lightly loaded or at no-load conditions, i.e., large load resistance or open-circuited. In some WPT applications where loads may enter into sleep mode, the power demands of these loads are usually very low which requires the duty cycle D_S to be small for maintaining low output power. In this situation, apart from the switching losses associated with the hard-switching operation, the switch S_{SC} will need to take most of or the complete short-circuit current of the pickup according to $(1-D_S)$. As such, the pickups may require large heat sinks for dissipating significant conduction losses which result in larger size of the pickup.

Low immunity to parameter variations is another disadvantage of using the shorting control. As the design philosophy of shorting control is to obtain the maximum power transfer capacity for each pickup to cover the full-range of their load variations, parameter variations in operating frequency, magnetic coupling, tuning capacitance, etc., are either normally ignored or roughly estimated to be included in the covering range of shorting control, which limited the pickups to work in applications having fixed coupling geometry, e.g. monorail systems, and high-quality track current.

Moreover, a tuned circuit such as the conventional parallel-tuned power pickup has a circuit sensitivity problem under high Q operation. Using high Q in the pickups is very advantageous for some WPT applications since the pickup output voltage can be boosted to a sufficient level without the need of increasing the magnitude and frequency of track current or more number of turns in the pickup coil windings, which essentially reduces the size and cost of the overall system. However, the pickup circuit sensitivity increases with the Q factor and can cause a significant power fluctuation under the effects of circuit parameter variations. Figure 8.2 illustrates the output power behavior of a parallel-tuned pickup with different Q factors.

The center frequency drift is used here as an example to illustrate how the secondary output power gets affected by the parameter variations under different Q values. There are two major parameters that could cause the secondary power pickups to detune, and they are the secondary tuning capacitance and the primary operating frequency. Generally, the tuning capacitance can be affected by the changes in ambient temperature during operations and causes the center frequency of the secondary resonant tank to drift from the nominal frequency f_0 to f_d (or to similar position on the other side) [9]. The variations in primary operating frequency have similar detuning effect on the secondary pickups. However, instead of having the same nominal frequency f_0 and a drifted resonant tank frequency f_d, the actual nominal frequency f_0 may be shifted to f_d as shown in Figure 8.2 with faded dashed lines and left the pickups being detuned from the position of the original nominal operating frequency. Depending on the value of Q factor, the power fluctuation can be quite different. If the pickup is operating with low Q, the change in output power may be negligible, whereas the power drop becomes significant when the pickup is operating with high Q. In a practical WPT system design, the detuning of the pickup is unlikely to be avoided during operations even the tuning circuit is initially tuned to a constant primary operating frequency.

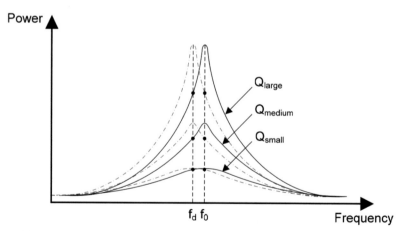

Figure 8.2   Output power behavior of parallel-tuned pickup using different Q factors under operating frequency variations.

Therefore, in order to prevent the system from having significant power fluctuation, the Q factor is normally kept below 10 or lower. Such a design allows the pickup to have better tolerance to some of the parameter variations but a reduced voltage boosting capability is inevitable as a trade-off.

8.1.2 Dynamic Tuning/Detuning Control

An alternate control method which has been investigated to improve the PFC of WPT systems is the dynamic tuning/detuning control technique proposed in [21–24]. Instead of fine-tuning the pickups to obtain the maximum power and then regulating it to meet the load requirements, this control strategy dynamically changes the power transfer capacity of the system to meet the load demands by deliberately mismatching the center frequency of the pickup tuning circuit with the operating frequency.

Figure 8.3 shows the general structure of a dynamic tuning-/detuning-controlled power pickup. It can be seen that a variable-tuning component, e.g., C_{S_v}, is adopted here to change the tuning condition of the pickup. The output voltage is used as a feedback signal for generating control signals to vary the capacitance C_{S_v}. The actual tuning capacitance C_S in the resonant tank is the summation of the fixed tuning capacitor C_{S_f} and the variable capacitance of C_{S_v}. Hence, by changing the value of C_{S_v}, the tuning condition of the pickup can be consequently changed according to $\omega_S = 1/\sqrt{L_S\left(C_{S\_f} + C_{S\_v}\right)}$, where ω_S is the center frequency of the pickup resonant tank. Figure 8.4 shows the relationship between the output power and center frequency variations of the pickup. From Figure 8.4, it can be seen that the maximum power is obtained

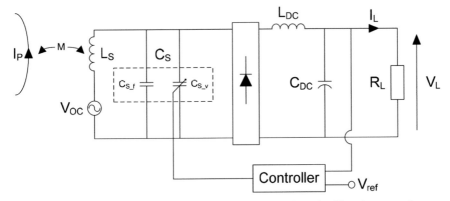

Figure 8.3 Parallel-tuned power pickup with dynamic tuning/detuning control.

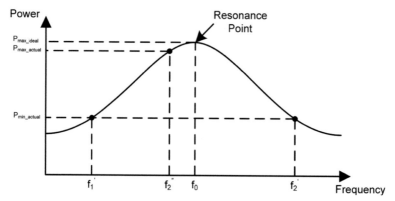

Figure 8.4   Relationship between pickup output power and center frequency variations of pickup tuning circuit.

when the pickup tuning circuit tunes to the nominal operating frequency f_0, and lower power is achievable by having a center frequency other than the nominal operating value. The center frequencies f_1' and f_2' are determined by the equivalent tuning capacitances $(C_{S\_f} + C_{S\_v})$ and C_{S_f}, respectively, and the output power is controllable within the range of these two frequencies [9, 25]. A simple PI controller is often employed to generate control signals for changing the equivalent tuning capacitance. However, since the operating frequency has a parabolic relationship with the power, the adopted PI controller would not be able to achieve full-range tuning control and constrained to only operate in one of the two tuning regions that are separated by the nominal frequency f_0. The frequency f_2'' shown in Figure 8.4 indicates the actual available maximum operating frequency instead of the ideal f_2'. Note that the frequency f_2'' is different from the nominal frequency f_0 by a safety margin, and this safety margin is used to prevent the PI tracking process from traversing between the two tuning regions and results in control failure.

In practice, prior knowledge to the operating region shown in Figure 8.5 is needed for selecting the values of C_{S_f} and C_{S_v}. By selecting one of the two operating regions from the above tuning curve and defining an operating range, a single-side tuning control can be achieved. However, since a normal PI controller is unable to perform tracking for the entire bell-shaped curve, the maximum power delivery is therefore impossible in this case. As a result, the directional tuning/detuning method is proposed to 1) adapt to the bell-shaped curve and gain the maximum power transfer capability when needed and 2) be able to track the closest reference point in either side of the tuning curve for achieving full-range tuning control.

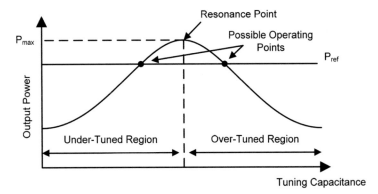

Figure 8.5 Relationship between tuning capacitance and output power of pickup.

8.2 Directional Tuning/Detuning Control (DTDC)

To overcome the problems associated with the existing power pickup control methods such as shorting control, dynamic tuning/detuning control, a control algorithm named directional tuning/detuning control (DTDC) has been developed to enable the pickup to achieve full-range tuning control [26–28].

8.2.1 Fundamentals of DTDC

In order to develop a controller which is capable of regulating the pickup output voltage under different operating conditions, all possible correct tracking processes on the tuning curve need to be thoroughly studied and compared.

The tuning curve therefore has been divided into four different operating regions such as A, B, C, and D as shown in Figure 8.6. Each operating region of the tuning curve is defined as follows:

- Region A—$V(t_n) < V_{ref}$, and the pickup is operating in the under-tuned region.
- Region B—$V(t_n) > V_{ref}$, and the pickup is operating in the under-tuned region.
- Region C—$V(t_n) > V_{ref}$, and the pickup is operating in the over-tuned region.
- Region D—$V(t_n) < V_{ref}$, and the pickup is operating in the over-tuned region.

Note that the operating points shown in Figure 8.6 correspond to different values of tuning capacitance, and by changing the tuning capacitance, the

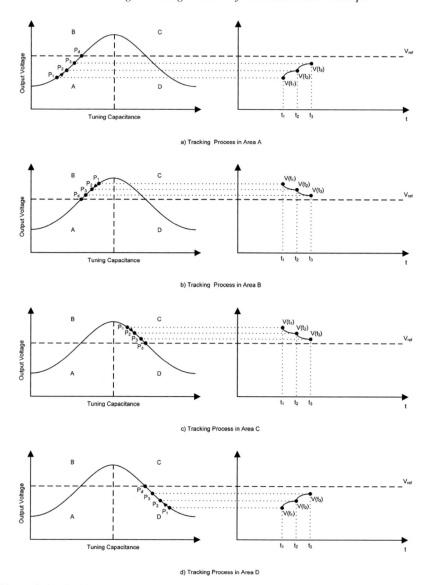

Figure 8.6 Tracking process of coarse-tuning stage in different areas of tuning curve and corresponding time-domain results.

operating point can be shifted which would result in different output voltages. For example, by shifting operating point from P_1 to P_2, the output voltage will change from $V(t_1)$ to $V(t_2)$.

The tracking process in these regions can then be divided into stages of coarse tuning and fine-tuning. The coarse-tuning stage generally takes place immediately after the initialization of the controller (start-up of the pickup circuit) or when the output voltage is considerably deviated from the reference voltage. And once the output voltage reaches close to the reference voltage, the circuit will be fine-tuned to approach it.

8.2.2 Coarse-Tuning Stage

8.2.2.1 Coarse tuning in region A

The tracking process in the region A can be clearly observed from the time-domain illustration of Figure 8.6a. It can be seen that between any of the two successful sampling instances, if the voltage lies exclusively in one of the operating regions, the sign of the error remains unchanged. For example, if between two successful sampling instances t_1 and t_2, the voltage $V(t_1)$ and $V(t_2)$ are both in the operating region A, then the error remains positive. Therefore, the error signal defined as:

$$e(t_n) = V_{ref} - V(t_n) \qquad (8.5)$$

where t_n is the sampling instance, is always positive in the region A, and this error goes on decreasing as time progresses. In addition, the rate of error defined as:

$$\dot{e}(t_n) = e(t_n) - e(t_{n-1}) \qquad (8.6)$$

is always negative in this region. To obtain a correct tracking in the region A by brining the operating point P_1 to P_d, the tuning parameter, e.g., tuning capacitance (TC) should be increased from the initial starting point to the value that corresponds to the desired output voltage.

8.2.2.2 Coarse tuning in region B

In the region B, it can be seen from Figure 8.6b that the error signal is always negative and decreases as the time progresses. The rate of error in this case is always positive. To perform a correct tracking in the region B, the TC needs to be decreased from point P_1 to P_d.

8.2.2.3 Coarse tuning in region C

In the region C, the error signal which has been observed from Figure 8.6c is always negative and decreases as the time progresses. The rate of error in this case is always positive. It can be seen that the error and rate of error are identical in both regions B and C, and the only difference between them is that to perform a correct tracking in the region C, the TC needs to be increased from point P_1 to P_d.

8.2.2.4 Coarse tuning in region D

In the region D, the error signal which has been observed from Figure 8.6d is always positive and decreases as the time progresses. The rate of error in this case is always negative. It can be seen that the error and rate of error are identical in both regions A and D, and the only difference between them is that to perform a correct tracking in the region D, the TC needs to be decreased from point P_1 to P_d.

8.2.3 Fine-Tuning Stage

Figure 8.7 shows where the fine-tuning stage happens in the tuning curve and how it appears in the time domain. As it can be seen, the fine-tuning process can occur between any two possible operating points shown in the figure. Note that the coarse-tuning stage and fine-tuning stage are represented by the shaded and unshaded area €, respectively. The area € is a variable whose position and size are dependent on the positions of P_1 and P_2.

8.2.3.1 Fine-tuning between regions A and B

The term "fine-tuning stage" here means the controller is at a phase where it has just ended the tracking process of coarse tuning (shaded area) and started to fine-tune the circuit to reach the reference point within the area €. In the tracking process of fine-tuning between the regions A and B, there are two possible situations that need to be considered, i.e., whether the operating point is approaching toward the reference from region A or from region B. If the two consecutively sampled voltages are taken from region A across to region B, a negative error and a positive error are always obtained at the two successful sampling instances t_n and t_{n-1}, respectively. The rate of error under such a condition is always negative for both the illustrations shown in Figure 8.7a and b between t_1 and t_2. The correct tuning action for this particular case is to decrease TC. On the other hand, if the two consecutively sampled voltages are taken from region B across to region A as shown in Figure 8.7a and b between t_1' and t_2', a positive error and a negative error are always obtained at the two successful sampling instances t_n and t_{n-1}, respectively. The rate of error under such a condition is always positive. The correct tuning action for this particular case is to increase TC.

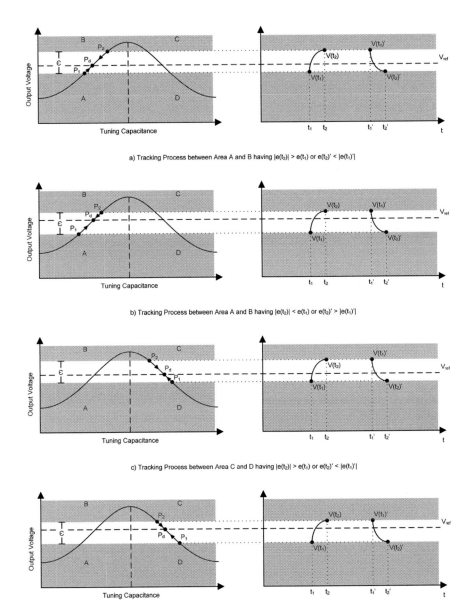

a) Tracking Process between Area A and B having $|e(t_2)| > e(t_1)$ or $e(t_2)' < |e(t_1)'|$

b) Tracking Process between Area A and B having $|e(t_2)| < e(t_1)$ or $e(t_2)' > |e(t_1)'|$

c) Tracking Process between Area C and D having $|e(t_2)| > e(t_1)$ or $e(t_2)' < |e(t_1)'|$

d) Tracking Process between Area C and D having $|e(t_2)| < e(t_1)$ or $e(t_2)' > |e(t_1)'|$

Figure 8.7 Tracking process of fine tuning stage in different areas of tuning curve and corresponding time-domain results.

8.2.3.2 Fine-tuning between regions C and D

In the tracking process of fine-tuning stage between the regions C and D, it can also be separated into two different situations, depending on which direction the operating points are approaching the reference. If the two consecutively sampled voltages are taken from region D across to region C, a negative error and a positive error are always obtained at the two successful sampling instances t_n and t_{n-1}, respectively. The rate of error under such a condition is always negative for both the illustrations shown in Figure 8.7c and d between t_1 and t_2. The correct tuning action for this particular case is to increase TC. On the contrary, if the two consecutively sampled voltages are taken from region C across to region D as shown in Figure 8.7c and d between t_1' and t_2', a positive error and a negative error are always obtained at the two successful sampling instances t_n and t_{n-1}, respectively. The rate of error under such a condition is always positive. The correct tuning action for this particular case is to decrease TC.

Table 8.1 summarizes the results in the tracking process of coarse and fine-tuning stages. Comparisons have been made between similar operations as follows: region of A and D, region of B and C, region of A-to-B and D-to-C, and region of B-to-A and C-to-D. It has been found that the error in the regions A and D is always positive, whereas it is negative in regions B and C. In addition, the rate of error in the regions A and D is always negative, whereas it is positive in regions B and C. It is therefore impossible to distinguish the difference between pickup being operated in regions A and

Table 8.1   Results in tracking process of coarse and fine-tuning stages

Coarse Tuning	$e(t_n) >$ $e(t_{n-1})$	$\dot{e}(t_n)$	$V(t_n) >$ $V(t_{n-1})$	$TC(t_n) >$ $TC(t_{n-1})$
Region A	0	Negative (−)	1	1
Region B	1	Positive (+)	0	0
Region C	1	Positive (+)	0	1
Region D	0	Negative (−)	1	0
Fine-Tuning				
Region A-to-B Outcome1	0	Negative (−)	1	0
Region B-to-A Outcome1	1	Positive (+)	0	1
Region C-to-D Outcome1	1	Positive (+)	0	0
Region D-to-C Outcome1	0	Negative (−)	1	1

D or regions B and C by using the error signal alone. However, by including the immediate past tuning result, it then becomes possible to distinguish the exact operating region of the pickup.

The following conditions are given for determining the operating region of the pickup:

- Region A: $V(t_n) < V_{ref}$ and $TC(t_n) > TC(t_{n-1})$ and $V(t_n) > V(t_{n-1})$, or $V(t_n) < V_{ref}$ and $TC(t_n) < TC(t_{n-1})$ and $V(t_n) < V(t_{n-1})$.
- Region B: $V(t_n) > V_{ref}$ and $TC(t_n) > TC(t_{n-1})$ and $V(t_n) > V(t_{n-1})$, or $V(t_n) > V_{ref}$ and $TC(t_n) < TC(t_{n-1})$ and $V(t_n) < V(t_{n-1})$.
- Region C: $V(t_n) > V_{ref}$ and $TC(t_n) > TC(t_{n-1})$ and $V(t_n) < V(t_{n-1})$, or $V(t_n) > V_{ref}$ and $TC(t_n) < TC(t_{n-1})$ and $V(t_n) > V(t_{n-1})$.
- Region D: $V(t_n) < V_{ref}$ and $TC(t_n) > TC(t_{n-1})$ and $V(t_n) < V(t_{n-1})$, or $V(t_n) < V_{ref}$ and $TC(t_n) < TC(t_{n-1})$ and $V(t_n) > V(t_{n-1})$.

From the above conditions, it can be seen that the under-tuned and over-tuned regions are differentiated by the immediate past tuning result in terms of the tuning direction of the TC and moving direction of the output voltage. In addition, the upper and lower regions of each tuning region are differentiated by their relative position to the reference voltage. To correctly track the desired output voltage from different operating regions of the tuning curve, the possible operating conditions are given as:

- Condition 1: If $V(t_n) < V_{ref}$ and $TC(t_n) < TC(t_{n-1})$ and $V(t_n) < V(t_{n-1})$, then $TC(t_{n+1})$ should be increased.
- Condition 2: If $V(t_n) < V_{ref}$ and $TC(t_n) > TC(t_{n-1})$ and $V(t_n) < V(t_{n-1})$, then $TC(t_{n+1})$ should be decreased.
- Condition 3: If $V(t_n) < V_{ref}$ and $TC(t_n) < TC(t_{n-1})$ and $V(t_n) > V(t_{n-1})$, then $TC(t_{n+1})$ should be decreased.
- Condition 4: If $V(t_n) < V_{ref}$ and $TC(t_n) > TC(t_{n-1})$ and $V(t_n) > V(t_{n-1})$, then $TC(t_{n+1})$ should be increased.
- Condition 5: If $V(t_n) > V_{ref}$ and $TC(t_n) < TC(t_{n-1})$ and $V(t_n) < V(t_{n-1})$, then $TC(t_{n+1})$ should be decreased.
- Condition 6: If $V(t_n) > V_{ref}$ and $TC(t_n) > TC(t_{n-1})$ and $V(t_n) < V(t_{n-1})$, then $TC(t_{n+1})$ should be increased.
- Condition 7: If $V(t_n) > V_{ref}$ and $TC(t_n) < TC(t_{n-1})$ and $V(t_n) > V(t_{n-1})$, then $TC(t_{n+1})$ should be increased.
- Condition 8: If $V(t_n) > V_{ref}$ and $TC(t_n) > TC(t_{n-1})$ and $V(t_n) > V(t_{n-1})$, then $TC(t_{n+1})$ should be decreased.

By summarizing the above possible operating conditions and categorizing the results once again into the coarse and fine-tuning stages, a truth table for determining the tuning direction of TC can be given as shown in Table 8.2.

Instead of depending solely on the error signals to generate the control signals as done by the conventional PI controllers, the DTDC includes the validity of the previous control action into considerations. To further simplify the results of Table 8.2, a Boolean expression can be derived as:

$$S_4 = S_3 \left(S_1 \oplus S_2 \right) + \overline{S_3} \left(S_1 \equiv S_2 \right) \tag{8.7}$$

where S_4 is the signal in the control algorithm for determining the tuning direction with logic 1 to increase or logic 0 to decrease the TC. The actual output signal of the controller can therefore be expressed by:

$$U(t_n) = U(t_{n-1}) + (-1)^{S_4+1} \cdot \Delta H(t_n) \tag{8.8}$$

where $U(t_n)$ is the present-state control signal, $U(t_{n-1})$ is the previous-state control signal, and $\Delta H(t_n)$ is the step-size of the adjustment in the present state.

8.2.4 Design and Performance Considerations of DTDC

After each tuning action being taken, the pickup tuning circuit would require a certain period of time (time constant of pickup circuit) for the output voltage to stabilize. Since the complete effect of each control action on the output voltage is required to be fully observed for a proper validity check before the controller can take the next step, the controller therefore has to wait for the

Table 8.2 Truth table for determining tuning direction signal S_4

Coarse Tuning	$V(t_n) > V_{ref}$ $S_1(t_n)$	$V(t_n) >$ $V(t_{n-1})$ $S_2(t_n)$	$TC(t_n) >$ $TC(t_{n-1})$ $S_3(t_n)$ or $S_4(t_{n-1})$	$TC(t_{n+1}) >$ $TC(t_n)$ $S_4(t_n)$
A (Cond. 4)	0	1	1	1
B (Cond. 5)	1	0	0	0
C (Cond. 6)	1	0	1	1
D (Cond. 3)	0	1	0	0
Fine-Tuning				
A-to-B (Cond. 8)	1	1	1	0
B-to-A (Cond. 1)	0	0	0	1
C-to-D (Cond. 2)	0	0	1	0
D-to-C (Cond. 7)	1	1	0	1

pickup circuit to reach its steady state. This makes the selection of sampling frequency for the pickup output voltage very important as it may significantly affect the controller performance. However, as the time constant of different pickups with different circuit configurations and parameters may be different, the selection of sampling frequency will need to be individually determined case by case, either through simulations or experiments.

Another important design factor that affects the performance of the controller is the tuning step-size. From Figure 8.8, it can be seen that the output voltage can be controlled by operating the pickup either in the under-tuned or in the over-tuned region from the initial operating point A or E, which corresponds to the initial TC of TC_0 or TC_0'. For the DTDC to perform an efficient control so the operating point can be stabilized to the desired voltage with the shortest tuning distance, the operating points of the circuit should not jump across the two regions during any two consecutive samples. At the initial controller startup, this can be ensured if the value of each tuning adjustment does not exceed a maximum TC variation range of Δh_m. This can vary, depending on the circuit characteristics and required dynamic performance. But its theoretical maximum should be confined by half of the distance between the initial operating point (A or E) and the maximum point (C).

Once the maximum tuning step-size is determined, the controller can use it for coarse-tuning adjustment, but to achieve satisfactory control result with

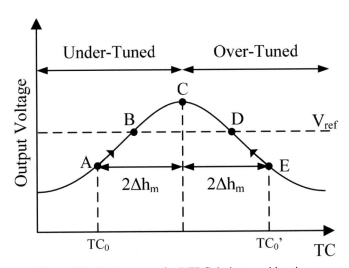

Figure 8.8 Tuning curve for DTDC design considerations.

stable and error-free output, the value of Δh during operations needs to be variable for fine tuning process. This is due to the fact that a large fixed value of Δh can allow the system to have fast response but may cause the output voltage to have large oscillations around the reference voltage, whereas a small fixed value of Δh makes the system sluggish but leads to a more stable output. A judicious compromise between these two is therefore needed and can be obtained by algorithmically changing Δh at each sampling instant t_n as follows:

$$\begin{cases} \Delta h(t_n) = \Delta h_m = \alpha \cdot \Delta H_m & \text{Coarse tuning}, |e(t_n)| > \varepsilon \\ \Delta h(t_n) = \Delta h = \alpha \cdot [\Delta H(t_{n-1}) - \beta \cdot \Delta H_m] & \text{Fine tuning}, |e(t_n)| \leqslant \varepsilon \end{cases} \quad (8.9)$$

where α is a scaling factor (depending on the physical controller design) between the step-size of control signal ΔH and the physical tuning step-size Δh, and β is a scaling factor less than one.

The above-introduced method is named Simple Step-Size Adjustment (SSSA), and it can be modified or substituted by different algorithms to fulfil the desired output requirements [29]. The determination for the value of β is however difficult since the behavior of pickup circuit may be constantly changing due to variations of the circuit parameters. Therefore, it can only be designed through heuristic method if the SSSA is used.

The decision on whether the controller should be performing coarse tuning or fine-tuning is purely dependent on S_1 and S_2 at any two consecutive sampling instances. If the controller is in the coarse-tuning stage, then the logic signal $S_1 S_2$ (from Table 8.2) should either be 01 or 10 at any two consecutive sampling instances. On the contrary, if the controller is in the fine-tuning stage, then $S_1 S_2$ should be switching in-between 00 and 11 at any two consecutive sampling instances. Under such an assumption, the control result of applying SSSA can lead to two different outcomes which are the ideal and the indefinite control result. Figure 8.9 shows the two possible outcomes when the SSSA is applied. It can be seen from Figure 8.9a that if an ideal control result is obtained, the controller would bring the output voltage gradually toward V_{ref} and eventually reaches V_{ref} by reducing the tuning step-size according to Equation (8.9). The reduction in the tuning step-size occurs at each time when the output voltage traverses the voltage reference. The second possible outcome is shown in Figure 8.9b. The controller in this case reduces the tuning step-size (Δh_m to Δh) at t_2 since the output voltage has entered the fine-tuning stage, which allows the output voltage

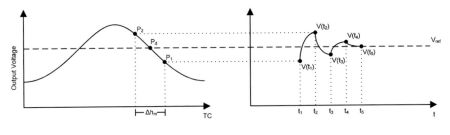

a) Ideal control result after applying SSSA

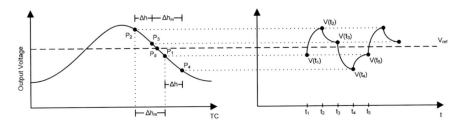

b) Indefinite control result after applying SSSA

Figure 8.9 Possible control results of applying SSSA.

at the next time instance move toward the voltage reference (P_2 to P_3). However, as the output voltage at t_3 does not traverse the voltage reference but remains above it, the controller switches back to the coarse-tuning mode and uses Δh_m as the tuning step-size. The mode switching then results in an infinite looping between the coarse-tuning and fine-tuning stages of the control algorithm and causes the output voltage to never reach the desired value, which eventually leads to an indefinite control result. In order to solve this problem, a prior knowledge of whether the output voltage has entered the fine-tuning stage is required so that the tuning step-size can be adjusted correctly.

Table 8.3 shows the signals for determination of Δh adjustment. From Table 8.3, all possible cases can be classified into one of the three different categories as below:

8.2.4.1 Category I
This includes case 1, 2, 15, and 16. Regardless of the result of S_6, the signal S_1, S_2, and S_5 have implied that the tuning direction from the previous control action is incorrect and therefore requires Δh to have the coarse-tuning value (represented by logic 1) in the present state.

Table 8.3   Truth table for Δh adjustment determination (S_7)

Possible Cases	$V(t_n) > V_{ref}\ S_1(t_n)$	$V(t_n) > V(t_{n-1})\ S_2(t_n)$	$V(t_{n-1}) > V_{ref}\ S_5(t_n)$	$V(t_{n-2}) > V_{ref}\ S_6(t_n)$	$S_7(t_n)$
Case 1	0	0	0	0	1
Case 2	0	0	0	1	1
Case 3	0	0	1	0	0
Case 4	0	0	1	1	0
Case 5	0	1	0	0	1
Case 6	0	1	0	1	0
Case 7	0	1	1	0	N/A
Case 8	0	1	1	1	N/A
Case 9	1	0	0	0	N/A
Case 10	1	0	0	1	N/A
Case 11	1	0	1	0	0
Case 12	1	0	1	1	1
Case 13	1	1	0	0	0
Case 14	1	1	0	1	0
Case 15	1	1	1	0	1
Case 16	1	1	1	1	1

8.2.4.2 Category II

This includes case 3, 4, 13, and 14. Regardless of the result of S_6, the signal S_1, S_2, and S_5 have implied that the output voltage has entered the fine-tuning stage and therefore requires Δh to be reduced (represented by logic 0) in the present state.

8.2.4.3 Category III

This includes case 5, 6, 11, and 12. The signal S_1, S_2, and S_5 in this category have shown that the tuning direction from the previous control action is correct, and whether the controller is in the coarse-tuning or fine-tuning stage is completely dependent on the location of $V(t_{n-2})$. As a result, both the case 5 and 12 belong to tracking stage and requires Δh to have the coarse-tuning value. Conversely, the case 6 and 11 belong to the fine-tuning stage and requires Δh to be reduced.

Note that the result of case 7, 8, 9, and 10 are marked with N/A, since these cases are impossible to happen. By summarizing the obtained result in Table 8.3, a Boolean expression for determining whether the pickup should be coarse-tuned or fine-tuned (logic 1 for coarse tuning and logic 0 for fine-tuning) can be derived as:

$$S_7 = (S_1 \equiv S_2 \equiv S_5) + (S_1 \equiv S_5 \equiv S_6) \qquad (8.10)$$

8.2.5 Standard Procedure of DTDC

To effectively control the pickup output voltage using DTDC, a standard procedure for executing the algorithm has been recommended as shown in Figure 8.10. The procedure starts with the initialization of the algorithm. In this process, the controller initializes the settings according to the user specifications and the system characteristic that the controller is applied to. These include sampling time of the controller, initial state of each processing block (shown in the square blocks), maximum increment (or decrement) level of the controller's output, and the hysteresis band around the voltage reference.

After the initialization, the first sampled output voltage $V(t_n)$ would be stored in the memory blocks with an unit delay in each one of them before outputting to the next processing block, and this allows the controller to have the memory of present output voltage after two executions of the algorithm. For example, $V(t_n)$ will become $V(t_{n-2})$ in the first memory block after the first execution and then become $V(t_{n-2})$ in the second memory block after the second execution. Note that in order to perform a proper reference tracking at the initial stage, the controller should not provide tuning adjustments to the pickup anytime before the third execution of the algorithm since the knowledge of $V(t_{n-2})$ and $V(t_{n-2})$ are still lacking in the first and second executions.

With the values of $V(t_{n-1})$ and $V(t_{n-2})$ being obtained, the present sampled output voltage $V(t_n)$ can then be compared to V_{ref} and $V(t_{n-1})$ for generating the logic signals of $S_1(t_n)$ and $S_2(t_n)$ respectively, and $V(t_{n-1})$ and $V(t_{n-2})$ can be compared to V_{ref} for generating the logic signals of $S_5(t_n)$ and $S_6(t_n)$ respectively. The results of $S_1(t_n)$, $S_2(t_n)$, $S_5(t_n)$, and $S_6(t_n)$ would then be used in the two separate branches of the algorithm. In the left branch where the tuning direction of the pickup is determined, the signals of $S_1(t_n)$, $S_2(t_n)$, and $S_3(t_n)$ are employed to obtain the result of $S_4(t_n)$ which has an output of 1 for increasing or 0 for decreasing the value of the controlled tuning component. Note that $S_3(t_n)$ is based on the previous result of $S_4(t_n)$ which has been stored in the memory block and delayed by one unit. In the right branch where the value of $\Delta H(t_n)$ is determined, the signals of $S_1(t_n)$, $S_2(t_n)$, $S_5(t_n)$, and $S_6(t_n)$ are employed to obtain the result of $S_7(t_n)$ which has an output of 1 for coarse or 0 for fine-tuning the pickup.

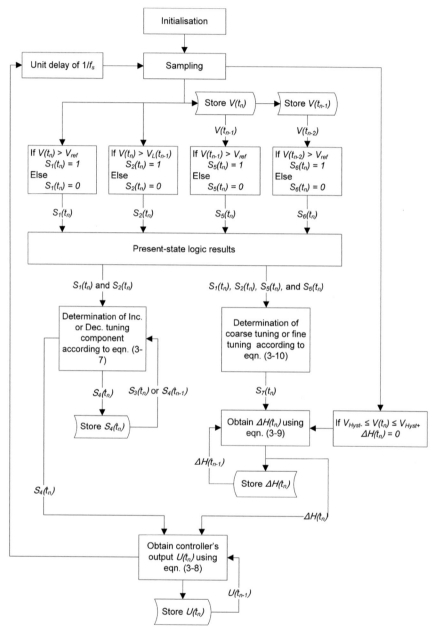

Figure 8.10   Flowchart of standard procedure of DTDC.

Based on the result of $S_7(t_n)$, the actual value of $\Delta H(t_n)$ can then be calculated according to Equation (8.9). Note that the value of $\Delta H(t_n)$ may be based on $\Delta H(t_{n-1})$ depending on the result of $S_7(t_n)$. Therefore, $\Delta H(t_n)$ also needs to be stored in the memory block and delayed by one unit for the next calculation.

It is worth noting that a subroutine is introduced in the standard DTDC to avoid the controller to take further control actions after the output voltage reaches a satisfactory value; for example, if the sampled output voltage lies outside the hysteresis band, then $\Delta H(t_n)$ would still need to be calculated by the main algorithm, and on the contrary, if the output voltage lies inside the hysteresis band, then the value of $\Delta H(t_n)$ would be set to zero.

Finally, by combining the result of $S_4(t_n)$ and $\Delta H(t_n)$, the output $U(t_n)$ of the controller can be calculated using Equation (8.8), and it would be stored in the memory block with an unit delay of $1/f_s$ (f_s represents the sampling frequency of the controller) for the next iteration.

8.3 DTDC-Controlled Parallel-Tuned LC Power Pickup

The parallel-tuned LC circuit is the most commonly seen tuning configuration in the wireless pickups of WPT systems and therefore used here as an example for the DTDC practical implementation.

8.3.1 Fundamentals of Parallel-Tuned LC Power Pickup

The simplified circuit of a parallel-tuned LC power pickup, without considering the ac-dc rectification, is shown in Figure 8.11. As can be seen from the figure, the LC power pickup consists of a secondary pickup coil with a self-inductance of L_S, a tuning capacitance C_S, and a load resistance

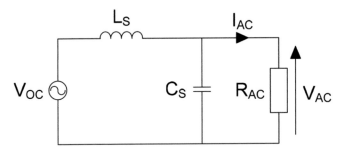

Figure 8.11 Simplified model of parallel-tuned LC power pickup circuit.

R_{AC} to form a simple second-order system. The resistance R_{AC} is the ac-equivalent resistance of the actual dc load resistor R_L, and the conversion ratio between these two resistances under the dc inductance L_{DC} being continuous conducting is $R_{AC}/R_L = \pi^2/8$ [30, 31].

The voltage transfer function of the pickup can be determined from:

$$H_V(s) = \frac{V_{AC}(s)}{V_{OC}(s)} = \frac{\frac{1}{L_S C_S}}{s^2 + \frac{1}{R_{AC}C_S}s + \frac{1}{L_S C_S}} \tag{8.11}$$

Considering $I_{AC}(s) = V_{AC}(s)/R_{AC}$ and $I_{SC}(s) = V_{OC}(s)/sL_S$, the current transfer function of the pickup is obtained from:

$$H_I(s) = \frac{I_{AC}(s)}{I_{SC}(s)} = \frac{\frac{1}{R_{AC}C_S}s}{s^2 + \frac{1}{R_{AC}C_S}s + \frac{1}{L_S C_S}} \tag{8.12}$$

Equations (8.11) and (8.12) can also be expressed in the frequency domain using rectangular form as:

$$H_V(j\omega) = \frac{R_{AC}^2(1 - \omega^2 L_S C_S) - j\omega R_{AC} L_S}{R_{AC}^2(1 - \omega^2 L_S C_S)^2 + \omega^2 L_S^2} \tag{8.13}$$

$$H_I(j\omega) = \frac{\omega^2 L_S^2 + j\omega R_{AC} L_S(1 - \omega^2 L_S C_S)}{R_{AC}^2(1 - \omega^2 L_S C_S)^2 + \omega^2 L_S^2} \tag{8.14}$$

Under fully tuned condition where ω_0 is the nominal operating frequency and $\omega_0^2 L_S C_S = 1$, Equations (8.13) and (8.14) can further be reduced to:

$$H_V(j\omega_0) = \frac{-jR_{AC}}{\omega_0 L_S} \tag{8.15}$$

$$H_I(j\omega_0) = 1 \tag{8.16}$$

The absolute value of Equation (8.15) is also known as the quality factor $Q_{S\text{-}p}$ of the parallel-tuned LC power pickup. From Equation (8.16), it can be seen that the output current is equal to the short-circuit current of the pickup coil under fully tuned condition.

8.3.2 Controllable Power Transfer Capacity of Parallel-Tuned LC Power Pickup

Considering the normalized adjusting ratio of the tuning capacitance is presented by:

$$r_{adj} = \frac{C_S}{C_{S\_\omega_0}} \tag{8.17}$$

where C_S is the actual equivalent tuning capacitance and $C_{S\_\omega_0}$ is the tuning capacitance under fully tuned condition which is equal to $1/\omega_0^2 L_S$, the magnitude of the output voltage and current can be obtained as:

$$V_{AC} = \frac{R_{AC}\sqrt{R_{AC}^2(1-r_{adj})^2 + \omega_0^2 L_S^2}}{R_{AC}^2(1-r_{adj})^2 + \omega_0^2 L_S^2} \cdot V_{OC} \tag{8.18}$$

$$I_{AC} = \frac{\omega_0 L_S \sqrt{R_{AC}^2(1-r_{adj})^2 + \omega_0^2 L_S^2}}{R_{AC}^2(1-r_{adj})^2 + \omega_0^2 L_S^2} \cdot I_{SC} \tag{8.19}$$

By substituting $Q_{S\_p} = R_{AC}/\omega_0 L_S$ into Equations (8.18) and (8.19), the equations can be further simplified as:

$$V_{AC} = K_V \cdot V_{OC} = B_C \cdot Q_{S\_p} V_{OC} \tag{8.20}$$

$$I_{AC} = B_C \cdot I_{SC} \tag{8.21}$$

where K_V is the operational voltage boosting factor of the tuning circuit and equals to $B_C \cdot Q_{S\_p}$. B_C is a newly introduced variable which represents the controllable boosting coefficient and can be determined from:

$$B_C = \frac{\sqrt{Q_{S\_p}^2(1-r_{adj})^2 + 1}}{Q_{S\_p}^2(1-r_{adj})^2 + 1} \tag{8.22}$$

Note that the controllable boosting coefficient has a maximum of unity when r_{adj} equals to 1 and a minimum of 0 when r_{adj} approaches infinity.

Considering the output power of the pickup is $P_{AC} = V_{AC} I_{AC}$, the variable power transfer capacity of the parallel-tuned LC power pickup can therefore be determined from:

$$P_{AC} = \frac{Q_{S\_p}}{Q_{S\_p}^2(1-r_{adj})^2 + 1} \cdot V_{OC} I_{SC} \tag{8.23}$$

8.3.3 Effects of Parameter Variations on Output Voltage of Parallel-Tuned LC Power Pickup

To compensate for the effects of system parameter variations on the pickup output voltage, possible variations need to be identified and taken into considerations of the DTDC design. The most common parameter variations that can be seen in the wireless power pickup include, but not limited to, the variations in the operating frequency, magnetic coupling, and load. In order to compensate for these variations by using the variable-tuning capacitor C_S, this section focuses on the analyses of the relationships between the variable tuning capacitance and the pickup output voltage under these parameter variations.

8.3.4 Operating Frequency Variation

As it has been discussed before, variations in the operating frequency can cause the pickup to detune as well as change the magnitude of open-circuit voltage of the pickup coil. Considering these two factors, the output voltage of the pickup under the variations of the operating frequency can be determined from:

$$V_{AC} = \frac{\sqrt{Q_{S\text{-}p}^2 \left(\frac{1}{\alpha_f} - \alpha_f r_{adj}\right)^2 + 1}}{Q_{S\text{-}p}^2 \left(\frac{1}{\alpha_f} - \alpha_f r_{adj}\right)^2 + 1} \cdot Q_{S\text{-}p} V_{OC} \qquad (8.24)$$

where α_f is the normalized frequency variation index, i.e., α_f equals to 1.1 if the nominal operating frequency ω_0 is +10% higher. From Equation (8.24), it can be seen that the maximum pickup output voltage remains unchanged, but the adjusting ratio r_{adj} for achieving the fully tuned condition has been shifted to:

$$r_{adj\_fvr} = \frac{1}{\alpha_f^2}$$

Figure 8.12 shows the behavior of the pickup output voltage under variations of the operating frequency. It can be seen that the entire tuning curve would be shifted to the left if the operating frequency is increased from its nominal value and to the right if it is decreased from its nominal value.

In order to maintain the output voltage to a constant desired value V_{ref}, the adjusting ratio of the tuning capacitance has to be varied accordingly. By rearranging Equation (8.24) into a quadratic equation with respect to the ratio r_{adj}, solving the equation gives:

$$r_{adj\_fv} = \frac{1}{\alpha_f^2} \cdot \left[1 \pm \frac{\alpha_f \sqrt{\frac{1}{r_k^2} - 1}}{Q_{S\text{-}p}}\right] \qquad (8.25)$$

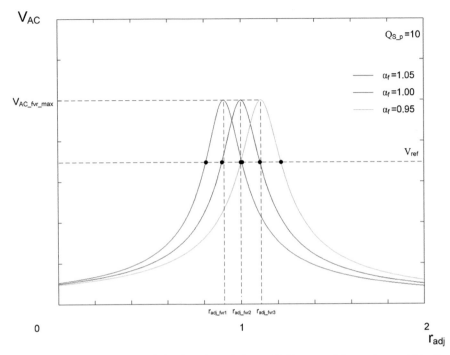

Figure 8.12   Output voltage behavior of parallel-tuned LC power pickup using variable C_S under operating frequency variations.

where r_k is the ratio between the desired voltage V_{ref} and the nominal maximum pickup output voltage under fully tuned condition, and it can be expressed as:

$$r_k = \frac{V_{ref}}{V_{AC\_max}} = \frac{V_{ref}}{Q_{S\_p}V_{OC}} \tag{8.26}$$

8.3.5 Magnetic Coupling Variation

Variations in the magnetic coupling between the primary track and the secondary pickup can also affect the power being delivered to the load. Considering that the normalized magnetic coupling variation index is α_{oc}, and the affected open-circuit voltage is equal to $\alpha_{oc}V_{OC}$, the output voltage of the pickup under the variations of the magnetic coupling can be expressed as:

$$V_{AC} = \frac{\alpha_{oc}\sqrt{Q_{S\_p}^2(1-r_{adj})^2+1}}{Q_{S\_p}^2(1-r_{adj})^2+1} \cdot Q_{S\_p}V_{OC} \qquad (8.27)$$

Figure 8.13 shows the output voltage behavior of the pickup using variable tuning capacitor under magnetic coupling variations. From the figure it can be seen that the shape of the tuning curve and the position of the tuned point remain unchanged, but the level of the tuning curve is vertically shifted by a factor of α_{oc}, and therefore, the maximum output voltage $V_{AC\_ocvr\_max}$ under the magnetic coupling variations is changed according to $\alpha_{oc}Q_{S\_p}V_{OC}$.

   To keep the output voltage constant at a desired level, the required adjusting ratio of the variable tuning capacitor can be obtained from:

$$r_{adj\_ocv} = 1 \pm \frac{\alpha_{oc}\sqrt{\frac{1}{r_k^2}-\frac{1}{\alpha_{oc}^2}}}{Q_{S\_p}} \qquad (8.28)$$

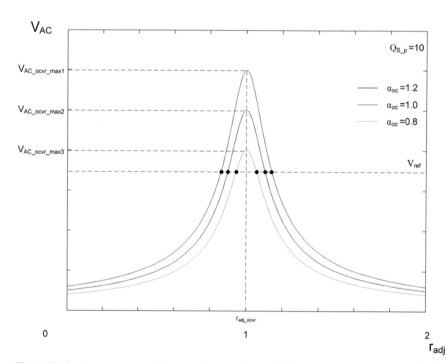

Figure 8.13   Output voltage behavior of parallel-tuned LC power pickup using variable C_S under magnetic coupling variations.

Note that the valid solutions in Equation (8.28) can only be obtained when $r_k \leq \alpha_{oc}$. This means that to allow the controller to successfully maintain the output voltage to be constant, the desired output voltage has to be less than or equal to the minimum of the maximum output voltage ($V_{AC\_ocvr\_max3}$ shown in Figure 8.13) obtained at the tuned point.

8.3.6 Load Variation

Load change is yet another commonly seen parameter variation in many systems. To observe the effects of load variations on the output voltage of the pickup, the normalized load variation index α_r is used here. The output voltage under the load variations can be obtained from:

$$V_{AC} = \frac{\sqrt{Q_{S\_p}^2 \left(1 - r_{adj}\right)^2 + \frac{1}{\alpha_r^2}}}{Q_{S\_p}^2 \left(1 - r_{adj}\right)^2 + \frac{1}{\alpha_r^2}} \cdot Q_{S\_p} V_{OC} \qquad (8.29)$$

Figure 8.14 shows the output voltage behavior of the pickup under load variations. From Figure 8.14, it can be seen that the output voltage behavior in this case is similar to the result of magnetic coupling variations.

The required adjusting ratio for maintaining the desired output voltage under the load variations can be determined from:

$$r_{adj\_rv} = 1 \pm \frac{\sqrt{\frac{1}{r_k^2} - \frac{1}{\alpha_r^2}}}{Q_{S\_p}} \qquad (8.30)$$

Notice that the valid solutions in Equation (8.30) can only be obtained when the condition of $r_k \leq \alpha_r$ is met, which requires the desired output voltage to be less than the minimum of the maximum output voltage ($V_{AC\_rvr\_max3}$ shown in Figure 8.14) obtained at the tuned point.

8.3.7 Operating Range of Variable C$_S$

To fully compensate for the above parameter variations, the integrated effect of these parameter variations on the output voltage needs to be investigated. The pickup output voltage under such a consideration can be expressed by:

$$V_{AC} = \frac{\alpha_{oc} \sqrt{Q_{S\_p}^2 \left(\frac{1}{\alpha_f} - \alpha_f r_{adj}\right)^2 + \frac{1}{\alpha_r^2}}}{Q_{S\_p}^2 \left(\frac{1}{\alpha_f} - \alpha_f r_{adj}\right)^2 + \frac{1}{\alpha_r^2}} \cdot Q_{S\_p} V_{OC} \qquad (8.31)$$

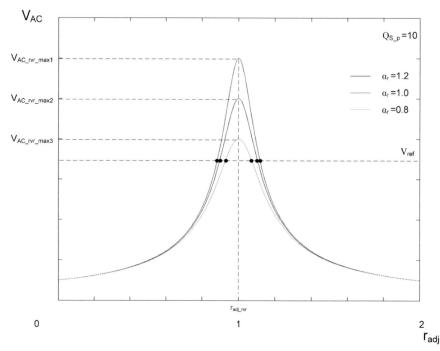

Figure 8.14   Output voltage behavior of parallel-tuned LC power pickup using variable C_S under load variations.

Equation (8.31) can also be rearranged to obtain the required adjusting ratio r_{adj_pv} for achieving the desired output voltage under the integrated effect of the parameters variation, and the required ratio can be determined from:

$$r_{adj\_pv} = \frac{1}{\alpha_f^2} \cdot \left[1 \pm \frac{\alpha_f \alpha_{oc} \sqrt{\frac{1}{r_k^2} - \frac{1}{(\alpha_{oc}\alpha_r)^2}}}{Q_{S\_p}} \right] \tag{8.32}$$

To have valid solutions in Equation (8.32), the following condition has to be met:

$$r_k \leq \alpha_{oc}\alpha_r$$

In order to obtain the operating range of the tuning capacitance C_S, the worst-case scenario (extreme operating condition) of the practical operations has to be considered for obtaining the minimum and the maximum of the adjusting ratio r_{adj_pv}. Based on the results shown in Figures 8.12–8.14, it can be seen that the minimum and maximum of r_{adj_pv} can be calculated by using Equation (8.32) with the following conditions.

8.3.7.1 Maximum required ratio ($r_{adj\_pv\_max}$)

- The parallel-tuned LC power pickup is operating in the over-tuned region of the tuning curves.
- The operating frequency is at *nominal value – maximum allowable tolerance*, and the magnetic coupling and load variations are at *nominal value + maximum allowable tolerance*.

8.3.7.2 Minimum required ratio ($r_{adj\_pv\_min}$)

- The parallel-tuned LC power pickup is operating in the under-tuned region of the tuning curves.
- The considered parameter variations are all at *nominal value + maximum allowable tolerance*.

8.3.8 Implementation of DTDC Controlled Parallel-Tuned LC Power Pickup

The basic structure of the DTDC-controlled parallel-tuned LC power pickup with switch-mode variable-tuning capacitor is shown in Figure 8.15. It consists of the main circuit of a parallel-tuned LC power pickup and a control circuitry to form a complete secondary system. The tuning capacitance C_S here is divided into two parts: the first part is a fixed value capacitor C_{S1} which is used for starting up the pickup circuit, and the second part consists

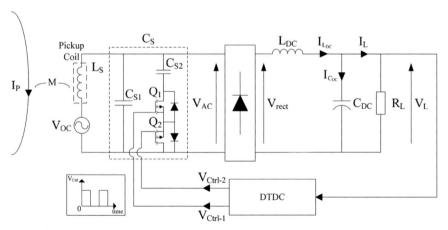

Figure 8.15 Structure of DTDC-controlled parallel-tuned LC power pickup with switch-mode variable-tuning capacitor.

of a capacitor C_{S2} and two MOSFET switches Q_1 and Q_2, which functions as a switch-mode variable capacitor for changing the tuning condition of the pickup. The output voltage V_L is used as a feedback signal to the DTDC for producing switching signals V_{Ctrl_1} and V_{Ctrl_2} with controlled duty cycles so that Q_1 and Q_2 can be turned on/off accordingly for obtaining the desired equivalent capacitance. This eventually allows the pickup to deliver the power as required by the load.

8.3.8.1 Selection of C_{S1} and C_{S2}

To have sufficient power for starting up the controller at the initial stage, the capacitance of C_{S1} is required to be selected according to the location shown in Figure 8.16.

The voltage $V_{startup}$ is the voltage required to startup the control circuitry. The value of C_{S1} can be determined by using Equation (8.32), with the operating frequency, magnetic coupling, and load variations being considered at their minimum tolerable values. This would allow the pickup to start up the control circuitry under the worst-case scenario of the parameters variation at the initial stage.

The operating range of C_S to fully compensate for the parameters variation has been determined previously; however, since the capacitance obtained from the minimum required ratio $r_{adj\_pv\_min}$ may not be equal to the actual value of C_{S1}, the previously calculated operating range can only be regarded as the minimum required operating range. Hence, the actual operating range of C_{S2} is expressed as:

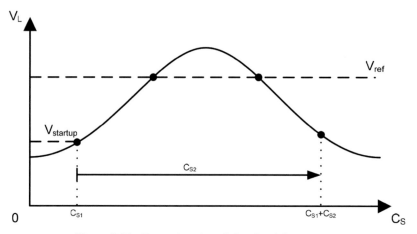

Figure 8.16   Proper location of C_{S1} for pickup startup.

$$C_{S2} = C_{pv\_max} - C_{S1} \qquad (8.33)$$

where C_{pv_max} is the capacitance obtained from the maximum required ratio $r_{adj\_pv\_max}$.

8.3.8.2 Equivalent Capacitance of C_{S2}

For the resonant circuit using an ac-switched capacitor, the waveform of the voltage in the ac tank and across the capacitor can be represented by V_{AC} and V_C respectively, as shown in Figure 8.18. From Figure 8.17, it can be seen that the signal V_{Ctrl-1} and V_{Ctrl-2} control the on/off period of the capacitor in the positive and the negative cycles, respectively. When the capacitor is switched off (at θ), the voltage across the capacitor would be capped at V_{C_off} since there is no more current flowing through the capacitor. The capacitor only gets discharged when V_{AC} is lower than V_C (at $\pi-\theta$). Such a technique can control the amount of electric charges accumulated inside the capacitor and hence achieves a variable equivalent capacitance for changing the tuning condition of the pickup [9, 25].

The relationship between the switching angle and the voltages can be expressed as:

$$\theta = \sin^{-1}\left(\frac{V_{C\_off}}{\hat{V}_{AC}}\right) \qquad (8.34)$$

The switching angle can also be directly related to the duty cycles of the control signal and has a linear relationship of:

$$\theta = \frac{\pi}{100}(D - 50) \qquad (8.35)$$

where D is the duty cycle of the control signal V_{Ctrl-1} and V_{Ctrl-2} in percentage. Considering the electric charges stored inside an ac-switched capacitor are equal to that of using an equivalent capacitance C_{eq}, an equation can be obtained as follows:

$$\int_0^\pi C_{S2}V_C \, d(\omega t) = \int_0^\pi C_{eq}\hat{V}_{AC}\sin(\omega t) \, d(\omega t) \qquad (8.36)$$

By expanding Equation (8.36) into segments according to V_C as shown in Figure 8.17, the following equation can be obtained:

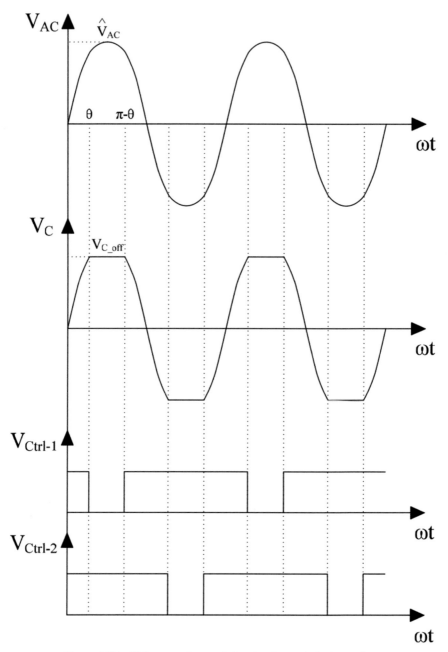

Figure 8.17   Voltages and control signals of ac-switched capacitor.

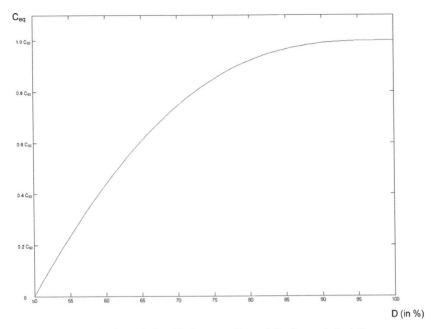

Figure 8.18 Relationship between C_{eq} and D of ac-switched C_{S2}.

$$C_{eq} = \frac{1}{2\hat{V}_{AC}} \left[\int_0^\theta C_{S2}\hat{V}_{AC}\sin(\omega t)\ d(\omega t) + \int_\theta^{\pi-\theta} C_{S2}V_{C\_off}\ d(\omega t) \right.$$
$$\left. + \int_{\pi-\theta}^\pi C_{S2}\hat{V}_{AC}\sin(\omega t)\ d(\omega t) \right]$$

$$(8.37)$$

Solving Equation (8.37), the variable equivalent capacitance which is obtained by varying the duty cycle of control signals is determined from:

$$C_{eq} = C_{S2}\left[1 - \cos\left(\frac{\pi(D-50)}{100}\right) + \pi\left(1 - \frac{D}{100}\right)\sin\left(\frac{\pi(D-50)}{100}\right)\right]$$

$$(8.38)$$

where D has a variation range of 50–100% since the control signals V_{Ctrl-1} and V_{Ctrl-2} are responsible for each 50% of the complete cycle. Figure 8.18 shows the relationship between the equivalent capacitance and the duty cycle. As can be seen from the figure, the variable equivalent capacitance has a capacitance equals to C_{S2} when both switches are fully turned on and zero capacitance when both switches are at 50% duty cycle.

8.3.8.3 Integration of Control and ZVS Signals for Q$_1$ and Q$_2$

To improve the system efficiency and minimize the heat generation from the control circuitry, soft-switching techniques such as zero voltage switching (ZVS) need to be implemented.

Figure 8.19 shows the conditions of achieving ZVS in Q$_1$ and Q$_2$. The conditions of Figures 8.19a and c show that Q$_1$ and Q$_2$ can be turned off anywhere in between θ_1 to θ_2 and θ_4 to θ_5 as shown in Figure 8.20, respectively, to achieve ZVS given the voltage across the switch is negligible under full-conduction state. To achieve ZVS in the conditions of Figures 8.19b and d where the forward voltage drop of the body diode is neglected, Q$_1$ and Q$_2$ can be switched on in-between θ_3 to θ_5 and θ_6 to θ_7, respectively.

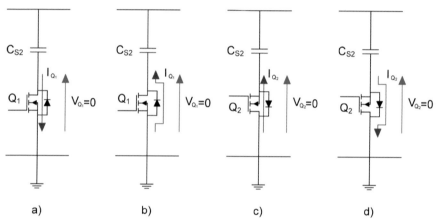

| | | | |
| a) | b) | c) | d) |

Figure 8.19   Conditions of achieving ZVS in Q$_1$ and Q$_2$.

Table 8.4   Truth table for $V_{\text{Ctrl}-1}$ and $V_{\text{Ctrl}-2}$ generation

Possible Cases	$V_{sq} > 0 \ (x_1)$	$S_{Q_1} > V_{\text{trig}}$ (x_2)	$S_{Q_2} > V_{\text{trig}}$ (x_3)	$V_{\text{Ctrl}-1}$	$V_{\text{Ctrl}-2}$
Case 1	0	0	0	1	0
Case 2	0	0	1	N/A	N/A
Case 3	0	1	0	1	0
Case 4	0	1	1	1	1
Case 5	1	0	0	1	1
Case 6	1	0	1	N/A	N/A
Case 7	1	1	0	0	1
Case 8	1	1	1	0	1

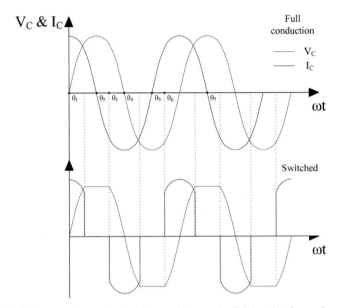

Figure 8.20 Voltage and current waveforms of C_{S2} under full conduction and switch mode.

In order to achieve ZVS while giving the correct switching signals to Q_1 and Q_2, a control signal conversion method is used here. Figure 8.21 shows the signal waveforms that are required for achieving such a task. The voltage signal V_{sq} is obtained by comparing V_{AC} with ground reference through a comparator and used as an indicator for the positive ($V_{sq} > 0$) and the negative ($V_{sq} < 0$) cycles of V_{AC}. The voltage signal V_{trig} is obtained by integrating V_{sq} through a passive integrator and used as a reference signal with which the control signals S_{Q_1} and S_{Q_2} are compared. The signals S_{Q_1} and S_{Q_2} have a relationship of:

$$S_{Q_1} = -S_{Q_2} \tag{8.39}$$

where S_{Q_1} is equal to the output signal $U(t)$ of the DTDC.

Table 8.4 summarizes the waveform of signals shown in Figure 8.21. The result of Table 8.4 can further be simplified using Boolean expressions. Hence, the control signals V_{Ctrl-1} and V_{Ctrl-2} can be obtained as:

$$V_{\text{Ctrl}-1} = \bar{x}_1 + \bar{x}_2 \tag{8.40}$$

$$V_{\text{Ctrl}-2} = x_1 + x_3 \tag{8.41}$$

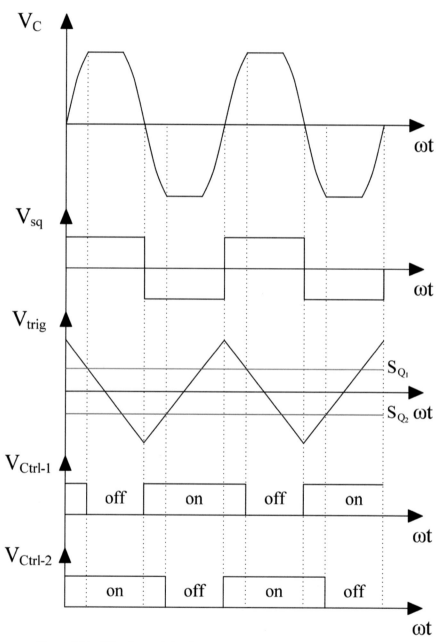

Figure 8.21   Waveform of signals used for generating V_{Ctrl-1} and V_{Ctrl-2}.

8.4 Conclusions

A novel algorithm named directional tuning/detuning control is developed to enable the full-range tuning of wireless pickup for the power flow regulation of wireless power transfer systems. Integrated effects of some of the parameter variations that are commonly seen in the wireless power transfer systems have been investigated and used to determine the operating range of the variable tuning capacitance, which has been employed to change the tuning condition of a parallel-tuned LC power pickup according to the load demand. To improve the steady-state performance of the controller, a simple algorithm for tuning step-size automation has been incorporated into the DTDC design. A parallel-tuned LC power pickup based on DTDC has been implemented, and a thorough description of using switch-mode variable-tuning capacitor with ZVS has been given in this chapter.

8.5 Problems

P8.1: A parallel-tuned wireless power pickup has a rated output voltage of 12V. If the output voltages in two consecutive sampling instances are 14.5V and 13V and the tuning capacitance has been increased, determine the operating region and tuning stage of the pickup.

P8.2: A WPT system has an ac track current with a constant operating frequency of 38.4 kHz. The pickup coil of the system has a self-inductance of 19.8uH with a measured open-circuit voltage of 2V. If a dc load resistance of 50Ω is connected to the output of the pickup, determine the value of the tuning capacitance required for the pickup output to have a rated voltage of 12V.

P8.3: Continuing from problem P8.3, the system is designed to have a variation range of ±1%, ±20%, and ±40%, in the operating frequency, open-circuit voltage of the pickup coil, and load, respectively. Find the minimum required operating range of the tuning capacitance.

P8.4: Continue from the problem P8.3. An embedded microprocessor is used as the controller of the system and requires a minimum dc supply voltage of 5V. Calculate the values of C_{S1} and C_{S2} for performing proper startup and control of the pickup under the worst-case scenario of the parameters variation.

References

[1] Bieler, T., M. Perrottet, V. Nguyen, and Y. Perriard. "Contactless Power and Information Transmission." *IEEE Transactions on Industry Applications* 26, no. 5: 1266–1272 (2002).

[2] Kim, C.-G., D.-H. Seo, J.-S. You, J.-H. Park, and B.H. Cho. "Design of a Contactless Battery Charger for Cellular Phone." *IEEE Transactions on Industrial Electronics* 48, no. 6: 1238–1247 (2001).

[3] Hu, A.P., I.L.W. Kwan, C. Tan, and Y. Li. "A Wireless Battery-Less Computer Mouse with Super Capacitor Energy Buffer." in *ICIEA 2007 2nd IEEE Conference on Industrial Electronics and Applications.* May 23–25 (2007).

[4] Feezor, M.D., F.Y. Sorrell, and P.R. Blankinship. "An Interface System for Autonomous Undersea Vehicles." *IEEE Journal of Oceanic Engineering* 26, no. 4: 522–525 (2001).

[5] Wang, L., M. Chen, and D. Xu. "Increasing Inductive Power Transferring Efficiency for Maglev Emergency Power Supply." in *PESC 2006 37th IEEE Power Electronics Specialists Conference.* June 18–22 (2006).

[6] Egan, M.G., D.L. O'Sullivan, J.G. Hayes, M.J. Willers, and C.P. Henze. "Power-Factor-Corrected Single-Stage Inductive Charger for Electric Vehicle Batteries." *IEEE Transaction on Industrial Electronics* 54, no. 2: 1217–1226 (2007).

[7] Harrison, R.R. "Designing Efficient Inductive Power Links for Implantable Devices." in *ISCAS 2007 IEEE International Symposium on Circuits and Systems.* May 27–30 (2007).

[8] Gao, J. "Inductive Power Transmission for Untethered Micro-Robots." in *IECON 2005 32nd IEEE Annual Conference of Industrial Electronics Society.* November 6–10 (2005).

[9] Kwan, L.I. "Battery-Less Wireless Computer Mouse." Master thesis, The University of Auckland, December (2004).

[10] Jackson, D.K., S.B. Leeb, and S.R. Shaw. "Adaptive Control of Power Electronic Drives for Servomechanical Systems." *IEEE Transactions on Power Electronics* 15, no. 6: 1045–1055 (2000).

[11] Chao, Y.-H., J.-J. Shieh, C.-T. Pan, W.-C. Shen, and M.-P. Chen. "A Primary-Side Control Strategy for Series-Parallel Loosely Coupled Inductive Power Transfer Systems." in *ICIEA 2007 2nd IEEE Conference on Industrial Electronics and Applications.* May 23–25 (2007).

[12] Covic, G.A., J.T. Boys, M.L.G. Kissin, and H.G. Lu. "A Three-Phase Inductive Power Transfer System for Roadway-Powered Vehicles."

IEEE Transactions on Industrial Electronics 54, no. 6: 3370–3378 (2007).

[13] Wang, C.-S., O.H. Stielau, and G.A. Covic. "Load Models and Their Application in the Design of Loosely Coupled Inductive Power Transfer Systems." in *PowerCon 2000 International Conference on Power System Technology*. December 4–7 (2000).

[14] Chen, H., A.P. Hu, and D. Budgett. "Power Loss Analysis of a TET System for High Power Implantable Devices." in *2nd IEEE Conference on Industrial Electronics and Applications, ICIEA 2007*, pp. 240–245 (2007).

[15] Si, P., A.P. Hu, J.W. Hsu, M. Chiang, Y. Wang, S. Malpas, and D. Budgett. "Wireless Power Supply for Implantable Biomedical Device Based on Primary Input Voltage Regulation." in *2nd IEEE Conference on Industrial Electronics and Applications, ICIEA 2007*, pp. 235–239 (2007).

[16] Si, P., A.P. Hu, S. Malpas, and D. Budgett. "A Frequency Control Method for Regulating Wireless Power to Implantable Devices." *IEEE Transactions on Biomedical Circuits and Systems* 2, no. 1: 22–29 (2008).

[17] Green, A.W., and J.T. Boys. "10 kHz Inductively Coupled Power Transfer-Concept And Control." in *5th International Conference on Power Electronics and Variable-Speed Drives*, pp. 694–699 (1994).

[18] Xu, Y.X., J.T. Boys, and G.A. Covic. "Modeling and Controller Design of ICPT Pick-Ups." in *Proceedings of International Conference on Power System Technology, PowerCon 2002*, pp. 1602–1606 (2002).

[19] Boys, J.T., G.A. Covic, and A.W. Green. "Stability and Control of Inductively Coupled Power Transfer Systems." *IEE Proceedings of Electric Power Applications* 147, no. 1: 37–43 (2000).

[20] Keeling, N.A., J.T. Boys, and G.A. Covic. "Unity Power Factor Inductive Power Transfer Pick-Up for High Power Applications." in *34th Annual IEEE Conference of Industrial Electronics, IECON 2008*, pp. 1039–1044 (2008).

[21] Hu, A.P., and S. Hussmann. "Improved Power Flow Control for Contactless Moving Sensor Applications." *IEEE Power Electronics Letters* 2, no. 4: 135–138 (2004).

[22] James, J., J. Boys, and G. Covic. "A Variable Inductor Based Tuning Method for ICPT Pickups." in *7th International Power Engineering Conference, IPEC 2005*, pp. 1142–1146 (2005).

[23] Eghtesadi, M. "Inductive Power Transfer to an Electric Vehicle-Analytical Model." in *40th IEEE Vehicular Technology Conference*, pp. 100–104 (1990).

[24] Si, P., A.P. Hu, S. Malpas, and D. Budgett. "Switching Frequency Analysis of Dynamically Detuned ICPT Power Pick-Ups." in *International Conference on Power System Technology, PowerCon 2006*, pp. 1–8 (2006).

[25] Si, P. "Wireless Power Supply for Implantable Biomedical Devices." PhD thesis, the Department of Electrical and Computer Engineering, University of Auckland, Auckland (2008).

[26] Hsu, J.-U.W., A.P. Hu, A. Swain, D. Xin, and S. Yue. "A New Contactless Power Pick-Up with Continuous Variable Inductor Control Using Magnetic Amplifier." in *International Conference on Power System Technology, PowerCon 2006*, pp. 1–8 (2006).

[27] Hsu, J.-U.W., A.P. Hu, and A. Swain. "A Wireless Power Pickup Based on Directional Tuning Control of Magnetic Amplifier." *IEEE Transactions on Industrial Electronics* 56, no. 7: 2771–2781 (2009).

[28] Hsu, J.-U.W. "Full-Range Tuning Power Flow Control of IPT Power Pickups." PhD thesis, the Department of Electrical and Computer Engineering, University of Auckland, Auckland (2010).

[29] Hsu, J.U.W., A.P. Hu, and A. Swain. "Fuzzy Based Directional Tuning Controller for a Wireless Power Pick-Up." in *IEEE Region 10 Conference, TENCON 2008*, pp. 1–6 (2008).

[30] Hu, A.P. "Selected Resonant Converters for IPT Power Supplies." PhD thesis, the Department of Electrical and Electronic Engineering, University of Auckland, Auckland (2001).

[31] Boys, J.T., G.A. Covic, and Y. Xu. "DC Analysis Technique for Inductive Power Transfer Pick-Ups." *IEEE Power Electronics Letters* 1, no. 2: 51–53 (2003).

9

Technology Overview and Concept of Wireless Charging Systems

Pratik Raval, Dariusz Kacprzak and Aiguo Patrick Hu

Department of Electrical and Computer Engineering,
The University of Auckland, New Zealand

This chapter overviews current technologies of near-field inductive wireless power transfer systems. One application is related to charging low-power electronics. For this purpose, magnetic structures are developed to demonstrate the concept of two-dimensional and three-dimensional wireless low-power transfer systems. The presented development is aided by utilizing state-of-the-art finite-element-method simulation software packages.

9.1 Introduction

The foundation of the operating principles of near-field magnetic induction (NFMI) trace back to the early nineteenth century. More specifically, there are two laws that lay the theoretical foundation of the near-field magnetic flux coupling form of wireless power transfer. These laws are termed Ampere's law and Faraday's law [1]. In simplified terms, Ampere's law implies that an electric current produces a magnetic field. This magnetic field may be determined using Ampere's circuital law which relates the integrated magnetic field around a closed loop to the electric current passing through the loop. Faraday's law implies that an electromotive force (EMF) is induced in any closed circuit through a time-varying magnetic flux through the circuit. In fact, the EMF generated is proportional to the rate of change of the magnetic flux. Therefore, these two laws are the building blocks of NFMI technology.

As such, the technique of using near-field magnetic field coupling as a link of power transfer is not new and has been used in low-frequency, such as 50 Hz or 60 Hz, applications in transformers and induction machines for centuries. However, it is also well-known that power transfer across a large air gap at such low frequencies is practically impossible. Furthermore, the earlier power semiconductor switching devices, such as diode, thyristor, triac and Gate Turnoff Thyristor (GTO), are mainly designed for low-switching frequencies of 50/60 Hz, up to 1 kHz. However, beyond 1970 the innovation of modern power electronic semiconductor-switching materials and devices such as Metal Oxide Silicon Field Effect Transistor (MOSFET), Bipolar Junction Transistor (BJT) and Insulated Gate Bipolar Transistor (IGBT) has led to practically feasible near-field and loosely coupled magnetic induction applications. This is mainly due to these devices higher switching frequencies along with lower switching losses that are now attainable through resonant soft-switching power-conversion techniques. The range of high frequencies in early NFMI systems was between 10 and 100 kHz. Nowadays, it is not uncommon to find NFMI systems operating in the 1–10 MHz range. In addition, recent developments of software packages that can accurately compute the numerical technique termed finite element method have enabled great improvements in the magnetic structures of NFMI systems. Such software works by segmenting the system model into elements. The associated approximate solution to the partial differential and integral equations of each element is then solved for minimum potential energy. The resulting simulation outputs provide a qualitative and quantitative visualization and frequency analysis of various magnetic parameters related to NFMI systems. This chapter begins with an overview of the state-of-the-art wireless power-transfer technologies and applications before presenting the development of a wireless charging system.

9.2 System Technology

In order to understand the technologies involved in a typical NFMI charging system, the basic structure is shown in Figure 9.1.

The system includes three main parts: the primary circuitry, magnetic coupling and secondary circuitry. Typically, the power supply may be a three-phase or single-phase input voltage applied to the primary circuitry. In the primary circuitry, the power electronics includes the power conversion (DC–AC) for generating a high-frequency time-varying magnetic flux for voltage induction. Next, the compensator networks are used to provide resonance or maximize power transfer and efficiency. Following this, the magnetic coupling is achieved through proper core and coil design. On the primary side, this is commonly referred to as the charging surface or primary track loop. On the secondary side, it is referred to as the power pickup. The primary track loop is used to carry an AC current to provide a power transfer platform. This may

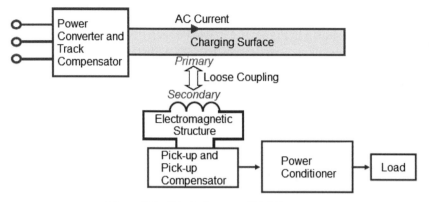

Figure 9.1 Block diagram of NFMI system.

be an elongated track, lumped coil, rail-mounted, overhead, flat, or spiral PCB traces as required by various electronic charging systems. The power pickup is generally loosely coupled to the primary platform for voltage induction. This could simply be another lumped coil similar to the primary structure or a customized one with ferrite cores to maximize power transfer capabilities. It is the power-coil(s) of the pickup that induce an EMF, which acts as an AC voltage source to the secondary circuitry. Due to the loosely coupled nature of the system, the induced voltage may be very weak and unstable. As such, the induced voltage is resonated in the secondary circuitry by using a compensation network. Following this, the induced AC voltage is then power-conditioned to produce a regulated output voltage suitable to drive a load.

9.2.1 Power Converter

The power converter inverts the DC input to produce a high-frequency AC current in the primary track loop. Such power converters are typically categorized as linear or switch-mode circuits. Switch-mode power converters are currently the most widely used attributable to a higher power efficiency and power density. The switching techniques in switch mode power conversion can be categorized as hard switching and resonant soft switching.[1]

Since 1970s, a conventional example of hard switching is pulse width modulation. In this technique, the fully on and fully off transitions of the

---

[1]Soft switching is often called resonant soft switching, because it incorporates a resonant tank in a power converter to create oscillatory voltage or current waveforms so that the zero-voltage-switching and zero-current-switching conditions can be created for switching the semiconductors on and off.

switching devices occur at nonzero voltage or current instances. The trajectory of hard switching is shown in Figure 9.2. It can be seen that during the turn-on and turn-off processes, the power devices have to withstand high voltage and current simultaneously, resulting in high-switching losses and stress. In many cases, dissipative passive snubbers are added to power circuits to reduce the rate of change of voltage or current of the power devices to reduce the power losses. Overall, hard switching inherently results in high-switching losses, low power efficiency, high electromagnetic interference (EMI) and poor waveform generation as exaggerated in Figure 9.2. To largely overcome these drawbacks, soft switching is commonly used in ICPT applications. The trajectory of soft switching is also shown in Figure 9.2. Evidently, the switching devices encounter much lower stress due to approximate zero voltage or current during switching. In this technique, the fully on and fully off switching instances are controlled to occur at zero voltage or zero current instances [2]. This is commonly termed zero-voltage-switching (ZVS) or zero-current-switching (ZCS), which occur under the following conditions:

1. ZVS is approximately achieved by controlling switching transitions to occur at zero voltage instances across a capacitor.
2. ZCS is approximately achieved by controlling switching transitions to occur at zero current instances through an inductor.

ZVS and ZCS conditions can be summarized as conventional square wave power conversion during the switches on time with resonant switching transitions. Overall, resonant soft switching results in lossless switching, high

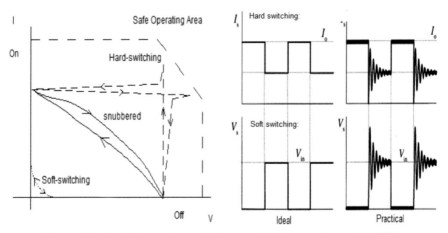

Figure 9.2   Switching trajectory (left) and switching waveforms (right) [2].

efficiency, reduced transient switching voltage or currents, reduced EMI and reduced gate drive requirements.

Inverters may generally be classified as either voltage-fed (V-fed) or current-fed (I-fed). Each may be classified as either full-bridge or half-bridge configurations. Generally, the power capability increases from half bridge to full bridge [3].

A voltage-fed inverter, as the name suggests, is driven by a voltage source. The full-bridge topology shown in Figure 9.3 consists of four switches. This topology outputs a voltage range equal to twice the input voltage. In contrast, the half-bridge topology also shown in Figure 9.3 consists of two switches and two suitably large capacitors. This topology outputs a voltage range equals to the input voltage. These two configurations however have major disadvantages: The voltage and current waveforms contain harmonic components, and they require an external controller to maintain a constant track current. As such, voltage-fed configurations are more suitable for medium- to high-power level applications, and not for charging personal electronics, at low-power levels, such as on a consumer electronics charging platform.

A current-fed inverter, as the name suggests, is driven by a current source. This current source is often achieved by placing a large inductor in series with an input voltage source. Analogous to the voltage-fed case, the full-bridge topology shown in Figure 9.4 consists of four switches. This topology outputs an AC current equal to the magnitude of the current source. In contrast, the half bridge topology also shown in Figure 9.4 or commonly termed push-pull topology consists of two switches and a phase-splitting transformer. The phase-splitting transformer divides the DC current in half and consequently, the output current is half the magnitude of the current source and output voltage doubles. This push-pull topology is advantageous in that the switching transitions only occur at the ZVS and the inverter can be designed for automatic startup and self-sustaining operation without external controllers [3], thereby reducing switching and conduction power losses.

9.2.2 Compensation Networks

Compensation networks are required on both the primary and secondary sides. The main purpose of primary compensation is to reduce the input VA rating of the power supply. Inherently, the primary track inductance results in an increase in the power requirement of the power supply, by adding a reactive (complex) component to total power [4]. This is understood by Equation (9.1), where the total power S is a sum of the real power P and the imaginary power Q.

$$S = P + jQ \tag{9.1}$$

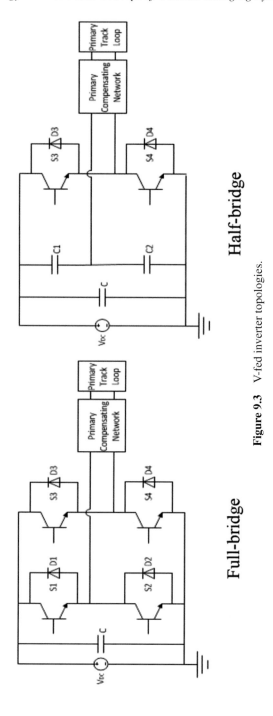

Figure 9.3   V-fed inverter topologies.

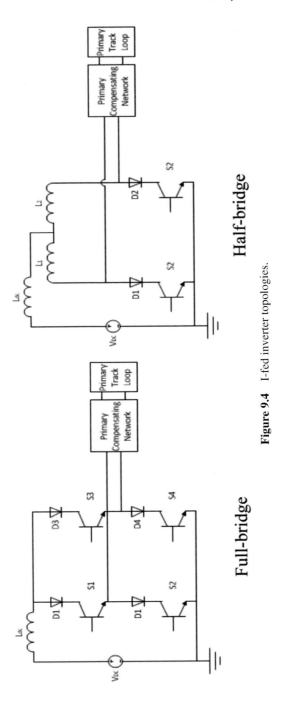

Figure 9.4 I-fed inverter topologies.

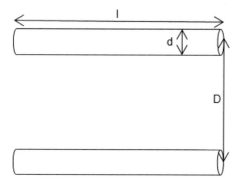

Figure 9.5   Two parallel track conductors.

This is undesirable as it means a phase displacement between the primary track voltage and current. In these systems, it is ideal to operate at zero-phase angle (ZPA) in order to reduce the apparent power required. Therefore, track inductance, L_P, must be compensated through a capacitor of suitable value, C_P, at the operating angular frequency ω. In other words, the finite inductance of the track wire essentially places an upper limit on the track length. The design of the compensating network may be understood through an example.

Example:

A parallel cable of length 1 m is in air with a spacing of 0.5 m between the parallel conductors of diameter 0.105 m is to be used as the primary track loop. Your task is to design a resonant compensator network for an operating frequency of f = 13.56 MHz.

Solution:

Firstly, the track inductance L_P that is to be compensated must be calculated. This is calculated from the equation as derived for the inductance L_P for parallel cables, of length l, diameter d and a spacing of D between the parallel cables expressed meters, and relative permeability $\mu_r = 1$ for air, as formulated below.

$$L_p = \frac{\mu_r \mu_0 l}{\pi} \left[\frac{1}{4} + \ln\left(\frac{2D}{d}\right) \right]$$

$$= \frac{1 \times 4 \times \pi \times 10^{-7} \times 1}{\pi} \left[\frac{1}{4} + \ln \left(\frac{2 \times (0.5)}{0.105} \right) \right] \approx 1 \, (\mu H)$$

The compensating capacitance then can be calculated at the resonant frequency.

$$C_P = \frac{1}{L_P \omega^2}$$

$$= \frac{1}{1 \times 10^{-6} \times (2 \times 3.14159265 \times 13.56 \times 10^6)^2} \approx 138 \, (pF)$$

As the result shows, the track inductance is roughly 1 μH per 1 m length of wire. The primary track compensation enables longer track lengths to be driven for a given voltage rating. Another reason for primary compensation is to form a resonant circuit. Resonant circuits have the advantage of reducing switch losses and EMI by enabling soft-switched operation of semiconductor switches which is vital in high-frequency operation of resonant soft-switched converters. The most common forms of primary compensation are series and parallel configurations as shown in Figure 9.6. Such configurations are often required for resonant soft-switching inverters.

For V-fed inverters, compensation consists of placing a resonating capacitor in series with the primary track loop inductance as shown in Figure 9.6. Such series compensation presents a current source nature from the primary inductor and matches the voltage source input for the inverter. If the inductive and capacitive impedances are denoted by Z_L and Z_C then the complex impedance Z of the series compensated circuit is given by Equation (9.2). In

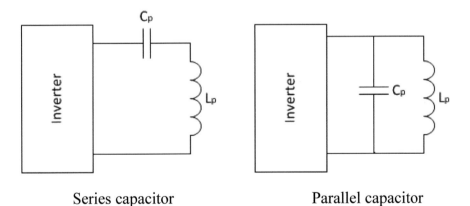

Series capacitor   Parallel capacitor

Figure 9.6 Common forms of primary compensation.

contrast, the I-fed inverter compensation consists of parallel resonant capacitor also shown in Figure 9.6. This forms a voltage source from the tuning capacitor C_P and matches the current source of the inverter. Similarly, the complex impedance of the parallel compensated circuit is given by Equation (9.3). Notably, if the resonant tank is tuned at the resonant angular frequency given by Equation (9.3), the circuit impedances of Equations (9.2–9.3) become zero.

$$Z(j\omega) = Z_L + Z_C = \frac{j(\omega^2 L_P C_P - 1)}{\omega C_P} \tag{9.2}$$

$$Z(j\omega) = \frac{Z_L Z_C}{Z_L + Z_C} = \frac{-j\omega L_P}{\omega^2 L_P C_P - 1} \tag{9.3}$$

$$\omega = \frac{1}{\sqrt{L_P C_P}} \tag{9.4}$$

Alternatively, the state-of-the-art systems may compensate for V-fed inverters through a composite LCL form [13] that prevents the formation of voltage sources in parallel. In comparison, an I-fed inverter may be compensated through a composite CCL form [13], in order to prevent the formation of current sources in series. It is important to note that tuning configurations influence power supply properties such as reactive power flow, power factor, track current sensitivity, system stability and efficiency. As such, an inevitable trade-off exists between those parameters as compared in [4, p. 19].

The main purpose of pickup compensation is to maximize the output power capability. This may be achieved by a resonant series or parallel capacitor, similar to primary compensation. The resulting effect of pickup compensation is graphically illustrated in Figure 9.7, which shows a series tuned pickup can supply an infinite current at a given voltage; a parallel compensated pickup can supply an infinite voltage at a given current; and no compensation results in a trade-off between load voltage and current.

9.2.3 Electromagnetic Structures

NFMI systems are generally loosely coupled, so the magnetic cores of the primary and secondary, the primary track loop and pickup, need to be separated. As such, the configuration of the electromagnetic structures is specific and highly dependent on the application requirements. For instance, in some applications the secondary load is to remain fixed or largely motionless— such as a stationary electric car charging platform or electric toothbrush while in other applications the secondary load is allowed to move freely

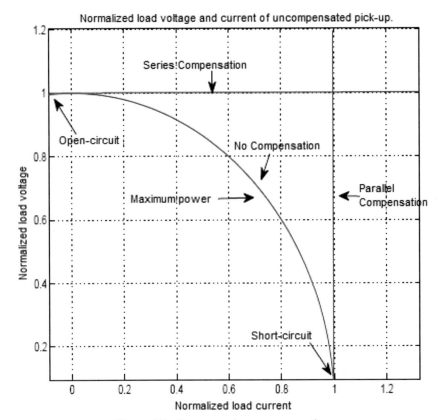

Figure 9.7 Common primary compensation.

relative to the primary as in monorail systems or a three-dimensional charging electronics box. It is also possible that multiple secondary load(s) or pickup(s) are required in such systems. As such, the electromagnetic cores need to work synergistically to maximize the performance of the wireless power transfer system. More specifically, various parameters including power distribution and uniformity, core volume, material, shape, weight, cost and efficiency must be considered when designing the primary track loop and pickup.

The main purpose of the primary track loop is to generate the required magnetic flux distribution or power transfer window. To date, various state-of-the-art designs have been proposed for two-dimensional (2D) and three-dimensional (3D) charging surfaces. Various planar configurations are shown in Figure 9.8. Once the resulting magnetic flux distribution is visualized in single-layer configurations, the main limitation is a lack of magnetic flux

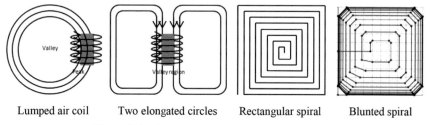

| Lumped air coil | Two elongated circles | Rectangular spiral | Blunted spiral |

Figure 9.8 2D spiral primary track configurations.

uniformity. For instance, earlier innovations simply used a lumped air coil. In this configuration, the MMF peak is located near the perimeter of the coil and weakest around the middle of the coil, progressively weakening as the size of the coil is increased. Another configuration considers elongated circles. This has a destructive-zone in the location adjacent to the circles, making uniform power induction across the surface difficult. Generally, in spiral configurations the magneto motive force (MMF) peaks are located near the centre of the coil and is weakest at the edges of the coil. This restricts the free-positioning or movement of any load(s) on the charging platform. To overcome this problem, two notable concepts have previously been proposed [5, 19]. The first concept is still based on a single-layer rectangular configuration consisting of blunted edges and non-uniform wire spacing in the spiral. That is, the density of the number of turns is greater at the outer edges and gradually declines towards the centre. This was shown in [19] to generate a relatively uniform MMF for a single layer. The second concept extends the single-layer spiral concept to introduce multi-layer array winding matrices. One such three-layer configuration is illustrated in Figure 9.9. This three-layer concept has a distance-phase displacement between each layer, so that the central peak in MMF of one layer can be made to coincide with the outer valley of another layer so as to produce a uniform MMF distribution. Overall, such a design has been simulated and experimentally verified in [5] to show a substantially uniform magnetic flux over a major part of the primary surface. In fact, the load power ranged from 1.3 to 1.58 W over the surface. In this way, precise position and orientation of the chargeable device on the platform is not critical.

The pickup is an essential part of the NFMI system. The sole purpose of a pickup is to provide the induced power, from the primary circuit, to the secondary circuit in a satisfactory manner. In order to do this, the two most important factors to consider are the accessibility to the primary track loop and the power or coupling requirement of the particular application.

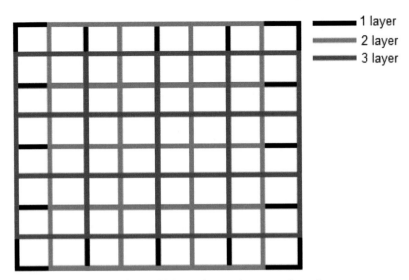

Figure 9.9 Multi-layer planar primary configuration.

Firstly, the pickup accessibility factor is readily achieved by using a variety of shapes and core materials to suit the particular application. Some common pickup configurations include the following: E-, toroidal- and planar-pickup types as shown in Figure 9.10. These cores are usually made from a highly permeable magnetic material (such as soft ferrite) to help confine and guide the magnetic field towards to secondary circuit. Practical examples of these configurations include E-type cores that are commonly used for monorail trolleys and materials handling systems where freedom of mechanical movement is important, toroidal-type cores being used in a current transformer with a primary and moveable secondary circuit [3, p. 30], and planar- or flat-type cores are typically used when the primary track loop is a flat coil, as is the case in PCBs and in transportation applications such as electric vehicles where the

E-core   Toroid   Planar

Figure 9.10 Various pickup configurations.

geometry of the pickup allows larger vertical displacement. More recently, the S-type pickup configuration has been proposed in [6] with no practical application as yet. Secondly, the pickup must also induce enough power to meet the power requirement for the particular application. The coupling factor is largely a function of the geometry and core materials being used. As example values, toroidal type typically has coupling at 0.9, E types about 0.6–0.8 and the flat type has the lowest at 0.4 or less [3]. Generally, higher coupling factors are generally achieved by greater use of ferrite cores at the expense of increased weight and cost [3].

9.2.4 Power Conditioner

The purpose of the power conditioner is to improve the quality of the induced power before it is delivered to drive a load. In a general sense, the parameters defining the quality of power include variation in voltage magnitude, transient voltages and currents and harmonic content in the waveforms. This stage is commonly referred to as voltage regulation. Voltage regulators fall into two categories, linear and switch-mode.

A linear regulator operates by using a voltage-controlled current source to enforce a fixed voltage to appear at the output of the regulator. Typically, a control or sense circuitry is required to adjust to current source so as to maintain a constant output voltage. This often involves the use of an active device such as BJT, FETs, or vacuum tube. Common linear regulators include the standard regulator, low-dropout regulator and quasi low-dropout regulator, as shown in Figure 9.11. These are often connected to a differential amplifier that monitors the output determined by a potential divider. In this way, the desired voltage level is essentially maintained by dissipating excess power as heat.

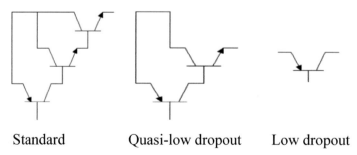

       Standard   Quasi-low dropout   Low dropout

Figure 9.11   Linear regulators.

In contrast, switch-mode regulators maintain the desired voltage by using switches to switch between ideal storage elements. Common switch-mode regulators used in many inductively coupled power transfer systems include switch-controlled boost and buck regulators as shown in Figures 9.12 and 9.13.

Boost regulators are commonly used with parallel compensated pickups. The output of this configuration is a boosted voltage which appears at the input of the rectifier. However, a disadvantage of such a parallel-tuned case is that it reflects both the real and reactive loads back onto the track power supply [7] as derived in [8].

In contrast, buck regulators are commonly used with series compensated pickups. The output of this configuration is a boosted current which appears at the input of the rectifier. This circuit has the advantage of no reactive loading on the power supply [8]. However, at start-up, resonant-tuned voltages are difficult to control [8, 9].

9.3 Applications

In the early 1990s, the Power Electronics Research Group of the University of Auckland developed a prototype near-field wireless power transfer system rated at 180 W, suggesting potential industrial applications. The first major

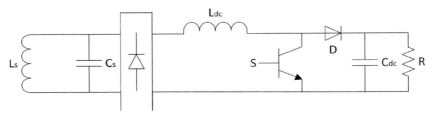

Figure 9.12   Switch-mode boost regulator.

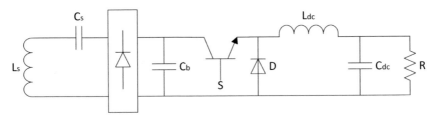

Figure 9.13   Switch-mode buck regulator.

industrial application of this technology was found at Daifuku Ltd., a material-handling system manufacturer in Osaka, Japan, in its Ramrun electrified monorail system. Ramrun is widely used in assembly and processing lines, particularly in the automotive industry. Typically, the track conductors are powered in 100 m section lengths and may have a current of 60 A at 10 kHz. Due to such large track length and the loosely coupled nature of the system, multiple monorail trolleys can be used. In such a case, each trolley can typically receive 750 W of power. This application includes a customized pickup constructed from ferromagnetic material termed E-core pickup. Such a magnetic structure is essential for efficient power transfer. Other applications have now developed at Daifuku Ltd., including clean-rooms, silicon chip manufacture, automobile manufacturing and assembly.

In 1997, NFMI technology had an impact on road lighting in the form of road studs [10]. The idea behind this was to extract power inductively from a buried wire and drive a number of light-emitting diodes. This was done by connecting a 2-kW, 20-A @ 20-kHz power converter to a long closed track loop buried under the road. Such a design successfully improved reliability, robustness and most importantly visibility of road-studs. A New Zealand company, Harding Traffic Ltd, developed this lighting technology called Smart-Studs. Smart-Studs are rapidly being installed in Europe, Asia and the United States for use in tunnels, bridges, roads, walkways and pedestrian crossings. In New Zealand, this technology is famously used in Terrace tunnel in Wellington.

ICPT technology spread further in people mover or transportation applications. In 1997, Wampfler AG, Germany, and the University of Auckland successfully implemented an electric vehicle using a high-power ICPT electric power charger [11]. An application of this vehicle is found in a Whakarewarewa Geothermal Park [12] in Rotorua, New Zealand. In fact, at the end of 2009 contactless charging of electric vehicles was standardized. Further applications include inductive loading of busses in Turin and Genoa, Italy, with a charging power of 60 kW and more recently inductive supply of trams in Augsburg, Germany.

In addition to the applications mentioned, ICPT has also found applications in kid-karts [3, p. 5], machine tools [3, p. 5], cordless power stations [3, p. 5], underwater power plugs [3, p. 5], biomedical implantation [3, p. 5], aircraft entertainment systems [3, p. 5] and electronics charging applications [13]. In summary, NFMI technology has found applications in a vast range of industries.

9.4 Development of Wireless Low-Power Transfer System

The motivation for charging the battery cells via inductive coupling rather than direct electrical connection arises to remove the inconveniences caused from physical wires and the need to purchase and replace new cells, as well as supporting for free-positioning of any load(s) while translating other non-contact power transfer benefits to the particular application such as safety, longevity and low maintenance. The two main magnetic structures of the NFMI system are the primary track loop and pickup. These are to be developed through the use of modern research and analysis tools in this section.

9.4.1 Methodology

In general terms, analysis of a problem can be broken down into two steps. The first step is often termed modelling. This is where all aspects of interest are identified and selected. The second step is often termed discretization. This is where the most suitable numerical method to solve the problem is selected. The most common methods [14] for this processare as follows:

1. Finite Difference Method (FDM)
2. Boundary Element Method (BEM)
3. Moments Method (MM)
4. Monte Carlo Method (MCM)
5. Finite Element Method (FEM)

The FDM, BEM, MM and MCM are most suitable for linear problems with regular geometry. FDM is often used in time-dependent problems. BEM, MM and MCM may not generate a mesh in the complete geometry or may only be used in linear cases with simple geometries. In contrast, the FEM method may be used in linear and non-linear problems without restrictions on the geometry. So, the FEM method is most suitable for the development of the magnetic structures of an NFMI system. The FEM method may be defined as a mathematical method for solving ordinary and elliptic partial differential equations via a piecewise polynomial interpolation scheme. In other words, the FEM evaluates the partial differential and integral equations by using a number of polynomial curves arising from a plane strain triangular quadratic element to approximate the shape of a more complex function. The implementation of FEM is termed Finite Element Analysis (FEA). Nowadays, this is commonly done through powerful software packages. FEA software works by segmenting the system model into elements. This process is often

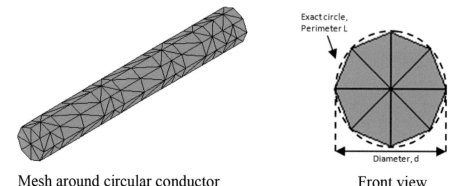

Mesh around circular conductor   Front view

Figure 9.14   Elements inside a circle.

described as the generation of the mesh. This process may be understood by evaluating the value of π, by considering a triangular mesh generated using FEA software for a circular conductor as illustrated in Figure 9.14. The elements form a polygon inside the exact circle. The polygon sides are termed elements and the vertices are termed nodes. Each generic element has a length dependent on the number of elements n. This suggests the accuracy of the calculated length increases with increasing the number of elements. To demonstrate this, the approximate value of π is to be calculated as a function of the perimeter L of the exact circle of diameter d as in Equation (9.5).

$$\pi = L/d = n\sin(\pi/n) \tag{9.5}$$

Upon considering integer values of n, the approximate calculated value of π along with the associated error is tabulated in Table 9.1. Notably, the error is significantly improved with a large number of elements.

9.4.1.1 Finite element formulation

Fundamentally, let us consider the wave equation derived from Maxwell's equations in Cartesian coordinates (x, y) for a wave[2] of electric field E, in medium of conductivity σ, permittivity ε, permeability μ at angular frequency ω.

---

[2]As the magnetic field intensity, H, may also be written as a form of Helmholtz's equation, a very similar analysis to the one presented can be done in terms of H.

Table 9.1 Approximate calculated value of π

N	π	Error	Error (%)
2	2.0000000000	1.1415926536	36.33802
4	2.8284271247	0.3131655288	9.968368
8	3.0614674589	0.0801251947	2.550464
16	3.1214451523	0.0201475013	0.641315
32	3.1365484905	0.0050441630	0.160561
64	3.1403311570	0.0012614966	0.040155
128	3.1412772509	0.0003154027	0.01004
256	3.1415138011	0.0000788524	0.00251
512	3.1415729404	0.0000197132	0.000627
1024	3.1415877253	0.0000049283	0.000157
2048	3.1415914215	0.0000012321	3.92E-05

$$\frac{\partial}{\partial x}\left(\frac{1}{\mu}\frac{\partial E}{\partial x}\right) + \frac{\partial}{\partial y}\left(\frac{1}{\mu}\frac{\partial E}{\partial y}\right) - (j\omega\sigma - \omega^2\varepsilon)E = 0 \qquad (9.6)$$

Analytically, there is no simple exact solution to Equation (9.6). Hence, the FEM may be used as a tool to find approximate solutions to the partial differential equation. For thisformulation, let us consider just one triangular element with three nodes. The weighted residual method by application of the Galerkin approach is to be applied to Equation (9.6) TO derive equations governing each element. Accordingly, this may be written in compact matrix form, as in Equation (9.7).

$$[M + K]\{E\} = 0 \qquad (9.7)$$

The matrices M and K are derived over element domain $d\Omega$ below.

$$M = (\omega^2\varepsilon - j\omega\sigma)\int_{d\Omega} N_i N_j \, d\Omega \qquad (9.8)$$

$$K = \frac{1}{\mu}\int_{d\Omega}\left(\frac{\partial N_i}{\partial x}\frac{\partial N_j}{\partial x} + \frac{\partial N_i}{\partial y}\frac{\partial N_j}{\partial y}\right) d\Omega \qquad (9.9)$$

where N_i and N_j are polynomial element shape functions and $i = j = [1, 3]$ are an element of integers representing the three nodes, a, b and c, of the triangular element. Equation (9.8) may be expanded to yield Equation (9.10) and then evaluated over the area element A to give Equation (9.11). Similarly, Equation (9.9) may be evaluated to give Equation (9.12).

$$M = (\omega^2\varepsilon - j\omega\sigma) \int_{d\Omega} \begin{bmatrix} N_1 \\ N_2 \\ N_3 \end{bmatrix} [N_1 N_2 N_3]\, d\Omega \qquad (9.10)$$

$$M = \frac{(\omega^2\varepsilon - j\omega\sigma)A}{12} \begin{bmatrix} 2 & 1 & 1 \\ 1 & 2 & 1 \\ 1 & 1 & 2 \end{bmatrix} \qquad (9.11)$$

$$K = \frac{1}{4\mu A} \begin{bmatrix} b_i b_i + c_i c_i & b_i b_j + c_i c_j & b_i b_k + c_i c_k \\ & b_j b_j + c_j c_j & b_j b_k + c_j c_k \\ Sym & & b_k b_k + c_k c_k \end{bmatrix} \qquad (9.12)$$

Notably, for one triangular element containing three nodes, the expression of the FEM approximation is a 3×3 matrix. This implies that a system with n nodes will consist of a nxn matrix. With the use of modern FEA software's such calculations for magnetic frequency analysis may be automated. One such state of the art FEA software, JMAG Designer 10.0, is capable of accurately performing magnetic frequency analyses to within 10%. So, the development of the magnetic structures of the proposed system is aided by this software.

9.4.2 D Planar Wireless Power Transfer System

9.4.2.1 Primary track loop

The main purpose of the primary track loop is to provide a power transfer window for the pickup to induce an EMF. With regard to a 2D planar electronics charging system, a spiral rectangular primary track loop is considered because of ease of manufacturing, commonly implemented using Litz wire or on printed circuit boards, and support of scalability as additional track loops are easily fitted to the four adjacent spaces. As part of the simulation, one of the most important parameters to consider is the magnetic flux density distribution. Generally, it is important to consider all components including the vertical or normal, tangential or horizontal, and magnitude components. This enables detailed analysis and determination of the dominant component for pickup design. The primary track loop is simulated with unity AC current at 25 kHz and the resulting tangential, normal and magnitude components of the magnetic flux density was analyzed by considering the average componential inter-conductor cancellation loss factor as in [15]. Upon computing the average magnetic flux density in both components for near-field planes above a rectangular spiral configuration, the resulting averaged magnitude magnetic flux density value yields information on the strength of the magnetic flux in the powering volume. In this case, the normal component of magnetic flux density is almost 57% stronger throughout various planes in the near-field region than the tangential component. As such, the normal component is chosen as the target component of the pickup. However, the research challenge identified is with regard to the non-uniformity of the normal magnetic flux. That is, the magnetic flux peak occurs towards the centre of the rectangular configuration and gradually declines towards the edges. The result is that each central peak (maximum) is surrounded by 4 valleys (minima). The particular challenge related to charging systems is with respect to free-positioning of the charging device—as the magnetic flux density is not uniform, the positioning of the charging device is critical. In order to somewhat overcome this challenge and to consider scalability of the platform, arrays of spirals are considered. For the purposes of designing, two different cases of current polarity are considered: 'Case A' and 'Case B'. Both cases are illustrated in Figure 9.15. In 'Case A' each spiral is excited with current in the same direction to adjacent spirals to produce an array of north (positive) poles. The result is that the net magnetic flux from adjacent spirals destructively interferes at the spiral edges, further restricting power transfer capabilities. In contrast, in 'Case B' each spiral is excited with a polarity of current that is of opposite direction to each adjacent spiral to produce a combination of north and south (positive and negative) poles. This is to create a channel that aids the flow of net magnetic flux between adjacent spirals via magnetic attraction as illustrated in Figure 9.15.

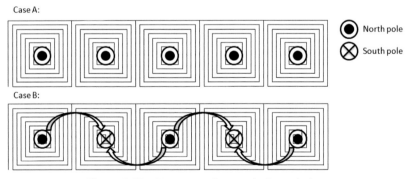

Figure 9.15 Two cases of magnetic flux polarity.

For an array of spirals considering the magnetic flux polarities of 'Case A' and 'Case B', the dominant normal component of magnetic flux density, along a plane on top of the spirals is shown in Figure 9.16. Both cases are analyzed in more detail in [15, 16] which considers the averaged normal magnetic flux density across in the near-field above the platform in relation to the number of turns in the spiral. Upon visualization, the magnetic flux density of 'Case B' may be used advantageously by design of a suitable smart pickup. That is, a pickup design that aids the flow of magnetic flux from the positive to negative regions can be designed to provide a relatively uniform power induction irrespective of the valley zones.

9.4.2.2 Pickup

In NFMI systems, the magnetic flux resulting from the primary track loop must be directed and concentrated to the pickup secondary circuitry. This is commonly done by the pickup. The pickup is usually constructed by using a material that supports a high degree of magnetization in response to the magnetic field generated by the primary circuitry—such as ferrite. This means that using such a material would provide a conductive pathway for the magnetic flux to be directed towards to secondary circuit. With regard to the developed scalable primary track configuration, one of the main objectives for pickup design is to provide relatively uniform power induction irrespective of pickup location on the platform. In order to satisfy this objective, it is important

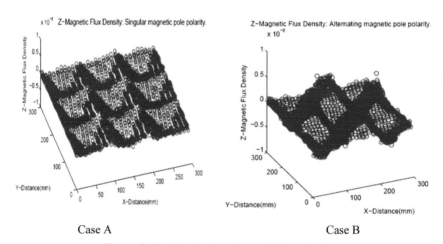

Case A   Case B

Figure 9.16   Magnitude of magnetic flux density.

to first understand how various pickup parameters are mathematically related to the induced power.

Firstly, the pickup parameters are to be mathematically related to the induced power. This power induced directly through the pickup is termed the uncompensated power. This uncompensated pickup power [13] can be expressed in terms of the parameters relevant to the pickup by using the definitions of the secondary open-circuit voltage (V_{OC}) and short-circuit current (I_{SC}) as in [13] stated here as Equations (9.13, 9.14).

$$V_{OC} = NBA_{B\perp}\omega \qquad (9.13)$$

where N is the number of coil turns, B is the average value of the magnetic flux density in the ferrite under the coil, $A_{B\perp}$ is the cross section of the ferrite perpendicular to the vectors of the magnetic flux density, and ω is the operating angular system frequency.

$$I_{SC} = JA_{I\perp}/N \qquad (9.14)$$

where J is the average value of the current density in the coil and $A_{I\perp}$ is cross-section of solid coil perpendicular to the flow of current.

The uncompensated power S_U can then be written as the product between the secondary open-circuit voltage and short-circuit current as in Equation (9.15).

$$S_U = BA_{B\perp}JA_{I\perp}\omega \qquad (9.15)$$

For pickup design purposes, it can be assumed that the operating frequency, track current and therefore the resulting magnetic field are fixed. This means that only two critical parameters, $A_{B\perp}$ and $A_{I\perp}$, are to be optimized during pickup design. Furthermore, the optimization of those critical parameters is justified from the point of view of the magnetic flux flow paths. Inherently, the magnetic field causes magnetic flux to follow the path of least magnetic reluctance. Upon using the U and E core pickups as a starting point [13], the proposed pickup shape was developed as shown in Figure 9.17.

The proposed pickup shape is designed to provide a low magnetic reluctance pathway to induce the dominant normal component of magnetic flux density. For the purposes of design, the dimensions of the proposed pickup shape must be maximized to yield greater power transfer capabilities. This is often done through FEA visual analysis of the magnetic density vector plots. As visualized in the magnetic flux density vector plot of Figure 9.18, a strong interception of the normal component of magnetic flux by the pickup core ensures satisfactory voltage induction. In this way, FEM software is useful to

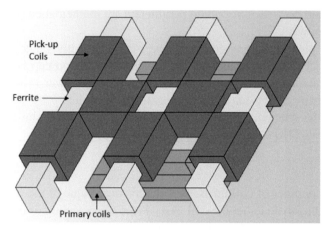

Figure 9.17   Proposed pickup shape.

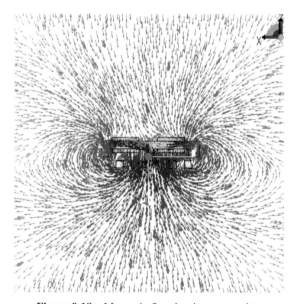

Figure 9.18   Magnetic flux density vector plot.

optimize the relative dimensions of the magnetic structures of NFMI systems to get enhanced power transfer capabilities.

In order to demonstrate the concept of scalability, various pickup positions must be considered on arrays of spiral loops as shown in Figure 9.19. With reference to the simulated magnetic flux density distributions, it can be

Figure 9.19 Various pickup positions on platform.

seen that moving the pickup along the platform or positions 1–5 results in coincidence of central portion of pickup core with magnetic flux peak zones. Position 6 results in the top portion of the pickup core incident on magnetic flux peak zones, while position 7 results in the outer portion of the pickup core incident on the magnetic flux peak zones. Finally, the rotated case as in position 8 also results in the interception of a portion of the pickup core with magnetic flux peak zones. The alternating polarity of magnetic flux is expected to aid the flow of flux through the pickup core. In fact, the variation in the experimentally measured total open-circuit voltage induced in the coils was 32.1% [15, 16].

Current researchers should be aware that present research challenges of such systems include:

1. *Electromagnetic interference*: The portion of the primary circuitry not in use by any load(s) remains switched on not only producing potentially harmful electromagnetic interference but also wasting considerable power. As such, an efficient receiver detection feature is required to

detect the presence of any load(s) for selective switching of the primary circuitry.

2. *Efficiency*: The efficiency, defined as the ratio of power received by the receiver to the power transmitted by the transmitter, is one of the biggest drawbacks of NFMI systems. The most common efficiency of such systems fall within the range of 50–80%. This is in contrast to conventional adapters which have efficiencies close to 100%. As a wireless power transfer technique, the efficiency is a function of two aspects. The first is the efficiency associated with variable load(s) and the second is the efficiency associated with transmitter to receiver load distance.

3. *Energy consumption*: This is related to the usage of power received by the receiver in an efficient manner to operate any load(s). For instance, NFMI systems suffer from increased resistive heating in comparison with direct contact. This is a result of additional NFMI components such as the magnetic structures, drive electronics and power coils.

9.4.3 Wireless Power Transfer System

To date, the research challenge of developing three-dimensionally radiating inductive power transfer systems remains. As the magnetic field arising from a coil immersed in a linear medium is unidirectional at any location, at least three distinct primary coils are required for the generation of three orthogonal magnetic field components. Notably, the magnetic field components are vectorial quantities with a single magnitude and direction in the space domain. Ideally, the receiver or pickup axis must be aligned parallel to the magnetic field to maximize the mutual inductance and consequently power transfer. However, since the magnetic field decays as an inverse square law with distance, the curvature nature of the magnetic field poses many challenges for generating three-dimensionally radiating primary surfaces. That is, the research challenge remains to develop a primary surface capable of uniformly distributing an omnidirectional power transfer window for an arbitrarily oriented pickup in the powering volume to induce sufficient magnetic flux to satisfactorily operate any load(s). In summary, current 3D primary links may be classified as continuous or discontinuous.

9.4.3.1 Continuous mode of operation

Upon exciting an AC sinusoidal current in all three orthogonally oriented coils simultaneously, the resulting magnitude of the magnetic field, $|H|$, traced out at any point in space is proportional to the sinusoidal current in the three components, as shown in Equation (9.16).

$$|H| \propto \sqrt{x(t)^2 + y(t)^2 + z(t)^2} \tag{9.16}$$

where $x(t)$, $y(t)$ and $z(t)$ are sinusoidal currents of amplitude I_0 and carrier angular frequency ω_C, given below in Equations (9.17–9.19).

$$x(t) = I_0 \sin(\omega_C t + \phi_1) \tag{9.17}$$

$$y(t) = I_0 \sin(\omega_C t + \phi_2) \tag{9.18}$$

$$z(t) = I_0 \sin(\omega_C t + \phi_3) \tag{9.19}$$

On one hand, if the phases of all currents are equal ($\phi_1 = \phi_2 = \phi_3 = \phi$), then Equation (9.16) reduces to Equation (9.20). As such, this is not an omnidirectional system but a bidirectional system [17, p. 31].

$$|H| \propto \sqrt{3} I_0 \sin(\omega_C t + \phi) \tag{9.20}$$

On the other hand, if the phases of the currents are not equal ($\varphi_1 \neq \varphi_2 \neq \varphi_3$), then Equation (9.16) may be expanded to yield Equation (9.21). This consists of two time dependent terms scaled by functions of the current phases [17, p. 32]. Essentially, the result is a rotating field vector with an elliptical path. However, there are still orientations where power is absent. As such, some form of amplitude, frequency, or phase modulation is required.

$$|H| \propto \sqrt{\sum_{i=1}^{3} [I_0 \sin(\omega_C t) \cos(\phi_i) + I_0 \cos(\omega_C t) \sin(\phi_i)]^2} \tag{9.21}$$

Amplitude modulation can be applied to the primary exciting currents. One such technique is to simultaneously amplitude modulate two orthogonal axes, at constant phase operation ($\varphi_1 = \varphi_2 = \varphi_3 = 0$). In this case, two of the applied sinusoidal source currents are scaled by the modulating signal with modulating angular frequency ω_M so that the input currents are given by Equations (9.22–9.24). The resulting field is given by Equation (9.25).

$$x(t) = I_0 \sin(\omega_C t) \sin(\omega_M t) \tag{9.22}$$

$$y(t) = I_0 \sin(\omega_C t) \cos(\omega_M t) \tag{9.23}$$

$$z(t) = I_0 \sin(\omega_C t) \tag{9.24}$$

$$|H| \propto I_0 \sqrt{\sin^2(\omega_C t) \sin^2(\omega_M t) + \sin^2(\omega_C t) \cos^2(\omega_M t) + \sin^2(\omega_C t)}$$
$$(9.25)$$

The resulting solution forms a spherical omnidirectional field with constant magnitude traced around the outline. By analogy to Double Sideband Suppressed Carrier (DSSC) in radio transmission, the Fourier transform of the resulting field occurs at ω_C and $\omega_C \pm \omega_M$. Notably, the latter frequencies have half the field magnitude compared to the former.

Frequency modulation can be implemented by applying alternate frequencies to one or more primary current coil(s). In this case, if the carrier frequency along the z-direction is ω_{FM} such that $\omega_{FM} \neq \omega_C$, the input currents are given by Equations (9.26–9.28). The resulting field is given by Equation (9.29).

$$x(t) = I_0 \sin(\omega_C t) \tag{9.26}$$

$$y(t) = I_0 \cos(\omega_C t) \tag{9.27}$$

$$z(t) = I_0 \sin(\omega_{FM} t) \tag{9.28}$$

$$|H| \propto \sqrt{I_0^2 (1 + \sin^2(\omega_{FM} t))} \tag{9.29}$$

This can lead to an omnidirectional field [17, p. 33] albeit at multiple frequencies. In terms of NFMI systems, this is problematic as pickup compensation is ideally chosen at a single frequency. As a result, such a scheme would restrict the quality factor compromising power transfer capabilities. However, in order to avoid this multi-tuning problem, the frequency separation can be made small. However, if this separation is too small, then the beat frequency is too low and can lead to null power transfer [17, p. 33]. The other form of modulation, phase modulation, can be regarded as frequency modulation and will not be discussed further.

9.4.3.2 Discontinuous mode of operation

The sequential application of current to three orthogonally oriented primary coils to align with the pickup axis is the simplest form of discontinuous operation. This mode has the advantage of being relatively uncomplicated and easy to implement. As this case is just an extension of the single-plane solution [17, p. 33], the field vectors are identical to those created in a bidirectional system except that the plane of rotation must be taken into account. This technique has been applied previously in biomedical implantation omnidirectional links [17, p. 34] and various approaches exist. One approach is to periodically switch the plane of field rotation based on the threshold level of power induced by a pickup. This is readily achieved by applying current in phase quadrature to the x and y coils and then to the y and z coils and finally to the z and x coils as shown in Figure 9.20. This makes it necessary to take into account the plane of field rotation with respect to the pickup axis.

In fact, in such a case it is unlikely that the pickup axis will always be aligned parallel to the resulting flux from the primary coil. This makes it difficult to maximize power transfer. Furthermore, this approach is difficult to implement with multiple pickups. Another approach is to operate the coils on a time-sharing basis [17]. The idea is to excite each coil for one-third of a periodic duration. However, this is difficult when the pickup axis orientation is also rotating in anti-phase with the flux being generated. Furthermore, [18] showed that the maximum possible attainable power is one-third the magnitude compared to the case when the field is perfectly aligned with the pickup.

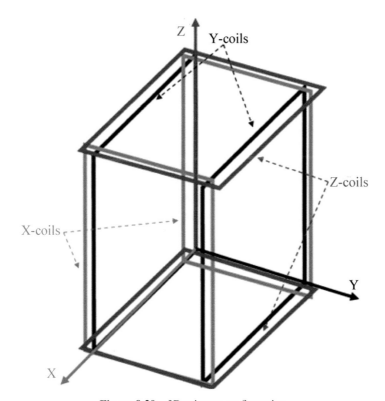

Figure 9.20   3D primary configuration.

This not only compromises power-transfer capabilities but also efficiency. In summary, there exist various continuous and discontinuous techniques [17] to supply power and the most suitable mode of operation must be chosen based on the application requirements.

9.4.3.3 Development

During development of the primary magnetic structure, the first design consideration is often the powering volume. In the case of a 3D powering volume, the air-region including normal and tangential displacements lays the foundation of the basic design shape—a 3D box. In order to prevent magnetic flux leakage outside the inner powering volume of the 3D box, ferrite is used as casing of the box. This will also act to enhance the magnetic field and provide a low-reluctance magnetic flow path. One such primary coil configuration that can form a box-like structure is to orient two parallel rectangular coils along two orthogonal axes with opposite polarity of current $I(t)$ for which there are two modes of operation: continuous and discontinuous.

The case of a continuous mode of operation is illustrated in Figure 9.21. In this case, all 4 coils are excited with current in-phase but of opposite polarity along both orthogonal axes. As this design is essentially produc- ing a net magnetic flux along two orthogonal axes, there is going to be interference. This interference results in two diagonally oriented regions of particularly low magnetic flux density defined as the 'Weak Zone' illustrated in Figure 9.21.

The case of discontinuous mode of operation aims to overcome this problem, as illustrated in Figure 9.22. In this case, the solution is to have a 90-degree phase displacement between the currents in the coils oriented along the horizontal axis to the coils oriented along the vertical axis. Such

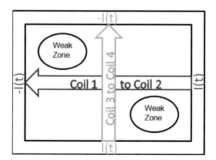

Figure 9.21   3D box (top view): Magnetic flux pathway in continuous mode of operation.

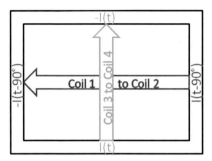

Figure 9.22 3D box (top view): Magnetic flux pathway in discontinuous mode of operation.

form of operation will ensure that the coils oriented along the horizontal axis turn on or off when the coils oriented along the vertical axis turn off or on. The net result is a rotation of the magnetic field as a function of time within the box.

The resulting magnetic structure termed 'Design 1' is shown in Figure 9.23. As it is not visible, it must be noted that identical coils are modelled on the opposite faces of the ferrite. In this design, the number of rectangular turns in all four track coils is constant. After FEA analysis on this design, it is evident that the central region of the box encounters a weak magnetic flux density also shown in Figure 9.23. This is in contrast to the 2D design which encounters a central peak in magnetic flux density.

In order to improve the central weak region, it is proposed to have an increasing number of coil turn density towards the central region of the rectangular loop. The resulting design, 'Design 2', is shown in Figure 9.24. The increased number of coil turns in the central region of the box is expected to improve the magnetic flux density in the weaker central region. This may be verified by simulating a magnetic frequency analysis on the developed model for a current of 4.5 A at 155 kHz. The resulting magnitude of magnetic flux density through a central plane covering the air-region within the box of area 130 × 90 mm is also shown in Figures 9.23 and 9.24. As the result shows, the case of constant turn density has magnetic flux peaks that gradually decline towards the central region of the box. In contrast, 'Design 2' has successfully improved the magnitude of magnetic flux within the central weak region by increasing coil turn density towards the low-power zone.

Another aspect of discussion with regard to such a 3D-radiating source is the level of magnetic field uniformity. It can be seen that the magnetic flux density peaks at the central edges of the box, while the top-most and

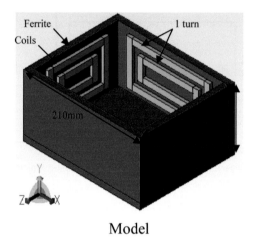

Model

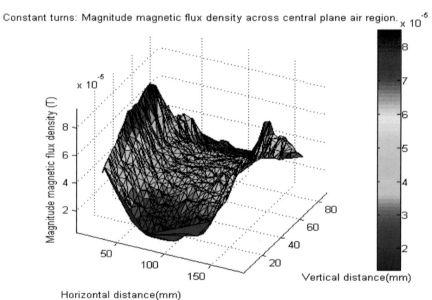

Magnetic flux density distribution

Figure 9.23 Design 1: Constant coil turn density.

bottom-most regions of the box encounter a weaker net magnetic flux density. On one hand, while a pickup is oriented at a region of maximum magnetic flux, it is able to induce a higher EMF to satisfactorily operate any load(s).

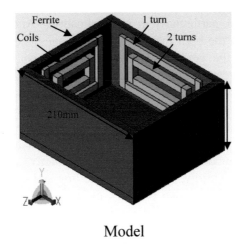

Model

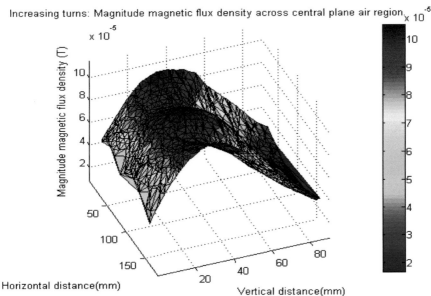

Figure 9.24 Design 2: Increasing coil turn density.

On the other hand, if the pickup is oriented at a region of minimum magnetic flux, it may not induce a sufficiently high EMF to satisfactorily operate any load(s). As such, the magnetic flux density variation within the powering volume must be minimized for satisfactory uniform power induction. This is a current research challenge and a significant limitation in the design of many NFMI systems, particularly in 3D applications. Another limitation in such a 3D design is the high magnetic flux leakage. The magnetic flux leakage is defined as the portion of magnetic flux outside the intended powering volume with respect to the total magnetic flux. This is an important parameter for consideration during design of the magnetic structure and particularly the use of ferrite. In the developed system, the ferrite can be seen to have three functions. Firstly, the ferrite enhances the magnetic field due to the high relative permeability of the ferromagnetic material. Secondly, the ferrite acts as a magnetic circuit to provide a conductive pathway for the resulting magnetic flux. Thirdly, by satisfying the first and second functions, it can be said that the ferrite also confines the magnetic field to provide electromagnetic shielding. However, the top of the box remains unshielded and is prone to causing magnetic flux leakage. In fact, the magnetic flux density at a plane of 4 cm on top of the box or intended powering volume was averaged to be 6.12 μT. This may be considerably high and can result in potentially harmful EMI with radiation sources nearby. As such, another key research area in relation to the design of 3D NFMI systems is often associated with electromagnetic shielding effects to comply with regulatory requirements.

9.5 Conclusions

This chapter has provided a general overview of the state-of-the-art near-field magnetic induction technologies and a summary of example applications. The chapter has also presented the development of the conceptual magnetic structures of a two-dimensional and three-dimensional wireless charging system. This was aided by modern finite-element-method simulation software to demonstrate the state-of-the-art research and design techniques. These designs also helped to demonstrate the current research states and challenges with regard to inductive power transfer charging systems and technologies.

9.6 Problems

P9.1: What are the advantages of zero-voltage switching during power conversion?

P9.2: Consider a planar charging system. The spiral primary track loop has an inductance of 1.268 μH, equivalent series resistance of 283.8 mΩ and an

AC track current of 1 A at 200 kHz. What is the power rating requirement on the input power supply?

P9.3: Does the system of Question 2 satisfy the zero-phase angle (ZPA) requirement? If not, how can the ZPA condition be achieved?

P9.4: Consider a series compensated primary circuit operating at 1 MHz with a track inductance of 1 μH and compensating capacitance 10 nF. Is this circuit tuned? If not what is the complex input impedance?

P9.5: What is the value of compensating capacitor required to tune the circuit of the Question 4?

P9.6: Describe the steps in the finite element formulation used to transform the wave equation into a compact matrix form.

P9.7: Sketch the magneto-motive force (MMF) distribution over a spiral primary track loop.

P9.8: The open-circuit voltage and short-circuit current on pickup coil has been measured to be 5 V and 0.5 A. What is the induced uncompensated power?

References

[1] Griffiths, D. *Introduction to Electrodynamics*, 3rd edition. Prentice Hall, New Jersey, ISBN 0-13-805326-X (1998).
[2] Andreycak, B. *Zero Voltage Switching Resonant Power Conversion, Application Note*. Unitrode Corp., slus138 (1999).
[3] Hu, A.P. Selected Resonant Converters for ICPT Power Supplies. PhD thesis, Department of Electrical and Computer Engineering. University of Auckland, October (2001).
[4] Tan, C. *Enhanced ICPT System for Wireless Computer Mouse*. ME thesis, Department of Electrical and Computer Engineering, University of Auckland, February (2006).
[5] Hui, S.Y.R., and W.C. Ho. *A New Generation of Universal Contactless Battery Charging Platform for Portable Consumer Electronic Equipment*, 35th Annual IEEE Power Electronics Specialists Conference, Aachen, Germany (2004).
[6] Kung, G., D. Kacprzak, G. Covic, and J. Boys. "A study of pick-ups performance for inductively coupled power transfer systems" *Application of Electromagnetic Phenomena in Electrical and Mechanical Systems*, Auckland, New Zealand (2004).

[7] Keeling, A.K., A.G. Covic, and T.J. Boys. A Unity-Power-Factor ICPT Pickup for High-Power Applications, *IEEE Transactions on Industrial Electronics*, Raleigh, USA, February (2010).

[8] Boys, J.T., G.A. Covic, and A.W. Green. "Stability and control of inductively coupled power transfer systems." *Electric Power Applications, IEE Proceedings* 147, no. 1, pp. 3743 (2000).

[9] Covic, G.A., J.E. James, and J.T. Boys. Analysis of a series tuned ICPT pickup using DC transformer modelling methods. In *Proceedings of the 6th International Power Engineering Conference,* pp. 5152, Singapore (2003).

[10] Boys, J.T., and A.W. Green Boys. "Intelligent road-studs–lighting the paths of the future, Intelligent road-studs–lighting the paths of the future." *IPENZ Transactions* 24, no. 1: 33–40 (1997).

[11] Covic, G.A., G. Elliott, O.H. Stielau, and R.M. Green. "The Design of Contactless Energy Transfer System for a People Mover System." *Proceedings of International Conference on Power System Technology*, Perth, Australia, pp. 79–94, December (2000).

[12] Covic, G.A., G. Elliott, O.H. Stielau, R.M. Green, and J.T. Boys. "Design of the contact-less battery charging system for the people movers at the Wakarewarewa Geothermal Park.: *Power System Technology* (2000).

[13] Raval, P. "Scalable Inductively Coupled Power Transfer Platform." ME thesis, Department of Electrical and Computer Engineering, University of Auckland, December (2009).

[14] Silvester, P.P., R.L. Ferrari. *Finite Elements for Electrical Engineers*, 3rd edition. Cambridge: Cambridge University Press (1996).

[15] Raval, P., D. Kacprzak P. Hu. "Scalable Inductively Coupled Power Transfer Platform." *Proceedings of 17th Electronics New Zealand Conference* pp. 57–62, Hamilton, New Zealand, November (2010).

[16] Raval, P., D. Kacprzak, A.P. Hu. "A Wireless Power Transfer System for Low Power Electronics Applications." *Proceedings of 6th Industrial Electronics and Applications Conference*, Beijing, China, June (2010).

[17] McCormick, J.D. "Omnidirectional Inductive Power Transfer from a Surface." PhD thesis, Department of Electrical and Computer Engineering, University of Auckland, New Zealand, March (2011).

[18] Ping, S., A.P. Hu, S. Malpas, and D. Budgett. "A Frequency Control Method for Regulating Wireless Power to Implantable Devices, Biomedical Circuits and Systems." *IEEE Transactions on Biomedical Circuits and Systems* 2: 22–29 (2008).

[19] Casanova, J.J., Z.N. Low, J. Lin, and R. Tseng. "Transmitting Coil Achieving Uniform Magnetic Field Distribution for Planar Wireless Power Transfer System." *Proceedings of the 4th International Conference on Radio and Wireless Symposium.* San Diego, CA, USA (2009).

10

Wireless Power Transfer in On-Line Electric Vehicle

10.1 Introduction

This chapter provides a general overview of a wireless power transfer system using magnetic field and the application toon-line electric vehicles. The magnetic field shape design and shielding technologies are introduced and discussed. The simulation and measurement results supporting the design methods are explained.

10.1.1 Wireless Power Transfer Technology

Since the invention of long-distance radio transmission by Guglielmo Marconi (1874–1937), the wireless telecommunication technology came to public and the wireless communication technologies rolled into our lifestyle. The wireless communication became a vital part of our everyday lives, and most modern people cannot even imagine a life without a cellular phone or a laptop computer. In the innovation of wireless communication, the technological improvement of energy storage system is also credited for its realization. People want more time to be free from the wire. However, they want higher-performance electronic devices at the same time. The huge necessity of mobile electronics with high computing power induced the development of high energy density lithium battery.

However, here comes a movement of a new revolution in wireless technology to remove the last wire for power transmission. The dream of wireless power transfer originates from the experiment of Tesla Tower (Figure 10.1) by Nikola Tesla (1856–1943) who sought to transmit power through a waveguide between the earth's surface and the ionosphere in the early 1900s [1].

Even though the experiment was not very successful at that time, researches on wireless power transfer technology have been continuing and there were several applications as shown in Figure 10.2 [2]. There were small

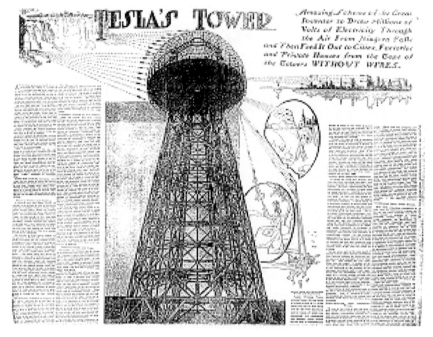

Figure 10.1 Tesla Tower constructed in 1904 [1].

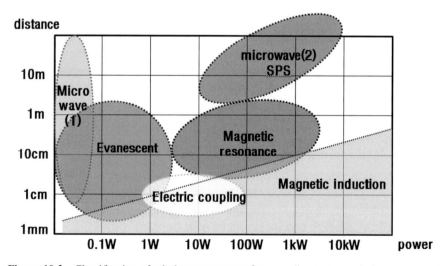

Figure 10.2 Classification of wireless power transfer according to transmission power and distance [2].

power microwave application, solar power satellite (SPS), magnetic induction for electric brush and cooker, and so on.

In the early 2000s again, the researches on wireless power transfer technologies became one of the hottest issues. As people want to pay more money for the convenience and safety, the necessity of wireless power transfer technology increased, and more engineers and scientists started to verify the mechanism and improved the technology for the application which end users finally want.

Although the research started from the universities and research institute, now prototypes have been developed by companies and some companies already introduced commercial products. Generally, this technology includes the removal of the cable used to recharge small electronic devices and the removal of power cord for home appliances.

10.1.2 Wireless Power Transfer System in the Market

Even though this technology is known for years, the portion of electronic devices with wireless power transfer technology was extremely small. After the publication of a paper on the Science magazine by a research group in Massachusetts Institute of Technology, lots of research on wireless power transfer technology has been initiated and more interests on this technology [3]. Recently, there were some movements of commercialization in a few years by some companies with the idea that the receiver coils can be embedded in consumer applications such as mobile phones, MP3 players, digital cameras, and laptop computers (Figure 10.3). However, the power regulation for the danger of over-powered devices and permanent damage, communication and control protocol, and interoperability with current market product were big challenges [4].

According to the research by Cahners In-Stat and iSuppli, the market of wireless charging will be increased from 0.38 billion dollars in 2010 to 4.3 billion dollars in 2014 [5, 6]. Forty percentage of the customers interviewed are willing to buy the wireless charger if the price is around 50 dollars. So, it is expected that the wireless charging system will become popular in the near future when the cost of devices will be lowered.

After the wireless charging system is on the market (Figure 10.4), the necessity for the standardization has been increased. The wireless power consortium was founded in December 2008, and the Qi interface, which defines the interoperability between the power transmitter and power receivers, was established to guarantee product interoperability for wireless battery charging,

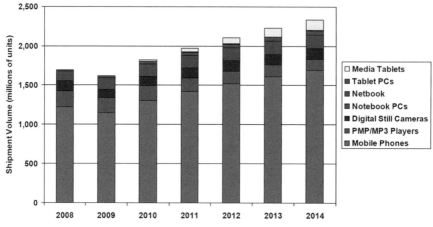

Figure 10.3   Market overview for key verticals [4].

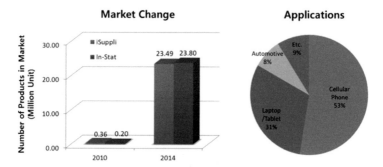

Figure 10.4   Market trend of wireless charger [4, 5].

give design freedom, allow product differentiation, and provide access to essential patents [7].

However, most of the commercial products in the market adopted inductive coupling, and the application has the limitation of distance from the transmitter to the receiver. Although the electronic device and charging station are not electrically connected with some electrodes exposed outside, the housing of charging station and mobile devices are still contacted or almost contacted. To enhance the distance and make evolutionary convenience, we need to apply the magnetic coupling with resonance with higher Q-factors.

10.1.2.1 Application to automobiles

The starting point of development was from the small portable consumer electronics. But if we change our sight to other electronic systems, we can find many more applications that use wireless power transfer technology. One of the biggest applications is the automobiles. As the automotive vehicles use more and more electronic devices inside, wireless power transfer technology can be applied to transfer power without wires. Also, the consumer electronic devices used inside the vehicle such as cellular phones or laptop computers can be recharged wirelessly.

However, the most challenging technology is to transfer the power to the vehicle. General electronic devices have the power of less than 100 W so it is relatively easy to implement and less challenging to enter the market. If we are going to apply the wireless power transfer technology to supply the power for the vehicle, the first thing we have to consider is to increase the scale of power to more than kilowatt. Increasing the power source is not a simple problem. When we increase the power, the problem extends to areas of high-power semiconductor devices, ferromagnetic materials, magnetic field safety, power transfer efficiency, thermal issues, weight and size, and cost. These issues can be significant for small-power applications; however, it becomes much more critical for high-power applications.

Even though there are many difficulties in the implementation of wireless power transfer system in electric vehicles, more attention is attracted because of the necessity of electric vehicles. In 1994, a Partners for Advanced Transit & Highways (PATH) project research group led by UC Berkeley derived research results on wireless power transfer technology [7].

They have proved that the power of 10 kW per module can be transferred with power efficiency of 60% from the infrastructure constructed under the road (Figure 10.5). The air gap which is defined as the distance between the road surface and the bottom of the vehicle was 7.5 cm when they used 2000 A for supply current. The vehicle is designed to get the power wirelessly no matter the vehicle is stopped or in motion. It was a reasonable approach and a motivation of the following researches; however, the small air gap was not enough to guarantee the comfort in driving on the real road. For example, all the vehicles have the air gap at least larger than 12 cm, and most of the vehicle has larger air gap because of speed bumps. Moreover, 2000 A of current was large to induce strong magnetic field to transfer power wirelessly, and this was at the same time a problem to exceed the magnetic field regulation which is adopted in many countries. Cost reduction and audible noise problems were also left to solve.

There were more research groups in automobile companies, universities, and research institutes, who were trying to use this technology to the electric vehicle applications in different manners. Figure 10.6 shows the

Figure 10.5   Wireless power transfer system for electric bus implemented by the PATH project team in 1994 [7].

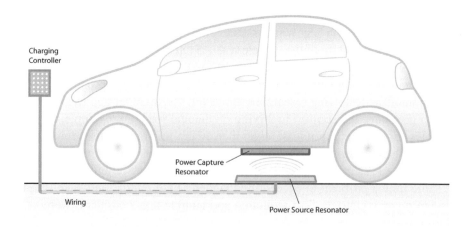

Figure 10.6   The wireless charging system for automobiles involving no plugs or charging cords [8].

wireless charging system for automobiles without any plugs or charging cords. Electricity flows upward from a power source resonator to a power capture resonator installed on the vehicle [8].

10.2 Mechanism of Wireless Power Transfer

10.2.1 Electric Field and Magnetic Field

The power can be transferred in several ways. The electric power is generally transferred by the movement of electrons in a conductor or semiconductor when the transmitter and receiver are electrically connected. To transfer the power wirelessly when transmitter and receiver are electrically separated, the electromagnetic field can be the medium of transmission. According to Maxwell's equations (Equation 10.1), time-varying electric field generates magnetic field, and time-faring magnetic field generates electric field, and the ratio of electric field to magnetic field is defined as the wave impedance. So, if we measure the magnitude of electric field and magnetic field at the observation point which is far from the source, the ratio of electric field to magnetic field will converge to a constant value of wave impedance.

$$
\begin{aligned}
\nabla \times E &= -\frac{\partial B}{\partial t} \\
\nabla \times H &= J + \frac{\partial D}{\partial t} \\
\nabla \cdot D &= \rho \\
\nabla \cdot B &= 0
\end{aligned}
\tag{10.1}
$$

The characteristics of a field are determined by the source, the media surrounding the source, and the distance between the source and the point of observation. At a point close to the source, the field properties are determined primarily by the source characteristics, and far from the source, the properties of the field depend mainly on the medium through which the field is propagating. Therefore, the space surrounding a source of radiation can be broken into two regions. If the observation point is close to the source, the field is in near-field region. If the observation point is far from the source, the field is in far-field region. The region is a transition region if

$$
\text{Distance} = \frac{\lambda}{2\pi}
\tag{10.2}
$$

As the value of lambda is proportional to the wavelength, the near-field region means the region where the distance is much smaller than the wavelength [9].

As shown in Figure 10.7, if the normalized distance is much larger than 1, it is in the far-field region, and wave impedance converged to around 377 ohm. If the normalized distance is much smaller than 1, it is in the near-field region, and the region can be categorized into two conditions: electric field predominant and magnetic field predominant. If dipole antenna is used for the generation of electromagnetic field, electric field is predominant and the wave impedance is higher than 377 ohm. If loop antenna is used, magnetic field is predominant and the wave impedance is lower than 377 ohm. In most cases of wireless charging for electric vehicle, the near-field region in magnetic field predominant condition is used.

Both electric field and magnetic field can be used of power transmission. However, most of the application adopted magnetic field transmission because magnetic field has a characteristic that it has less attenuation and reflection due to high dielectric material because magnetic field is less affected by permittivity. Also, magnetic coupling structure is easier to implement than

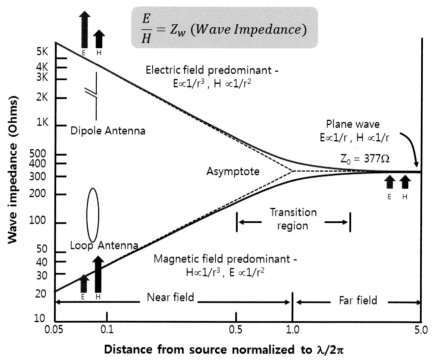

Figure 10.7   Wave impedance depending on the distance from the source [9].

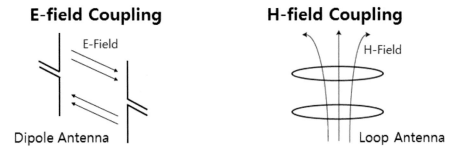

Figure 10.8 Transmission and reception of electric field and magnetic field using dipole antenna and loop antenna.

electric field coupling. As shown in Figure 10.8, the magnetic field can be generated and received by a loop antenna, and large wires can transmit and receive high power by magnetic field. In electric field coupling, however, large capacitance between transmitter and receiver is required, and this requires large plates in transmitter and receiver rather than dipole antenna with wires.

10.2.2 Inductive Coupling and Resonant Magnetic Coupling

Wireless power transmission by induction has been used for a long time in the form of electric toothbrushes. The electric toothbrushes have to use inductive coupling technology as a means of recharging their batteries because they are constantly exposed to water. This type of wireless charging system can be categorized to inductive coupling. The inductive coupling generally implies the wireless power transfer in a few millimeters of distance from the transmitter coil to the receiver coil (Figures 10.9 and 10.10).

As the coils have a short distance, the magnitude of leakage of magnetic field is very small, and the efficiency is very high in the range of 60–90 %. The limitations of the inductive coupling are that the distance is difficult to be increased and that the alignment of the transmitter and receiver should be accurate. Not fulfilling these conditions means the degradation of the transfer power [10].

The limitations of distance and alignment accuracy can be solved by applying the resonance circuit of the transmitter and receiver systems which is also called resonant magnetic coupling, and the theory is clearly explained in [3] using coupled-mode theory. After that, some researches show that this idea is also valid for different frequency ranges and power [11].

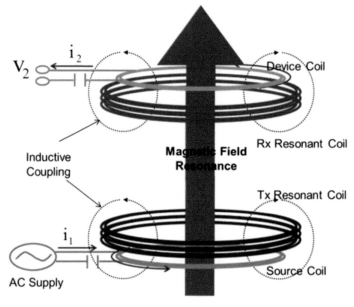

Figure 10.9 Magnetic field generation [10].

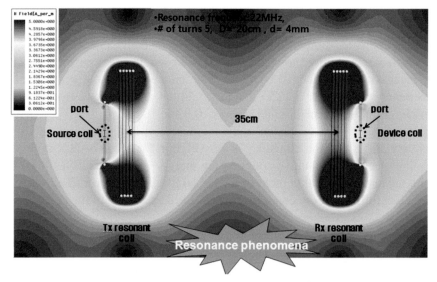

Figure 10.10 Example of simulated resonant magnetic coupling [11].

10.2.3 Topology Selection and Coil Design

The core of wireless power transfer system consists of a power source, transmitter coil, receiver coil, and load as shown in Figure 10.11. The design of each component is significant to implement a wireless power transfer system because the coil is tightly related to the dimensions, circuit design, transfer power, efficiency, magnetic field shape, and cost. So, source types, coil topologies, feeding types, and load types should be carefully determined at the beginning of the design.

Figure 10.12 shows the feeding types of a transmitter coil and receiver coil. The left figure shows indirect-fed coils where the source coil and transmitter coil are separated. In this case, the transmitter coil and receiver coil have high Q-factor values because the resistance of source and load is not included, and hence the distance of the wireless power transfer can be increased. However, just like the other system with high Q-factor, this system can be very sensitive to design parameters such as inductance value and resonance frequency. So, direct-fed wireless power transfer system can be useful when stability is more important than the transfer distance [12].

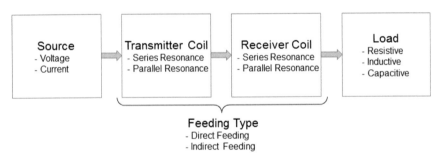

Figure 10.11 Overall structure of a wireless power transfer system.

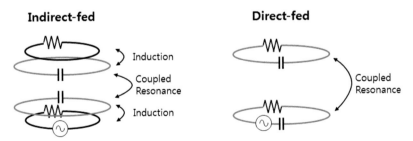

Figure 10.12 Two types of feeding for transmitter and receiver coils: indirect-fed coil and direct-fed coil.

In the design of coils of wireless power transfer system, the Q-factor significantly affects the power transfer efficiency. In Figure 10.13, coils with higher Q-factor can have higher power transfer efficiency if the frequency of the transmitted power and the resonance frequency of the system are exactly the same. However, if there are some changes in the resonance frequency due to process variations, inaccuracy of lumped component values, or dimension change by temperature increase, the efficiency can be changed. So, for more stable operation of the wireless power transfer system considering the manufacturing issues, the proper Q-factor should be selected first depending on the requirement of the application.

Figure 10.14 shows two types of transmitter coil and receiver coil with different resonance types. The left figure shows the self-resonant coil where the spiral wire has its own resonance due to the inductance of the wire and capacitance between each turn of wire. The right figure shows the LC

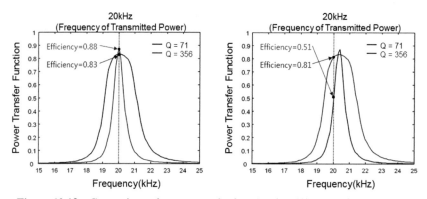

Figure 10.13   Comparison of power transfer function for different Q-factor of coils.

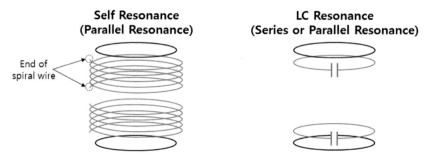

Figure 10.14   Two resonance types of transmitter and receiver coils: self-resonance coil and LC resonance coil.

resonant coil where the inductance of coil and lumped capacitor generates series resonance. The lumped capacitor component can make more loss than the fringing capacitance of self-resonant, but it is more controllable and easier to manufacture. Similar to Figure 10.12, the selection of resonance type determines the loss of transmitter and receiver coils and makes different coupling factor and distance.

To implement a wireless power transfer system with longer distance and with higher efficiency, indirect-fed, self-resonant coil can be a better choice; however, when we focus on the design of the wireless power transfer system for electric vehicle, the robustness and the less sensitivity to design parameter are important for direct-fed, LC resonant coils. Figure 10.15 shows the example of basic topologies of direct-fed, LC resonant coils for wireless power transfer system. There can be 24 combinations of the system with two source types, two transmitter coil types, two receiver coil types, and three load types. In selecting resonance types between series resonance and parallel resonance, the impedance of the load should be considered. For maximum power transfer, the impedance of the receiver coil should be matched with the load. So, if the load has higher impedance, the receiver coil with parallel resonance should be selected and for lower load impedance, the receiver coil with series resonance should be selected [13].

10.3 Design of On-Line Electric Vehicle

10.3.1 Necessity of On-Line Electric Vehicle

The development of online electric vehicle (OLEV) started in 2009 as an effort to address the two major problems the world is facing: resource exhaustion and environmental pollution. While intensive research has been performed on fully electric vehicles for a long time, there remain serious problems to be solved in the battery-powered electric transportation system (Figure 10.16). It is well known that electricity is a cheaper alternative to petroleum resources and is also preferable in terms of preventing air pollution. Many major automotive companies have developed electric cars, but issues of enlarged size and weight and high battery cost must be addressed. Moreover, diminished stocks of lithium could cause increasingly high prices and force electric vehicles out of the automotive market. Even if battery capacity is increased, other problems of long recharging time, safety concerns, limited availability of charging service points, and construction cost will have to be confronted.

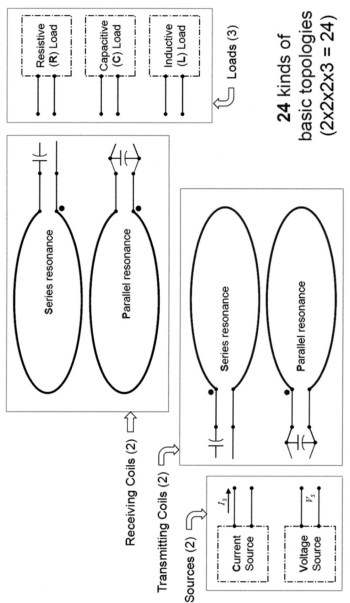

Figure 10.15   Example topologies of direct-fed, LC resonant coils for wireless power transfer system [13].

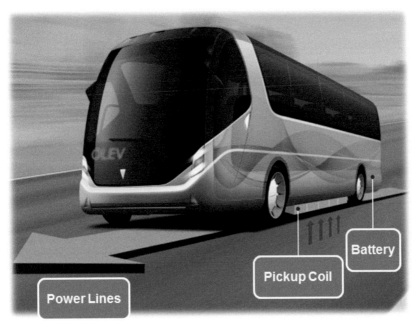

Figure 10.16 Concept of on-line electric vehicle. An electric vehicle can have minimum batter inside vehicle because of the wireless power transfer system between infrastructure and vehicle [13] (http://olev.kaist.ac.kr).

In order to resolve the battery problems, Korea Advanced Institute of Science and Technology (KAIST) invented the OLEV, a new type of electric vehicle that is powered from road-embedded infrastructure using wireless charging technology. The pickup coils (receiver coils) in OLEV constantly receives and recharges its power from power lines (transmitter coils) embedded under the road. As it has the reduced battery capacity of about 20 % compared to conventional battery-powered electric vehicles, it can consequently minimize the weight and the price of the vehicle as well as the construction cost of power stations.

The concept of a wireless power transfer system applied for electric vehicles, where energy is transferred continuously while the vehicle is stopped or in motion, was introduced as far back as 20 years ago [7]. However, the realization of a wirelessly powered electric vehicle as a substitute for fuel engine vehicles has proved extremely challenging because of demanding requirements.

10.3.2 Challenges

One of the greatest challenges in the development of OLEV was economic feasibility in design and manufacturing. The total cost of each component used only in OLEV and its infrastructure such as inverter operating in 20 kHz, cable for power lines (transmitter coil) and pickup coils, ferromagnetic material for magnetic field shaping, rectifier, regulator, and construction cost should be minimized to satisfy the economic competitiveness.

To drive a conventional bus, the transfer power should be around 100 kW, and the transmitting and receiving coil should be large enough to endure that amount of power transfer. When the magnitude of the current in power lines reaches 200 A, the voltage could be 500 V. The power transfer efficiency should be more than 80 % for economical usage, and the air gap should be at least 20 cm for general driving. It was not easy to satisfy this condition. As shown in Figure 10.17, the air gap of the other wireless charging electric vehicles is less than 10 cm, which is not applicable to wireless charging when the vehicle is moving.

Moreover, electromagnetic field (EMF) regulation for protection of the human body from exposure of the magnetic field from the OLEV system is one of the important criteria in terms of safety. Just like electromagnetic

Institute, Nation	Vehicle		Air gap	Efficiency	Out put	Charging
KAIST, Korea	On-line Electric Bus		20cm	Above 80%	75kW	Driving/ Stopping
HINO, Japan	Electric Bus		3cm	95%	–	Stopping
Auckland Univ., New Zealand	Passenger Car	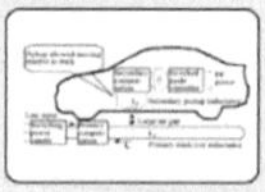	4.5cm	93%	32kW	Stopping
PATH Program, USA	Mini Electric Bus		7.5cm	Below 60%	6 ~ 10kW per module	Driving/ Stopping

Figure 10.17 Comparison of some electric vehicles with wireless power transfer.

interference (EMI) regulation, which is a critical requirement for electronic devices, EMF regulation is defined in most of the countries. In the application of mobile devices, the magnitude of current is in the range of a few amperes; however, the vehicle requires tens or hundreds of amperes and sometimes the coil with that current has several turns resulting in thousands of ampere-turns. However, shielding of the leakage magnetic field is a difficult problem when the frequency of the transfer power is as low as kHz and magnetic field is predominant. As the frequency is 20 kHz, the wavelength is about 15 km which is much longer than the dimension of the cable or vehicle, and therefore it is in the range of near-field. The shielding of magnetic field would be easier if the magnetic field is generated from a fixed structure; however, it is more difficult to block the magnetic field from the vehicle because it has to move with an air gap and the magnetic field passes through that air gap.

Before 2009, there have been no significant research results to satisfy the aforementioned requirements, and thus at the beginning of the project, few engineers believed this could be achieved successfully solving these challenging problems.

10.3.3 Topology Analysis

To simplify the theory and to explain with simple equivalent circuit, two LC resonant circuits with mutual inductance are popularly used as shown in Figure 10.18. The transmitter coil and the receiver coil have the mutual inductance (M) for wireless power transmission, while they also have the self-inductances. By adding the capacitances (C1 and C2) in the circuits of transmitter and receiver, the self-inductance of the inductance is cancelled by the capacitance, and then, the total impedance of each loop is decreased. To minimize the impedance maximally, the resonance frequency of the LC resonance should be the same as the frequency of the source (I1). As a result,

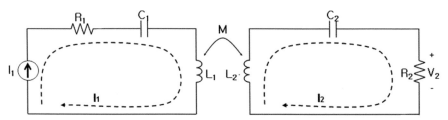

Figure 10.18 Simplified equivalent circuit of current-source wireless power transfer system with resonant magnetic field.

the imaginary part of the loop is minimized and the efficiency can be increased even if the distance is longer than inductive coupling. The calculation of the power transfer efficiency in the circuit is shown in Equation (10.4).

$$\text{Induced voltage}: V_2 = j\omega M i_1 \tag{10.3}$$

$$\text{Efficiency}: K = \frac{\omega^2 M^2}{R_1 R_2 + \omega^2 M^2} = \frac{1}{1 + \dfrac{R_1 R_2}{\omega^2 M^2}} = \frac{1}{1 + \dfrac{1}{Q_{M1} Q_{M2}}} \tag{10.4}$$

10.3.4 Coil Design for Electric Vehicle

Figures 10.19 and 10.20 show the vertical magnetic flux type of power lines and pickup coils [14]. There are two power lines with opposite current directions underneath the road surface forming a current loop. Due to the current in the power lines, magnetic flux is induced around each power line. The magnetic fluxes from the two power lines are added at the place between the power lines, and the magnetic fluxes are cancelled outside of the power lines. The pickup coil catches the vertical magnetic flux through copper coil loops around the ferrite core. The ferrite core is the most significant material

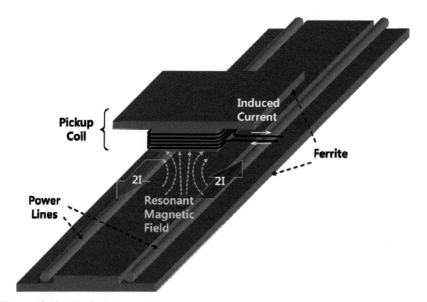

Figure 10.19 Vertical magnetic flux type power lines and pickup coil in perspective view.

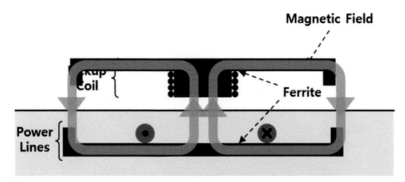

Figure 10.20   Vertical magnetic flux type power lines and pickup coil in cross-sectional view.

to apply the shaped magnetic field in resonance (SMFIR) technology, which is the key of wireless power transfer design in OLEV to maximize the transfer power and to minimize EMF. This type of power line and pickup coil pair has the advantage of efficient power transfer because the direction of the magnetic flux from the power lines is the same as the direction of the flux to the pickup coil.

Figures 10.21 and 10.22 show the horizontal flux type of power lines and pickup coil. In this type, there are four power lines forming two current loops. As two center conductors are placed close together, there are three magnetic flux loops in Figure 10.22. The copper pickup coils around the horizontal ferrite core catches the horizontal magnetic flux to generate power for vehicles. This type has a disadvantage in that the transfer power is less than the vertical flux type because induced magnetic field directions of power lines and pickup coils are perpendicular. The difficulty of increasing pickup coil loop size is another disadvantage because the increase of pickup coil loop size means the total thickness of the pickup coil. However, this type of power line and pickup coil pair has the current distribution generating weaker magnetic field at the side of the vehicle.

10.3.5 Electromagnetic Field Reduction Technology

Figure 10.23 shows the magnetic flux density distribution of OLEV. In the case of the vertical magnetic flux type, there is one magnetic flux path between the power lines and pickup coils where the power is transferred. The return flux comes back to the power lines via the sides of the main flux path. The horizontal magnetic flux type has two magnetic flux paths. The side power lines of this type have return flux paths on the side of the main flux path. The

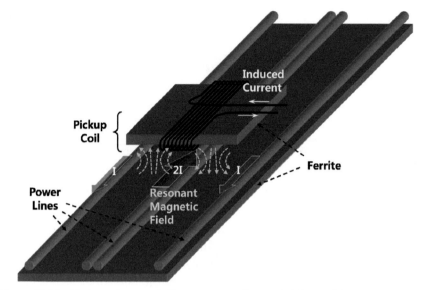

Figure 10.21 Horizontal magnetic flux type power lines and pickup coil in perspective view.

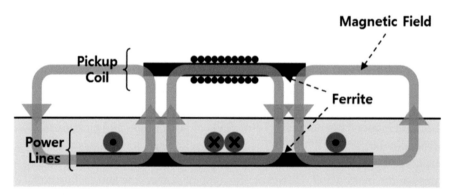

Figure 10.22 Horizontal magnetic flux type power lines and pickup coil in cross-sectional view.

return flux path creates the fringing magnetic flux, and this flux is measured as the EMF level of OLEV. In this work, the target EMF level of OLEV is 6.25 μT according to the regulation of Korea Communications Commission which follows the ICNIRP design guideline shown in Table 10.1 [15].

As the power supply system of OLEV generates large amounts of magnetic field to transfer 100 kW of power which is necessary for the vehicle, thousands μT of magnetic flux between the power lines and pickup coils beneath the

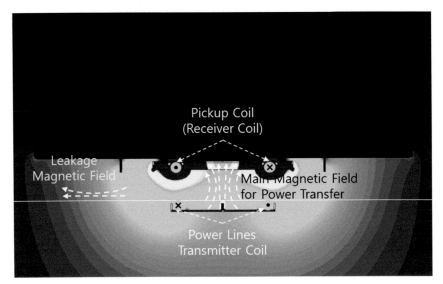

Figure 10.23 Distribution of magnetic field for OLEV.

Table 10.1 Distribution of magnetic field for OLEV

Frequency range	B-field (μT)
up to 1 Hz	4×10^4
1–8 Hz	$4 \times 10^4/f^2$
8–25 Hz	$5,000/f$
0.025–0.8 kHz	$5/f$
0.8–3 kHz	6.25
3–150 kHz	6.25
0.15–1 MHz	$0.92/f$
1–10 MHz	$0.92/f$
10–400 MHz	0.092
400–2,000 MHz	$0.0046f^{1/2}$
2–300 GHz	0.20

vehicle while power is transferred. So, if even 0.1 % of leakage magnetic field comes out from the OLEV system, the EMF level could exceed the regulation of 6.25 μT.

Basically, passive shielding using metal plates is applied to the first test version of OLEV for the reduction of electromagnetic field. For protection of passengers from magnetic field, a metal plate is applied to the bottom of the vehicle. As the power lines are the source of magnetic field, vertical plate shields are applied as shown in Figure 10.24 [16]. To improve the shielding effectiveness of the passive shield, soft contacts are additionally applied between the bottom plate and vertical ground plate by metal brushes as shown in Figure 10.25. The metal brush is a bundle of thin metal wires attached beneath the bottom plate and connects the current path between the vehicle body and ground plate underneath the road surface (Figure 10.26). The photograph of the implemented metal brush is shown in Figure 10.29. The number of connections using metal brushes is a significant factor to improve

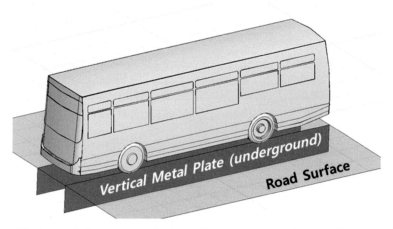

Figure 10.24 Vertical ground plate buried underground for the reduction of magnetic field.

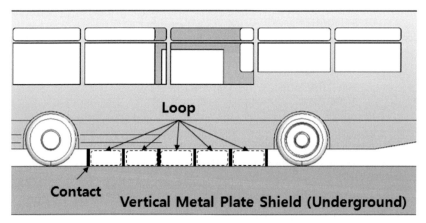

Figure 10.25 Connections between vehicle body and underground vertical metal plate for passive shield.

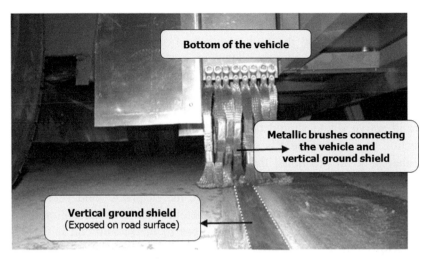

Figure 10.26 The implemented metal brush in passive shield at the bottom of OLEV.

the shielding effectiveness of the passive shielding. The EMF level has been decreased from 14.4 to 3.5 μT when the number of connections using metal brushes is increased from 2 to 8 as shown in Figure 10.27.

However, this can be applied only in some limited applications because the metal brush will be worn out due to friction, and the metal ground shield at the road surface can be covered with non-metal material disturbing the connection between the vehicle body and ground plate.

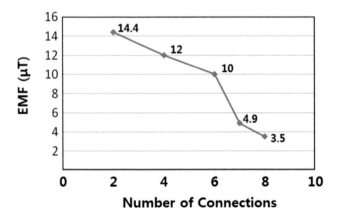

Figure 10.27 Effect of the number of connections between the metallic vehicle body and the horizontal ground shield.

The EMF can be minimized by active shielding with or without passive shields independently, and the basic concept of active shield is shown in Figure 10.28. Similar to power lines, the active shield is also a metal wire which carries the same freency with current but the phase is the opposite of the current in the pickup coil [17].

In the design of active shield, the directions of magnetic fields by the source and active shield should be carefully considered. In Figure 10.29, the direction of magnetic field is shown. To make the EMF level less than the regulation at all positions, the magnetic field from the active shield should be almost the same as that from pickup coils at all positions. At the position above 20 cm from road surface, the magnetic field vector is parallel to the metal plate because of the metallic shield at the bottom of the vehicle. So, to place the active shield close to the pickup coil is more effective. However, if the active shield goes closer to the pickup coil, the current of the active shield should be larger. For this reason, the placement of the active shield is compromised considering the shielding effectiveness and current magnitude. At the optimal value of current, the magnetic flux density is reduced to 1/10 of the density without the active shield as depicted in Figure 10.30.

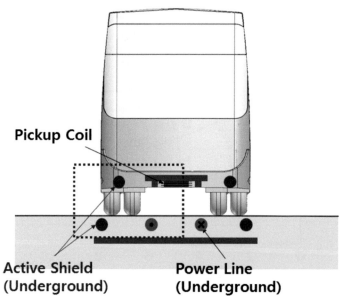

Figure 10.28   Concept of active shield for OLEV.

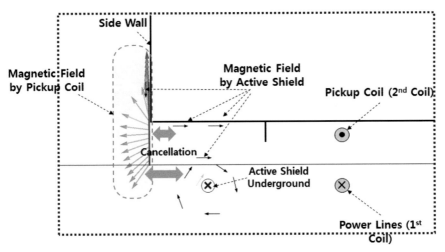

Figure 10.29 Direction of magnetic field from pickup coils and active shield.

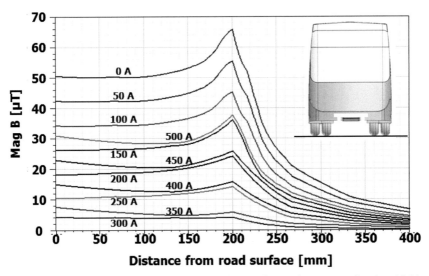

Figure 10.30 Simulated change of EMF level according to the current of active shield.

Figure 10.31 shows that the measured EMF after active shield is applied to OLEV. In the measurement, IEC 62110 standard which defines the magnetic field measurement of electric systems has been selected as a standard measurement method [18]. According to IEC 62110, the average value

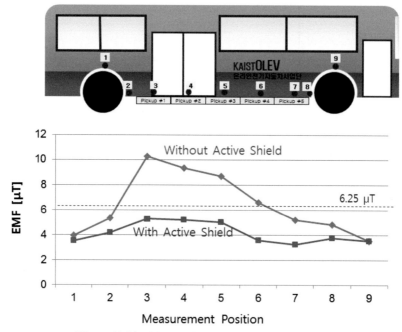

Figure 10.31   Measured EMF level applied to OLEV.

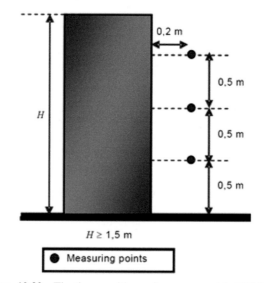

Figure 10.32   The three positions of measurement in IEC 62110.

of measurement at three points is reported as a final measurement. By using active shield, the leakage magnetic field has been reduced from 10.3 to 5.3 μT.

10.3.6 Design Procedure and Optimization

In the design of the power lines and the pickup coil structure for OLEV system, three criteria are mainly considered for the electrical performance of the wireless power transfer system: power transfer capability, power transfer efficiency, and leakage from the electromagnetic field [19].

The power transfer capability implies the maximum power that can be transferred from the power lines under the road to the load in the vehicle, which consequently determines the maximum speed and battery recharging time of the vehicle. From the simplified equivalent circuit model of the wireless power transfer system with two series resonant coils as shown in Figure 10.18, the power at the load R_L is calculated to be proportional to the frequency, mutual inductance, and magnitude of source current assuming that the system is operating at the resonance frequency as shown in Equation (10.5).

$$P_C \cong \frac{\omega^2 M^2}{(R_2 + R_L)^2 + \left(\omega L_2 - \frac{1}{\omega C_2}R_L\right)} I_1^2 R_L \cong \frac{\omega^2 M^2}{R_L} I_1^2 \quad (10.5)$$

$$P_C \cong \frac{\omega^2 M^2 R_L}{(R_2 + R_L)^2 + \omega^2 M^2(R_2 + R_L)} I_1^2 R_L \cong \frac{1}{1 + \frac{R_1 R_L}{\omega^2 M^2}} \quad (10.6)$$

The power transfer efficiency is also an important factor for commercialization, and it should be reasonably high compared with the efficiency of other types of vehicles. To increase the efficiency, we need to minimize the loss at each stage of the power system of OLEV. With the development of power components operating at 20 kHz, which was not available tens of years ago, the efficiency of the inverter is significantly increased. Also, the mutual inductance should be increased, and the parasitic resistances R_1 and R_2 which are the loss from these resistances should be decreased as derived in Equation (10.6) to increase the efficiency even more.

The third criterion of leakage EMF is simply proportional to the magnitude of the current and inversely proportional to the distance between the current position and measurement position without a shield. However, as the application of passive and active shields significantly changes the magnitude of EMF, the design of the EMF should be performed separately.

The previous design procedure for the wireless power transfer system for OLEV is shown in Figure 10.33. At the early stage of design, we have to determine the topology and outline of the dimensions for the physical structures such as the number of coils, coil size and dimension, and the position of the ferrite core because the mutual inductance is roughly determined when the physical dimension is fixed and it is hard to change the value significantly in the later stage. So, the design freedom is limited, and the final design parameter is difficult to be the optimal. For optimal design of wireless power transfer system in OLEV, several parameters should be considered at the same time in design stage. So, the optimization stage is necessary after the design parameters and performance parameters are defined as shown in Figure 10.34.

Table 10.2 Shows the result of simulated sensitivity analysis of transferred power for the change of main design parameters which is the reference for the optimization of the design. At each design stage, a sensitivity analysis on the effect of each design parameters has been performed using simulation with a 3-dimensional field solver.

To optimize the design, we first formulate a parameter optimization problem such that the transferred power to the pickup P_C which is consumed at the load R_L of Figure 10.18 is maximized while EMF level and power transfer efficiency K satisfy the requirements. We assume that the power

Table 10.2   Sensitivity analysis of transferred power for the change of design parameters

Design parameters		Change of Parameters			
		−20%	−10%	+10%	+20%
Dimension parameters	Air gap	+46.3%	+20.1%	−15.9%	−33.9%
	Number of turns in pickup coil	−44.0%	−21.0%	+21.0%	+44.0%
	Dist. between rail wires	−40.0%	−18.6%	+17.1%	+35.1%
	Pickup coil width	−24.2%	−9.5%	+6.9%	+12.2%
Material parameters	permeability (μ)	−1.0%	−0.4%	+0.4%	+0.72%
	Permittivity (ε)	0%	0%	0%	0%
	Conductance (σ)	0%	0%	0%	0%
	Frequency	−44.1%	−20.7%	+21.3%	+45.2%
Electrical parameters	Current	−44.0%	−21.0%	+21.0%	+44.0%
	Frequency	−44.1%	−20.7%	+21.3%	+45.2%

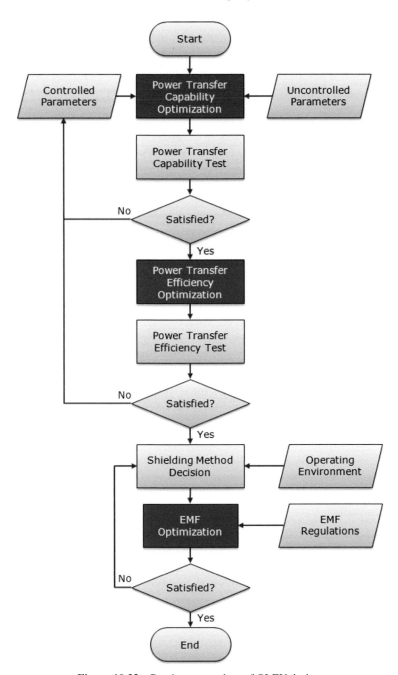

Figure 10.33   Previous procedure of OLEV design.

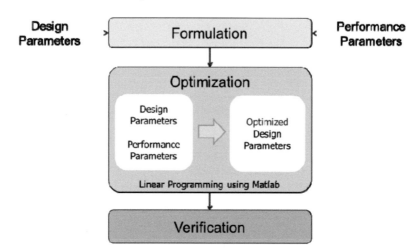

Figure 10.34   Optimized procedure for OLEV design.

transfer efficiency should be greater than or equal to 0.8, and the leakage EMF should be less than or equal to 6.25 μT.

Table 10.3 shows system parameters, which are divided into two categories: constant system parameters and variable system design parameters. We assume that the air gap between power lines and pickup coils, resonance frequency, parasitic resistance of power lines, parasitic resistance of pickup coil, and load resistance are given as in Table 10.3. We can change three system

Table 10.3   System parameters

Constant system parameters	
Air gap	g_{Air} (=20 cm)
Resonance frequency	f (=20 kHz)
Parasitic resistance of power lines	R_1 (=0.1 Ω)
Parasitic resistance of pickup coil	R_2 (=0.1 Ω)
Load resistance	R_L (= 10 Ω)
System design parameters	
Width of pickup coil	W_C
Current of power lines	I_S
Number of turns in pickup coil	N

design parameters: width of pickup coil W_C, current of power lines I_S, and number of turns in pickup coil n.

Accordingly, we formulate our optimization problem as follows:

$$
\begin{aligned}
&\text{variables} : W_C, \ n, \ I_S \\
&\text{maximize } P_C \\
&\text{such that} \\
&EMF \leq 62.5(mG), \\
&K \geq 0.8, \\
&0 \leq W_C \leq W_{C,\max}, \ 0 \leq n \leq n_{\max}, \ 0 \leq I_S \leq I_{S,\max}.
\end{aligned}
\tag{10.7}
$$

where $W_{C,\max}$, $n_{\max}$, and $I_{S,\max}$ are the allowable maximum values of W_C, n, I_S, respectively. To solve our problem, we need to express P_C, K, EMF in terms of W_C, n, I_S. Since $V_C = j(2\pi f)MnI_S$, the induced voltage V_C is proportional to f, n, and I_S, and the EMF is proportional to n and I_S. From the simulation and modeling, we can obtain the approximate expressions for the relationship between V_C and EMF as follows:

$$
|V_C| \approx c_1 f n I_S \sqrt{W_C} \tag{10.8}
$$

$$
EMF \approx c_2 n I_S W_C^2 \tag{10.9}
$$

where C_1 and C_2 are constants. Then, transfer power P_C and total power P_{Total} at resonant frequency can be represented as follows:

$$
P_C = \frac{V_C^2}{R_C} \approx \frac{c_1^2}{R_C} f^2 n^2 I_S^2 W_C, \tag{10.10}
$$

$$
P_{\text{Total}} \approx R_1 I_S^2 + \frac{c_1^2}{R_C} f^2 n^2 I_S^2 W_C. \tag{10.11}
$$

Therefore, the power transfer efficiency is

$$
K = \frac{P_C}{P_{\text{Total}}} \approx \left(1 + \frac{R_1 R_L}{c_1^2 f^2 n^2 W_C} \right)^{-1} \tag{10.12}
$$

From Equations (10.9), (10.10), and (10.12), we can express the optimization problem in Equation (10.5) as follows:

$$
\underset{n, I_S, W_C}{\text{maximize }} \alpha_1 f^2 n^2 I_S^2 W_C
$$

$$
\text{such that}
$$

$$
n I_S W_C^2 \leq \alpha_2
$$

$$f^2 n^2 W_C \geq \alpha_3 \tag{10.13}$$
$$0 \leq W_C \leq W_{C,max}, 0 \leq n \leq n_{max}, 0 \leq I_S \leq I_{S,max}$$

where

$$\alpha_1 = \frac{c_1}{R_C}, \quad \alpha_2 = \frac{62.5}{c_2}, \quad \alpha_3 = \frac{R_1 R_L}{c_1^2 f^2 \left(\frac{1}{0.8} - 1\right)} \tag{10.14}$$

Let

$$x = \log(n), \quad y = \log(I_S), \quad z = \log(W_C). \tag{10.15}$$

Then, the optimization problem in Equation (10.8) can be restated as:

$$
\begin{aligned}
&\underset{x,y,z}{\text{maximize}} \quad 2x + 2y + z + \beta_1 \\
&\text{such that} \\
&\qquad x + y + 2x \leq \beta_2 \\
&\qquad 2x + z \geq \beta_3 \\
&\qquad x \leq x_{max}, \; y \leq y_{max}, \; z \leq z_{max}
\end{aligned}
\tag{10.16}
$$

where

$$
\begin{aligned}
&\beta_i = \log(\alpha_i), i = 1, 2, 3, \quad x_{max} = \log(n_{max}), \\
&y_{max} = \log(I_{S,max}), \quad z_{max} = \log(W_{C,max}).
\end{aligned}
\tag{10.17}
$$

Note that the problem (10.16) is a form of typical linear programming (LP) problem.

In the process of finding optimal design parameters, the parameters which maximize the transfer power are determined. The width of pickup coil should be minimized because it increases EMF more significantly than current and number of turns. Similarly, the current and the number of turns should be increased unless it violates the boundary conditions. The boundary conditions on the power transfer efficiency affect the design parameters when the frequency is low or mutual inductance is small. Once the product of frequency and mutual inductance is large enough, the EMF is the only boundary condition, and then the combination of the design parameters is determined to make the EMF 6.25 μT, which is the maximum value allowed in the optimization. In this EMF boundary, the current and number of turns are maximized until they reach the maximum value we set as $W_{C,max}$, n_{max}, $I_{S,max}$ in Equation (10.13). Finally, two maximum values of n_{max} and $I_{S,max}$ determine the transferred power because the number of turns and current should reach the maximum value for maximum power.

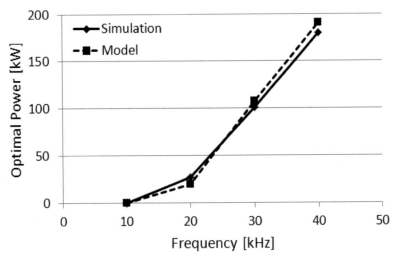

Figure 10.35 Optimal transferred power for different values of frequency.

Now, we obtain the optimal solution for problem (10.16) and compare it with the simulation results to investigate the validity of the approximation for LP formulation. Figure 10.35 shows the optimal transfer power P_C and the variation of constraints such as EMF and K for different values of the frequency f. The optimal power increases as the frequency increases because frequency simply increases the transfer power and has no effect on EMF. The efficiency and EMF should be maintained at the specific level. We can find that the simulation results are similar to the LP solution, which means that the approximation for LP formulation is reasonable. More accurate results can be obtained by applying more complex numerical models in Equations (10.9) and (10.10), which describes the voltage and EMF more accurately.

10.4 Conclusions

It is the time when the secrets of wireless power transfer technologies are revealed one by one and some commercial products using inductive coupling technology can be found in the market. However, application of wireless power transfer system to the automobile is now on the first step of development by universities, research institutes, and companies. The OLEV was the first product commercially used with the technology of wireless power transfer for a distance of 20 cm. In this chapter, we have overviewed from the background technologies to the implementation of OLEV with consideration of power, efficiency, and EMF.

10.5 Problems

P10.1: Explain the difference between electric field coupling and magnetic field coupling?

P10.2: What is the main factor to increase the transfer power and its efficiency?

P10.3: What is the effect of lossy cable used for the coils in transmitter and receiver?

P10.4: What is the disadvantage of conventional battery-powered electric vehicle?

P10.5: What difficulties are there in designing wireless power transfer system for automobile application?

P10.6: Discuss future applications of wireless power transfer applications.

References

[1] Tesla Society. http://www.teslasociety.com/tesla_tower.htm.
[2] Hidetochi Matsuki. "Frontier of wireless electric power transmission," August (2009).
[3] Kurs, A., A. Karalis, R. Moffatt, J.D. Joannopoulos, P. Fisher, and M. Soljačić. "Wireless Power Transfer via Strongly Coupled Magnetic Resonances." *Science* 317, 5834: 83–86 (2007).
[4] Tina Teng. "*Wireless Charging Eliminates Tangled Cords—Market Forecast Wireless Charging.*" iSuppli Corporation, January (2011).
[5] Cahners In-Stat. "*Cut the Cord: Wireless Charging Systems Analysis and Forecast,*" August (2010).
[6] iSuppli. "*Wireless Charging Market Set to Expand by Factor of Nearly 70 by 2014,*" June (2010).
[7] Wireless Power Consortium. http://www.wirelesspowerconsortium.com.
[8] PATH. "*Roadway Powered Electric Vehicle Project Track Construction and Testing Program,*" February (1994).
[9] Delphi. http://delphi.com.
[10] Henry W. Ott. "*Noise Reduction Techniques in Electronic Systems.*" Second Edition.

[11] Kong, Sunkyu, Myunghoi Kim, Kyoungchoul Koo, Seungyoung Ahn, Bumhee Bae and Joungho Kim. *"Analytical Expressions for Maximum Transferred Power in Wireless Power Transfer Systems."* IEEE Electromagnetic Compatibility Symposium, pp. 379–383, August 14–19 (2011).

[12] Park, Youngjin. *"Design Issues of Wireless Power Transmission Based on Magnetic Resonance."* Asian-Pacific Electromagnetic Compatibility Symposium, May (2011).

[13] Hirayama, Hiroshi. *"Basic Theory of Magnetic-Coupled Resonant Wireless Power Transfer and Recent Progress."* Asian-Pacific Electromagnetic Compatibility Symposium, May (2011).

[14] Kim, Jonghoon, Hongseok Kim, In-Myoung Kim, Young-il Kim, Seungyoung Ahn, Jiseong Kim, and Joungho Kim. *"Comparison of Series and Parallel Resonance Circuit Topologies of Receiving Coil for Wireless Power Transfer."* International Forum on Electric Vehicle, pp. 123–141, November 18 (2011).

[15] Ahn, Seungyoung, Jun So Pak, Taigon Song, Heejae Lee, Junggun Byun, Deogsoo Kang, Cheol-Seung Choi, Yangbae Chun, Chun taek Rim, Jae-Ha Yim, Dong-Ho Cho, and Joungho Kim. *"Low Frequency Electromagnetic Field Reduction Techniques for the On-Line Electric Vehicle (OLEV)."* IEEE Electromagnetic Compatibility Symposium, pp. 625–630, July (2010).

[16] ICNIRP Guidelines. "Guidelines for Limiting Exposure to Time-Varing Electric, Magnetic, and Electromagnetic Fields (UP TO 300 GHz)." *Health Physics* 74, 4 (1998).

[17] Hasselgren, L., and J. Luomi. "Geometrical Aspects of Magnetic Shielding at Extremely Low Frequencies." *IEEE Transactions on Electromagnetic Compatibility* 37, 3: 409–420 (1995).

[18] Buccella, C., M. Feliziani, and V. Fuina. *"ELF Magnetic Field Mitigation by Active Shielding."* IEEE International Symposium on Industrial Electronics, vol. 3, pp. 994–998, November (2002).

[19] IEC 62110. *"Magnetic Field Levels Generated by A.C. Power Systems— Measurement Procedures with Regard to Public Exposure,"* Ed. 1. July (2009).

[20] Ahn, Seungyoung, Yangbae Chun, Dong-Ho Cho, and Joungho Kim. "Wireless Power Transfer Technology in On-Line Electric Vehicle." *Journal of the Korean Institute of Electromagnetic Engineering and Science* 11, 3 (2011).

11

Wireless Powering and Propagation of Radio Frequencies through Tissue

Eric Y. Chow[1], Chin-Lung Yang[2] and Pedro P. Irazoqui[3]

[1]Cyberonics Inc., Houston, TX 77058 USA
[2]Department of Electrical Engineering in National Cheng Kung University, Tainan 70101, Taiwan
[3]Center for Implantable Devices, Weldon School of Biomedical Engineering, Purdue University, West Lafayette, IN 47907 USA
E-mail: eric.chow@cyberonics.com; cyang@mail.ncku.edu.tw; pip@purdue.edu

Wireless powering has benefits in a variety of different areas but has the potential to revolutionize the field of medical devices. With the advantages of this technology come complexities which become exemplified when applied to an implantable setting in biomedical applications. This chapter provides an overview of some of the applications of wireless powering and focuses on far-field remote powering for implantable applications using radiative electromagnetic fields. Analytical and simulation models are presented to help illustrate and gain an intuitive understanding of RF propagation through tissue. *Ex vivo* experiments and *in vivo* studies are performed to assess the validity and accuracy of the models, and these empirical results are discussed. Lastly, the electrical components, circuit design, and system integration required for transcutaneous wireless power transfer are described. There are significant benefits of the radiative wireless powering technique, as compared to its inductive coupling counterpart, but with these advantages come extremely complex challenges.

11.1 Introduction

Wireless power transmission can be traced back to the early work of Heinrich Hertz and Nikola Tesla in the 1880s and is documented in Tesla's 1891 patent "System of Electric Lighting" where he mentions the transfer of energy across

Wireless Power Transfer 2nd Edition, 421–454.

an air gap [1]. The vast applications of power transmission via radio frequency waves have ranged all the way from a microwave-powered helicopter, first developed by William Brown and John Burgess in 1963, to a satellite that would capture solar energy and wirelessly transmit the energy to the earth, a concept introduced by Peter Glaser in 1968 [2, 3]. For biomedical applications, wireless power transfer has also been around since the 1960s, with Robert Fischell inventing the first rechargeable cardiac pacemaker during that time; however, for these implantable applications, a near-field inductive coupling technique is used [4, 5]. Transcutaneous powering is traditionally done using near-field inductive coupling because the body tissue poses significant issues on the propagation of radiating fields, but is essentially transparent to a magnetic field that can generate a coupling current. Power transmission via radiative far-field power delivery is a relatively new concept in the field of implantable devices due to this transcutaneous boundary, but a thorough understanding of the effects of the body tissue, accurate simulation models, and proper design holds the key to utilization of this technique for this application.

11.2 Comparison of Transcutaneous Powering Techniques

The methods to transfer power through skin and body tissue are via conduction, capacitive coupling, induction, and radiative fields.

Conduction in biological tissue is carried out via ions (Na^+, K^+, Cl^-, Ca^{++}, etc.), whose concentration determines its conductivity. To transfer power from an electrical source, an electrochemical reaction is necessary at the tissue contact interface. The contact is connected to the electrical power source and is typically a metallic surface referred to as an electrode. The main issues with this method include the poor conductivity of some of the tissue layers (dry skin, fat, etc.), any potential air gaps, and the current density limits [6, 7]. This technique suffers from reliability and is not really wireless as it does not allow for any air distance between the source and the load.

The capacitive coupling technique is based on permittivity, as opposed to conductivity in the previous case, which along with area and distance determines the capacitance from source to load. Considering the current density limits, a decently sized capacitor value is required to transfer a useful power level amount [6, 7]. Unfortunately, given realistic areas and the permittivity of body tissues, reasonable separation distances cannot be used to achieve these necessary capacitance values. Thus, this method is also not really wireless since any real air gap would make the capacitive power coupling unusable.

The most commonly used approach for transferring power through the biological tissue is through magnetic induction. This is typically done using 2 coils, where a current running through the primary source coil generates a perpendicular magnetic field which passes through the nearby secondary coil where it induces a current flowing in the opposing direction. Since there is no real permeability mismatch between the body tissue and air, the magnetic field does not encounter any boundary conditions. Conductive surfaces, however, which are directly in the path of the magnetic field or even nearby, will alter the flux due to eddy current generation which will create opposing magnetic fields. The conductivities of biological tissues are much smaller than those of the metal used for the coils, and thus the eddy current generation in the tissue will be insignificant. Unfortunately, implantable devices typically utilize a metal enclosure to ensure hermeticity, longevity, and biocompatibility, and this would significantly affect the inductive coupling especially since it is typically desirable to have the coil within the metallic casing. Careful selection of low conductivity metals, case geometries, and primary–secondary coil optimizations allows for usable power transfer across a short distance, typically up to a couple of inches. The main drawbacks of magnetic coupling are that its efficiencies are extremely sensitive to primary–secondary coil alignment and distance beyond the near-field region, where the magnetic field quickly diminishes to insignificant levels.

A much less prevalent technique for transcutaneous power transfer is through the use of radiative propagating RF electromagnetic fields. This approach allows for the possibility of far-field power transfer and potentially some orientation robustness. Unfortunately, using this method in an implantable application to achieve reasonable efficiencies and useable power levels is far from trivial. A major obstacle stems from the dielectric relaxation loss of biological tissue, which results in interactions between the tissue and EM fields, which become drastically more significant with increasing frequencies. This characteristic becomes a major challenge when considering the fact that medical implants have stringent size requirements, which limits the antenna size and thus pushes the optimal radiating frequency to upper MHz and even GHz ranges. The remainder of this chapter will focus on the optimization of radiative far-field powering through an analysis of RF propagation through tissue, state-of-the-art simulation and modeling, and proper design of the electrical components.

11.3 Analysis

To achieve wireless radiative power transfer to an implantable device, the RF field must propagate through biological tissue, which is typically heterogeneous and consists of multiple layers. The following are the major considerations for this transcutaneous propagation:

- Impedance mismatches at the layer boundaries, causing reflections of the incident waves,
- Attenuation due to tissue absorption,
- Energy spreading, also referred to as free-space path loss,

Starting with some fundamentals for a lossy source-free medium, Maxwell's equations provide the following relationships for time-harmonic fields [8, 9]:

$$\nabla \times \overrightarrow{E} = -\frac{d\overrightarrow{B}}{dt} - \overrightarrow{M} = -j\omega\mu\overrightarrow{H} \qquad (11.1)$$

$$\nabla \times \overrightarrow{H} = j\omega\varepsilon\overrightarrow{E} + \sigma\overrightarrow{E} + \overrightarrow{J} = (j\omega\varepsilon + \sigma)\overrightarrow{E} \qquad (11.2)$$

The wave equation for electric (E) fields in this medium is given by

$$\nabla^2\overrightarrow{E} - \gamma^2\overrightarrow{E} = 0 \qquad (11.3)$$

where

$$\gamma^2 = j\omega\mu\,(j\omega\varepsilon + \sigma) \qquad (11.4)$$

Wave impedance is defined as the ratio of the transverse E-field to the transverse magnetic (H) field and using Maxwell's relationships and Equation (11.4):

$$\eta = \sqrt{\frac{j\omega\mu}{j\omega\varepsilon + \sigma}} = \frac{j\omega\mu}{\gamma} \qquad (11.5)$$

Considering a simplified case of wave propagation in only the z-direction and E-field, which must be perpendicular to the propagation, only in the x-direction, Equation (11.1) is simplified as follows

$$\begin{vmatrix} \hat{x} & \hat{y} & \hat{z} \\ dx & dy & dz \\ E_0 e^{-\gamma z} & 0 & 0 \end{vmatrix} = -j\omega\mu\,\overrightarrow{H}$$

$$\Rightarrow -\gamma E_0 e^{-\gamma z}\hat{y} = -j\omega\mu\,\overrightarrow{H}$$

$$\Rightarrow \overrightarrow{H} = \frac{\gamma}{j\omega\mu}E_0 e^{-\gamma z}\hat{y}$$

$$\qquad (11.6)$$

and the wave equation in Equation (11.3) is simplified to

$$\frac{d^2\overrightarrow{E_x}}{dz^2} - \gamma^2\overrightarrow{E_x} = 0 \qquad (11.7)$$

The dielectric properties (permittivity, permeability, and conductivity) of the media are needed to utilize these equations. For biological tissue types, the permeability is typically always equal to that of free-space and thus from now on, the free-space permeability of μ_0 will be used for μ. The permittivity and conductivity values will depend on the particular media. For human media, Camelia Gabriel and Sami Gabriel's work in 1996 provides empiricaldata for a variety of tissues, and values for some of the more common tissue types for a few popular frequencies are tabulated in Table 11.1 [10–13]:

11.3.1 Reflections at an Interface

At the interface between two different dielectric layers, the E-field boundary conditions dictate that the sum of incident and reflection waves on one side of the interface is equal to the transmission wave on the other side of the interface:

$$\overrightarrow{E}^i + \overrightarrow{E}^r = \overrightarrow{E}^t \tag{11.8}$$

Now, consider an example where a plane wave propagating in the z-direction in an air medium is incident on an air–skin interface. Through solutions to the wave equation in Equation (11.3), the E-field of each wave in Equation (11.8) is equated to:

$$
\begin{aligned}
\overrightarrow{E}^i &= \hat{x} E_0 e^{-\gamma_{1(air)} z} \\
\overrightarrow{E}^r &= \hat{x} R E_0 e^{-\gamma_{1(air)} z} \\
\overrightarrow{E}^t &= \hat{x} T E_0 e^{-\gamma_{2(skin)} z}
\end{aligned}
\tag{11.9}
$$

where R and T, the reflection and transmission coefficients respectively, are constants that will be derived in the subsequent steps.

First, looking at H-field of the incident wave and using Equation (11.6):

$$\overrightarrow{H}^i = \frac{\gamma}{j\omega\mu_0} E_0 e^{-\gamma_{1(air)} z} \hat{y} \tag{11.10}$$

Table 11.1 Measured conductivity and relative permittivity values of common human tissue types for popular frequency bands, Camelia Gabriel and Sami Gabriel's work in 1996 [10–13]

Frequencies	433 MHz	900 MHz	2.4 GHz	5.8 GHz
Tissue Type	$(\sigma(\text{S/m}), \varepsilon_r)$	$(\sigma(\text{S/m}), \varepsilon_r)$	$(\sigma(\text{S/m}), \varepsilon_r)$	$(\sigma(\text{S/m}), \varepsilon_r)$
Skin (dry)	0.702, 46.08	0.867, 41.41	1.441, 38.06	3.717, 35.11
Skin (wet)	0.681, 49.42	0.845, 46.08	1.562, 42.92	4.342, 38.62
Fat	0.042, 5.567	0.051, 5.462	0.102, 5.285	0.293, 4.955
Muscle	0.805, 56.87	0.943, 55.03	1.705, 52.79	4.962, 48.49
Bone (cortical)	0.094, 13.07	0.143, 12.45	0.385, 11.41	1.154, 9.674

Manipulating this equation using Equation (11.4):

$$\overset{\rightarrow i}{H} = \sqrt{\frac{j\omega\mu_0\left(j\omega\varepsilon_{1(air)} + \sigma_{1(air)}\right)}{(j\omega\mu_0)^2}} E_0 e^{-\gamma_{1(air)}z}\hat{y}$$

$$= \sqrt{\frac{j\omega\varepsilon_{1(air)} + \sigma_{1(air)}}{j\omega\mu_0}} E_0 e^{-\gamma_{1(air)}z}\hat{y} \qquad (11.11)$$

Using Equation (11.5) to express the incident wave's H-field as a function of the wave impedance:

$$\overset{\rightarrow i}{H} = \hat{y}\frac{E_0}{\eta_{1(air)}}e^{-\gamma_{1(air)}z} \qquad (11.12a)$$

Similarly for the reflection and transmission waves' H-fields:

$$\overset{\rightarrow r}{H} = -\hat{y}\frac{RE_0}{\eta_{1(air)}}e^{-\gamma_{1(air)}z} \qquad (11.12b)$$

$$\overset{\rightarrow t}{H} = \hat{y}\frac{TE_0}{\eta_{2(skin)}}e^{-\gamma_{2(skin)}z} \qquad (11.12c)$$

Defining the z-axis origin at the air–skin interface:

$$\overset{\rightarrow i}{E} + \overset{\rightarrow r}{E} = \overset{\rightarrow t}{E} \overset{z=0}{\longrightarrow} 1 + R = T \qquad (11.13)$$

$$\overset{\rightarrow i}{H} + \overset{\rightarrow r}{H} = \overset{\rightarrow t}{H} \overset{z=0}{\longrightarrow} \frac{1}{\eta_{1(air)}}\left(1 - R\right) = \frac{T}{\eta_{2(skin)}} \qquad (11.14)$$

Now solving for the reflection and transmission coefficients:

$$R_{a2s} = \frac{\eta_{2(skin)} - \eta_{1(air)}}{\eta_{2(skin)} + \eta_{1(air)}} \qquad (11.15a)$$

$$T_{a2s} = \frac{2\eta_{2(skin)}}{\eta_{2(skin)} + \eta_{1(air)}} \qquad (11.15b)$$

Typically in RF applications, the effect on the power of the wave is of most importance. The Poynting vector, which represents the power density in W/m^2, is given by

$$\vec{S} = \vec{E} \times H$$
$$\Rightarrow |S| = |E|\,|H| \qquad (11.16)$$

Therefore, the incident and reflection power densities at the interface on the air-side are:

$$\left|S_{1(air)}{}^{i}\right| = \frac{E_0^2}{\eta_{1(air)}}, \qquad \left|S_{1(air)}{}^{r}\right| = \frac{R_{a2s}{}^2 E_0^2}{\eta_{1(air)}} \qquad (11.17a)$$

The transmission power density at the interface on the skin-side is:

$$S_{2(skin)}{}^{t} = \frac{T_{a2s}{}^2 E_0^2}{\eta_{2(skin)}} \qquad (11.17b)$$

Therefore, the ratios of transmission-to-incident power ($P_{t/i}$) and reflection-to-incident power ($P_{r/i}$) are:

$$P_{r/i} = \frac{\left|S_{1(air)}{}^{r}\right|}{\left|S_{1(air)}{}^{i}\right|} = R_{a2s}{}^2, \qquad P_{t/i} = \frac{\left|S_{2(skin)}{}^{t}\right|}{\left|S_{1(air)}{}^{i}\right|} = \frac{T_{a2s}{}^2 \eta_{1(air)}}{\eta_{2(skin)}}$$

$$(11.17c)$$

11.3.2 Attenuation Due to Tissue Absorption

As a wave travels through a lossy medium, it encounters loss and thus decreases in magnitude. The solution of the second-order differential wave equation in Equation (11.7) indicates that this decrease is an exponential decay:

$$E_x = E_0 e^{-\gamma z} \qquad (11.18)$$

Representing the coefficient as a sum of real and imaginary components:

$$\gamma = \alpha + j\beta \qquad (11.19)$$

Using Equation (11.19) in Equation (11.4) and expanding:

$$\gamma^2 = j\omega\mu \left(j\omega\varepsilon + \sigma\right) = (\alpha + j\beta)^2$$
$$\Rightarrow -\omega^2\mu\varepsilon + j\omega\mu\sigma = \alpha^2 - \beta^2 + j2\alpha\beta \qquad (11.20)$$

Equating the coefficients of the real components in Equation (11.20):

$$-\omega^2\mu\varepsilon = \alpha^2 - \beta^2 \qquad (11.21)$$

Equating the coefficients of the imaginary components in Equation (11.20) and solving for:

$$\omega\mu\sigma = 2\alpha\beta$$
$$\Rightarrow \beta = \frac{\omega\mu\sigma}{2\alpha} \qquad (11.22)$$

Using Equation (11.22) in Equation (11.21) and organizing into the quadratic equation format:

$$\alpha^2 + \omega^2\mu\varepsilon - \left(\frac{\omega\mu\sigma}{2\alpha}\right)^2 = 0$$
$$\Rightarrow \alpha^4 + \omega^2\mu\varepsilon\alpha^2 - \frac{(\omega\mu\sigma)^2}{4} = 0 \tag{11.23}$$

Solving for the roots of the quadratic formula to get α^2 and taking the square root to get α:

$$\alpha = \pm\sqrt{\frac{-\omega^2\mu\varepsilon \pm \sqrt{(\omega^2\mu\varepsilon)^2 + (\omega\mu\sigma)^2}}{2}} \tag{11.24}$$

Since α represents the coefficient of the purely real component, the "+" under the radical must be a "+". Further simplification of Equation (11.24) is as follows:

$$\Rightarrow \alpha = \pm\sqrt{\frac{1}{2}\left(-\omega^2\mu\varepsilon + \sqrt{\omega^4\mu^2\varepsilon^2 + \omega^2\mu^2\sigma^2}\right)}$$

$$\Rightarrow \alpha = \pm\sqrt{\frac{1}{2}\left(-\omega^2\mu\varepsilon + \omega\mu\sqrt{\omega^2\varepsilon^2 + \sigma^2}\right)}$$

$$\Rightarrow \alpha = \pm\sqrt{\frac{1}{2}\left(-\omega^2\mu\varepsilon + \omega^2\mu\varepsilon\sqrt{1 + \frac{\sigma^2}{\omega^2\varepsilon^2}}\right)} \tag{11.25}$$

$$\Rightarrow \alpha = \pm\sqrt{\frac{\omega^2\mu\varepsilon}{2}\left(\sqrt{1 + \frac{\sigma^2}{\omega^2\varepsilon^2}} - 1\right)}$$

$$\Rightarrow \alpha = \pm\omega\sqrt{\mu\varepsilon}\sqrt{\frac{1}{2}\left(\sqrt{1 + \left(\frac{\sigma}{\omega\varepsilon}\right)^2} - 1\right)}$$

Now to solve for β, Equation (11.22) is first rearranged as:

$$\alpha = \frac{\omega\mu\sigma}{2\beta} \tag{11.26}$$

Now using Equation (11.26) in Equation (11.21) and organizing into the quadratic equation format:

$$\beta^2 - \omega^2\mu\varepsilon - \left(\frac{\omega\mu\sigma}{2\beta}\right)^2 = 0$$
$$\Rightarrow \beta^4 - \omega^2\mu\varepsilon\beta^2 - \frac{(\omega\mu\sigma)^2}{4} = 0 \tag{11.27}$$

Equation (11.27), in terms of β, is nearly identical to Equation (11.23) except that the middle coefficient (*b* component of the quadratic formula) has a negative sign. Therefore, carrying this extra negative sign through the derivation results in:

$$\beta = \pm\omega\sqrt{\mu\varepsilon}\sqrt{\frac{1}{2}\left(\sqrt{1+\left(\frac{\sigma}{\omega\varepsilon}\right)^2}+1\right)} \tag{11.28}$$

Considering the case of a source-free lossy medium, that is, biological tissue, the E-field can only be decreasing and therefore in the E-field expression (11.18) where

$$E_x = E_0 e^{-\gamma z} = E_0 e^{-\alpha z} e^{-j\beta z} \tag{11.29}$$

α and β must be positive:

$$\alpha = \omega\sqrt{\mu\varepsilon}\sqrt{\frac{1}{2}\left(\sqrt{1+\left(\frac{\sigma}{\omega\varepsilon}\right)^2}-1\right)} \tag{11.30}$$

$$\beta = \omega\sqrt{\mu\varepsilon}\sqrt{\frac{1}{2}\left(\sqrt{1+\left(\frac{\sigma}{\omega\varepsilon}\right)^2}+1\right)} \tag{11.31}$$

As seen in Equation (11.29), the α coefficient corresponds to the exponential decay of the E-field's magnitude and is thus often referred to as the attenuation constant and has units of m^{-1}. The distance, z, at which the amplitude of the wave decreases to e^{-1}, is typically referred to as the penetration depth or skin depth, has units of m, and is given by:

$$\delta = \frac{1}{\alpha} = \frac{1}{\omega\sqrt{\mu\varepsilon}\sqrt{\frac{1}{2}\left(\sqrt{1+\left(\frac{\sigma}{\omega\varepsilon}\right)^2}-1\right)}} \tag{11.32}$$

Therefore, as a wave travels through a lossy medium, the amplitude of the E-field will exponentially decrease as:

$$E_x = E_0 e^{-\alpha z} \tag{11.33}$$

Relating this to power levels, using the Poynting vector expression in Equation (11.16), the magnitude of the power through a lossy medium will exponentially decrease as:

$$P_x = P_0 e^{-2\alpha z} \tag{11.34}$$

11.3.3 Energy Spreading (Free-Space Path Loss)

Due to the finite size of a receiving antenna's aperture, only a portion of the transmitted energy can be picked up a certain distance away. Due to the spreading of a propagating wave's energy, which is typically approximated by a spherical expansion, this portion of received energy will decrease with the square of the distance. This energy spreading is not unique to this transcutaneous application and is typically referred to as free-space path loss. Given a certain receive antenna size, the ratio of received power to transmitted power as a function of distance, assuming the far-field approximation, is:

$$\frac{P_{Rx}}{P_{Tx}} = \frac{A_{eff}}{4\pi d^2} \tag{11.35}$$

where A_{eff} is the effective receive antenna aperture area and d is the distance between the antennas. For an isotropic antenna, this area is $\lambda^2/4\pi$, but for an implanted antenna, determination of this wavelength value (λ) is not trivial as it is dependent on the dielectric properties of the surrounding biological tissues.

Incorporating the gains of the receive and transmit antennas results in Friis transmission equation:

$$\frac{P_{Rx}}{P_{Tx}} = G_{Tx} G_{Rx} \left(\frac{\lambda_{eff}}{4\pi d} \right)^2 \tag{11.36}$$

where λ_{eff} is the effective wavelength of the wave that is seen at the receive antenna and equals the wavelength in free space divided by the square root of the effective permittivity (ε_r).

11.3.4 Expanding to Multiple Layers and Interfaces

Combining the effects of boundary reflections, tissue absorption, and free-space path loss, the derivations in Sections 11.3.1, 11.3.2, and 11.3.3 can be extended to any number of tissue layers and interfaces.

Consider first an ideal plane wave propagation (thus ignoring energy spreading) from air, to an air–skin interface, then through skin, then to a skin–fat interface.

From Equations (11.17c) and (11.15b), the magnitude of the power transmitted out of the air–skin interface, $P_0{}^t$, is:

$$P_0{}^t = \frac{T_{a2s}{}^2\eta_{air}}{\eta_{skin}}P_0i = \left(\frac{2\eta_{skin}}{\eta_{skin} + \eta_{air}}\right)^2 \frac{\eta_{air}}{\eta_{skin}}P_0{}^i$$

$$\Rightarrow P_0{}^t = \frac{4\eta_{skin}\eta_{air}}{(\eta_{skin} + \eta_{air})^2}P_0{}^i \qquad (11.37)$$

Now after propagating through the lossy skin, using Equation (11.34), the power that is incident, from that component, at the skin–fat interface is:

$$P_1^i = P_0^t e^{-2\alpha_{skin}z_{skin}} = \frac{4\eta_{skin}\eta_{air}}{(\eta_{skin} + \eta_{air})^2}P_0^i e^{-2\alpha_{skin}z_{skin}} \qquad (11.38)$$

Note that the reflected component at the skin–fat interface also propagates backward, reflects again at the air–skin interface (Figure 11.1), and then propagates forward again, which adds to the power incident at the skin–fat interface. After two reflections and the additional traveled distance through the lossy medium, the incident power from this component, $P_3{}^i$, is:

$$P_3^i = \left(\frac{\eta_{fat} - \eta_{skin}}{\eta_{fat} + \eta_{skin}}\right)^2 \left(\frac{\eta_{air} - \eta_{skin}}{\eta_{air} + \eta_{skin}}\right)^2 P_1{}^i e^{-4\alpha_{skin}z_{skin}}$$

$$\Rightarrow P_3^i = \left(\frac{\eta_{fat} - \eta_{skin}}{\eta_{fat} + \eta_{skin}}\right)^2 \left(\frac{\eta_{air} - \eta_{skin}}{\eta_{air} + \eta_{skin}}\right)^2 \frac{4\eta_{skin}\eta_{air}}{(\eta_{skin} + \eta_{air})^2}P_0^i e^{-6\alpha_{skin}z_{skin}}$$

$$(11.39)$$

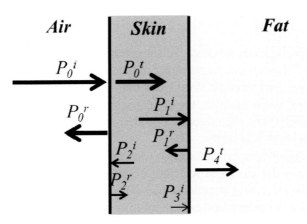

Figure 11.1 Air–skin–fat layers.

Lastly, the power transmitted out of the skin–fat interface is:

$$P_3 = \frac{4\eta_{fat}\eta_{skin}}{(\eta_{fat} + \eta_{skin})^2}\left(P_1^i + P_3^i\right)$$

$$\Rightarrow P_3 = \frac{4\eta_{fat}\eta_{skin}}{(\eta_{fat} + \eta_{skin})^2}\frac{4\eta_{skin}\eta_{air}}{(\eta_{skin} + \eta_{air})^2}P_0^i e^{-2\alpha_{skin}z_{skin}}$$

$$\left(1 + \left(\frac{\eta_{fat} - \eta_{skin}}{\eta_{fat} + \eta_{skin}}\right)^2\left(\frac{\eta_{air} - \eta_{skin}}{\eta_{air} + \eta_{skin}}\right)^2 e^{-4\alpha_{skin}z_{skin}}\right) \qquad (11.40)$$

Now, taking away the ideal plane wave approximation and considering the electromagnetic propagation from a transmit antenna to a receive antenna positioned in the fat layer but right by the skin–fat interface:

$$P_{Rx} = P_{Tx}G_{Tx}G_{Rx}\left(\frac{\lambda_{eff}}{4\pi\left(d_{total}\right)}\right)^2\frac{P_3}{P_0^i}$$

$$\Rightarrow P_{Rx} = P_{Tx}G_{Tx}G_{Rx}\left(\frac{\lambda_{eff}}{4\pi\left(d_{total}\right)}\right)^2\frac{4\eta_{fat}\eta_{skin}}{(\eta_{fat} + \eta_{skin})^2}$$

$$\frac{4\eta_{skin}\eta_{air}}{(\eta_{skin} + \eta_{air})^2}e^{-2\alpha_{skin}z_{skin}}$$

$$\times\left(1 + \left(\frac{\eta_{fat} - \eta_{skin}}{\eta_{fat} + \eta_{skin}}\right)^2\left(\frac{\eta_{air} - \eta_{skin}}{\eta_{air} + \eta_{skin}}\right)^2 e^{-4\alpha_{skin}z_{skin}}\right)$$

$$(11.41)$$

This expression now includes antenna gains and the free-space path loss due to distance traveled in air from a transmitting antenna source to the skin layer and the distance traveled through skin. Note that even though the receive antenna is positioned in the fat layer, the dielectric properties of the fat alone cannot be used to determine the effective wavelength as that will depend on the surrounding dielectrics as well.

The analysis performed in this section only consists of 3 layers and two interfaces and as the layer and interface count increases, the analytical calculations become extremely complex rather quickly. Simulation modeling and EM software are necessary to facilitate more comprehensive modeling.

11.4 Simulation Modeling

Full-wave 3D electromagnetic simulation software can help provide a good picture of the interaction between the human body and the propagation of RF signals. Full-wave simulation tools consider all the time-varying components in Maxwell's equations and do not make assumptions about field relationships, as opposed to quasi-static solvers which solve for only certain modes (typically just transverse electromagnetic) and approximates the time derivatives (typically just setting them to zero). Furthermore, full-wave solvers can accurately calculate losses due to radiation, which is useful in simulations for biomedical implantable applications.

There are several methods to achieve full-wave simulation capabilities, which include method of moments (MoM), the finite element method (FEM), and finite-difference time domain (FDTD). MoM is not really useful for accurately modeling complex dielectric structures such as the human body because it uses a surface mesh. FEM and FDTD both use volume meshing. FEM can more easily handle complicated geometries and solves in the frequency domain. FDTD is limited to more rectangular shapes but solves in the time domain and is thus good for capturing large frequency ranges.

A variety of simulation software packages exist for both FEM and FDTD techniques. Two popular EM simulators are Computer Simulation Technology (CST) Microwave Studio (MWS), which uses FDTD, and ANSYS High Frequency Structural Stimulator (HFSS), which uses FEM. Both these companies have human body models available. CST's human body model, called HUGO, is a model of slices consisting of several thousand layers, 32 tissue types, and down to a 1-mm resolution. ANSYS's human body model consists of 300+ objects (bones, muscles, organs), generates about half a million tetrahedral, and also has accuracies down to the millimeter level. In February of 2011, the Federal Communications Commission (FCC) ruled that FEM is valid for simulating the EM characteristics of medical implants and their surroundings [14].

The EM software packages can support numerous different levels of simulation fidelity, a few examples are shown below in Figure 11.2. The desired level of modeling depends on a trade-off between accuracy and simulation time.

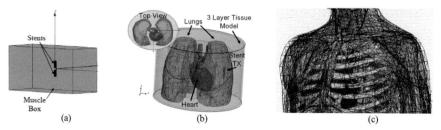

Figure 11.2   (a) simple model, (b) moderate model, (c) comprehensive model.

11.5 Empirical Studies

For the most accurate results, empirical measurements using biological tissue in *ex vivo* and *in vivo* studies should be taken.

Ex vivo studies can be a quick method to achieve approximations of the implanted setting. An extremely simple measurement can be taken by just connecting an RF transmitter to an antenna and measuring the received power a certain distance away with and without a piece of excised tissue placed over the antenna. This basic measurement is taken in [15] for a transcutaneous application at 6.7 GHz using an SMA cable to connect the transmitter to a patch antenna, placing a 4-mm section of porcine tissue over the patch, as seen in Figure 11.3a, and showed about a 15-dB power reduction due to the tissue. That setup could be okay for somewhat large antennas and patch-type antennas which have a ground plane to prevent stray fields from coupling onto the large SMA ground shield, but for more miniature antennas, the cable should be eliminated. For an ocular implant application, *ex vivo* studies done in [16] implanted a miniature 2.4-GHz transmitter and antenna in an excised porcine eye, as shown in Figure 11.3b. Measurements showed a power reduction of up to 6 dB, due to boundary reflections and attenuation, when implanted 1.5 mm beneath the surface of the eye, behind the cornea and scleral tissue.

The most accurate method to quantify the effects of surrounding biological tissue on RF propagation is through *in vivo* studies. Compared to the *ex vivo* studies described above, the setup for *in vivo* experiments is vastly more involved. For cardiovascular application, surgeries have been done on live porcine subjects, shown in Figure 11.4 [17–19]. Testing in an *in vivo* setting is highly regulated and for these porcine surgeries, a Purdue Animal Care and Use Committee (PACUC) approved protocol (PACUC No. 08-019) was strictly followed. On the day of the operation, the subject is first anesthetized with a Telazol (250 mg tiletamine and 250 mg zolazepam), ketamine (250x mg),

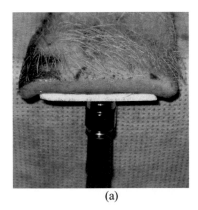 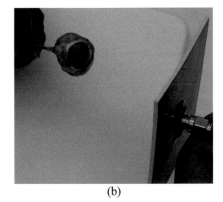

(a)   (b)

Figure 11.3 *Ex vivo* studies.

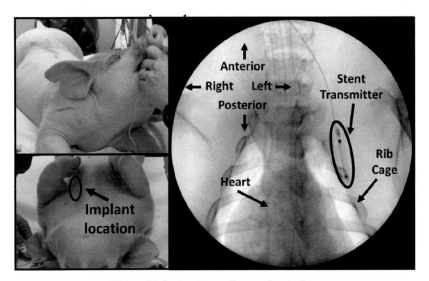

Figure 11.4 *In vivo* cardiovascular studies.

and xylazine (250 mg) combination. Anesthesia maintenance is accomplished with an isoflurane (1.5%–4.0% oxygen) inhalation anesthetic administered from a machine with vaporizer and waste gas ventilation system. The surgery involved implantation of a stent-based antenna, integrated with a miniature 2.4-GHz transmitter, at a depth of 3.5 cm. The results of these *in vivo* studies showed a power loss of 33–35 dB due to boundary reflections and tissue attenuation effects [17–19].

For ocular implant applications, studies have been done in live New Zealand white rabbit subjects, shown in Figure 11.5, chosen for their relatively large and representative eye size [20, 21]. Again, due to the strict regulations surrounding *in vivo* studies, a Purdue Animal Care and Use Committee (PACUC) approved protocol (PACUC No. 08–004) was strictly followed. The rabbits are first caged, maintained, and monitored for one week prior to surgeries. On the day of the surgery, anesthetic induction is done using a ketamine–xylazine injection in the leg muscle and maintained with an intravenous Propofol drip through a vein in the ear and a proparacaine eye drop applied every 10 minutes. For these RF propagation studies, a miniature device with a 2.4-GHz transmitter and antenna is implanted in the suprachoroidal space, about a millimeter beneath the surface of the eye. Measurements showed that the power of the 2.4-GHz RF signal reduced by about 4–5 dB due to tissue absorption and boundary reflections [20, 21].

To evaluate the accuracy of the simulation models, the *in vivo* results are compared back with those done using the EM simulation software. The simplest simulation model, shown above in Figure 11.2a, is just a box with the dielectric properties of muscle and has been shown to be accurate to within 10 dB when compared with the porcine *in vivo* measurements [17]. A more advanced model, shown in Figure 11.2b, incorporates several components of ANSYS's human body model and has been shown to be accurate to

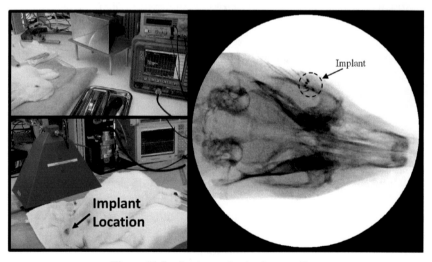

Figure 11.5   *In vivo* ocular implant studies.

within 3–5 dB when compared with *in vivo* porcine studies [18]. The stent-based antenna was near-omnidirectional in free-space, but in an implanted setting, the antenna pattern was altered and simulation results of this affect matched rather closely with *in vivo* measurements, which were taken at discrete points in the front hemisphere of the porcine test subject, as shown in Figure 11.6.

11.6 Antenna Design and Frequency Band Selection

After the radiated RF energy reaches the implant, it needs to be received to be used for powering. The design of this implanted receive antenna is very application and situation specific. Different implant locations can have drastically different surroundings, and typically medical devices have very specific size and geometry requirements. Typically for implantable applications, the area is extremely limited, which can drastically limit the efficiency. The transmit frequency should be chosen based on an analysis between electrically small antenna efficiency and tissue-induced power loss.

Optimal antenna size decreases with increasing frequency, and thus for ultra-miniaturization, a push for higher frequencies is desirable from this standpoint. Unfortunately, as described above, the higher the frequency, the more interactions with biological tissue and power loss, due to the attenuation and boundary reflections, increases drastically. Furthermore, the Federal

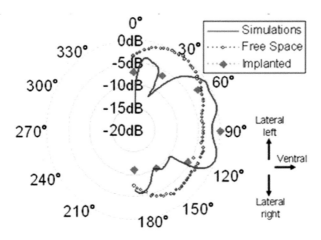

Figure 11.6 *In vivo* porcine antenna pattern plot compared with simulated results.

Communications Commission (FCC) places certain restrictions on the use of frequency bands as well as limits on the output power levels. A useful frequency spectrum category for this RF radiating-field-based wireless powering application is the industrial, scientific, and medical (ISM) radio bands [22]. These bands are considered unlicensed, and thus any device operating in these frequency regions must tolerate interference from other devices. For wireless powering applications, "data" integrity is not important, and thus this ISM radio band category is reasonable. ISM bands cover a wide range of frequencies ranging from low MHz to hundreds of GHz. The various ISM bands can be sorted into 3 categories with large frequency separations in between. The subcategories of bands are as follows: 6.7 MHz up to 40.7 MHz, 433 MHz to 5.8 GHz, and 24 GHz to 246 GHz. The size constraints of implantable devices make operating via RF radiating fields at the 6.7 MHz to 40.7 MHz band nearly impossible, as the miniature antennas cannot really radiate such low frequency signals. Conversely, the 24 GHz to 246 GHz range would allow for ultra-small-scale antenna operation, but the tissue interactions at those frequencies are overwhelming and pretty much all but completely block RF propagation. As a result, the ISM allocations between 433 MHz and 5.8 GHz, tabulated in Table 11.2, are ideal for this implantable application using RF radiating-field-based wireless powering.

In general, the maximum output allowed from the transmitter output is 30 dBm (1 watt) as described in section 47CFR15.247.b.1. Antenna directivity can improve performance significantly, but there are specific restrictions that must be followed. For the 433-MHz and 915-MHz bands, the maximum effective isotropic radiated power (EIRP) is 36 dBm (4 watts), as described in section 47CFR15.247.b.3. For 2.4-GHz operation, above 36 dBm EIRP, section 47CFR15.247.b.3.i describes that the transmitter output power only needs to be decreased by 1 dBm for every 3 dBi increase in antenna directivity/gain. According to section 47CFR15.247.b.3.ii, operation in the 5.8-GHz band does not have restrictions on antenna gain, and thus EIRP, and only has to abide by the maximum 30-dBm transmitter output power. These regulations

Table 11.2   Several ISM frequency band allocations [22]

Center Frequency (MHz)	Frequency Range (MHz)
433.92	433.05–434.79
915	902.00–928.00
2450	2400–2500
5800	5275–5875

make the 2.4-GHz and 5.8-GHz bands more appealing, but again, these higher frequencies will face more tissue losses and thus the specific application and implant environment must be evaluated to determine the optimal operating band.

Different antenna types can be explored to optimize the performance for a given geometry. To show a comparison between antenna type and size, the table below (Table 11.3) summarizes some of the published miniature antenna designs, not specifically for medical applications and targeted for ultra-wide-band use:

As shown in Table 11.3, monopole-types are typically smaller than dipoles and loop-type antennas in terms of area. An interesting fractal-type monopole is described in [35] and is reported to achieve an extremely small footprint. Thus, for a given area, a monopole-type antenna could better take advantage of the size constraint; however, note that the relatively high loss tangent of the surrounding tissue can result in significant attenuation of the electric field. A loop structure may consume more area, but it promotes magnetic field storage, resulting in less loss in the implanted environment. Fractal antenna design is a unique art that allows for many degrees of freedom for an involved optimization process; however, for biomedical implant applications, the surrounding environment is so intricate, complex, and variable that fractal design may not be the best approach. In [36], patch, spiral, and serpentine micro-strip antennas are compared and results from simulations done in a two-thirds muscle block showed the spiral antenna to be the most area efficient for a given frequency, followed closely by the serpentine antenna, with the worst being

Table 11.3   Comparison of sizes between antenna-type designs

Antenna	Size (λ_0^2)	Year	Reference
Monopole	0.038	2011	[23]
Monopole	0.049	2011	[24]
Monopole	0.057	2010	[25]
Monopole	0.062	2007	[26]
Monopole	0.098	2009	[27]
Dipole	0.100	2010	[28]
Dipole	0.103	2008	[29]
Loop	0.085	2005	[30]
Loop	0.101	2007	[31]
Slot	0.032	2006	[32]
Slot	0.078	2008	[33]
Slot	0.193	2008	[34]
Fractal	0.013	2011	[35]

the patch-type. Simulations showed that the spiral antenna had strong coupling only at the center of the antenna, while the serpentine antenna had coupling to its adjacent arms which makes it electrically shorter. Even though for a given frequency, the spiral antenna could be slightly smaller, it is sometimes desirable to still use a serpentine-type structure especially for a dual-band design [37]. Therefore, the optimal antenna type and design will depend on the particular use-case, application, size constraints, and implant location.

Design of the antenna should start in the simulation environment, but due to variations between simulation modeling and the real environment, an iterative design approach is necessary. Furthermore, during the iterations, the matching network and antenna should be simultaneously optimized. For optimizing the size of the antenna, a biocompatible substrate should have the largest possible permittivity, while having low conductivity. Due to the iterations required, a biological phantom material can be used to facilitate the design process. The OET Bulletin 65 written in June 2011, [38], provides good guidelines for RF phantom recipes, for a variety of frequency bands, using mixtures of water, salt, sugar, hydroxyethyl cellulose (HEC), bactericide, Triton X-100, and diethylene glycol butyl ether (DGBE).

11.7 Power Conversion Circuitry

The next important component for wireless power transfer in implantable applications is a power conversion circuit that takes the received signal from the antenna and converts it to a type that can be used for the subsequent structures. This converter is most often a rectification circuit which converts this received AC signal into a DC supply. Typically, in medical implants, the device itself has relatively low power consumption requirements but still requires a sufficient voltage level to bias its structures. As a result, a Cockcroft–Walton multiplier is a good rectification technique for these applications. This topology produces sufficiently high voltages with relatively low input power levels when compared to other structures such as the PMOS voltage multiplier, full-wave diode rectifier, and gate cross-connected bridge rectifier [39]. Techniques can be used to further optimize this output voltage to input power conversion such as using a complementary topology, which doubles the voltage, and implementing additional stages. The complementary architecture also yields additional advantages such as smaller input impedance and less reflected harmonics due to the symmetry. The two-stage complementary Cockcroft–Walton multiplier circuit is shown in Figure 11.7.

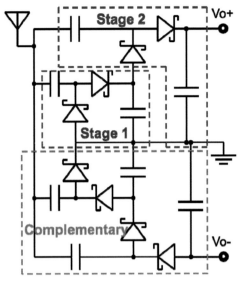

Figure 11.7 Two-stage complementary Cockcroft–Walton multiplier topology.

The overall efficiency of this multiplier-type rectification circuit is given by:

$$\eta = \frac{V_U I_L}{2N I_S B_1 \left(\dfrac{V_0}{V_T}\right) \exp\left(-\dfrac{V_U}{2N V_T}\right)} \qquad (11.42)$$

where V_U is the output voltage, I_s is the diode saturation current, I_L is the output current, $B_1(x)$ is the first-order modified Bessel function, N is the number of stages, and V_T is the thermal voltage for the diodes.

One technique to optimize power conversion efficiency for a given average power limit is through a pulsed/impulsive powering technique. Due to the finite forward bias of the rectification components and their non-linear character-istics, these rectification circuits typically have improved efficiencies when exposed to higher power levels. In medical applications, transmission power at RF frequencies through human tissue is constrained by the exposure limit guidelines for time-varying electromagnetic fields. According to the International Commission on Non-Ionizing Radiation Protection (ICNIRP) and the IEEE C95.1 standard, the whole-body average specific energy absorption rate (SAR), for frequencies between 10 MHz and 10 GHz, must be below 0.08 W/kg, averaged over a 6-minute period [40]. For an average adult human weighing about 80 kg, the 6-minute average absorption must be under

6.4 W [41]. For these electrical components, the response time is much shorter, when compared with biological structures, and thus a bursting technique can be used to expose the rectification structure to a higher instantaneous power while maintaining a low average power to meet the IEEE C95.1 SAR limits.

For an ocular implant application, *in vivo* studies are done on rabbit subjects in [21], and impulsive powering through ocular tissue was quantified. A miniature ocular implant is inserted into the eye into the suprachoroidal space, and an external transmitter and antenna is used to generate an RF field about 15 cm away from the rabbit subject. The radiated RF field is received by the implant and converted to a DC supply to charge a 20-uF capacitive load, which is required for using an impulsive powering technique. The duty cycle, or pulse width, of the transmitted signal varies from continuous wave (100%) down to very short instantaneous bursts (2%). The charge delivered to the capacitor as a function of time is quantified. The resulting current measurement is multiplied with the voltage, and the subsequent power received is analyzed as a function of the average power transmitted from the source. The results, shown in Figure 11.8a on a dB scale and Figure 11.8b on a linear scale, show significantly greater performance as the duty cycle of the transmitted RF signal is decreased.

To further optimize this pulsed powering technique, a monocycle pulse generator can be used. The circuitry in the proposed monocycle pulse generator is laid out with three parts: a Gaussian pulse generator, a pulse shaping circuit, and an RC differentiator as shown in Figure 11.9.

The Gaussian pulse is generated by the SRD accompanied by a second-order RLC transient circuit. Unlike pulses based on active transistors, the SRD requires no extra bias, and functions as a switch with extremely short transition time. The voltage across the serial inductor L_1 changes with the time-varying current through L_1. The voltage generates an instant, sharp spike at the moment that the shunt SRD switch turns off. The change in voltage on this inductor L_1 can be calculated from the product of L_1 and the time variation of the current through the inductor. Through the second-order transient circuit response of $R_1 L_2 C_1$, a sub-nanosecond Gaussian pulse can be created. Through Kirchhoff's rules:

$$L_1 \frac{di(t)}{dt} = \frac{1}{C_1} \int i(t)\, dt + L_2 \frac{di(t)}{dt} + R_1 i(t) \qquad (11.43a)$$

Expressing $i(t)$ as:

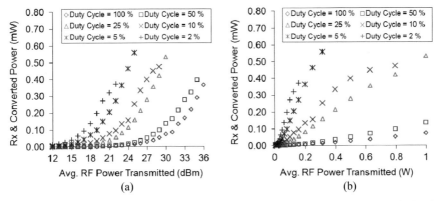

Figure 11.8 Impulsive powering results from *in vivo* rabbit studies where a device is implanted in the eye and power transfer across 15 cm, through radiating RF fields and conversion to DC, is quantified and plotted (a) as a function of transmitted power on a dB scale and (b) as a function of that on a linear scale.

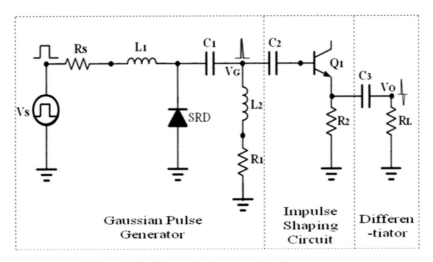

Figure 11.9 Schematic of monocycle pulse generator.

$$i(t) = ke^{st} \tag{11.43b}$$

and substituting into Equation (11.43):

$$s = -\frac{R_1}{2\left(L_2 - L_1\right)} \pm j\sqrt{\frac{1}{\left(L_2 - L_1\right)C_1} - \left(\frac{R_1}{2\left(L_2 - L_1\right)}\right)^2} \tag{11.44}$$

Defining the following:

$$\frac{R_1}{2\left(L_2 - L_1\right)} \equiv \varsigma \tag{11.45}$$

$$\sqrt{\frac{1}{\left(L_2 - L_1\right)C_1}} \equiv \omega_0 \tag{11.46}$$

where ω_o is the resonant frequency and ζ is the damping coefficient. Using Equations (11.45) and (11.46) in Equation (11.44) results in:

$$s = -\varsigma \pm j\sqrt{\omega_0^2 - \varsigma^2} \tag{11.47a}$$

where the negative and positive solutions for s can be defined in s_1 and s_2 and thus $i(t)$ can be expressed as:

$$i(t) = k_1 e^{-s_1 t} + k_2 e^{-s_2 t} \tag{11.47b}$$

A change in damping coefficient ζ can be classified according to three types of second-order transient responses: under-damping ($\zeta < \omega_o$), critical-damping ($\zeta = \omega_o$), and over-damping ($\zeta > \omega_o$). To reduce ringing in the monocycle pulse, the over-damping transient condition was selected as the principal component in the design of the Gaussian pulse (on node V_G). A wideband BJT (Q_1) and a simple RC differentiator were used to convert the Gaussian pulse into a bipolar monocycle pulse. A passive wideband BJT with floating collector was intentionally selected to function as a diode (base-emitter junction) in shaping Gaussian pulses. Additional DC offset is not required in our design. A fast transitional behavior of this BJT in diode mode can be achieved because of the shallow emitter and the narrow base width. Therefore, a fast transition is available by utilizing its EB junction diode. The collector-base can be shortened further to ensure that the junction remains reverse biased. When the Gaussian pulse is on the rising edge, the BJT (Q_1), functioning as a diode, turns on to charge the capacitor C_3. Conversely, when the Gaussian pulse is on the falling edge, the capacitor C_3 provides a reverse bias to the BJT (Q_1) to shut down Q_1 and discharges through R_2. The width of the Gaussian pulse can be further compressed after shaping. Finally, the Gaussian pulse is differentiated into a monocycle pulse through a simple RC differentiator. The simulation results of Gaussian pulse and monocycle pulse waveforms are shown in Figure 11.10.

A pulse width of 280 ps was generated; however, the amplitude was reduced considerably from 1.25 V to 0.235 V, due to loss in the differentiator.

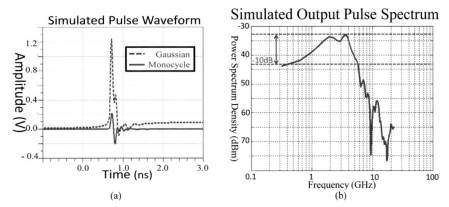

Figure 11.10 (a) Simulated waveforms of proposed Gaussian pulse and monocycle pulse using over-damping response in time domain; (b) The response of the simulated monocycle pulse in frequency domain.

Figure 11.9b is the Fourier transform of the simulated pulse in the frequency domain, implying an ultra-wide bandwidth of nearly 5.7 GHz (220 MHz– 5.9 GHz). Ringing exists in all practical impulse circuits; however, compensation can be used to reduce the effects of ringing. The ringing ratio can be reduced by exploiting its symmetrical waveform and if positive and negative cycles have precisely the same pulse width and opposite phase ripple, then the ringing can be reduced to the minimal level. Ringing is further optimized through the second-order transient response, the shaping network, and the RC differentiator.

11.8 Benefiting Applications and Devices

Wireless powering based on far-field radiating RF fields has revolutionary benefits to the world of biomedical implantable applications. Wireless powering and recharging is a growing need in current medical devices, and this technique allows for incredible size reductions and operation over vastly greater distances, which is beneficial in a variety of areas including neural, ocular, and cardiovascular applications.

For neural prostheses, there are a variety of devices that monitor and record while there are also those that stimulate. Companies such as NeuroVista and NeuroPace have developed implants that can record directly from the brain [42, 43]. Boston Scientific, St. Jude Medical (after acquisition of Advanced

Neuromodulation Systems) and Cyberonics are a few of the larger medical companies that have commercially available neurostimulation devices, while there are over 20 start-up companies entering the field, indicating the huge growth in this market [44–46]. Although size may not be as critical in some of these neural devices, miniaturization is always desirable and would also facilitate surgical procedure and potentially allow for flexibility in placement. Radiative-based powering could reduce or even eliminate the battery requirements through use of only a small-scale antenna and allow for operation across reasonable distances.

Cardiovascular applications have predominantly focused on pacemakers and implantable defibrillators, which are typically relatively large devices implanted in the chest with electrodes that reach the heart, although a new "mini-pacemaker" about the size of a vitamin is under development by Medtronic, showing the drive for miniaturizing devices in this area [47]. Other areas of cardiac applications include monitoring of physiological statistics such as blood pressure, hemodynamics, and glucose, which have been striving for miniaturization for ease of surgery and flexibility of placement. CardioMEMS developed a Champion Heart Failure Monitoring system, which is based on a passive inductor-capacitor (LC) resonator circuit device and can be implanted in the patient's heart chambers or into the pulmonary artery, but unfortunately, after a 550-patient study, the FDA did not vote in favor of this device [48–50]. ISSYS has introduced a pressure monitoring system, still under development, that is comprised of two parts: an implantable, battery-less, telemetric sensor and a companion handheld reader [51, 52]. The implantable device is to be anchored in the left atrium and contains a MEMS pressure transducer, electronics, and an antenna which transmits data via magnetic coupling [53]. St. Jude Medical has been performing studies on their HeartPOD^{TM} device, which is used to directly monitor left atrial pressure. The 3×7 mm device, to be implanted in the atrial septum, consists of an implantable sensor leadcoupled with a subcutaneous antenna coil, a titanium pressure-sensing membrane, and^{a} circuitry for measuring and communicating LAP, temperature, and intracardiac electrogram [54]. A cardiac pressure sensor, using far-field radiative powering, is described in [55], where the device is integrated with an FDA-grade medical stent, as shown in Figure 11.11, for use as an antenna to receive RF energy to power the device. The device includes an application-specific integrated circuit (ASIC) with on-chip electronics to process measurements from a micro-electromechanical systems (MEMS) pressure sensor and wirelessly transmit out the digitized data.

Figure 11.11 Stent-based cardiac pressure–sensing implant radiograph, unpackaged device, and final sealed package.

The field of ocular biomedical implants is a relatively less developed area in part due to the strict size limitations in the available implantation locations; however, substantial work is still being done for sight restoration and treatment and monitoring of other ocular conditions such as glaucoma [56–59]. Due to the relatively small scale of the eye, extreme size minimization is required, making this far-field radiative RF powering technique an enabling technology for powering active ocular implants. A miniature device described in [60] takes advantage of this wireless powering technique to achieve a fully active intraocular pressure monitoring device with an integrated MEMS pressure sensor, capacitive power storage array, ASIC, and antenna all assembled into a biocompatible package. Two versions of this device are shown in the figure below where the implant in Figure 11.11a measures about 3×6 mm with the antenna embedded directly into a ceramic substrate, and that shown in Figure 11.12b has a head that measures about 3×6 mm and a 30-mm tail, which is used to house an antenna, and is packaged in a flexible liquid-crystal-polymer material.

(a) (b)

Figure 11.12 Ocular implants for monitoring intraocular pressure. (a) 3×6 mm version on a ceramic substrate with antenna embedded in the substrate layers. (b) Flexible liquid-crystal-polymer version (showing both sealed and unsealed pictures) with a 3×6 mm head and 30-mm tail (antenna).

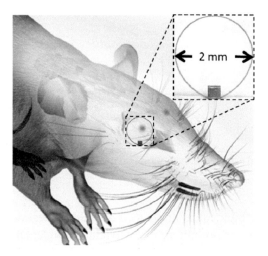

Figure 11.13 Device under development for intraocular pressure monitoring in mice.

The system provides an active intraocular pressure monitoring system that can operate over a 24-hour period, recording a digitized pressure measurement every 5 min, and storing into on-board memory. The intent is that at the end of each day, the patient holds an external device nearby to download the pressure data and recharge, via far-field wireless power transfer, the onboard power storage on the implant.

This work has been further extended to research applications on the genetic studies of intraocular pressure on mice where devices, as shown in Figure 11.13, are being developed to monitor the pressure within the eye of a mouse [61, 62]. Obviously, there are incredible size restrictions for this application which will benefit from the radiative powering technique.

11.9 Conclusions

This chapter on radiative wireless powering through tissue has derived an analytical model of RF propagation through biological tissue to help in the understanding of the electromagnetic considerations that need to be taken in an implantable setting. Computer-based simulation tools and models have also been described as a method to help in the design of the implantable antenna and system. Empirical studies in both *ex vivo* and *in vivo* settings have been presented to validate the models and give an idea of the experimental setup and requirements. Some antenna design guidelines and comparisons with existing work were also discussed. RF rectification circuitry was also described, and novel techniques to help improve performance were presented. Finally, applications of this technology were discussed, and some state-of-the-art biomedical implants were described.

11.10 Problems

P11.1: Consider a 2.4-GHz plane wave propagating in an air medium that is incident on an air–skin(dry) interface. How much of the power is reflected back into the air medium and how much power is transmitted into the skin layer?

P11.2: How much does the magnitude of the power attenuate as a 2.4-GHz wave propagates through a 3-mm dry skin?

P11.3: Considering two isotropic antennas implanted deep in a homogeneous human muscle media and separated by a distance of 1 meter, how much path loss (in decibels), due to just energy spreading, does a 2.4 GHz wave traveling between the antennas encounter?

P11.4: Considering two isotropic antennas implanted deep in a homogeneous human muscle media and separated by a distance of 2 cm, how much loss (in decibels) due to both energy spreading and tissue absorption (attenuation) and assuming the far-field approximation, does a 2.4-GHz wave traveling between the antennas encounter?

P11.5: Consider a 433-MHz plane wave propagating through 3 tissue layers. The wave initially has a magnitude of 1 W in wet skin and first travels through 3 mm of this wet skin, then 2.5 cm of fat, and finally through 5 cm of muscle. What is the final power magnitude (after the 5 cm of muscle)?

P11.6: Consider two antennas communicating at 433 MHz through 3 tissue layers. The transmit antenna is a small isotropic radiator in wet skin and the path of its output signal first travels through 3 mm of this wet skin, then 2.7 cm of fat, and finally through 7 cm of muscle before its incidence on the receive antenna. Assume the receive antenna has a gain of 1 and an effective area of 10 cm^2. What is the ratio of received power to transmitted power?

References

[1] Tesla, N. "System of Electric Lighting." U.S.P. Office, Ed. United States (1891).

[2] Brown, W.C. "The History of Power Transmission by Radio Waves." IEEE *Transactions on Microwave Theory and Techniques* 32: 1230–1242 (1984).

[3] Glaser, P. "Satellite Solar Power Station and Microwave Transmission to Earth." *Journal of Microwave Power* 5 (1970).

[4] Fischell, R.E. "The Retrospectroscope-the Invention of the Rechargeable Cardiac Pacemaker: vignette #9." *Engineering in Medicine and Biology Magazine IEEE* 9: 77–78 (1990).

[5] Fischell, R.E. "Fixed Rate Rechargeable Cardiac Pacemaker." U.S. Patent, Ed.: The Johns Hopkins University (1971).

[6] ICNIRP. "Guidelines for Limiting Exposure to Time-Varying Electric, Magnetic and Electromagnetic Fields (up to 200 GHz)." *Health Physics* 74: 494–522 (1998).

[7] ICNIRP. "Guidelines for Limiting Exposure to Time-Varying Electric and Magnetic Fields (1 Hz to 100 kHz)." Health Physics 99: 818–836 (2010).

[8] Balanis, C.A. *Advanced Engineering Electromagnetics.* New York: Wiley (1989).

[9] Ramo, S., J.R. Whinnery, and T.V. Duzer. *Fields and Waves in Communication Electronics*, 3rd ed. New York: Wiley (1994).

[10] Gabriel, C., S. Gabriel, and E. Corthout. "The Dielectric Properties of Biological Tissues: I. Literature survey." 41: 2231 (1996).

[11] Gabriel, S., R.W. Lau, and C. Gabriel. "The Dielectric Properties of Biological Tissues: II. Measurements in the Frequency Range 10 Hz to 20 GHz." 41: 2251 (1996).

[12] Gabriel, S., R.W. Lau, and C. Gabriel. "The Dielectric Properties of Biological Tissues: III. Parametric Models for the Dielectric Spectrum Of Tissues." 41: 2271 (1996).

[13] Gabriel, C. "Compilation of the Dielectric Properties of Body Tissues at RF and Microwave Frequencies." *Report N.AL/OE-TR- 1996–0037, Occupational and environmental health directorate, Radiofrequency Radiation Division, Brooks Air Force Base*, Texas, USA (1996).

[14] Knapp, J.P. "ANSYS Inc. Request for Waiver of 47 C.F.R. § 1.1307(b)(2) of Commission Rules, DA 11-192." *Federal Communications Commission* (2011).

[15] Chow, E.Y., A. Kahn, and P.P. Irazoqui. "High Data-Rate 6.7 GHz Wireless ASIC Transmitter for Neural Prostheses." in *Engineering in Medicine and Biology Society, 2007. EMBS 2007. 29th Annual International Conference of the IEEE*, pp. 6580–6583 (2007).

[16] Chow, E.Y., Y. Chin-Lung, A. Chlebowski, W.J. Chappell, and P.P. Irazoqui. "Miniature Antenna for RF Telemetry Through Ocular Tissue." in *Microwave Symposium Digest*, 2008 IEEE MTT-S International, pp. 1309–1312 (2008).

[17] Chow, E.Y., B. Beier, O. Yuehui, W.J. Chappell, and P.P. Irazoqui. "High frequency transcutaneous transmission using stents configured as a dipole radiator for cardiovascular implantable devices." in Microwave Symposium Digest, 2009. MTT '09. *IEEE MTT-S International*, pp. 1317–1320 (2009).

[18] Chow, E.Y., O. Yuehui, B. Beier, W.J. Chappell, and P.P. Irazoqui. "Evaluation of Cardiovascular Stents as Antennas for Implantable Wireless Applications." *IEEE Transactions on Microwave Theory and Techniques* 57: 2523–2532 (2009).

[19] Chow, E.Y., B.L. Beier, A. Francino, W.J. Chappell, and P.P. Irazoqui. "Toward an Implantable Wireless Cardiac Monitoring Platform Integrated with an FDA-Approved Cardiovascular Stent." *Journal of Interventional Cardiology* 22: 479–487 (2009).

[20] Chow, E.Y., Y. Chin-Lung, A. Chlebowski, M. Sungwook, W.J. Chappell, and P.P. Irazoqui. "Implantable Wireless Telemetry Boards for *In Vivo* Transocular Transmission." *IEEE Transactions on Microwave Theory and Techniques* 56: 3200–3208 (2008).

[21] Chow, E.Y., Y. Chin-Lung, O. Yuehui, A.L. Chlebowski, P.P. Irazoqui, and W.J. Chappell. "Wireless Powering and the Study of RF Propagation Through Ocular Tissue for Development of Implantable Sensors." *IEEE Transactions on Antennas and Propagation* 59: 2379–2387 (2011).

[22] "Part 15 - Radio Frequency Devices (47 CFR 15), Title 47 of the Code of Federal Regulations." *Federal Communications Commission*, current as of December 22 (2011).

[23] Ojaroudi, M., S. Yazdanifard, N. Ojaroudi, and M. Naser-Moghaddasi. "Small Square Monopole Antenna With Enhanced Bandwidth by Using Inverted T-Shaped Slot and Conductor-Backed Plane." *IEEE Transactions on Antennas and Propagation 59*: 670–674 (2011).

[24] Ryu, K.S., and A.A. Kishk. "UWB Dielectric Resonator Antenna Having Consistent Omnidirectional Pattern and Low Cross-Polarization Characteristics." *IEEE Transactions on Antennas and Propagation* 59: 1403–1408 (2011).

[25] Thomas, K.G., and M. Sreenivasan. "A Simple Ultrawideband Planar Rectangular Printed Antenna With Band Dispensation." *IEEE Transactions on Antennas and Propagation* 58: 27–34 (2010).

[26] Zhi Ning, C., S.P.S. Terence, and Q. Xianming. "Small Printed Ultrawideband Antenna With Reduced Ground Plane Effect." *IEEE Transactions on Antennas and Propagation* 55: 383–388, (2007).

[27] Ching-Wei, L., and C. Shyh-Jong. "A Simple Printed Ultrawideband Antenna With a Quasi-Transmission Line Section." *IEEE Transactions on Antennas and Propagation* 57: 3333–3336 (2009).

[28] Nazli, H., E. Bicak, B. Turetken, and M. Sezgin. "An Improved Design of Planar Elliptical Dipole Antenna for UWB Applications." Antennas and Wireless Propagation Letters, IEEE 9: 264–267 (2010).

[29] Jin-Ping, Z., X. Yun-Sheng, and W. Wei-Dong. "Microstrip-Fed Semi-Elliptical Dipole Antennas for Ultrawideband Communications." *IEEE Transactions on Antennas and Propagation* 56: 241–244 (2008).

[30] Yazdanboost, K.Y., and R. Kohno. "Ultra Wideband L-loop Antenna." in *IEEE International Conference on Ultra-Wideband, ICU 2005*, pp. 201–205 (2005).

[31] Joshi, R.K., and A.R. Harish. "Printed Wideband Variable Strip Width Loop Antenna." in *Antennas and Propagation Society International Symposium, 2007* IEEE, pp. 4793–4796 (2007).

[32] Yi-Cheng, L., and H. Kuan-Jung. "Compact Ultrawideband Rectangular Aperture Antenna and Band-Notched Designs." *IEEE Transactions on Antennas and Propagation* 54: 3075–3081 (2006).

[33] Jia-Yi, S., and S. Jen-Yi. "Design of Band-Notched Ultrawideband Square Aperture Antenna With a Hat-Shaped Back-Patch." *IEEE Transactions on Antennas and Propagation* 56: 3311–3314 (2008).

[34] Shi, C., P. Hallbjorner, and A. Rydberg. "Printed Slot Planar Inverted Cone Antenna for Ultrawideband Applications." *Antennas and Wireless Propagation Letters, IEEE 7*: 18–21 (2008).

[35] Oraizi, H., and S. Hedayati. "Miniaturized UWB Monopole Microstrip Antenna Design by the Combination of Giusepe Peano and Sierpinski Carpet Fractals." *Antennas and Wireless Propagation Letters, IEEE* 10: 67–70 (2011).

[36] Soontornpipit, P., C.M. Furse, and C. You Chung. "Design of Implantable Microstrip Antenna for Communication with Medical Implants." *IEEE Transactions on Microwave Theory and Techniques* 52: 1944–1951 (2004).

[37] Karacolak, T., A.Z. Hood, and E. Topsakal. "Design of a Dual-Band Implantable Antenna and Development of Skin Mimicking Gels for Continuous Glucose Monitoring." *IEEE Transactions on Microwave Theory and Techniques* 56: 1001–1008 (2008).

[38] Means, D.L., and K.W. Chan. "Evaluating Compliance with FCC Guidelines for Human Exposure to Radiofrequency Electromagnetic Fields." *OET Bulletin 65 (Edition 97-01) Supplement C (Edition 01-01), Federal Communications Commission Office of Engineering & Technology* (2001).

[39] Sun, X., C. Zhang, Y. Li, Z. Wang, and H. Chen. "Design of Several Key Circuits of UHF Passive RFID Tag." *China Integrated Circuit* 16 (2007).

[40] ICNIRP. "Guidelines for Limiting Exposure to Time-Varying Electric, Magnetic, and Electromagnetic Fields (up to 300 GHz)." *Health Phys.* 74: 494–522 (1998).

[41] Ogden, C.L., C.D. Fryar, M.D. Carroll, and K.M. Flegal. "Mean Body Weight, Height, and Body Mass Index, United States 1960–2002." *Centers for Disease Control and Prevention* 347 (2004).

[42] "NeuroVista Corporation." http://www.neurovista.com/.

[43] "NeuroPace, Inc." http://www.neuropace.com.

[44] "Boston Scientific Corporation." http://www.bostonscientific.com.

[45] "St. Jude Medical, Inc." http://www.sjm.com/.

[46] "Cyberonics, Inc." http://www.cyberonics.com.

[47] "Medtronic, Inc." http://www.medtronic.com/.

[48] Fonseca, M., M. Allen, D. Stern, J. White, and J. Kroh. "Implantable wireless sensor for pressure measurement within the heart." A61B005/0215 ed United States: Cardiomems Inc (US) (2005).

[49] Allen, M., M. Fonseca, and J. White, et al. "Implantable Wireless Sensor for Blood Pressure Measurement with an Artery." Cardiomems Inc (US) (2005).

[50] Joy, J., J. Kroh, and M. Ellis, et al. "Communicating with Implanted Wireless Sensor." United States: CardioMEMS, Inc. (2007).

[51] Najafi, N., and C.A. Rich. "Method for Monitoring a Physiologic Parameter of Patients with Congestive Heart Failure." US: Integrated Sensing Systems, Inc., Filed (2006).

[52] Najafi, N., and A. Ludomirsky. "Initial Animal Studies of a Wireless, Batteryless, MEMS Implant for Cardiovascular Applications" *Biomedical Microdevices* 6: 61–65 (2004).

[53] Schneider, R.L., N. Najafi, and D.J. Goetzinger. "Anchor for Medical Implant Placement and Method of Manufacture." United States: Integrated Sensing Systems, Inc. (2008).

[54] Ritzema, J., I.C. Melton, and M.A. Richards, et al. "Direct Left Atrial Pressure Monitoring in Ambulatory Heart Failure Patients." Circulation 116: 2952–2959 (2007).

[55] Chow, E.Y., A.L. Chlebowski, S. Chakraborty, W.J. Chappell, and P.P. Irazoqui. "Fully Wireless Implantable Cardiovascular Pressure Monitor Integrated with a Medical Stent." *IEEE Transactions on Biomedical Engineering* 57: 1487–1496 (2010).

[56] Humayun, M.S., J.D. Weiland, G.Y. Fujii, R. Greenberg, R. Williamson, J. Little, B. Mech, V. Cimmarusti, G. Van Boemel, G. Dagnelie, and E. de Juan Jr. "Visual perception in a blind subject with a chronic microelectronic retinal prosthesis." *Vision Research* 43: 2573–2581 (2003).

[57] Margalit, E., M. Maia, J.D. Weiland, R.J. Greenberg, G.Y. Fujii, G. Torres, D.V. Piyathaisere, T.M. O'Hearn, W. Liu, G. Lazzi, G. Dagnelie, D.A. Scribner, E. de Juan Jr, and M.S. Humayun. "Retinal Prosthesis for the Blind." Survey of Ophthalmology 47: 335–356 (2002).

[58] Yanai, D., J.D. Weiland, M. Mahadevappa, R.J. Greenberg, I. Fine, and M.S. Humayun. "Visual Performance Using a Retinal Prosthesis in Three Subjects With Retinitis Pigmentosa." *American Journal of Ophthalmology* 143: 820–827.e2 (2007).

[59] "SOLX, Inc." http://www.solx.com/.

[60] Chow, E.Y., A.L. Chlebowski, and P.P. Irazoqui. "A Miniature-Implantable RF-Wireless Active Glaucoma Intraocular Pressure Monitor." *IEEE Transactions on Biomedical Circuits and Systems* 4: 340–349 (2010).

[61] Chow, E.Y., D. Ha, L. Tse-Yu, W.N. deVries, S.W.M. John, W.J. Chappell, and P.P. Irazoqui. "Sub-cubic millimeter intraocular pressure monitoring implant to enable genetic studies on pressure-induced neurodegeneration." in *Engineering in Medicine and Biology Society (EMBC), 2010 Annual International Conference of the IEEE*, pp. 6429–6432 (2010).

[62] Chow, E.Y., S. Chakraborty, W.J. Chappell, and P.P. Irazoqui. "Mixed-Signal Integrated Circuits for Self-Contained Sub-Cubic Millimeter Biomedical Implants." in *Solid-State Circuits Conference Digest of Technical Papers (ISSCC), 2010 IEEE International*, pp. 236–237 (2011).

12

Microwave Propagation and Inductive Energy Coupling in Biological Human Body Tissue Channels

Current literature shows that consideration of the presence of skin and tissue in the magnetic channel linking an inductive transmitter and receiver is often glossed over during system design and analysis. In doing so, pertinent system performance degradations are left unaccounted for. In this chapter, we perform a rigorous analysis of the effects of body tissues and specifically skin, fat and muscle tissues on biomedical systems where energy is propagated or coupled from a source to a receiver. The two scenarios of electric field and magnetic flux coupling are described. Models of tissues including the skin are given and analysed in great detail. The effects of tissues on the induced magnetic field are shown to be mainly threefold: Ohmic heating in the tissue at both low and high frequencies which reduces the induced voltage available to the receiver and a phase change resulting from the relaxation time (characteristic frequency) of the tissue. At mid-frequency range, a gain factor is also manifested in the voltage at the output of the tissue impedance transfer function. We show that absorption of E-field in the tissue can be greatly minimised by using protective materials with left-handed materials which maximises reflection. We also show that for the same system and tissue characteristics, the specific magnetic field absorption rate is always smaller than the specific absorption rate for the electric field, suggesting preference for inductive coupling in embedded medical applications.

12.1 Introduction

Probing of biological tissues and the behaviour of biological tissues in the presence of electromagnetic fields has been the subject of a significant number of publications and for various reasons. In the medical field, apart from the use of x-rays, two methods of probing biological tissues are prominent, either using electromagnetic propagation or inductive techniques. This includes magnetic resonance imaging, tomography, tissue stimulation and ablation. These are for unique purposes. A popular recourse however is through the application of bio-impedance which permits studying how the body reacts to electromagnetic radiation. Like all other materials in nature, human tissues have their electrical properties through which electromagnetic propagation in them can be studied and understood. The general area of bio-impedance and its application is too wide to review in a paper like this. Therefore, only a select number of papers which are in line with this investigation will be reviewed.

All naturally occurring materials can and have been modelled as either combinations of resistive and capacitive (RC) circuits for pure dielectrics or as combinations of RC and inductive (RLC) circuits when the materials also have significant inductive behaviour in the frequencies of interest. This modelling approach includes biological tissues. A great deal of work has been published on the dielectric properties of biological tissues [3, 10–14]. An interesting and highly informative summary of the dielectric models of biological objects and methods of computation is given by Gabriel [14].

Apart from medical applications, the need to monitor as much as possible body signals for increased health-care delivery, entertainment and individual security has increased interest in data communication through the use of body area networks. Body area networks mostly employ electromagnetic wave propagation with limiting consequences due to the nature of signal degradation in the human body channel. The human body is composed of tissues which behave as lossy dielectric materials. Hence, uniform plane waves propagating in such tissues will satisfy Maxwell's equations.

Electromagnetic wave degradation around the human body can be as large as 40 to 100 dB at 2.4 GHz depending on the body movements. This is because the human body behaves as both a dielectric and conductive material. The permittivity of body tissues consists of real and imaginary parts. The imaginary part accounts for the electric and magnetic field losses. Also, because many of the tissues including the skin can be modelled as dielectric slab layers, propagation of transverse electric and magnetic waves can result in reflection, refraction and transmission at the interfaces and through the materials. Because of these properties, the human body renders itself to microwave probing with numerous health-related applications including radio-frequency (RF) ablation and cancer treatment. RF ablation is used for the treatment of tumours as a minimal invasive alternative to surgery [10]. In RF ablation, a tumour is exposed to an electric field through electrodes placed on or near it. Typically, RF electrodes can ablate about 5 cm diameter of tumour within 20 minutes [10]. The heating effects resulting from the imaginary part (conductive dielectric) provide the motivation for this technology. The heat delivered to the tumour is proportional to the current density. The current delivered to the tissue via an electrode and return path is up to 4 ground pads. The pads are each typically 10×10 cm and placed on the thighs [10]. The high currents result in high specific absorption rates (SAR). The total current density is due to both conduction and displacement currents given by the expression $\mathbf{J} = (\sigma + j\omega\varepsilon)\mathbf{E}$, where $\mathbf{J}$ is the current density, $\mathbf{E}$ is the electric field intensity at frequency w, σ is conductivity and ε is the dielectric constant of the tissue. The heat created by the current in the tissue is used to kill cancer cells.

Apart from the direct medical application, an area not completely unrelated to the medical field where electromagnetic propagation in tissues is of interest is body area networks. The effects of tissue between a transmitter and a receiver are essential in most applications to quantify their performance and also to ensure exposures are not above the legal limits. The effects of a tissue between a transmitter and a receiver in an embedded bi omedical system can be quantified. See Figure 12.5 in which a tissue material comes in between the transmitter and receiver coils. This situation which exists in all embedded medical devices practically means that a thick slab or layers of dielectric slabs have been placed in the magnetic channel linking the receiver. The effects are several. The tissue layers cause internal bending of the fields (refraction), reflection, absorptions of the magnetic and electric fields and power losses. The objective of this paper is to model such systems and the impact of the tissue material in the channel on the induced current (voltage) in the receiver.

The application of inductive coupling in medicine is diverse with the most prevalent being embedded medical devices [15], embedded drug release capsules [16], neural prosthesis [17], power transfer and regulation [18–20]. To date, attention has been mostly on the characterisation of the effects of electric fields on tissues [21–25], and little to no attention has been paid to the effects of tissues on the performance of inductive couplers (magnetic fields) in the near-field region. How magnetic field energy is coupled rather than propagated in inductive systems is as important as how they propagate in the far-field region. Most of the papers have assumed far-field propagation while in fact the system they described operates in the near-field region. Since most embedded systems operate in the near-field region (near-field region $<\lambda/2\pi$) where energy is mostly coupled and the radiation resistance of the antennas and tissues is very small, impact of the skin and other tissues on inductive coupling and propagation in the near-field is significantly important. For example, at 403 MHz, a frequency often quoted in papers in this area, the near-field edge is 11.85 cm (1.95 cm at 2.45 GHz), which means that antennas and coils located closer to the body than this value are actually within the near-field region and yet the analyses all assume far-field propagation. Thus, the tissue is located in the near-field region where the AC resistance of the antenna is only marginally responsible for radiation and is insignificant. In a nutshell, systems which seek to use propagation methods need to select frequencies which enable them to operate well in the far-field regions close to the body or use inductive coupling methods. Such propagation methods require very short wavelengths. The following sections present two methods of probing the human tissue, based on inductive coupling and propagation and we start with the latter.

12.2 Electromagnetic Wave Propagation in Tissues

Most communication systems operate in the radiating far-field of an antenna. In the radiating far-field, the electric and magnetic fields are orthogonal to each other and energy transfer is modelled by the Poynting vector equation (**P = ExH**) which shows the mutual travel of both the electric and magnetic fields and at the same speed as in Figure 12.1. In the near-field however, the H-field does not propagate but rather quasi-static.

Figure 12.2 illustrates the far-field scenario in which the two fields propagate in a medium, in this case a tissue and the tissue separates the transmitter from the receiver (it acts as the medium in the channel). The tissue is represented as lossy dielectric slabs. An incident TE or TM wave is applied to the topmost layer with properties ε_0, μ_0. The permittivity, conductivity and permeability of the other layers are $\varepsilon_i, \sigma_i, \mu_i$; $1 \leq i \leq 3$. Due to the nature of the system, the tissue presents complex impedance to fields. In the scenario in Figure 12.2, the tissue structure consists of the epidermis in contact with the air, dermis as layer two and the fat as layer three. Both TE and TM waves are considered concurrently. In this analysis, we are at first interested in the effects the tissue has on electromagnetic waves and later we will focus on the near-field with the analysis based on energy coupling as in [1].

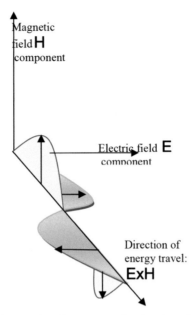

Figure 12.1 Propagating electromagnetic, electric and magnetic fields.

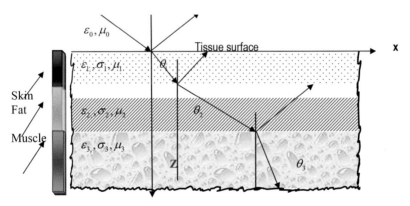

Figure 12.2 Oblique incident electric field on an obstructing tissue in the channel.

In Figure 12.2, the lossy dielectric tissue layers are irradiated by a plane wave at an oblique angle. The plane wave travels through it in the form shown and decays from $z = 0$ to $z = d$. The region $z < 0$ is considered to be free space. The effect of the tissue is to introduce impedance variations and also variations in the refractive index from $z = 0$ to $z > 0$. This variation in refractive index conditions the path taken by the wave. Similarly, the variation in the impedances of the layers causes impedance mismatches which result in reflections. Figure 12.3 shows what happens in each layer and how the signal propagates in it. It shows reflections, attenuation and phase change which are quantified in the following analysis.

First we introduce pertinent system variables. Plane waves propagating in lossy human tissues will satisfy Maxwell's equations. A wave propagating in a lossy tissue sets up both conductive and displacement currents given by the expressions

$$\mathbf{Jc} = \sigma \mathbf{E}$$
$$\mathbf{Jd} = j\omega \mathbf{D} = j\omega \varepsilon_d \mathbf{E}$$
(12.1)

The total current flowing in the tissue is thus the sum of both currents given by the expression

$$\mathbf{J_t} = \mathbf{J_c} + \mathbf{J_d} = (\sigma + j\omega \varepsilon_d) \mathbf{E} = j\omega \varepsilon_c \mathbf{E}$$
(12.2)

ε_c is the total effective complex permittivity of the tissue. Hence,

$$\sigma + j\omega \varepsilon_d = j\omega \varepsilon_c$$
$$\varepsilon_c = \varepsilon_d - \frac{j\sigma}{\omega}$$
(12.3)

Equation (12.3) predicts that conductivity increases with frequency, and permittivity decreases with frequency. These results have been validated by

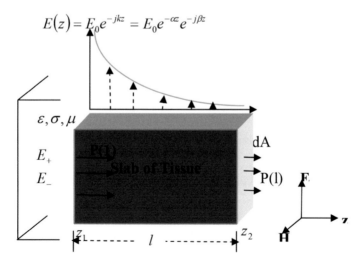

$$E(z) = E_0 e^{-jkz} = E_0 e^{-\alpha z} e^{-j\beta z}$$

Figure 12.3   Wave propagation in tissue.

several authors [21–28]. Note that both σ and ε_d are complex and frequency dependent. If we let the parameter

$$\varepsilon_d = \varepsilon_d' - j\varepsilon_d'', \tag{12.4}$$

then

$$\sigma + j\omega\varepsilon_d = j\omega\varepsilon_c$$
$$\varepsilon_c = \varepsilon_d' - j\left(\varepsilon_d'' + \frac{\sigma}{\omega}\right) = \varepsilon' - j\varepsilon''. \tag{12.5}$$

The imaginary component of the permittivity is non-zero and ε_d'' accounts for power losses in the dielectric material. Henceforth for the sake of generality, we will drop the subscripts 'c' and 'd' in the above equations and define propagation constants, attenuation and the refractive index of the tissues using the commonly known forms. They will be used in the analysis of the reflection coefficients and SAR. The following definitions apply:

$k = \omega\sqrt{\mu\varepsilon}$ is the complex wave number of the tissue,

$\eta = \sqrt{\mu/\varepsilon}$ is the intrinsic impedance of the tissue, and

$n = \sqrt{\mu_r\varepsilon_r}$ is the refractive index of the wave in the tissue. The refractive index determines the path the wave takes through the tissue. Tissues are generally non-magnetic ($\mu_r = 1$) lossy dielectrics. The refractive index of the material is complex and defined by the following expression:

$n = \sqrt{(\varepsilon' - j\varepsilon''/\varepsilon_0)} = n_r - jn_i$ is the complex refractive index of tissues.

Noting that the wave number can be written as $k = \beta - j\alpha = \omega\sqrt{\mu(\varepsilon' - j\varepsilon'')}$, as a complex wave number, we may therefore write the wave number in terms of the refractive index of tissues using the above results: $jk = \alpha + j\beta = jk_0n = jk_0(n_r - jn_i)$ or $\alpha = k_0n_i$; $\beta = k_0n_r$. A figure of merit for materials is their skin depth which defines how far the wave has penetrated the medium when its power has reduced to 37%. The skin depth of the tissue is given by $\delta = 1/\alpha$ representing the penetration depth of the wave into the tissue. These variables determine the nature of wave behaviour in the tissue, which we explore further in the next section. Thus, in Figure 12.2, the intrinsic impedance of each layer can be computed from the expression

$$\eta_i = \frac{\omega\mu_0}{k_i} = \frac{\eta_0}{\sqrt{\left(\varepsilon' - \frac{j\sigma}{\varepsilon_0\omega}\right)}} \cong \frac{376.7343}{\sqrt{\left(\varepsilon' - \frac{j\sigma}{\varepsilon_0\omega}\right)}} \text{ Ohms.} \tag{12.5a}$$

As a net effect, the wave sees impedance different from the impedance of free space.

12.2.1 Wave Reflections in Tissues

When a wave travels through a tissue, it experiences obstructions from the impedance of the tissue and it loses energy. Due to mismatches in the structure, wave reflections also take place. The reflection components reduce the power that is available to be delivered to the receiver. This section quantifies this. Consider Figure 12.3. At the tissue interfaces z = 0 and z = d, according to Maxwell's equations, the electric and magnetic fields are continuous:

$$E = E'; \quad H = H' \tag{12.6}$$

E' and H' are components of the electric and magnetic fields inside the tissue. Continuity of the fields holds inside and outside the tissue. Hence, the relationships between the fields can be used to model the elemental reflection and transmission coefficients. Generally, the waves in the material can be shown to be related by the following equations expressed in a matrix form:

$$\begin{bmatrix} E(z) \\ H(z) \end{bmatrix} = \begin{bmatrix} e^{-jkz} & e^{jkz} \\ e^{-jkz}/\eta & -e^{jkz}/\eta \end{bmatrix} \begin{bmatrix} E_{0+} \\ E_{0-} \end{bmatrix} \tag{12.7}$$

By defining the reflection coefficient $\Gamma = E_-/E_+$ as the ratio of incident and reflected electric fields and also the wave impedance $Z = E(z)/H(z)$ as the ratio of electric field to the magnetic field, we can write the system matrix equation at any point z in the medium as:

$$\Gamma(z) = \left[\frac{Z(z) - \eta}{Z(z) + \eta} \right] \quad \text{or } Z(z) = \eta \left[\frac{1 + \Gamma(z)}{1 - \Gamma(z)} \right] \tag{12.8}$$

These relationships hold for the reflection coefficients and impedance in the tissue.

The relationship between the impedance at the two interfaces shows how the impedance is propagated in the material as follows:

$$Z_1 = \eta \left[\frac{Z_2 + j\eta \tan kl}{\eta + jZ_2 \tan kl} \right] \tag{12.9}$$

This impedance is a function of the intrinsic impedance of the medium and the wave number. At the two interfaces, the electric and magnetic fields experience reflections, and the impedances at the two interfaces are similarly related through the coefficients of reflection as

$$Z_1 = \eta \left[\frac{1 + \Gamma_1}{1 - \Gamma_1} \right] = \eta \left[\frac{1 + \Gamma_2 e^{-j2kl}}{1 - \Gamma_2 e^{-j2kl}} \right] \qquad (12.9a)$$

These relationships describe propagation of EM waves in most dielectric media and apply equally well to bio-tissues. They show that the waves experience attenuation, reflection, transmission and in multiple layer tissues, they experience refraction as well.

12.2.2 Matlab Simulations

The intrinsic impedance was evaluated at several frequencies of interest. The data in Table 12.1 used for the simulations were taken from [9, 26, 27]. While the data from [9] and [27] appear to agree significantly, the data on the conductivity and relative permittivity from [26] do not. The trend in conductivity and permittivity values from [26] does not agree well with the rest. Six frequencies were considered between 403 and 4500 MHz. The 900 and 916.5 MHz fall within the GSM band. The wireless local area network (WLAN) frequency of 2450 is also considered. These are frequencies employed in ubiquitous devices (mobile phones and access points). In each row, in columns two and three, the three sets of values are for skin, fat and muscle in that order.

The results in Table 12.1 (columns 4, 5 and 6) show that body tissues have complex impedance values imposed by the complex permittivity. The tissue impedance increases with frequency. The fat tissue has very high impedance compared to the other layers of tissues near the skin with the impedance of the muscle being the smallest. Conductivity increases with frequency but permittivity evidently decreases with frequency.

The next section discusses their propagation regimes and how electromagnetic waves propagate in them. In the next section, we employ the intrinsic variables of the tissue to model its impedance and hence its impact on inductive energy coupled to an implanted device. In Figure 12.4, the blue curves are for TM and the green ones for TE.

12.3 Applications

In Figure 12.4, the reflection coefficients from skin and muscle are very high. This may be desirable in applications where the objective is to prevent RF penetration into the tissue such as in absorbers and in health. If the

Microwave Propagation and Inductive Energy Coupling

Table 12.1 Tissue impedance at various frequencies

Frequency (MHz)	Conductivity (skin, fat, muscle); σ	Relative Permittivity (skin, fat, muscle); ε_r	Impedance (η_1) skin	Impedance (η_1) fat	Impedance (η_1) muscle
236	0.61; 0.07; 0.75	53; 11.9; 59.3	41.999 + j15.8029	102.02 + j21.811	38.5038 + j15.5292
450	0.71; 0.08; 0.81	45.8; 11.6; 56.8	49.369 + j14.048	107.63 + j14.554	45.043 + j11.9294
900	0.87; 0.11; 0.94	41.4; 11.3; 55.0	55.117 + j11.098	110.53 + j10.645	48.7513 + j8.0915
1000 [34]	1.1; 0.1; 1.4	44; 6.0; 50.5	53.037 + j11.3695	148.94 + j21.831	48.8202 + j11.4904
1500	1.3; 0.11; 1.4	43.5; 6.0; 50.0	54.6055 + j9.4831	151.11 + j16.404	51.1979 + j8.3605
1800	1.18; 0.19; 1.34	38.9; 11.0; 53.6	58.4540 + j8.6594	112.35 + j9.6186	50.3072 + j6.1849
2450	1.46; 0.27; 1.74	38.0; 10.8; 52.7	59.3923 + j8.2111	113.22 + j10.298	50.8001 + j6.0653
3000	2.4; 0.17; 2.2	42; 6.0; 47	55.4414 + j9.1052	152.17 + j12.825	53.4187 + j7.3493
5000	4.0; 0.22; 4.4	40.5; 5.5; 44	56.6325 + j9.7559	159.41 + j11.403	54.2767 + j9.4601
5800	3.72; 0.83; 4.5	35.1; 9.86; 48.5	61.2022 + j9.7940	117.06 + j15.018	52.5132 + j7.4003
7000	5.0; 0.33; 6.7	39; 5.0; 42	58.0518 + j9.3102	166.70 + j14.027	54.8668 + j10.8023
8.5	7.0; 0.37; 8.3	37; 4.5; 40	58.601 + j11.2878	175.62 + j15.155	55.7874 + j11.7016
10	7.7; 0.43; 10.0	35.8; 4.4; 37.6	59.778 + j11.1534	177.57 + j15.478	56.9123 + j12.9046
11.5	8.8; 0.49; 11.7	34.4; 4.1; 35.5	60.7784 + j11.701	183.68 + j17.009	57.9379 + j14.0464

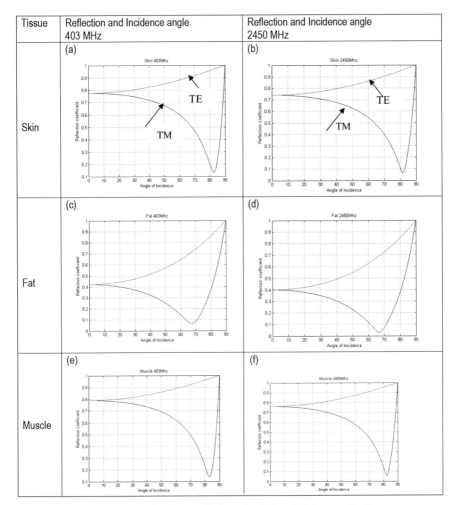

Figure 12.4   Reflection coefficients for TE and TM waves in tissue.

objective is communication at high signal-to-noise ratio and if the reflected
waves are used for the communication, then it is desirable to have high
reflection. Then, the waves should be incident at those angles which help
to maintain large reflections. This point is further illustrated in Figure 12.5 for
a left-handed metamaterial with $\varepsilon_r = -1$ at an air interface. The reflection
coefficient is extremely high for both TE and TM.

Nearly all the incident radiation is reflected at all oblique angles. On
the other hand, only skin nodes of body area network could benefit from

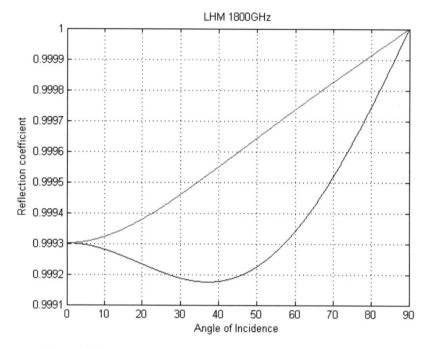

Figure 12.5   Extreme reflection by left-handed materials (metamaterials).

this arrangement. In many sensing applications however the objective is RF penetration into an embedded device to either monitor internal bio-signals, charge battery power and system reconfiguration of an embedded device large reflections are not desirable. Hence, incident angles which result in reduced reflections should be used.

Figure 12.5 shows that a left-handed material coated on a dielectric substrate or as part of the constituents of a material which sits on the surface of the skin can reflect most of the incident electromagnetic radiation away from the skin. Figure 12.6 is an illustration of the process of signal reflection by such a coating. A suitable application is an ear guard to prevent excessive penetration of mobile phone radiations into the head around the ear. Such a device could be worn as part of a daily clothing or as when needed. To ensure that the reflected field is not recirculated back towards the ear, the mobile phone piece should have an EM wave absorber covering as a thin shield. This ensures that any reflected field is absorbed into the phone covering fabric.

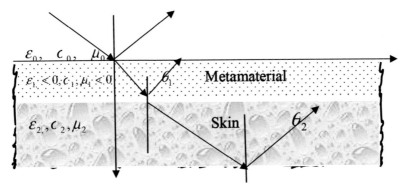

Figure 12.6 Application of left-handed metamaterial to increase electromagnetic wave reflection.

12.4 Inductive Energy Coupling Systems in Tissues

Near-field methods differ significantly from their far-field counterparts in that in the near-field region transmission is by inductive coupling rather than propagation. Although this makes them more secure from interceptors, it in effect limits the usable range of near-field systems to very short distances from the transmitter. The near-field region can be divided into the reactive and radiating near-field (intermediate region). This section focuses mainly on the former. RFIDs operate in this reactive region.

This study assesses the influence of skin on energy coupling in inductive embedded, on-body biomedical and body area network systems. The presence of a tissue introduces impedance (AC resistance in addition to a DC resistance) variations in the channel. The net effect of the changing impedance can be estimated by modelling the impedance of the tissue. The electric field propagation case described in Figure 12.2 in which an oblique field is incident on the tissue at an angle θ has been analysed in earlier sections of this paper.

The inductive coupling case is shown in Figure 12.7. The transmitter circuit consists of a coil modelled as a series of RLC combination with an input time–varying current source of i_1 Amperes and input impedance Z_s. This varying current sets up magnetic flux in the coil. Due to mutual inductance, the changing current in the transmitter coil sets up induced current i_{RX} in the receiver and a proportional voltage $jwMi_1$. The receiver coil also has a self-inductance L_2, radiation resistance and a loss resistance as shown in the right-hand half of the Figure 12.7. The output load impedance (resistance) of the receiver is Z_L. Normally, the radiation resistances are very small, a few

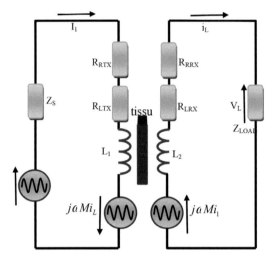

Figure 12.7   Implanted system with obstructing body tissue.

tens of Ohms. Therefore, both the radiation and loss resistors will be added and be considered as one resistor as:

$$R_{TX} = R_{RTX} + R_{LTX} \tag{12.10a}$$

$$R_{RX} = R_{RRX} + R_{LRX} \tag{12.10b}$$

Applying Kirchoff's voltage law to Figure 12.7 gives the voltages in the transmitter and receiver circuits as follows:

$$v_{TX} = (Z_S + R_{TX}) i_1 + j\omega L_1 i_1 - j\omega M i_L \tag{12.11a}$$

$$v_L = Z_L i_L = j\omega M i_1 - j\omega L_2 i_L - R_{RX} i_L \tag{12.11b}$$

These equations reveal that at the transmitter, the voltage resulting from the mutual inductance is a loss, whereas at the receiver it is a gain. Hence, these equations provide a means for assessing the level of induced voltages. The load voltage at the receiver is directly proportional to the mutual inductance from which we can derive some estimation of the effects of the coupling coefficient $k(x)$ on the communication link. This effect will now be quantified formally first without the biological tissue and then with a biological tissue in place. Before delving into that, a careful observation of Equations (12.10b) and (12.11a) shows that in the reactive near-field region when the radiation resistance is zero, more energy is coupled to the load because $(R_{RTX} = R_{RRX} = 0)$. The radiation resistance increases with the radius r of the loop antenna [1]. Hence, in the radiating far-field region, less power is transferred to the receiver load because it appears as a loss in Equation (12.11a).

The radiation efficiency of the loop antenna is therefore given by the ratio

$$\eta_{rad} = \frac{R_{rad}}{R_{rad} + R_{loss}} \tag{12.12}$$

Friis propagation equation applies only to communication when the receiver is located in the radiating far-field of an antenna. In the radiating near-field, signal propagation has been shown to not follow the Friis equation [2]. For this case, Yates [2] showed that when the input impedance Z_S of the transmitter is chosen to resonate with the transmitting antenna inductor L_1, maximum power is transferred to the receiver load. Hence, when

$$Z_S = \frac{1}{j\omega C} \text{ where } C = \frac{1}{\omega^2 L_1} \tag{12.13}$$

The transmitter power therefore is given by the expression

$$P_{TX} = I_{1rms}^2 R_{TX} \tag{12.14}$$

By using the Biot–Savart law, the resulting magnetic field induced to the receiver antenna is

$$H = \frac{I_1 N_1 r_1^2}{2\left(\sqrt{r_1^2 + x^2}\right)^3} \tag{12.15}$$

In this expression, x is the distance separating the transmitter and the receiver when they are aligned on a common axis. The transmitting coil has radius r_1 and N_1 turns. Using Faraday's law of induction, the induced voltage in the receiver coil (antenna) is given by the expression

$$V = N_2 \mu_0 A_2 . j\omega H \tag{12.16}$$

The induced voltage is proportional to the magnetic field that was set up by the time-varying current. Its magnitude is a function of frequency, the area and number of turns of the receiving coil. This voltage is available at the input of the receiver coil, and the equation assumes uniform flux linkage to the receiver coil when the distance between the transmitter and the receiver is much larger than the coil radii. In the equation, N_2 is the number of turns of the receiver antenna coil, A_2 is the area of the receiver antenna, and $\phi_2 = \mu_0 A_2 . j\omega H$ is the permeability. Clearly, Equation (12.16) shows there is advantage in well-designed coils that optimise the magnetic field H. This optimisation depends on the number of turns of the transmitter coil and the current flowing in them. It also depends on the nature of the coil as shown by the denominator of Equation (12.15). The radius of the transmitting coil is clearly important. In terms of the distance x, the shorter the receiver is to the transmitter, the larger the induced magnetic field. By substituting Equation (12.15) in Equation (12.16), the induced voltage at the receiver antenna is

$$V = \frac{j\mu_0 \omega I_1 N_1 N_2 A_2 r_1^2}{2\left(\sqrt{r_1^2 + x^2}\right)^3} \tag{12.17}$$

This voltage is available to a load in the receiver circuit. Maximum power transfer is achieved at the receiver by matching the receiver load impedance as a conjugate of the receiver antenna impedance. Therefore

$$Z_L = R_{RX} - j\omega L_2 \tag{12.18}$$

The power delivered to the load is maximum at resonance and is given by the expression

$$P_{RX} = \frac{V_{rms}^2}{4R_{RX}} = \frac{\left(\mu_0 \omega I_{1rms} N_1 N_2 A_2 r_1^2\right)^2}{16 R_{RX} \left(r_1^2 + x^2\right)^3} \tag{12.19}$$

The power transfer ratio given by the ratio of Equations (12.5) and (12.10) compares the received power to the transmitted power. This should be maximised in practice.

$$\frac{P_{RX}}{P_{TX}} = \frac{\left(\mu_0 \omega N_1 N_2 A_2 r_1^2\right)^2}{16 R_{TX} R_{RX} \left(r_1^2 + x^2\right)^3} \tag{12.20}$$

When the distance between the receiver and the transmitter is much larger than the radius of the transmitting coil, the power transfer ratio expression simplifies to

$$\frac{P_{RX}}{P_{TX}} \cong \zeta \cdot \frac{\left(\pi \mu_0 \omega N_1 N_2 r_2^2 r_1^2\right)^2}{16 R_{TX} R_{RX} x^6} \tag{12.21}$$

$A_2 = \pi r_2^2$, and ζ represents a factor for misaligning the antennas. Hence, the power transfer increases with the radii of the coils, frequency and the number of turns of the coils in the transmitter and receiver antennas. Unfortunately, when a tissue is in between the transmitter and receiver coils, the net effect is an introduction of a non-magnetic material in the path of the magnetic

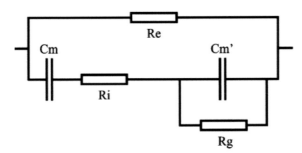

Figure 12.8   Bio-impedance model of biological tissues [5, 6].

flux and prevention of maximum power from being delivered to the receiver. The next section is an analysis of the effects of the introduced tissue. The next section presents an analysis of the effect of skin tissue when it is placed between the transmitter and receiver on the received power. To do that, the impedance of skin tissues is first modelled and used as part of the system transfer function between the output of the transmitter and the input of the receiver.

12.5 Bio-Impedance Models of Tissues

Implanted medical devices experience propagation problems from tissues in general and also the skin. This section provides two models, the first for tissues and the second for skin which will be used in later sections to analyse their impact on implanted devices. The body is electrically conducting due to conductive fluids, and this includes organs and tissues including the skin. Considering that the human skin consists of several layers such as the epidermis and dermis of varying thickness (from 0.03 to 1.4 mm for men and from 0.86 to 3 mm for women), it can be modelled as dielectric layers. The skin however also manifests anisotropic dielectric property due to the variation of the dielectric components of its constituent parts. The approximate thicknesses of the epidermis and dermis are 0.429 mm and 3 mm, respectively, with the palm having a skin thickness of about 3 mm. There is significant regional variability of skin impedance with the least values obtained in the palm due to the presence of more sweat glands there. Hence, modelling of human tissues and organs requires detailed attention to variations in properties and body functions. Different body functions result in different models.

In general, the impedance of a biological tissue consisting of intracellular (ICS) and extracellular structures (ECS) depends on the frequency of the signal applied to the tissue. To date, the range of input frequencies in use includes low frequencies, radio-frequencies, microwaves, infrared and optical range. Bio-impedance models are mostly combinations of resistive and capacitive (RC) circuits of various forms. Tissues are thus non-magnetic. A number of authors have proposed a model for tissues in terms of RC circuits with Figure 12.1 being one of the most elaborate [5, 6].

This model uses extracellular space (resistor R_e), transcellular pathways (capacitor C_m) and cytosol (resistor R_i) as means of modelling the electrical properties of tissues. Further parameters such as R_g and C'_m represent the gap junction resistance and adjacent membrane capacitor, respectively. Thus, the overall impedance is given by the combinations of C_m and R_i in series with R_g and C'_m. This series combination is in parallel with R_e. These RC components as we will show shortly induce characteristic frequencies representing the processes. The processes which lead to the low and high frequency responses are shown in Figure 12.9.

From Figure 12.9, the signal path at low frequencies is through the intercellular pathways, avoiding the cell membranes. At higher frequencies, the waves pass through the cell membranes as well as the intercellular pathways. Hence, the tissue model should account for this behaviour. In the tissue models, the resistors and capacitors are obtained from the expressions $R = A/(\sigma\, l)$ and $C = (\varepsilon_r \varepsilon_0 A)/l$, respectively. The cell membrane has length l, conductivity σ and area A. From Figure 12.4, the overall tissue impedance therefore can be estimated as follows.

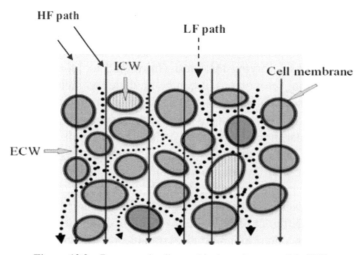

Figure 12.9   Processes leading to bio-impedance models [30].

Two different explanations are given in the current literature as to why the signal takes different paths at low and high frequencies [30–32]. Some authors have suggested that high frequency signals are able to penetrate objects more and hence pass through the cells [30, 31]. The second group of authors suggests the behaviour is due to changes in the dielectric properties of the tissue areas [32]. Clearly, the penetration depth of a radiation into a material is inversely proportional to the frequency of the wave. Hence, the penetration depth at high frequencies should be less than at low frequencies, suggesting that its explanation is in error. If they were right, low frequency signals should pass through the cells. The second explanation given is based on resistance which suggests that differences in the resistances of the cell (membrane and internal fluid) compared with the impedance of the extracellular pathways make the signal to pass in between the cells at low frequencies and to pass through the cells at high frequencies. Indeed, fundamental theory of how electromagnetic waves propagate in materials suggests that the paths taken by waves and signals are guided by the refractive index of the medium. We have shown earlier in this paper that the refractive index of a material is given by the expression $n = \sqrt{\mu_r \varepsilon_r}$, where $\mu_r = 1$ for biological tissues. Hence, the refractive indices of the cells and of the extracellular pathways are different and are functions of frequency. Clearly, at low frequencies, the intrinsic impedance of the pathways and of the cells given by $\eta = 1/\sqrt{\mu_r \varepsilon_r} = 1/\sqrt{\varepsilon_r}$ is different and must be low at low frequencies for the extracellular pathways compared with the path through the cells. Also at higher frequencies, the cell membranes and ICW present smaller intrinsic impedances to the current and hence become the path of least impedance. From basic circuit theory at a current junction in a circuit, more current always passes the path of least impedance (reactance) and in this case the path is through the cells at high frequencies. Remember the relative permittivity decreases with frequency (Equation (12.5a)) and hence η becomes larger and close to infinity (open circuit) at high frequencies for the extracellular pathways. Comparatively therefore, the intercellular pathways must have the larger intrinsic impedance (smaller relative permittivity) than the ICW, making the path through it the preferred current path. This proves the second set of authors theoretically right. Henceforth let

$$Z_{mi} = R_i + \frac{1}{j\omega C_m} \tag{12.22}$$

and

$$Z_{mg} = \frac{R_g(1/j\omega C'_m)}{R_g + 1/j\omega C'_m} = \frac{R_g}{1 + j\omega C'_m R_g} \tag{12.23}$$

Then, the total impedance of this model is as follows:

$$Z = \frac{R_e(Z_{mi} + Z_{mg})}{R_e + Z_{mi} + Z_{mg}} \tag{12.24}$$

Therefore

$$Z = \frac{R_e\left[\left(1 - \omega^2 C_m R_i C'_m R_g\right) + j\omega(C_m R_i + C'_m R_g + C_m R_g)\right]}{\left(1 - \omega^2 C_m R_i C'_m R_g - \omega^2 C_m R_e C'_m R_g\right)} \\ + j\omega\left(C_m R_i + C'_m R_g + C_m R_g + C_m R_e\right) \tag{12.25}$$

To emphasise the dependence of this model on several tissue parameters that are related to its state, we define the following time constants

$$\begin{aligned} \tau_{mi} = \tau_4 = C_m R_i; \quad \tau_{mg} = \tau_2 = C_m R_g \\ \tau_{m'g} = \tau_1 = C'_m R_g; \quad \tau_{me} = \tau_3 = C_m R_e \end{aligned} \tag{12.26}$$

By using these relaxation time constants, Equation (12.21) simplifies to the dispersion expression

$$Z_{tissue}(\omega) = \frac{R_e\left[\left(1 - \omega^2 \tau_4 \tau_1\right) + j\omega(\tau_4 + \tau_1 + \tau_2)\right]}{\left(1 - \omega^2 \tau_4 \tau_1 - \omega^2 \tau_3 \tau_1\right) + j\omega(\tau_4 + \tau_1 + \tau_2 + \tau_3)} \tag{12.27}$$

Hence, the tissue manifests real and imaginary impedance values. It can be shown quite easily that at DC, the dispersion equation reduces to $Z_{tissue,DC} = R_e$ and at infinite frequency, it reduces to the parallel combination

$$Z_{tissue,\infty} = \frac{R_e R_i}{R_e + R_i} \tag{12.28}$$

Alternatively, this may be viewed in terms of the characteristic frequencies of the tissue where $\omega_i = 1/\tau_i$

$$Z_{tissue}(\omega) = \frac{R_e\left[\left(1 - \frac{\omega^2}{\omega_1 \omega_4}\right) + j\omega\left(\frac{1}{\omega_1} + \frac{1}{\omega_2} + \frac{1}{\omega_4}\right)\right]}{\left(1 - \frac{\omega^2}{\omega_1 \omega_3} - \frac{\omega^2}{\omega_1 \omega_4}\right) + j\omega\left(\frac{1}{\omega_1} + \frac{1}{\omega_2} + \frac{1}{\omega_3} + \frac{1}{\omega_4}\right)} \tag{12.29}$$

The circuit for a simplified high frequency model is shown in Figure 12.10. At high frequency, the imaginary part of the impedance vanishes. The high frequency approximation of this model is a real value given by the expression

$$Z_{tissue}(\omega) \cong \frac{R_e}{\left(1 - \frac{\omega^2 \tau_1 \tau_4}{1 - \omega^2 \tau_1 \tau_3}\right)} \cong \frac{R_e}{\left(1 + \frac{\tau_4}{\tau_3}\right)} \tag{12.30}$$

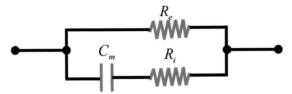

Figure 12.10 Simplified high frequency model of tissue.

This means that the high frequency approximation is well modelled by the tissue impedance circuit:

Thus, as shown in Figure 12.9 at high frequencies, the ICW path does not exist and hence the influences of R_i and C_m may be ignored ($C_m = 0$ and $R_i = 0$). Hence, $\tau_2 = 0$; $\tau_3 = 0$; $\tau_4 = 0$ and $Z_{tissue}(\omega) = R_e$. This is the same result as we got for DC. The low frequency impedance is dominated by the static impedance of the ECW.

Given the values of the parameters for this model from [33] as $C_m = 1\ \mu F$, $C'_m = 10\ \mu F$, $R_i = 470\ \Omega$, $R_e = 2600\ \Omega$ and $R_g = 10,000\ \Omega$, the system time constants are

$$\tau_1 = C'_m R_g = 100\ \text{ms};\ f_2 = 10.00\ \text{Hz};$$
$$\tau_2 = C_m R_g = 10\ \text{ms};\ \ f_3 = 100.00\ \text{Hz};$$
$$\tau_3 = C_m R_e = 2.6\ \text{ms};\ f_4 = 384.62\ \text{Hz};$$
$$\tau_4 = C_m R_i = 0.47\ \text{ms};\ f_1 = 2127.66\ \text{Hz}.$$

The frequency response has a notch around the high frequency point of 2127.66 Hz. The correction terms have affected the shape of the response. As expected, the resistance R_e affects only the magnitude of the impedance response. R_g affects the position of the notch. As its value is decreased, the response shifts towards higher frequencies. Also as the magnitudes of the capacitors are reduced, higher frequency responses are obtained.

Increasing R_g or decreasing C_m or C'_m has the same effect of shift in the position of the notch as in Figure 12.11b. A plot of the real versus imaginary components of the impedance response is given in Figure 12.12.

12.5.1 Skin Model

A large body of knowledge exists for the electrical impedance models of skin. This captures only what is needed for the analysis in this paper. The human skin is often modelled as a series-parallel RC circuit as shown in Figure 12.2 [3]. The deeper skin layers which receive blood from the

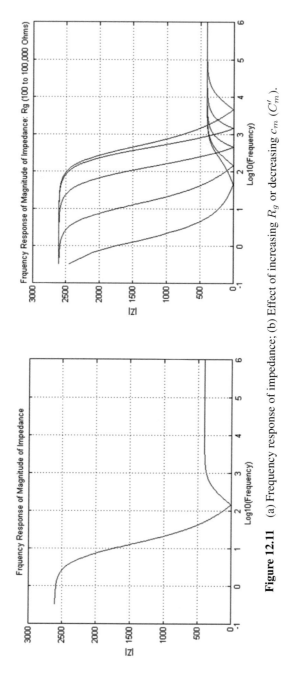

Figure 12.11   (a) Frequency response of impedance; (b) Effect of increasing R_g or decreasing c_m (C'_m).

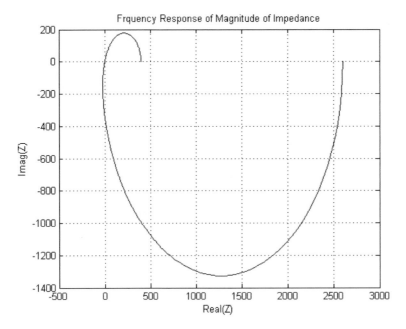

Figure 12.12   Normalised skin impedance (imaginary vs. real part).

capillary networks in the dermis are modelled with the resistor R_s. The superficial layers of the skin are modelled with the parallel RC circuit consisting of R_p and C_p.

Hence, the transfer function of a sample of section of skin is given by the expression:

$$Z_S = R_s + \frac{R_p/j\omega C_p}{R_p + \frac{1}{j\omega C_p}} = R_s + \frac{R_p}{1 + j\omega R_p C_p} \qquad (12.31)$$

This equation can be further simplified if we recognise the time constant $\tau_p = R_p C_p$ so that

$$Z_S = R_s + \frac{R_p}{1 + j\omega\tau_p} \qquad (12.32)$$

12.5.2 Matlab Simulations

The parallel resistance is typically around 4.5 Ω and 40 nF and the series resistance is approximately 700, which means 28×10^{-6} sec or the characteristic frequency is 5.56 MHz. The above model is identical to the Cole dispersion model provided $\alpha = 1$, $R_s = R_\infty$ and $R_p = R_0 - R_\infty$. Figure 12.11 shows the normalised impedance represented in Equation (12.32) as a function of frequency. The values are normalised to the maximum value of $Z_s = R_s + R_p$. Thus, the maximum Ohmic power to the skin when a current of amplitude I amperes is delivered to it is $P = \left(I_{rms}^2/2\right)\left(R_s + R_p\right)$.

The plot shown covers the frequency range of 1 Hz to 20 MHz. The impedance values of the impedance at the left edge and the right edge of the Re(Z) axes are determined by the resistance at DC and infinite frequencies. The frequency response of the impedance is shown in Figure 12.13. The horizontal axis uses a logarithmic scale.

This is in agreement with the analysis above that around DC, the impedance is purely resistive and equal to $Z_s = R_s + R_p = 704.5\ \Omega$.

12.6 Impact of Tissue Impedance on Inductive Coupling

From Equation (12.6), the impact of the tissue impedance on the inductive coupling system can be modelled. If the transmitting coil couples the magnetic field directly to the tissue as is in most applications, the field at the input of the tissue is equal to

$$H = \frac{I_1 N_1 r_1^2}{2\left(\sqrt{r_1^2 + x^2}\right)^3} \tag{12.33}$$

The transfer function for this system is shown in Figure 12.8.

With an induced input voltage, the current at the output of the tissue structure is

$$Y_0(s) = Z_{tissue}(s)\,H(s) \tag{12.34}$$

Equation (12.34) is also shown as Figure 12.16. The impulse response of the tissue therefore is

$$\frac{Y_0(s)}{H(s)} = G_{tissue}(s) = Z_{tissue}(s) \tag{12.35}$$

Figure 12.14 shows the normalized skin impedance as evaluated in Matlab. The frequency response of skin is given also in Figure 12.15.

Figure 12.13   A skin model.

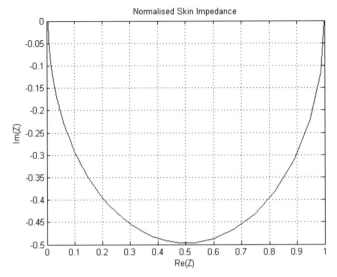

Figure 12.14   Normalised skin impedance (imaginary vs. real part).

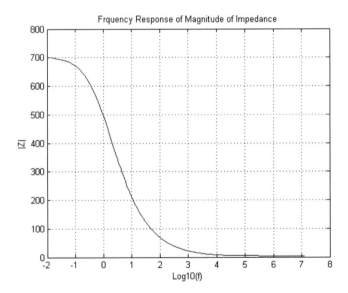

Figure 12.15   Frequency response of skin impedance.

Hence, the output signal from the tissue must be a voltage proportional to the input magnetic field. Technically, this is correct. Hence, analysis of the tissue impedance will reveal the impact on the induced voltage in the receiver coil. Hence, in the time domain, the convolution of the tissue impedance with the magnetic field strength gives the impact of the tissue. Only the current in Equation (12.33) is a function of time, the rest are space dependent and enable us to rewrite it as:

$$H = g(x)\, I_1(t) \tag{12.36}$$

$$g(x) = \frac{N_1 r_1^2}{2\left(\sqrt{r_1^2 + x^2}\right)^3} \tag{12.37}$$

In essence given a skin model

$$Z_S = R_s + \frac{R_p}{1 + j\omega\tau_p} \tag{12.38}$$

The induced voltage at the receiver is

$$V_r = Z_S \otimes H(x,t) = \left(R_s + \frac{R_p}{1 + j\omega\tau_p}\right) \otimes g(x)\, I_1(t) \tag{12.39}$$

This suggests two effects from the skin impedance. The first is obviously a heating effect due to R_s and the second appears to be both a gain and a phase change proportional to R_p and τ_p, respectively.

Alternatively, the magnetic energy absorbed in the near-field of a transmitter is given by the following Equations [28, 29]

$$SAR = \frac{\sigma}{\rho}\frac{\mu\omega}{\sqrt{\sigma^2 + \varepsilon^2\omega^2}}\left(1 + \delta_{corr}\Gamma\right) H_{rms}^2 \tag{12.40}$$

All the parameters in this equation retain their usual meanings except Γ which is the reflection coefficient of a plane wave in the tissue as defined in Equation (12.8) and evaluated in Figure 12.4 of this chapter; δ_{corr} is a correction term which accounts for the distances of the scatterer from the antenna and is defined as [28]

$$\delta_{corr} = \begin{cases} 1 & d \geq 0.08\lambda/\Gamma \\ \sin\left(\frac{\pi}{2}\frac{|\Gamma|}{0.08}\frac{d}{\lambda}\right) & d < 0.08\lambda/\Gamma \end{cases} \tag{12.41}$$

The reflection coefficient may be computed with the methods in Equation (12.8) or it may be estimated from the dielectric properties of the medium as given in [28] as follows:

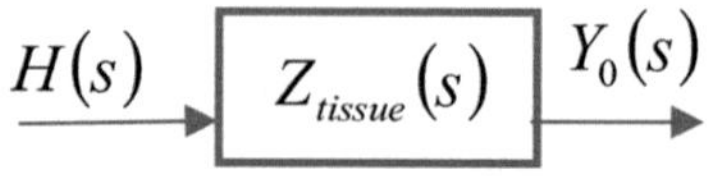

$$H(s) \longrightarrow \boxed{Z_{tissue}(s)} \xrightarrow{Y_0(s)}$$

Figure 12.16   Response of tissue structure to input inductive voltage.

$$\Gamma = \frac{2\left|\sqrt{\varepsilon'}\right|}{\left|\sqrt{\varepsilon'} + \sqrt{\varepsilon_0}\right|} - 1 \tag{12.42}$$

$\varepsilon' = \varepsilon - \sigma/j\omega$. By substituting for the H-field, SAR in the near-field is [29]

$$SAR_H = \frac{\sigma}{\rho} \frac{\mu\omega}{\sqrt{\sigma^2 + \varepsilon^2\omega^2}} \frac{I^2 l^2 \sin^2\theta}{(4\pi R^2)^2} \tag{12.43}$$

R is the distance between the receiver and the transmitter, and θ is the angular inclination between their antennas.

To appreciate the level of absorption by both the magnetic and electric fields, we compare them at the near-field edge where E-field propagation takes over from H-field coupling. Since SAR at the far-field is determined by the E-field and is given by the expression

$$SAR_E = \frac{\sigma\beta^2\eta^2}{\rho} \frac{I^2 l^2 \sin^2\theta}{(4\pi R^2)^2}, \tag{12.44}$$

at the edge of the near-field region, we expect normal RF propagation to begin, and the range R should be the same for both fields. The ratio of the absorption rates at the near-field edge where $R = \lambda/2\pi$, $\beta = 2\pi/\lambda$, $\mu_r = 1$, $\eta = \sqrt{\mu_r\mu_0/\varepsilon_r\varepsilon_0}$ and $Z_0 = \sqrt{\mu_0/\varepsilon_0}$ is

$$\frac{SAR_H}{SAR_E} = \frac{Z_0^2}{\eta^2} \frac{1}{\sqrt{\varepsilon_r^2 + (\sigma^2/\varepsilon_0^2\omega^2)}} = \frac{\varepsilon_r}{\sqrt{\varepsilon_r^2 + (\sigma^2/\varepsilon_0^2\omega^2)}} \tag{12.45}$$

Figure 12.11 is the absorption ratio for a skin tissue with relative permittivity 53 and conductivity 1.2 between the frequencies of 1 MHz and 2.45 GHz. At low frequencies, the magnetic field absorption rate is very small compared with the E-field absorption. At high frequencies, they are approximately equal. In the near-field region therefore, low frequency applications will lead to smaller absorption of coupled energy in the tissues.

The above expression is only an approximation because the conductivity of the medium is a function of frequency, but the equation assumes an average value. Hence, the more accurate expression for the ratio of SARs should account for this variation with frequency and is

$$\frac{SAR_H}{SAR_E} = \frac{\varepsilon_r}{\sqrt{\varepsilon_r^2 + (\sigma_i^2/\varepsilon_0^2\omega_i^2)}} \tag{12.46}$$

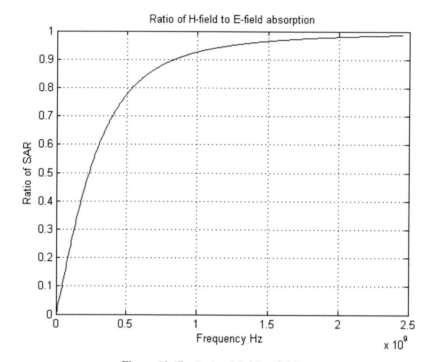

Figure 12.17   Ratio of SAR_H/SAR_E.

This ratio (Equation 12.46) is given as Figure 12.17. It shows the magnetic filed specific absorption ratio normalized with the electric field absorption ratio.

12.7 Circular Coil

The magnetic field created by a circular current carrying coil at distance x from the coil is

$$H = \frac{I_1 N_1 r_1^2}{2 \left(\sqrt{r_1^2 + x^2} \right)^3} \qquad (12.47)$$

When the receiver coil has radius r2 and N2 turns, the total field collected by the coil is

$$H_r = \frac{I_1 N_1 N_2 r_1^2 \pi . r_2^2}{2 \left(\sqrt{r_1^2 + x^2} \right)^3} \qquad (12.48)$$

The power absorbed in the near-field from the magnetic field is

$$SAR = \frac{\sigma}{\rho} \frac{\mu\omega}{\sqrt{\sigma^2 + \varepsilon^2\omega^2}} \left(1 + \delta_{corr}\Gamma\right) H_{rms}^2$$

$$= \frac{\sigma}{\rho} \frac{\mu\omega}{\sqrt{\sigma^2 + \varepsilon^2\omega^2}} \left(1 + \delta_{corr}\Gamma\right) \frac{1}{2} \left[\frac{I_1 N_1 N_2 r_1^2 \pi . r_2^2}{2\left(\sqrt{r_1^2 + x^2}\right)^3} \right]^2 \qquad (12.49)$$

12.8 Conclusions

This chapter has provided techniques for modelling human skin in the form of electric circuits. Due to the conductive nature of the skin and the fact that it absorbs and heats up when irradiated with RF signals, the skin is modelled by various forms of RC circuits. The skin is generally not magnetic and hence the dominant effects of RF signals appear to be more of electric field interactions with the skin. The chapter has also provided the specific absorption ratios for both magnetic and electric fields.

References

[1] Agbinya, Johnson I. *Principles of Inductive Near Field Communications for Internet of Things*, ISBN: 978-87-92329-52-3, River Publishers, Aalborg, Denmark (2011).

[2] Yates, David C., Andrew S. Holmes. and Alison J. Burdett. "Optimal Transmission Frequency for Ultralow-Power Short Range Radio Links." *IEEE Transaction On Circuits and Systems–I: Regular Papers* 51, no. 7: 1405–1413, July (2004).

[3] Dorgan, Stephen J., and Richard B. Reilly. "A Model for Human Skin Impedance During Surface Functional Neuromuscular Stimulation." *IEEE Transactions on Rehabilitation Engineering* [see also *IEEE Transaction on Neural Systems and Rehabilitation*] 7: 341–348, September (1999).

[4] Gersing, E. "Impedance Spectroscopy on Living Tissue for Determination of the State of Organs." *Bioelectrochemistry and Bioenergetics* 45: pp. 145–149 (1998).

[5] Mellert, Fritz, Kai Winkler, and Christian Schneider, Taras Dudykevych, Armin Welz, Markus Osypka, Eberhard Gersing, and Claus J. Preusse. "Detection of (Reversible) Myocardial Ischemic Injury by Means of

Electrical Bioimpedance." *IEEE Transaction on Biomedical Engineering* 58, no. 6: 1511–1518, June (2011).

[6] Hu, Yumu, Mohamud Suwan. "A Fully-Integrated Low-Power BPSK Based Wireless Inductive Link for Implantable Medical Devices." *The 47th IEEE International Mdwest Symposium on Circuits and System,* pp. III-25–III-28 (2004).

[7] Simard, Guillaume, Mohamad Sawan, and Daniel Massicotte. "High-Speed OQPSK and Efficient Power Transfer Through Inductive Link for Biomedical Implants." *IEEE Transaction on Biomedical Circuits and Systems* 4, no. 3: 192–200, June (2010).

[8] Zhen, Bin, Kenichi Takizawa, Takahiro Aoyagi, and Ryuji Kohno. "A body surface coordinator for implanted biosensor networks." in Proceedings of IEEE ICC (2009).

[9] Shultz, Kim, Pascal Stang, Adam Kerr, John Pauly and Greig Scott. "RF Field Visualisation of RF Ablation at the Larmor frequency." *IEEE Transactions on Medical Imaging* (2011).

[10] Seoane, Fernando, Ruben Buendia and Roberto Gil-Pita. "Cole Parameter Estimation from Electrical Bioconductance Spectroscopy Measurements." *32nd Annual International Conference of the IEEE EMBS Buenos Aires*, Argentina, pp. 3495–3498, August 31 – September 4, (2010).

[11] Hirschorn, Bryan, Mark E. Orazem, Bernard Tribollet, Vincent Vivier, Isabelle Frateur, and Marco Musiani. "Determination of Effective Capacitance and Film Thickness from Constant-Phase-Element Parameters." *Electrochimica Acta* 55, no. 21: 6218–6227, August (2010).

[12] White, Erick A., Mark E. Orazem, and Annette L. Bunge. "A Critical Analysis of Single-Frequency LCR Databridge Impedance Measurements Of Human Skin." *Toxicology in Vitro* 25: 774–784 (2011).

[13] Gabriel, C., S. Gabriel, and E. Corthout. "The Dielectric Properties of Biological Tissues: I-III." *Physics in Medicine and Biology* 41: 2231–2293 (1996).

[14] Theilmann, Paul T., and Peter M. Asbeck. "An Analytical Model for Inductively Coupled Implantable Biomedical Devices With Ferrite Rods." *IEEE Transaction on Biomedical Circuits and Systems* 3, no. 1: 43–52 (2009).

[15] Smith, S., et al. "Development of a Miniaturised Drug Delivery System with Wireless Power Transfer and Communication." *Nanobiotechnology* 1, no. 5: 80–86, October (2007).

[16] Coulombe, J., M. Sawan, and J.-F. Gervais. "A Highly Flexible System for Microstimulation of the Visual Cortex: Design and Implementation."

IEEE Transaction on Biomedical Circuits System 1, no. 4: 258–269, December (2007).

[17] Schuder, John C., Jerry H. Gold, and Hugh E. Stephenson, Jr. "An Inductively Coupled RE System for the Transmission of I kW of Power Through the Skin." *IEEE Transaction on Biomedical Engineering* BME-18, no. 4: 265–273, July (1971).

[18] Jow, Uei-Ming, and Maysam Ghovanloo. "Design and Optimization of Printed Spiral Coils for Efficient Transcutaneous Inductive Power Transmission." *IEEE Transaction on Biomedical Circuits and Systems* 1, no. 3: 193–202, February (2007).

[19] Chen, Qianhong, Siu Chung Wong, Chi K. Tse, and Xinbo Ruan. "Analysis, Design, and Control of a Transcutaneous Power Regulator for Artificial Hearts." *IEEE Transaction on Biomedical Circuits and Systems* 3, no. 1: 23–31, February (2009).

[20] Yusoff, N.I.M., S. Khatun, and S.A. AlShehri. "Characterization of Absorption Loss for UWB Body Tissue Propagation Model." *Proceedings of the 2009 IEEE 9th Malaysia International Conference on Communications*, Kuala Lumpur, Malaysia, pp. 254–258, 15–17 December (2009).

[21] Gavriloaia, Gheorghe, Gheorghe Serban, Emil Sofron, M.-R. Gavriloaia, and A.-M. Ghemiogean. "Evaluation of Microwave Electromagnetic Field Absorbed by Human Thyroid Gland." *2010 IEEE 16th International Symposium for Design and Technology in Electronic Packaging (SIITME)*, Pitesti, Romania, pp. 43–46, 23–26 September (2010).

[22] Chamaani, Somayyeh, Yuriy I. Nechayev, Peter S. Hall, and Seyed A. Mirtaheri. "In-body to Off-body Channel Modelling." *2010 Loughborough Antennas & Propagation Conference*, Loughborough, UK, pp. 609–612, 8–9 November (2010).

[23] Fujii, Masafumi, Ryo Fujii, Reo Yotsuki, Tuya Wuren, Toshio Takai, and Iwata Sakagami. "Exploration of Whole Human Body and UWB Radiation Interaction by Efficient and Accurate Two-Debye-Pole Tissue Models." *IEEE Trans on Antennas and Propagation* 58, no. 2: 515–524, February (2010).

[24] Lim, Hooi Been, Dirk Baumann, and Er-Ping Li. "A Human Body Model for Efficient Numerical Characterization of UWB Signal Propagation in Wireless Body Area Networks." *IEEE Transaction on Biomedical Engineering* 58, no. 3: 689–697, March (2011).

[25] Scanlon, William G., J. Brian Burns, and Noel E. Evans. "Radiowave Propagation from a Tissue-Implanted Source at 418 MHz and 916.5 MHz." *IEEE Transaction on Biomedical Engineering* 47, no. 4: 689–697, April (2000).

[26] Hinrikus, Hiie, Jaanus Lass, and Jevgeni Riipulk. "The sensitivity of living tissue to microwave field." *Proceedings of the 20th Annual International Conference of the IEEE Engineering in Medicine and Biology Society* 20, no 6: 3249–3252 (1998).

[27] Kuster, Niels, and Quirino Balzano. "Energy Absorption Mechanism by Biological Bodies in the Near Field of Dipole Antennas Above 300 MHz." *IEEE Transaction on Biomedical Engineering* 41, no. 1: 17–23, February (1992).

[28] Wang, Lujia, Li Liu, Chao Hu, and Max Q.-H Meng. "A Novel RF-based Propagation Model with Tissue Absorption for Location for the GI Tract." 32nd *Annual International Conference of the IEEE EMBS, Buenos Aires*, Argentina: 654–657, August 31–September 4 (2010).

[29] Al-Surkhi, Omar. I., P.J. Riu, Michel Y. Jaffrin. "Monitoring body fluid shifts during haemodialysis (HD) using electrical bioimpedance measurements." 1st *Middle East Conference on Biomedical Engineering (MECMBE)*: 108–113, February 21–24 (2011).

[30] Aberg, Peter. *Skin Cancer as Seen by Electrical Impedance.* ISBN: 91-7140-103-2, Karolinska Institutet, Stockholm, Sweden (2004).

[31] Halter, Ryan J., Alex Hartov, John A. Heaney, Keith D. Paulsen, and Alan R. Schned. "Electrical Impedance Spectroscopy of the Human Prostate." *IEEE Transaction on Biomedical Engineering* 54, no. 7: 1321–1327, July (2007).

[32] Gersing. E. "Impedance Spectroscopy on Living Tissue for Determination of the State of Organs." *Bioelectrochemistry and Bioenergetics* 45: 145–149 (1998).

[33] 1Lea, Andrew, Ping Hui, Jani Ollikainen, and Rodney G. Vaughan. "Propagation Between On-Body Antennas." *IEEE Transaction on Antennas and Propagation* 57, no. 11: 3619–3627, November (2009).

[34] Christ, Andreas, Anja Klingenbock, Theodoros Samaras, Cristian Goiceanu, and Kiels Kuster. "The Dependence of Electromagnetic Far-Field Absorption on Body Tissue Composition in the Frequency Range From 300 MHz to 6 GHz." *IEEE Transaction on Microwave Theory and Techniques* 54, no. 5: 2188–2195, May (2006).

13

Critical Coupling and Efficiency Considerations

13.1 Introduction

A great deal of behaviour change occurs in a wireless power transfer system before and after the critical coupling value. The critical coupling value is a function of distance because all the other variables in the equation for the coupling coefficient such as the radii of the coils remain fixed once they are created. At distances smaller than the critical coupling distance, the power transfer system is overcoupled. Also at distances larger than the critical coupling distance, the power transfer system is undercoupled. The critical coupling value is a transition point between the overcoupled regime and undercoupled regime. The critical coupling value (coefficient) corresponds to the greatest distance at which maximum power transfer efficiency is achieved. This point is the point at which the system achieves resonance and the frequency response manifests one peak. In the overcoupled regime, the frequency response shows several peaks at different frequencies as shown in [1, 2]. These peaks diverge or spread apart as the distance between the receiving and transmitting coils decreases below the critical coupling point. Frequency splitting will be discussed in another chapter.

It is noteworthy that in the overcoupled and undercoupled regimes, the power transfer system is still capable to transfer wireless power. However, in both regimes, the power transfer efficiency decreases below the value obtainable at critical coupling. Thus, in the over- and undercoupled regimes, the flux shared between the loops falls rapidly, well below the critical point value. This decrease is due to impedance mismatch between the transmitting and receiving circuits. Impedance matching is treated in another chapter. In the rest of the chapter, we will undertake analysis of how to obtain the value of the critical coupling coefficients for the two-coil and four-coil systems.

13.2 Two-Coil Coupling Systems

We have shown in [3] that the power transfer equation for the two-coil system is governed by the expression

$$P_{r,2n} = P_t(Q_1 \cdot Q_2 k^2(x))^n \left[\frac{2}{(1 + Q_1 \cdot Q_2 k^2(x))} \right]^{2n} \tag{13.1}$$

This expression is modified by the efficiencies of the transmitting and receiving coils which we will show for brevity as η_1 and η_2, respectively. Hence, the received power is reduced by them. In this section, the focus is for the case when $n = 1$, the so-called peer-to-peer system given by Equation (13.2).

$$P_{r,2} = P_t(Q_1 \cdot Q_2 k^2(x)) \left[\frac{2}{(1 + Q_1 \cdot Q_2 k^2(x))} \right]^2 \tag{13.2}$$

From observation, we note that the received power is a function of the quality factors and the coupling coefficient. In this section, we will focus on the quality factor but will derive the expression for critical coupling coefficient in terms of the quality factors. Equation (13.1) is highly simplified as it does not reveal the contributions from the source resistance and internal resistances of the transmitting and receiving loops. We will make adjustments for them. Wires used to wind the coils have their internal resistances. We will refer to them as the 'self-resistance' of the transmitting and receiving coils. Usually, their values are very small of the order of a several ohms and rarely more than 10 ohms. In most small loops, they are actually less than 1 ohm. Such low values create high Q resonators, and therefore, their impacts on power transfer need to be assessed. For the simplest peer-to-peer system at low values of the quality factors, the resulting power equation when the associated performance degradations in the transmitter and receiver are introduced is

$$P_r(\omega = \omega_0) = P_t Q_1 Q_2 \eta_1 \eta_2 k^2(x) = P_t Q_1 Q_2 \eta_1 \eta_2 \frac{r_1^3 r_2^3}{(x^2 + r_1^2)^3}; \quad r_1 \ll x \tag{13.3}$$

By defining the quality factors, we will demonstrate the effects of the source, self- and load resistances. Let

$$Q_1 = \frac{\omega L_1}{R_{L1} + R_S}; \quad Q_2 = \frac{\omega L_2}{R_{L2} + R_L}; \quad Q_{1,s} = \frac{\omega L_1}{R_{L1}}; \quad Q_{2,s} = \frac{\omega L_2}{R_{L2}} \tag{13.4}$$

By substituting Equation (13.4) in Equation (13.2), the following expression is obtained:

$$\frac{P_r}{P_t} = \eta = \left(Q_1 \cdot Q_2 k^2 \left(x\right)\right) \left[\frac{2}{\left(1+Q_1 \cdot Q_2 k^2 \left(x\right)\right)}\right]^2$$
$$\left(1 - \frac{Q_1}{Q_{1,s}}\right)\left(1 - \frac{Q_2}{Q_{2,s}}\right) \tag{13.5}$$

Several conclusions can be drawn from this equation. The internal resistances of the coils are usually very small, typically below 10 ohms. Hence, the self-quality factors $Q_{1,s}$ and $Q_{2,s}$ are a lot bigger than the loaded quality factors Q_1 and Q_2. Therefore, Equation (13.5) becomes identical with Equation (13.2). Hence, Equation (13.2) is approximately correct in strong coupling situations and for distances shorter than the critical coupling point. When the loaded quality factors are very high so that the ratios $(1 \approx> Q_1/Q_{1,s})$ and $(1 \approx> Q_2/Q_{2,s})$, maximum efficiency cannot be achieved ($\approx>$ means approximately higher).

13.2.1 Strong-Coupling Regime

Under this regime, the self-quality factors of the transmitting and receiving coils are comparatively very high so that $(Q_{1,s} \gg Q_1)$ and $(Q_{2,s} \gg Q_2)$. Hence, the expression (13.5) reduces to

$$\eta_{\text{strong}} = \left[\frac{4Q_1 \cdot Q_2 k^2 \left(x\right)}{\left(1 + Q_1 \cdot Q_2 k^2 \left(x\right)\right)^2}\right] \tag{13.6}$$

When $Q_1 \cdot Q_2 k^2 \left(x\right) \gg 1$, this equation predicts that the power transfer efficiency is approximately

$$\eta_{\text{strong}} = \left[\frac{4}{\left(Q_1 \cdot Q_2 k^2 \left(x\right)\right)}\right] \tag{13.7a}$$

The efficiency thus varies inversely with the coupling coefficient k. Also when $Q_1 \cdot Q_2 k^2 \left(x\right) \ll 1$, the efficiency is given by the expression:

$$\eta_{\text{strong}} = 4Q_1 \cdot Q_2 k^2 \left(x\right) \tag{13.7b}$$

where the position of the maximum efficiency is given by Equation (13.6). To obtain it, set the derivative of Equation (13.6) with respect to either Q_1 or Q_2 to zero. For example, let

$$\frac{d\eta_{\text{strong}}}{dQ_1} = 0$$

The maximum value of efficiency occurs when $Q_1 \cdot Q_2 k^2 (x) = 1$. By substituting this in Equation (13.5), the maximum efficiency is

$$\eta_{\text{strong, max}} = \left(1 - \frac{Q_1}{Q_{1,s}}\right)\left(1 - \frac{Q_2}{Q_{2,s}}\right) \qquad (13.8)$$

In a nutshell, the maximum efficiency is determined by the system resistors to be

$$\eta_{\text{strong, max}} = \left(\frac{R_{L1}}{R_{L1} + R_S}\right) \times \left(\frac{R_{L2}}{R_{L2} + R_L}\right) \qquad (13.8a)$$

This result was derived previously by the author in [4] and was presented as a product of the efficiencies of the transmitting and receiving sides:

$$\eta_{\text{strong, max}} = \eta_1 \eta_2 = \left(\frac{R_{L1}}{R_{L1} + R_S}\right) \times \left(\frac{R_{L2}}{R_{L2} + R_L}\right) \qquad (13.8b)$$

$\eta_1 = \left(\dfrac{R_{L1}}{R_{L1} + R_S}\right)$ and $\eta_2 = \left(\dfrac{R_{L2}}{R_{L2} + R_L}\right)$, respectively. The transmitter efficiency η_1 can be maximised for small source resistances. The receiver efficiency η_2 can be maximised for small loads. Often these conditions cannot be met in practice due to system limitations. Since for strong coupling the maximum efficiency occurs when $Q_1 \cdot Q_2 k^2 (x) = 1$, the strong-coupling coefficient is determined by the equation

$$k(x)_{\text{strong}} = \frac{1}{\sqrt{Q_1 \cdot Q_2}} \qquad (13.9)$$

13.2.2 Weak-Coupling Regime

In the weak-coupling regime, the value $k \ll 1$ and the denominator of the second term in Equation (13.5) is approximately unity.

$$\frac{P_r}{P_t} = \eta_{\text{weak}} = 4\left(Q_1 \cdot Q_2 k^2 (x)\right)\left(1 - \frac{Q_1}{Q_{1,s}}\right)\left(1 - \frac{Q_2}{Q_{2,s}}\right) \qquad (13.10)$$

By differentiating the efficiency expression with respect to Q_1 or Q_2, we obtain the criteria for maximum power transfer as $Q_1 = Q_{1,s}/2$ and $Q_2 = Q_{2,s}/2$. The efficiency of a weak-coupling wireless power transfer system is approximately:

$$\eta_{\text{weak}} = Q_1 \cdot Q_2 k^2 (x) \qquad (13.11)$$

13.3 Efficiency and Impedance Matching

Understanding of how wireless power transfer systems operate comes through understanding the efficiency of the system of power transfer loops. The efficiency of the system is often presented as a product of the transmitter and receiver circuits as in Equation (13.12):

$$\eta = \eta_t \eta_r \tag{13.12}$$

We have used this expression in our analysis so far. A better insight can be obtained by specifying the power transfer efficiency in terms of the voltage and current transferred to the receiver load. In terms of power transfer, we have

$$\eta = \frac{P_{\text{out}}}{P_{\text{in}}} = \frac{V_L I_r}{V_S I_t} = G_V G_I \tag{13.13}$$

The power transfer efficiency is the product of the voltage transfer and current transfer efficiencies. The efficiency may also be written exclusively in terms of the voltage or current efficiencies as

$$\eta = \frac{P_{\text{out}}}{P_{\text{in}}} = \frac{V_L^2 / R_L}{V_S^2 / R_S} = G_V^2 \frac{R_S}{R_L} \tag{13.14}$$

$$\eta = \frac{P_{\text{out}}}{P_{\text{in}}} = \frac{I_L^2 R_L}{I_S^2 R_S} = G_I^2 \frac{R_L}{R_S} \tag{13.15}$$

13.3.1 Efficiency of Peer-to-Peer WPT

Several factors affect the efficiency of wireless power transfer systems. The first factor that affects wireless power transfer is if there exists a mismatch between the input impedance of the transmitting circuit and the source impedance. To avoid signal reflections, the source input impedance should match the source impedance. This condition requires that the input impedance be the complex conjugate of the source impedance. When this condition is achieved the reflection coefficient at the input circuit becomes zero. Signal reflections cause reductions in the available signal to be transferred by removing the source of reflection, and we also boost the available signal from the transmitter. Also at the receiver, to ensure that maximum power is received, not only should resonance condition be created but the output impedance of the receiver must also match the receiver load. This matching of the receiver load to the output impedance eliminates signal reflections at the receiver load. Like in the transmitter, this helps to convey most of the available induced power to the load. A third reason why the system efficiency is less than expected is when there is frequency splitting. Often the receiver circuit is monitored at one frequency. However, at strong coupling, the system spectral response has two peaks at two different frequencies. Several authors [1–4] have explained and analysed this phenomenon well. Therefore, if the receiver power is monitored at only one of the split frequencies, the effect is a major reduction in efficiency. To enhance the efficiency, the receiver should be monitored at the split frequencies concurrently and the received power at the split frequencies be summed. This enhances the system efficiency considerably. In the rest of this chapter and the next, these three conditions which result in reduced efficiency are discussed in great details.

13.4 Impedance Matching and Maximum Power Transfer Considerations

A better understanding of the circuit theory model of wireless power transfer systems can be derived if the transmitting and receiving circuits are treated as two port networks with input and output impedances. The transmitting circuit is connected to the input port which is the output of the voltage source. The load impedance is connected to the output port.

The input impedance looking into the two points to which the input source is connected as shown in Figure 13.1 is influenced by the receiver circuit due to mutual inductance. This influence is modelled by the reflected receiver impedance $(\omega M)^2 / Z_2$ which affects the input impedance through the following equation:

$$Z_{\text{in}} = Z_1 + \frac{(\omega M)^2}{Z_2} = R_1 + j\omega L_1 + \frac{1}{j\omega C_1} + \frac{(\omega M)^2}{R_2 + R_L + j\omega L_2 + \frac{1}{j\omega C_2}}$$

$$(13.16)$$

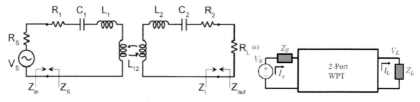

Figure 13.1   Maximum power transfer impedance relationships.

and

$$Z_{\text{out}} = Z_2' + R_1 + j\omega L_1 + \frac{1}{j\omega C_1} = R_2 + j\omega L_2 + \frac{1}{j\omega C_2}$$
$$+ \frac{(\omega M)^2}{R_1 + + R_S + j\omega L_1 + \frac{1}{j\omega C_1}} \qquad (13.17)$$

Equation (13.16), which depicts the input impedance of the WPT system, includes the load, while Equation (13.17) includes the source resistance. The reflected impedance is discussed in a later section of the chapter. At resonance, the reflected impedance is resistive. At any other frequencies, it adds phase changes to the signals at the transmitter and receivers because the two circuits can 'hear each other'.

13.4.1 Bi-Conjugate Matching

In bi-conjugate matching in wireless power transfer circuit, simultaneous matching of the input source and also conjugate matching at the load are required. In Figure 13.1, the source will deliver maximum power to the transmitter only when the input impedance is matched to the source impedance. This happens only when the input impedance is equal to the complex conjugate of the source impedance ($Z_{\text{in}} = Z_S^*$). This is *conjugate matching at the source*. At the output (receiver) circuit, maximum power can be delivered to the load only when the output impedance is also equal to the complex conjugate of the load impedance ($Z_{\text{out}} = Z_L^*$). This condition is called *conjugate matching at the load*. For any other impedance relationships, the impedance mismatches in the system cause voltage and current reflections in the circuits causing reduced power transfer and hence reduced power transfer efficiency as well. Therefore, for bi-conjugate matching, the two impedance matching equations therefore need to be considered as:

$$Z_{\text{in}} = Z_S^*$$
$$Z_{\text{out}} = Z_L^* \qquad (13.18)$$

Hence, at resonance, the system efficiency is

$$\eta = \frac{P_{\text{out}}}{P_{\text{source}}} \times 100\%$$
$$= \frac{\omega^2 M^2 R_L}{[R_1(R_2+R_L)^2 + \omega^2 M^2(R_2+R_L)]} \left(1 - |\Gamma_{\text{in}}|^2\right) \times 100\% \qquad (13.19)$$

where the reflection coefficient at the source is given by the expression

$$\Gamma_{\text{in}} - \frac{Z_{\text{in}} - Z_S}{Z_{\text{in}} + Z_S} \qquad (13.20)$$

To compare the efficiency in Equation (13.20) to what was previously obtained, the equation is expressed in terms of the quality factors of the coils given as:

$$\eta = \frac{k^2 Q_1 Q_2 R_2 R_L}{(R_2 + R_L)^2 \left[1 + \dfrac{k^2 Q_1 Q_2}{(R_2 + R_L)}\right]} \left(1 - |\Gamma_{\text{in}}|^2\right) \times 100\,\% \qquad (13.21)$$

When the following conditions are obeyed

$$Q_2' = Q_2 / (R_2 + R_L),$$

$$R_L' = (R_2 R_L) / (R_2 + R_L)$$

$$\Gamma_{\text{in}} = 0$$

When a wireless power transfer system is conjugate-matched, the system contains five major building blocks. These are the input source, the input matching network, the induction system, the receiver matching circuit and lastly the load. The efficiency expression when the source impedance is considered is given as:

$$\eta = \frac{k^2 Q_1 Q_2' R_L'}{[1 + k^2 Q_1 Q_2']} \times 100\,\% \qquad (13.22)$$

When conjugate matching exists in the system, we can show that

$$Z_S = \left(\frac{1}{j\omega C_1} - j\omega L_1\right) + \frac{R_1}{R_2}\sqrt{R_2^2 + (\omega M_{12})^2 \frac{R_2}{R_1}} \qquad (13.23)$$

and

$$Z_L = \left(\frac{1}{j\omega C_2} - j\omega L_2\right) + \sqrt{R_2^2 + (\omega M_{12})^2 \frac{R_2}{R_1}} \qquad (13.24)$$

At resonance, these two expressions reduce to

$$R_S = \frac{R_1}{R_2}\sqrt{R_2^2 + (\omega_0 M_{12})^2 \frac{R_2}{R_1}} \qquad (13.25)$$

and

$$R_L = \sqrt{R_2^2 + (\omega_0 M_{12})^2 \frac{R_2}{R_1}} \tag{13.26}$$

Equations (13.25) and (13.26) are two conditions for maximum power transfer when bi-conjugate matching is considered. By substituting Equation (13.27) into the efficiency Equation (13.19) when $|\Gamma_{in}| = 0$, we can show that the efficiency expression reduces to:

$$\eta = \frac{\omega_0^2 M_{12}^2 \sqrt{R_1 R_2^2 + (\omega_0 M_{12})^2 R_2}}{\left[\sqrt{R_1}\left(R_2 + \sqrt{R_2^2 + (\omega_0 M_{12})^2 R_2}\right)^2 + \omega_0^2 M_{12}^2 \left(R_2 \sqrt{R_1}\right.\right.}$$

$$\left.\left. + \sqrt{R_2^2 R_1 + (\omega_0 M_{12})^2 R_2}\right)\right] \times 100\% \tag{13.27}$$

Let

$$\alpha = \sqrt{R_1 R_2^2 + (\omega_0 M_{12})^2 R_2}$$

Then,

$$\eta = \frac{\omega_0^2 M_{12}^2 \alpha}{\left[\sqrt{R_1}\left(R_2 + \sqrt{R_2^2 + (\omega_0 M_{12})^2 R_2}\right)^2\right.}$$

$$\left. + \omega_0^2 M_{12}^2 \left(R_2 \sqrt{R_1} + \alpha\right)\right] \times 100\% \tag{13.28}$$

Further discussions on bi-conjugate matching of a four-node WPT system are given in [7].

13.5 Reflected Impedance

One of the unique features of inductive systems is that the various separate circuits interact with their neighbours through the reflected impedance. The presence of reflected impedances in each stage of the inductive system introduces phase changes, signal distortion and hence reduction of received signals. In the next subsections, we present brief overviews of reflected impedances for two-, three- and four-stage inductive systems.

13.5.1 Two-Coil Systems

In a two-coil system, the transmitter impedance affects the impedance of the receiver and also the receiver impedance affects the impedance of the transmitter. At the transmitter, the reflected impedance is

$$Z_{\text{Tref}} = \frac{(\omega M)^2}{Z_2} = \frac{\omega^2 M_{12}^2}{Z_2} = \frac{k^2 \omega^2 L_1 L_2}{Z_2} \tag{13.29}$$

where Z_2 is the receiver circuit impedance. Also the reflected impedance at the receiver circuit due to the influence of the transmitter circuit is

$$Z_{\text{Rref}} = \frac{(\omega M)^2}{Z_1} = \frac{\omega^2 M_{12}^2}{Z_1} = \frac{k^2 \omega^2 L_1 L_2}{Z_1} \tag{13.30}$$

where Z_1 is the transmitter circuit impedance. In a nutshell, the ratio of reflected impedances can be used to assess how strong the circuit impedances affect the other. This is given by the expression

$$\frac{Z_{\text{Tref}}}{Z_{\text{Rref}}} = \frac{Z_1}{Z_2} \tag{13.31}$$

When identical coils are used at the receiver and transmitter, the circuits affect each other equally through the reflected impedance. Therefore, to limit the influence of the transmitter impedance on the receiver impedance, the impedance of the transmitter should be smaller than that of the receiver. Also to remove any phase changes introduced by the reflected impedances in the circuits at any frequency other than resonance, conjugate cancellation of the reactive components of the impedances should be used.

13.5.2 Three-Coil Systems

In a three-coil system, the relay circuit impedance is influenced by both the transmitter and receiver impedances. Hence, the transmitter impedance is affected by also the impedance of the receiver through the expression:

$$Z_{\text{Tref}(2-1)} = \left(\frac{M_{12}}{M_{23}}\right)^2 R_L = \frac{k_{12}^2 L_1 R_L}{k_{23}^2 L_3} \tag{13.32}$$

where R_L is the load impedance at the receiver. Hence, for strong coupling of fluxes from transmitter to receiver, optimal choice of the inductors at the transmitter and receiver is highly essential.

13.5.3 Four-Coil Systems

A four-coil WPT system contains a driver coil, a primary coil, a secondary coil and load (receiver) coil. The driver and load circuits should be at resonance for maximum power transfer. However, at all other frequencies, the relationship between the recursive reflected impedances from the load coil to the driver coil can be shown to be given by the expression:

$$Z_{\text{Tref}(2-1)} = \frac{1}{R_L}\left(\frac{\omega M_{12} M_{34}}{M_{23}}\right)^2 = \frac{\omega^2 k_{12}^2 k_{34}^2 L_1 L_4}{k_{23}^2 R_L} \tag{13.33}$$

Hence, the impedance of the driver circuit depends on the coupling coefficients of the tandem connection between the driver, primary, secondary and load circuits.

13.6 Relating Reflected Impedance to Impedance Matching

The situation when many inductor stages or relays are used for extending the range of wireless power transfer, impedance matching at each stage could improve the system efficiency. An approach made popular in transmission lines was used in [8]. For this case, impedance inverters are used between two neighbouring stages as in Figure 13.2. This configuration represents a narrowband application.

The impedance inverters are T and pi sections formed from inductors and capacitors. Examples of T and pi sections are shown in Figure 13.3.

For the inductive inverters, the ABCD matrices for the T and pi sections are given by

$$\begin{bmatrix} 0 & -j\omega L \\ \frac{-j}{\omega L} & 0 \end{bmatrix} \text{ for the T section and } \begin{bmatrix} 0 & j\omega L \\ \frac{j}{\omega L} & 0 \end{bmatrix} \text{ for the pi section.}$$

When these matrices are used, the relationships between the input current (voltage) and output current (voltage) are given by the equation:

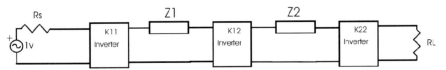

Figure 13.2 Impedance matching using inverters.

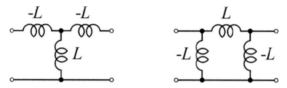

Figure 13.3   Inductive T and Pi sections.

$$\begin{bmatrix} V_1 \\ I_1 \end{bmatrix} = \begin{bmatrix} A & B \\ C & D \end{bmatrix} \begin{bmatrix} V_2 \\ I_2 \end{bmatrix} = \begin{bmatrix} 0 & -j\omega L \\ \dfrac{-j}{\omega L} & 0 \end{bmatrix} \begin{bmatrix} V_2 \\ I_2 \end{bmatrix} \quad \text{for the T section}$$

and $\begin{bmatrix} V_1 \\ I_1 \end{bmatrix} = \begin{bmatrix} A & B \\ C & D \end{bmatrix} \begin{bmatrix} V_2 \\ I_2 \end{bmatrix} = \begin{bmatrix} 0 & j\omega L \\ \dfrac{j}{\omega L} & 0 \end{bmatrix} \begin{bmatrix} V_2 \\ I_2 \end{bmatrix}$ for the pi

section. In Figure 13.4, capacitors are used to create the T and pi sections instead of inductors.

For the capacitive inverters, the ABCD matrices for the T and pi sections are given by

$$\begin{bmatrix} 0 & \dfrac{j}{\omega C} \\ j\omega C & 0 \end{bmatrix} \text{ for the T section and } \begin{bmatrix} 0 & \dfrac{-j}{\omega C} \\ -j\omega C & 0 \end{bmatrix} \text{ for the pi section.}$$

Consider one of these sections is used to match a WPT system as in Figure 13.5.

The impedance introduce by the T section at the source is derived in this section. At resonance, since $j\omega_0 L_0 - \dfrac{j}{\omega_0 C_0} = 0$ and $j\omega_0 L_{11} - \dfrac{j}{\omega_0 C_{11}} = 0$, the impedance introduced by the T section is given by the sum of $(-j\omega L_{m11})$ and the parallel combination of $-j\omega L_{m11}//R_{L1}$ and $(j\omega L_{m11})$. This impedance is:

$$Z_1 = -j\omega L_{m11} + \left[\frac{j\omega L_{m11} \times (R_{L1} - j\omega L_{m11})}{R_{L1}} \right] = \frac{\omega_0^2 L_{m11}^2}{R_{L1}} = \frac{K_{11}^2}{R_{L1}}$$

$$(13.34)$$

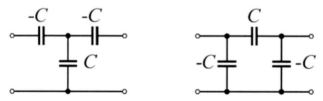

Figure 13.4   Capacitive T and Pi sections.

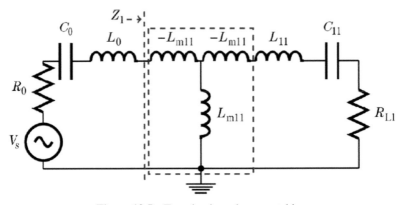

Figure 13.5 T section impedance matching.

The term K_{11} is called the characteristic impedance of the inverter. The inverter circuit impedance is equal to the square of the characteristic impedance of the inverter circuit divided by the resistance of the load. The equation for this impedance is useful for impedance matching in WPT systems. Note that the characteristic impedance is determined by the mutual reactance where

$$K_{11} = \omega_0 L_{m11} = \omega_0 k \sqrt{L_0 L_{m11}}$$

where L_{m11} is the mutual inductance of the peer-to-peer network. Following the above analysis, it can be shown that to match impedances between sections, the impedances of the peer-to-peer circuits and that from the mutual inductance should be used. For the circuit shown in Figure 13.6, the impedance matching equations are:

The full circuit for the SIMO configuration above is shown in Figure 13.7.

We can show from previous analysis that the required impedance transformation equations should be

$$Z_1 = \frac{K_{11}^2}{R_{L1}} \tag{13.34a}$$

$$Z_2 = \frac{K_{12}^2}{R_{L2}} \tag{13.34b}$$

We also have

$$K_{11} = \omega_0 k_{11} \sqrt{L_0 L_{11}} \tag{13.35a}$$

$$K_{12} = \omega_0 k_{12} \sqrt{L_0 L_{12}} \tag{13.35b}$$

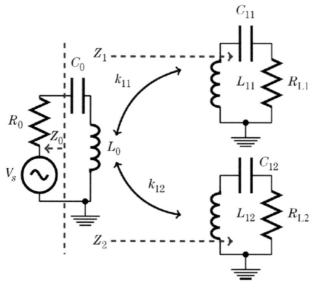

Figure 13.6   SISO system.

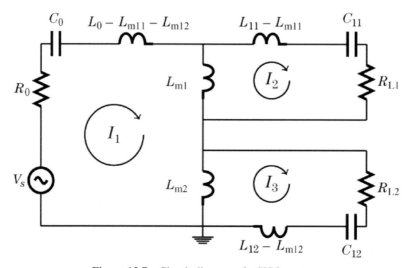

Figure 13.7   Circuit diagram of a SISO system.

13.6.1 Three-Coil Systems

The architecture of a three-coil system is given in Figure 13.8. The impedances looking into the inputs of previous sections are given by the following equations

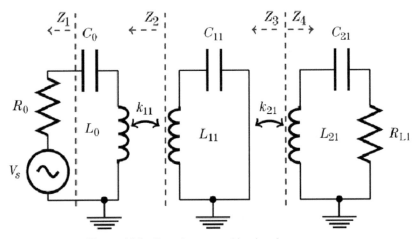

Figure 13.8 Impedance matching in relay systems.

$$Z_1 = R_0 \tag{13.36}$$

$$Z_2 = \frac{K_{11}^2}{Z_1} \tag{13.37}$$

$$Z_3 = \frac{K_{21}^2}{Z_2} \tag{13.38}$$

$$K_{11} = \omega_0 k_{11} \sqrt{L_0 L_{11}} \tag{13.39}$$

$$K_{21} = \omega_0 k_{21} \sqrt{L_{11} L_{21}} \tag{13.40}$$

$$Z_4 = Z_3 \tag{13.41}$$

References

[1] Hoang Nguyen, Johnson I. Agbinya, and John Devlin. "FPGA-Based Implementation of Multiple Modes in Near Field Inductive Communication Using Frequency Splitting and MIMO Configuration". *IEEE Transactions on Circuits and Systems*; Accepted for publication 7 September (2014).

[2] Hoa Doan Thanh, and Johnson I. Agbinya. "Investigation and Study of Mode Splitting in Near Field Inductive Communication Systems". *International Journal of Electronics and Telecommunications* 59, 2: 185–194. ISSN (Print) 0867-6747, doi:10.2478/eletel-2013-0022 (2013).

[3] Hoang Nguyen, Johnson I. Agbinya, and John Devlin. "Channel Characterization and Link Budget of MIMO Configuration in Near Field Magnetic Communication". *International Journal of Electronics and Telecommunications* 59, 3: 257–264. doi:10.2478/eletel2013-0030 (2013).

[4] Hoa Doan Thanh, and Johnson I. Agbinya. "Effects of Second Neighbour Interaction on Frequency Splitting in Near Field Inductive Communication Systems". In *Proceedings of the 7th IB2COM*, Sydney, Australia. ISBN: 978-0-9872129-1, pp. 190–195 November 5–8 (2012).

[5] Kurs, A., A. Karalis, R. Moffatt, J.D. Joannopoulos, P. Fisher, and M. Soljacic. "Wireless Power Transfer via Strongly Coupled Magnetic Resonances." *Science* 317: 83–86 (2007).

[6] Erin M. Thomas, Jason D. Heebl, Carl Pfeiffer, and Anthony Grbic. "A Power Link Study of Wireless Non-Radiative Power Transfer Systems Using Resonant Shielded Loops." *IEEE Transactions Circuits and Systems-I*. Regular Papers (2012).

[7] Matthew Chabalko, Eduard Alarcon, Elisenda Bou, and David S. Ricketts. "Optimisation of WPT Efficiency Using a Conjugate Load in Non-Matched Systems". Proceedings of AP-S, pp. 645–646 (2014).

[8] Marco Dionigi, Alessandra Costanzo, and Mauro Mongiardo. "Network Methods for Analysis and Design of Resonant Wireless Power Transfer Systems". In Ki Young Kim, ed. *Wireless Power Transfer—Principles and Engineering Explorations*, ISBN: 978-953-307-874-8. In Tech, Available from: http://www.intechopen.com/books/wire-less-power-transfer-principles-and-engineering-explorations/ networksmethods-for-the-analysis-and-design-of-wireless-power-transfer-systems (2012).

[9] Kim Ean Koh, Teck Chuan Beh, Takehiro Imura, and Yoichi Hori. "Impedance Matching and Power Division Using Impedance Inverter for Wireless Power Transfer via Magnetic Resonant Coupling". *IEEE Transactions on Industry Applications* 50, 3: 2061–2070 (2014).

14

Impedance Matching Concepts

14.1 Introduction

Impedance matching is defined as the process of making one impedance look like another using circuit design methods with the objective of achieving optimum system performance. Most of the time impedance matching reduces to matching load impedance to the internal impedance of a circuit driving the system or to source impedance. Often some authors use the term 'compensators' to refer to impedance matching. In practice, varied components and circuits can be used for impedance matching.

Several sources of system degradation exist in wireless power transfer systems. Two of the most prominent ones are frequency splitting and impedance mismatch. We have discussed one of them in terms of frequency splitting which occurs due to overcoupling of the transmitter to the receiver. When the receiver is tuned to the resonant frequency, the received power (efficiency) will be highly reduced. When it is tuned to any of the split frequencies, the achievable efficiency is still significantly less than that at resonance. The receiver needs to be tuned to all the split frequencies and the received power at the split frequencies summed to achieve better system efficiency. This means designing a better receiver. This complicates the power transfer system and increases the cost of production.

The efficiency of a wireless power transfer system is also reduced by the mismatch between the transmitter and receiver sections. This mismatch can be quantified as differences in the input impedances of the two circuits which result in reflected signals from the receiver back into the transmitter. A reflection coefficient is used to quantify these losses as in Equation (14.2).

The requirement for impedance matching in wireless power transfer systems can be appreciated by analysing the peer-to-peer system. We can show for a peer-to-peer resonant system, the efficiency when reflections exist in the system is given by the expression:

$$\eta = \frac{P_{out}}{P_{source}} = \frac{\omega^2 M_{12}^2 R_L}{\omega^2 M_{12}^2 (R_2 + R_L) + R_1(R_2 + R_L)^2} \left(1 - |\Gamma_{in}|^2\right) \quad (14.1)$$

From Equation (14.1), the reflection coefficient reduces the system efficiency and the reduction is due to the reflection coefficient given by the expression:

$$\Gamma_{in} = \frac{Z_{in} - Z_S}{Z_{in} + Z_S} \quad (14.2)$$

The reflection coefficient due to its complex nature introduces phase change. When the reflection coefficient is maximised, the system efficiency is zero, and when the reflection coefficient is zero, maximum power is transferred.

14.1.1 Rationale and Concept

The maximum power transfer theorem shows that to transfer maximum amount of power from a source to a load, the load impedance should match the source impedance Figure (14.1). This is basically a restatement of the fact that when there is impedance mismatch in sections of a circuit or system,

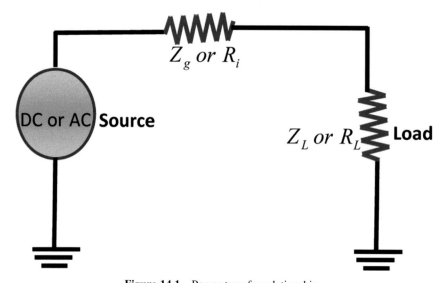

Figure 14.1   Power transfer relationship.

reflection of signals from the point of mismatch takes place. Reflections mean that a proportion of the power available to drive the load is reduced.

Impedance mismatch manifests itself as unusable reactive power and therefore needs to be removed from the system.

In the following simple analysis, we consider a general circuit. The circuit is powered by either a DC or AC source whose internal resistance is R_i. The configuration is shown in Figure 14.2. The source may also be a generator with output impedance Z_g. The driven load has impedance Z_L. In many cases this impedance is a resistance R_L, but in other applications it can be impedance with reactive (inductor and capacitor) components. The objective of the impedance matching (Figure 14.2) process is therefore to ensure that the impedances conjugate or compensate for each other, so that:

$$Z_L = Z_g^* \quad \text{or} \quad R_L = R_i \tag{14.3}$$

When the source impedance is not matched to the load, a matching circuit must be inserted to correct for the mismatch as in Figure 14.2.

Often impedance matching is for operations at a single frequency. Some applications may however require matching over a range of frequencies. For such applications, the bandwidth of the matching network is an important design parameter for the matching network. Applications in which the load impedance is a variable over a range require a matching network whose parameters can be adjusted or tuned.

For the matching networks discussed in this chapter, we use the terms characteristic impedance Z_0, input impedance Z_i or source impedance Z_g interchangeably.

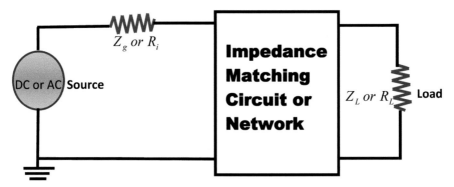

Figure 14.2 Impedance matching diagram.

Example 14.1:

Consider Figure 14.3 with a source voltage E_S. Let the voltage across the resistive load be designated as E_R. Determine the relationships between the source impedance and the load resistance for maximum power transfer when

(a) The source impedance is resistive
(b) The source impedance is inductive

Solution (a): The active power supplied to the load resistance may be obtained using the voltage divider rule. When the source is resistive ($Z_S = R_S$), the power is given by the expression:

$$P = \frac{E_R^2}{R} = \frac{\left(\frac{RE_S}{R+R_S}\right)^2}{R} = \frac{E_S^2 R}{(R+R_S)^2} \tag{14.4}$$

The best load that maximises the power transferred to the load is obtained by setting the derivative of P with respect to R to zero and solving the expression. That is

$$\frac{dP}{dR} = \frac{E_S^2}{(R+R_S)^2} - \frac{2E_S^2 R}{(R+R_S)^3} = 0 \tag{14.5}$$

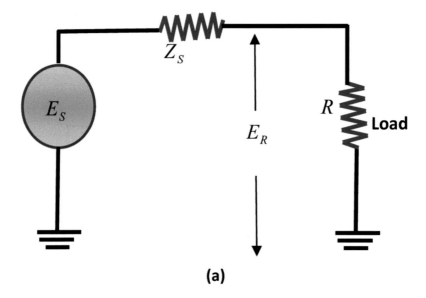

(a)

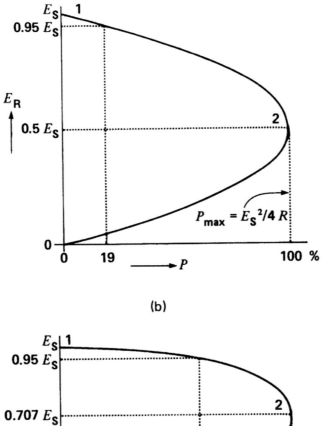

(b)

(C)

Figure 14.3 Figure example 1.

The solution shows that maximum power occurs at $R = R_S$, and the value of the maximum power is

$$P_{\max} = \frac{E_S^2 R_S}{(R_S + R_S)^2} = \frac{E_S^2}{4R_S} \tag{14.6}$$

This relationship is plotted in Figure 14.3b.

From the above analysis, the source delivers only half of its output power to the load when its impedance is matched to the load. Thus, equal power is dissipated in the load and the source. Therefore, the source should be capable of developing twice the power required by a load. Also the source must be able to dissipate half of that power as wasted power as well.

Solution (b): When the source is inductive, reactive power is created by the source inductance. The apparent power S developed by the source when the source is inductive $(Z_S = X_L)$ is given by the expression:

$$S = \frac{E_S^2}{R + jX_L} = \frac{RE_S^2}{R^2 + X_L^2} - \frac{jX_L E_S^2}{R^2 + X_L^2} \tag{14.7}$$

$$\frac{dS}{dR} = \frac{E_S^2}{R^2 + X_L^2} - \frac{2RE_S^2}{\left(R^2 + X_L^2\right)} = 0 \tag{14.8}$$

This solves to $R = X$. Maximum power is delivered to the load when its resistance is equal to the reactance of the source. The maximum real power supplied to the load is:

$$P_{\max} = \frac{X_L E_S^2}{X_L^2 + X_L^2} = \frac{E_S^2}{2X_L} \tag{14.9}$$

For this example, maximum power is transferred when $E_R = E_S/\sqrt{2}$.

14.1.2 Applications of Impedance Matching

Impedance matching is not unique to wireless power transfer systems. It is applicable to applications such as amplifiers, antenna systems connected to input amplifiers and coaxial cables. Ideally to cancel a positive reactance (inductive impedance), a negative capacitive reactance that is equivalent is required. Consider Figure 14.4 for which 'Amplifier 1' provides input power to 'Amplifier 2'.

The source or 'Amplifier 1' has inductive output impedance (an RL circuit). The input impedance of 'Amplifier 2' has a capacitive component (RC circuit).

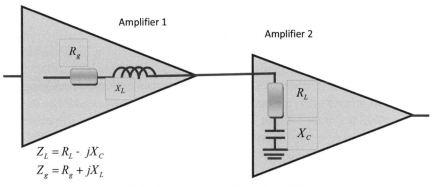

Figure 14.4 Impedance matching of amplifier stages.

For maximum power to be transferred from Amplifier 1 to Amplifier 2, the two impedances should match such that $Z_L = Z_g$. This situation occurs in transceivers when matching of the output impedance of one stage to the input impedance of the next stage is required for maximum power transfer between them.

A second example where impedance matching is routinely used is when the output stage (amplifier) of a transmitter needs to drive an antenna. Usually, it is required that the input impedance of an antenna be matched to output impedance of the driving amplifier stage. In Figure 14.5, the transmitter amplifier output impedance should be matched to the input impedance of the antenna ($R_L = R_g$).

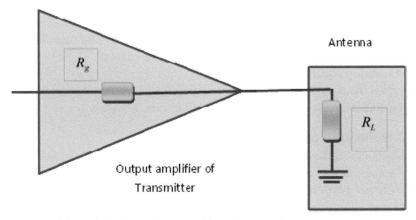

Figure 14.5 Impedance matching of a transmitter output stage.

14.1.3 Transmission-Line Impedance Matching

Impedance matching is used in electric power transmission and distribution systems. Power lines are basically transmission lines which can be modelled as RLC sections. Also from a practical point of view, a coaxial cable is considered to be a transmission line when its length is longer than $\lambda/8$ at an operating frequency f. In other words, the operating frequency determines whether the connecting cable is a transmission line or not. When the length of the cable is such that it becomes a transmission line, impedance matching is essential. This situation occurs in mobile phone base stations when connecting low-noise amplifiers and antennas.

Example 14.2:

At what length does interconnecting cables for a GSM900 become a transmission line?

Solution: For GSM900, the wavelength is 1/3 m, and thus, when the interconnecting cables are longer than 0.33/8 m or 4.17 cm, they become transmission lines and impedance matching becomes necessary.

14.1.3.1 Characteristic impedance

All transmission lines have characteristic impedances. The characteristic impedance Z_0 is a function of the inductance and capacitance per unit length of the line. It is given by the expression:

$$Z_0 = \sqrt{\frac{L}{C}}$$

The characteristic impedance is thus a function of the inductance and capacitance per unit length of the transmission line. This characteristic impedance must be matched to both the source and load impedances as shown in Figure 14.6 and in equations that follow.

Note from the equation of the characteristic impedance that it is resistive. Hence, the source and load impedances should also be resistive.

14.1.3.2 Reflection coefficient

When impedance mismatch occurs, the resulting problem is twofold: (a) standing waves are developed along the line due to reflections, and (b) maximum power is not transferred under such situation. Matching of the load to the line and source avoids the reflected wave and hence also the standing wave pattern.

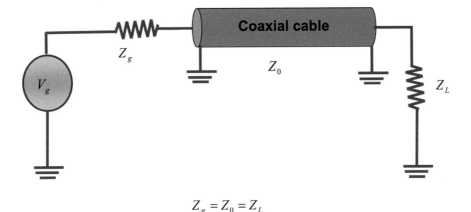

Figure 14.6 Coaxial cable transmission-line matching.

Standing waves are two waves of equal intensity and frequency travelling in opposite directions. Standing waves are thus confined within the transmission line. The ratio of the amplitudes of the wave at all points in time remains constant. Since some of the power is reflected from the load back into the source, maximum power is not delivered to the load. The reflected power if significant is dangerous to the source.

For the sake of illustration, consider the voltage and current Equation (14.10) in a transmission line of Figure 14.7:

$$V(z) = V_0^+ e^{-j\beta z} + V_0^- e^{j\beta z}$$

$$I(z) = \frac{V_0^+ e^{-j\beta z}}{Z_0} - \frac{V_0^- e^{j\beta z}}{Z_0} \tag{14.10}$$

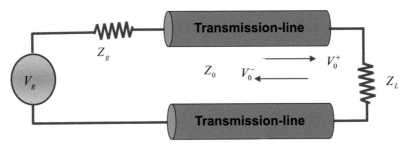

Figure 14.7 Transmission-line impedance matching.

The wave equations show that the incident wave propagates in the $(+z)$ direction with phase term $e^{-j\beta z}$ and the reflected wave propagates in the $(-z)$ direction with phase term $e^{j\beta z}$. The equations also show that the line impedance is complex and is a function of the length of the line. Hence, for a mismatched line, the generator sees a complex impedance which is a function of the length of the line. The load voltage and currents when $z = 0$ are given by the expressions

$$V_L(z = 0) = V_0^+ + V_0^-$$

$$I_L(z = 0) = \frac{V_0^+}{Z_0} - \frac{V_0^-}{Z_0} \tag{14.11}$$

Thus, we can write an expression for the load impedance to be

$$Z_L = \frac{V_L(z = 0)}{I_L(z = 0)} = \frac{\left(V_0^+ + V_0^-\right) Z_0}{\left(V_0^+ - V_0^-\right)} \tag{14.12}$$

The voltage amplitudes may therefore be determined with the expression

$$V_0^- = \frac{(Z_L - Z_0) V_0^+}{(Z_L + Z_0)} \tag{14.13}$$

The reflection coefficient is similarly determined as

$$\Gamma = \frac{V_0^-}{V_0^+} = \frac{(Z_L - Z_0)}{(Z_L + Z_0)} = |\Gamma| \, e^{j\theta} \tag{14.14}$$

Since the impedances are complex quantities, the reflection coefficient is also a complex quantity. Its net effect is to introduce a phase change and power loss. The amount of power lost depends on the reflection coefficient and the standing wave ratio. For perfect match, the reflection coefficient is zero. For imperfect match, the magnitude of the reflection coefficient lies between zero and one, $0 \leq |\Gamma| \leq 1$.

14.1.3.3 Standing wave ratio
The amount of power lost due to reflection is a function of the reflection coefficient (Γ) and the standing wave ratio (SWR). These are determined by the amount of mismatch between the source and load impedances. Two

expressions for the standing wave ratio are used. When the load impedance is larger than the characteristic impedance of the line, $Z_L > Z_0$, and when the load impedance is less than the characteristic impedance of the line, $Z_L < Z_0$:

$$\text{SWR} = S = \frac{Z_L}{Z_0} \ (\text{for } Z_L > Z_0)$$

$$S = \frac{Z_0}{Z_L} \ (\text{for } Z_L < Z_0) \tag{14.15}$$

Example 14.3:

The load connected to a transmission line is 75 ohms, and the line impedance is 50 ohms. Calculate the reflection coefficient and the standing wave ratio for the line.

Solution: The standing wave ratio is:

$$S = \frac{Z_L}{Z_0} \ (\text{for } Z_L > Z_0) = \frac{75}{50} = 1.5$$

The reflection coefficient is:

$$\Gamma = \frac{(Z_L - Z_0)}{(Z_L + Z_0)} = \frac{75 - 50}{75 + 50} = \frac{25}{125} = 0.2$$

The return loss in a transmission line is defined as the logarithm to base ten of the ratio of the input power to the reflected power or:

$$\rho_{\text{rloss}} = 10 \log_{10} \left(P_{\text{in}} / P_{\text{reflected}} \right) dB$$

where P_{in} is the line input power and $P_{\text{reflected}}$ is the reflected power. A large value of return loss or a small value of reflected power is desirable for efficient power transfer. Each transmission line also manifests inherent resistive loss, which is a function of the length of the line. Manufacturers' datasheets for coaxial cables should be consulted for values of the resistive losses per length of the cable.

14.1.4 Impedance Matching Circuits and Networks

So far we have discussed the general theory and reasons for impedance matching in electrical circuits. This is shown in Figure 14.2. The impedance matching circuit or network corrects for the mismatch between the source and the load impedances. This section will discuss only a few approaches for impedance matching as are relevant for wireless power transfer systems. We will discuss the use of transformers and Q-sections ($\lambda/4$ sections).

14.1.4.1 Ideal transformer model of WPT

A loosely coupled WPT may be considered as an ideal transformer. The inductors in ideal transformer have coupling coefficient $k = 1$ and zero self-resistances $(R_1 = R_2 = 0)$, and the inductances are infinite. Under these conditions, the impedance model of the LC-WPT reduces to Equation (14.16):

$$Z = j\omega L_1 + \frac{\omega^2 L_1 L_2}{R_L + j\omega L_2} \simeq \frac{\omega^2 L_1 L_2 R_L}{R_L^2 + \omega^2 L_2^2} \simeq \frac{L_1 R_L}{L_2} \tag{14.16}$$

By rewriting Equation (14.16), we obtain Equation (14.17):

$$Z \simeq \frac{L_1 R_L}{L_2} \tag{14.17}$$

Equation (14.17) is the familiar ideal transformer impedance relationship. Also the impedance is real, and provided the above assumptions hold, there are no reactive power losses. Hence, only real power is transferred to the load. The power transfer ratio favours increasing the transmitter inductance. The higher the transmitter inductance is compared to the receiver inductance the more real power is induced in the receiver and across the load.

14.1.4.2 Ideal transformer model

A transformer transforms one impedance to look like another by using the turns ratio. Let the turns ratio be:

$$\text{turns ratio} = a = \frac{N_p}{N_s} \tag{14.18}$$

In normal transformer parlance, the relationship 'a' is written as $N_p{:}N_s$, or N_p turns to N_s.

N_s is the number of turns in the secondary winding and N_p is the number of turns in the primary winding of the transformer. The primary and secondary impedances Z_p and Z_s, respectively, of the transformer are related to the turns ratio 'a', and it is given by the expression:

$$a^2 = \frac{Z_p}{Z_s} = \left(\frac{N_p}{N_s}\right)^2 \tag{14.19}$$

which means that the turns ratio becomes:

$$a = \sqrt{\frac{Z_p}{Z_s}} \qquad\qquad (14.20)$$

From Figure 14.8, Z_s is the load impedance Z_L.

Example 14.4

The source impedance of a wireless power transfer system is 400 ohms. Calculate the turns ratio of a transformer that is required to match this impedance to a 100 ohms receiver load.

Solution: The turns ratio for this transformer is given by the expression

$$a = \sqrt{\frac{Z_p}{Z_s}} = \sqrt{\frac{400}{100}} = 2$$

The turns ratio for this example is 2:1 with the primary side having twice the number of turns in the secondary.

Impedance matching with transformers allows for wide bandwidth applications and is applicable below 1 GHz.

The discussions so far focused on a constant impedance matching design. However, system loads in some cases are variable, and hence, adaptive design could enable varied applications. Adaptive designs use autotransformers either at the primary side or at the secondary side. When autotransformers are used at

Transformer

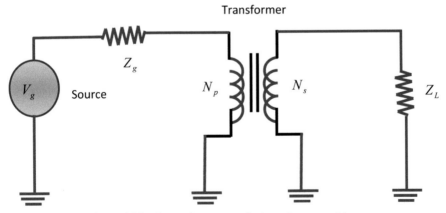

Figure 14.8 A transformer user for impedance matching.

the primary side, the source is connected directly to the winding. The secondary winding is tapped off the main winding as in Figure 14.9.

In Figure 14.9, a step-up impedance matching is used. The windings may be part of a resonant circuit. The governing relations are

$$a^2 = \frac{Z_p}{Z_s} = \left(\frac{N_p}{N_s}\right)^2; \quad a > 1 \tag{14.21}$$

In Figure 14.10, a step-down transformer impedance matching is shown. The governing equation is the same as in the step-up except that in this case $(N_s > N_p)$, and therefore, $a < 1$.

14.1.5 Q-Section Impedance Matching

Quarter wavelength (Q-section) impedance matching technique uses transmission lines to match the source to the receiver load and hence is popular in antenna applications where the antennas must be connected to transmitter output stages or receiver input stages. The matching circuit is a transmission line of length $\lambda/4$ as the operating frequency. For example, if the operating frequency is 10 MHz, then $\lambda = 30$ m and the length of the Q-section is 7.5 m. The impedance of this Q-section transmission line depends on the

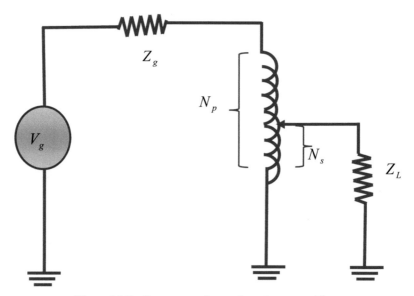

Figure 14.9   Step-up transformer impedance matching.

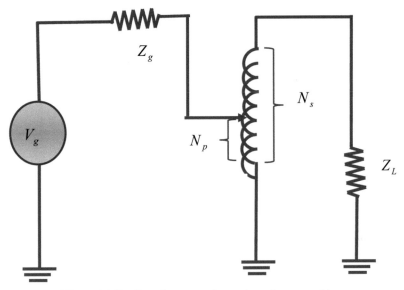

Figure 14.10 Step-down transformer impedance matching.

characteristics of the coaxial cable (resistance, inductance and capacitance per unit length of the cable).

The impedance of the Q-section is computed from well-known expressions. In Figure 14.11, a quarter wavelength section of characteristic impedance Z_ℓ is placed between a transmission line of characteristic impedance Z_0 and used to match a load impedance R_L. The input impedance looking into the input of the Q-section can be determined from well-known expressions. The input impedance of a Q-section is:

$$Z_{\text{in}} = Z_\ell \frac{R_L + jZ_\ell \tan \beta \ell}{Z_\ell + jR_L \tan \beta \ell} \tag{14.22}$$

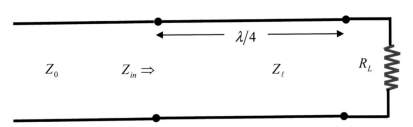

Figure 14.11 Quarter wavelength impedance matching.

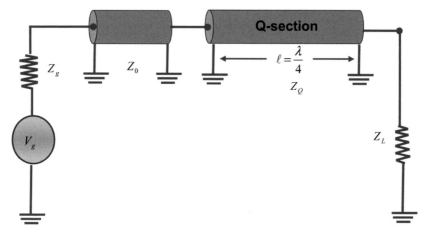

Figure 14.12   Example of impedance matching using Q-section.

The input impedance is real when the transmission line is a quarter of a wavelength long. The terminology 'electrical length' is often used to characterise the transmission line in terms of how much phase change it introduces. The electrical length of a cable is defined by the expression:

$$\text{Electrical length} = \beta\ell = 2\pi \cdot \frac{\ell}{\lambda} \qquad (14.23)$$

In other words, if we know the physical length of a transmission line and the operating frequency, we can determine the electrical length or the phase change caused by the line.

Electrical length may also be written as a function of time and frequency. Since the wavelength is determined by the wave velocity in the line and the operating frequency, we can rewrite the expression for the electrical length of the transmission line using:

$$v = f\lambda \qquad (14.24)$$

Thus, $\text{Electrical length} = \beta\ell = 2\pi \cdot \frac{f \cdot \ell}{v} = 2\pi f \cdot T_e$ and T_e is the time taken by the wave to propagate through the length of transmission line.

The electrical length of a Q-section is given by the expression:

$$\beta\ell = \frac{2\pi}{\lambda} \cdot \frac{\lambda}{4} = \frac{\pi}{2} \qquad (14.25)$$

At this electrical length, $\tan \beta\ell = \tan(\pi/2) \to \infty$, and the expression for the input impedance of the Q-section can be simplified to

$$Z_{\text{in}} = Z_\ell \frac{jZ_\ell \tan(\pi/2)}{jR_L \tan(\pi/2)} = \frac{Z_\ell^2}{R_L} \tag{14.26}$$

This allows us to arrive at another useful expression for a matched system when $Z_{\text{in}} = Z_0$. The expression is

$$Z_{\text{in}} = Z_0 = \frac{Z_\ell^2}{R_L} \tag{14.27}$$

and

$$Z_\ell = \sqrt{Z_0 R_L} \tag{14.28}$$

Example 14.5:

A quarter wavelength vertical ground plane antenna of impedance 36 ohms is to be driven by a transmitter operating at 10 MHz. Its source impedance $Z_g = 50$ ohms is connected to a matching transmission line of characteristic impedance $Z_0 = 50$ ohms. Calculate the characteristic impedance of the Q-section to be used for matching the transmitter to the load. Calculate also the physical length of the Q-section.

Solution: We are given $Z_g = 50$ ohms, $Z_0 = 50$ ohms and 36 ohms. The characteristic impedance of the Q-section is therefore given by

$$Z_Q = \sqrt{Z_0 R_L} = \sqrt{50 \times 36} \approx 42\Omega$$

At the operating frequency of 10 MHz, the physical length of the Q-section is:

$$l = \lambda/4 = 30/4 = 7.5\,\text{m}$$

Observe that the physical length computed in Example 14.5 is usually modified by the velocity factor of the transmission line. The refractive index of a medium determines the speed v with which waves travel through the medium. The refractive index n of a medium is a function of the relative permittivity ε_r and permeability μ_r given as

$$n = \sqrt{\varepsilon_r \mu_r} \quad \text{and} \quad n = c/v \tag{14.29}$$

In other words, if the refractive index $n > 1$, the medium will slow down waves propagating through it so that the speed of the waves is less than the speed of light c in vacuum ($v < c$).

Since the relative permeability of transmission-line materials such as aluminium, copper and Teflon is approximately 1, the velocity factor is therefore governed by the relative permittivity. The velocity factor (VF) of a transmission line is given by the reciprocal of the relative permittivity by the expression

$$VF = 1/\varepsilon_r \tag{14.30}$$

Therefore, the physical length of the Q-section computed is reduced by the VF. The new physical length is the computed length multiplied by the VF:

$$l = \ell \times VF \tag{14.31}$$

Since VF < 1, the practical length of the Q-section is less than the length computed without the VF.

Short circuit stub

$$Z_{\text{in}} = Z_\ell \frac{R_L + jZ_\ell \tan \beta\ell}{Z_\ell + jR_L \tan \beta\ell}$$

In a short circuit stub, the load impedance is a short circuit $R_L = 0$. Therefore, the input impedance reduces to

$$Z_{\text{in}} = jZ_\ell \tan \beta\ell \tag{14.32}$$

This impedance matches the characteristic impedance Z_0 when $Z_0 = Z_\ell$ at which point we have

$\tan \beta\ell = 1$ and $\beta\ell = \frac{\pi}{4}$; $\frac{2\pi}{\lambda}\ell = \frac{\pi}{4}$ or $\ell = \frac{\lambda}{8}$. The input impedance becomes an inductor with $jX_L = jZ_0$.

Open circuit stub

An open circuit stub has load impedance $R_L = \infty$. Its input impedance for the matching circuit $Z_0 = Z_\ell$ becomes

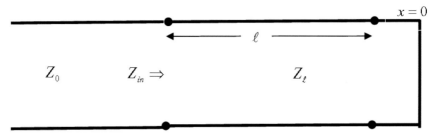

Figure 14.13   Short circuit stub.

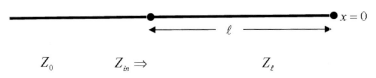

Figure 14.14 Open circuit stub.

$$Z_{\text{in}} = Z_0 \frac{R_L}{jR_L \tan \beta\ell} = -jZ_0 \cot \beta\ell \qquad (14.33)$$

When, $\ell = \frac{\lambda}{8}$ we obtain a shunt capacitor with reactance $jX_c = jZ_0$.

References

[1] Koh Kim Ean, et al. "Impedance Matching and Power Division Algorithm Considering Cross Coupling for Wireless Power Transfer via Magnetic Resonance" (2012).

[2] Nima Soltani, et al. "A High-Gain Power-Matching Technique for Efficient Radio-Frequency Power Harvest of Passive Wireless Microsystems" (2010).

[3] George L. Matthaei. "Microwave Filters Impedance-Matching Networks, And Coupling Structures". Artech House (1985).

[4] Koh Kim Ean, et al. "Novel Band-Pass Filter Model for Multi-Receiver Wireless Power Transfer via Magnetic Resonance Coupling and Power Division".

[5] Chwei-Sen Wang, and Grant A. Covic. "Power Transfer Capability and Bifurcation Phenomena of Loosely Coupled Inductive Power Transfer Systems" (2004).

15

Impedance Matching Circuits

15.1 Introduction

Impedance matching using transmission lines as discussed in Chapter 14 has several limitations. First, a cable of desired characteristic must be available. This is not always possible. Most cables sold in the market are limited to a few impedances such as 125, 93, 75 and 50 Ω. Secondly, we noticed from the analysis in the previous section that we need to factor in velocity factor or the nature of the material used for creating the cable. We also need to factor in the operating frequency so as to determine the quarter wavelength. For low frequencies, the operating wavelengths can be very long, and hence, it is not always advisable or feasible to use a Q-section that is going to be too long. At higher frequencies above 300 MHz where the wavelength is less than 1 m, using Q-sections becomes advantageous. Hence, Q-sections are useful at UHF and microwave frequencies using stripline or microstrip on PCBs.

In this section, the focus is on L-sections or inductor–capacitor (LC) circuits for matching electronic system loads to the output impedance of a generator. The analysis in this section is useful for applications involving radio frequency (RF) design and electronic circuits.

15.1.1 Series–Parallel Transformations

The equations which facilitate the design of matching L-sections are obtained from impedance transformations between series and parallel circuits. Consider the following circuits in Figure 15.1.

Consider that the series impedance and the parallel impedance of the two circuits in Figure 15.1 are equal. This allows us to derive the equations which can be used to convert between series and parallel impedances. The objective is first to isolate the real and imaginary parts of the impedances and to equate

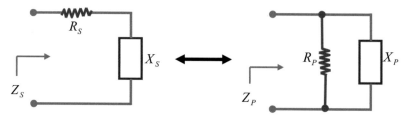

Figure 15.1   Series–parallel impedance transformation circuits.

them. We know for the series circuit that

$$Z_S = R_S + jX_S \tag{15.1}$$

We also know for the parallel equivalent circuit that

$$Z_P = \frac{X_P R_P}{R_P + jX_P} = \frac{R_P X_P^2}{R_P^2 + X_P^2} - \frac{jX_P R_P^2}{R_P^2 + X_P^2} \tag{15.2}$$

By comparing the real and imaginary parts of the Equations (15.1) and (15.2), we have

$$R_S = \frac{R_P X_P^2}{R_P^2 + X_P^2} = \frac{R_P}{\dfrac{R_P^2}{X_P^2} + 1} = \frac{R_P}{1 + Q_P^2} \tag{15.3}$$

In other words, the quality value of the parallel circuit is given by the relation

$$Q_P = \sqrt{\frac{R_P}{R_S} - 1} \tag{15.4}$$

Therefore, we can also write the expression for the parallel impedance in terms of the Q value as

$$R_P = R_S \left(Q_P^2 + 1 \right) \tag{15.5}$$

To find the relationship between the reactance of the series and parallel equivalent circuits, we also solve for them using Equations (15.1) and (15.2). Thus, we can write that

$$X_S = \frac{X_P R_P^2}{R_P^2 + X_P^2} = \frac{X_P}{1 + \dfrac{X_P^2}{R_P^2}} = \frac{X_P}{1 + \dfrac{1}{Q_P^2}} \tag{15.6}$$

Therefore, series reactance is given by the Q of the parallel circuit as

$$X_S = \frac{Q_P^2 X_P}{1 + Q_P^2} \tag{15.7}$$

Therefore, we arrive at the relationship:

$$X_P = X_S \frac{\left(1 + Q_P^2\right)}{Q_P^2} \qquad (15.8)$$

Note that in these derivations

$$X_P = \frac{1}{\omega C_P} \quad \text{and} \quad X_S = \frac{1}{\omega C_S} \qquad (15.9)$$

The Q of the parallel circuit is thus obtained as well as

$$Q_P = \frac{X_P}{R_P} = \frac{1}{\omega C_P R_P} \qquad (15.10)$$

15.1.2 Impedance Matching with L-Sections

Impedance matching is historically well established and understood. The primary purpose of this section is to provide a complimentary picture to the previous chapter and to illustrate the design of impedance matching circuits using L-sections. L-networks are used for impedance matching in RF circuits, transmitters and receivers. L-networks are also useful in matching one amplifier output to the input of a following stage. Another use of L-sections is matching antenna impedance to a transmitter output or a receiver input. Generally, applications using RF circuits covering narrow frequency bands are candidates for impedance matching with L-network. Figure 15.2 shows an L-section.

L-sections come with different configurations, how resistors, capacitors and inductors are arranged in the circuits as we will soon see. In the designs, the types and mix of the L and C components depend on whether the source impedance is less than or greater than the load impedance. For the examples given in this section, we will highlight this as we progress in the designs. There are four configurations in use. They consist of two low-pass and two high-pass configurations.

15.1.2.1 Low-pass sections

For the low-pass configurations in addition to impedance matching, they also attenuate higher frequency harmonics, noise and interference in the system. Two situations are considered for the low-pass section. For the first condition, the resistance of the source is smaller than the load resistance.

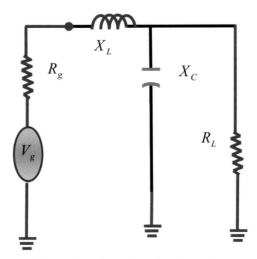

Figure 15.2   L-section when $R_g < R_L$.

Circuit components for low-pass sections are given by the expressions:

$$Q = \sqrt{\frac{R_L}{R_g} - 1}; \quad X_L = Q \cdot R_g \quad \text{and} \quad X_C = \frac{R_L}{Q} \tag{15.11}$$

The second case is when the source resistance is bigger than the load resistance as shown in Figure 15.3. The values of the circuit components are given by the expressions:

$$Q = \sqrt{\frac{R_g}{R_L} - 1}; \quad X_L = Q \cdot R_L \quad \text{and} \quad X_C = \frac{R_g}{Q} \tag{15.12}$$

15.1.2.2 High-pass sections
Figure 15.4 depicts two forms of the high-pass section. The two configurations of interest are shown in Figure 15.4a and b. In Figure 15.4a, the source resistance is smaller than the load resistance, and in Figure 15.4b, the source resistance is larger than the load resistance.

L-section circuits have their limitations. The first limitation is that Q-values of L-sections cannot be chosen or controlled. The Q is fixed by the impedances that are being matched, R_g and R_L. The range of impedances that can be matched with L-sections is limited. The second limitation is that it is possible to find values of inductances and capacitances that are either too large

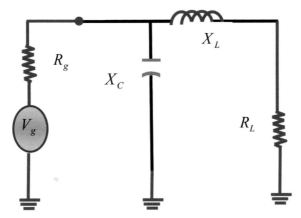

Figure 15.3   L-section when $R_g > R_L$.

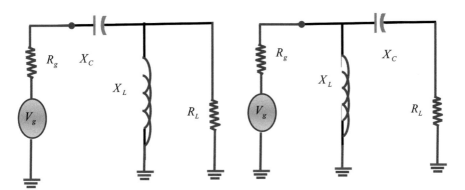

Figure 15.4   (a) and (b) L-sections.

or too small at some frequencies for practical use. If the calculated reactance is practically impossible, the design is useless. While switching between low-pass to high-pass versions may sometimes be helpful, at other times other design types must be used.

For these reasons, for applications which require control of bandwidth and Q, T or π-network is a better option.

Example 15.1

Design an L-section matching circuit for an RF transistor amplifier operating at 20 MHz. The load is 50 Ω, and the output impedance of the amplifier is 10 Ω. Calculate the values of the capacitor and inductor.

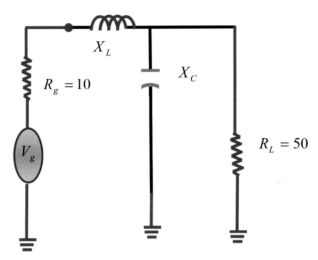

Figure 15.5   Matching a transistor source amplifier to a load.

Solution: We will use the equations for impedance conversion derived earlier. For this example, we know that $R_L > R_g$.

$$R_g = 10\,\Omega \text{ and } R_L = 50\,\Omega;\ \text{therefore,}$$

$$Q = \sqrt{\frac{R_L}{R_g} - 1}$$

Hence,

$$Q = \sqrt{\frac{50}{10} - 1} = 2$$

$$X_L = Q.R_g = 2\,(10) = 20\,\Omega$$

$$L = \frac{X_L}{\omega} = \frac{X_L}{2\pi f} = \frac{20\,\Omega}{2\pi \times 20 \times 10^6} = 15.9\,\text{nH} \approx 42\,\text{nH}$$

$$X_C = \frac{R_L}{Q} = \frac{50}{2} = 25\,\Omega$$

$$C = \frac{1}{\omega X_C} = \frac{1}{2\pi f X_C} = \frac{1}{2\pi \times 20 \times 10^6 \times 25} = 318.3\,\text{pF}$$

This solution omits any output impedance reactance such as transistor amplifier output capacitance or inductance and any load reactance that could be shunt capacitance or series inductance. When these factors are known, the computed values can be compensated.

The bandwidth (BW) of the circuit is relatively wide given the low Q of 2:

$$\text{BW} = \frac{f}{Q} = \frac{20 \times 10^6}{2} = 10\,\text{MHz}$$

To confirm that indeed the matching circuit has converted the parallel part of the circuit has been converted to a series circuit which also cancels the inductance, we show that the parallel combination results in a series $20\,\Omega$ reactance and $10\,\Omega$ resistance. We demonstrate this in two ways, first using the formulae and next using simple impedance analysis:

With formulae:

$$R_S = \frac{R_p}{Q^2 + 1} = \frac{50}{2^2 + 1} = 10\,\Omega$$

$$X_S = \frac{X_p}{(Q^2 + 1)/Q^2} = \frac{25}{(2^2 + 1)/2^2} = 20\,\Omega$$

With impedance analysis: Note that the conversion of the parallel side of the circuit may also be undertaken using the simple expression:

$$Z_S = \frac{-jX_C R_L}{(R_L - jX_C)} = \frac{-j25 \times 50}{(50 - j25)}$$

$$= \frac{50(1 - j2)}{5} = 10 - j20$$

This means that the series impedance (In Figure 15.6) is capacitive with $R_S = 10\,\Omega$ and $X_S = 20\,\Omega$. Note how the series equivalent capacitive reactance equals and cancels the series inductive reactance. Also the series equivalent load of $10\,\Omega$ matches the generator resistance for maximum power transfer.

15.1.3 Equivalent Circuits

Sometimes in the course of a design work it might be necessary to convert series RC or RL circuits to equivalent parallel RC or RL circuits or vice versa. These types of conversions are useful in RLC circuit analysis and design. Four groups of equivalent circuits are shown in Figure 15.7.

These equivalent circuits are useful for explaining how L-circuits work. The variables in the circuits have the following designations:

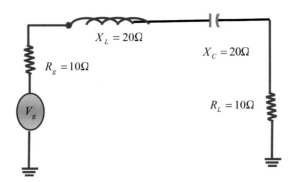

Figure 15.6   Matched circuit.

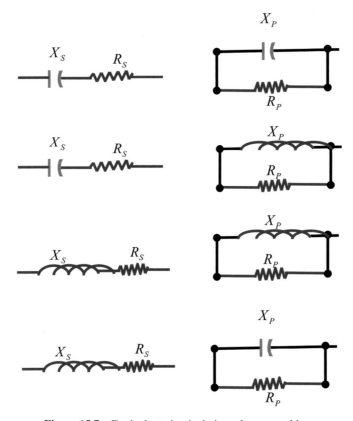

Figure 15.7   Equivalent circuits in impedance matching.

R_S = series resistance
R_P = parallel resistance
X_S = series reactance
X_P = parallel reactance

The conversion formulae relating the circuit elements are:

$$R_S = \frac{R_p}{(Q^2 + 1)} \tag{15.13}$$

$$X_S = \frac{Q^2 X_p}{(Q^2 + 1)} \tag{15.14}$$

$$R_p = R_S \left(Q^2 + 1\right) \tag{15.15}$$

$$X_p = \frac{X_S \left(Q^2 + 1\right)}{Q^2} \tag{15.16}$$

From Equations (15.13) to (15.16), it is apparent that high-Q circuits require small values of series resistance and high values for the parallel resistance. Using the inductive compensator $Q = X_L/R_S$ and the capacitive compensator with $Q = R_p/X_C$, when Q is greater than 5, a useful approximation in terms of the following equation may be used:

$$R_p = R_S Q^2 \quad \text{and} \quad X_p = X_S \tag{15.17}$$

Example 15.2

Match the output impedance of 50 Ω from a 20-MHz transmitter to a 10-Ω loop antenna impedance (Figure 15.8). Most loop antennas have low impedances traditionally.

From the values given, the following parameters may be calculated as follows

$$Q = \sqrt{\frac{R_L}{R_g} - 1} = \sqrt{\frac{50}{10} - 1} = 2$$

$$X_L = Q \cdot R_L = 2 \, (10) = 20 \, \Omega$$

$$L = \frac{X_L}{\omega} = \frac{X_L}{2\pi f} = \frac{20 \, \Omega}{2\pi \times 20 \times 10^6} = 159.2 \, \text{nH}$$

$$X_C = \frac{R_g}{Q} = \frac{50}{2} = 25 \, \Omega$$

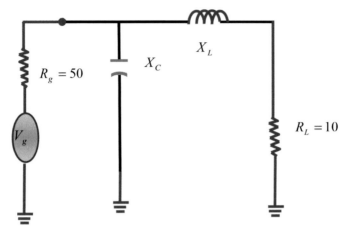

Figure 15.8   Pi-section matching circuit.

$$C = \frac{1}{\omega X_C} = \frac{1}{2\pi f X_C} = \frac{1}{2\pi \times 20 \times 10^6 \times 25} = 318.3\,\text{pF}$$

Wireless power transfer systems normally use two resonant circuits, one as transmitter and the other as receiver. In the following example, assume that the transmitter is resonating from a 50-Ω source. The transmitter is an RLC parallel resonant circuit. At resonance, the parallel circuit (Figure 15.9) behaves like an equivalent resistance of value:

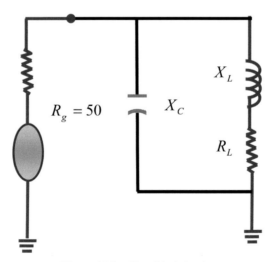

Figure 15.9   Simplified circuit.

$$R_o = \frac{L}{CR} \tag{15.18}$$

The resonant equivalent resistance may also be estimated with the expression:

$$R_0 = R\left(Q^2 + 1\right)$$

In the example here, we use the two expressions to obtain the following estimates. The first expression gives:

$$R_o = \frac{159.2 \times 10^{-9}}{318.3 \times 10^{-12} \times 10} = 50.1\,\Omega$$

The second expression gives:

$$R_0 = R\left(Q^2 + 1\right) = 10\left(2^2 + 1\right) = 50\,\Omega$$

From the two expressions, the parallel resonant load equivalent resistance is about 50 Ω. This is equal to the generator resistance and thus allows for maximum power transfer. Again, adjustments in these values should be made to include any load reactive component. The equivalent high-pass networks could also be used. One benefit is that the series capacitor can block dc if required.

15.2 Impedance Matching Networks

The Q values of L-networks are given directly by the load and generator resistances. Hence, they cannot be controlled nor varied. In some cases, the Q of the circuit may also not be suitable for the application in hand. For example, in wireless power transfer, high-Q circuits provide better power transfer efficiency. This limitation is removed with π-networks. Hence, for applications where the quality factors of the circuits need to be controllable or variable, π-networks provide options for doing so. In this section, how to use π-networks for controlling the quality factor is demonstrated.

High Q values mean that the circuit has small bandwidth which may or may not be suitable for an application. For example, in broadband inductive communication systems, a large bandwidth supports higher data rates and hence is desirable. Unfortunately, if the L-section is used for impedance matching in such situations, it limits the system bandwidth. Therefore, one of the main reasons for employing a T-network or π-network should be to get control of the circuit Q and hence also the system bandwidth. While limiting the bandwidth will help to reduce harmonics or filter out adjacent interfering signals without using additional filters, if this is not the main objective for the application then the use of appropriate matching circuit should be pursued. In inductive systems, the relationship between bandwidth and quality factor is important and is:

$$Q = \frac{f}{BW} \tag{15.19}$$

where f is the operating frequency and BW is the bandwidth. Using a T-network and π-network will provide enough freedom for applications requiring varied Q.

15.2.1 π-Networks

In addition to being able to control the Q of a π-network, its primary application is to match a high-impedance source to lower value load impedance. A π-network (Figure 15.10a) can also be used to match low source impedance to high load impedance.

In Figure 15.10c, two L-sections connected back-to-back are used to create a π-network. An intermediate virtual impedance R_V is assumed.

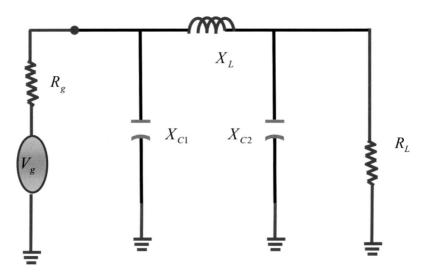

Figure 15.10a   Low-pass π-network.

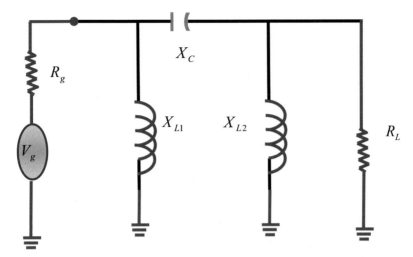

Figure 15.10b High-pass π-network.

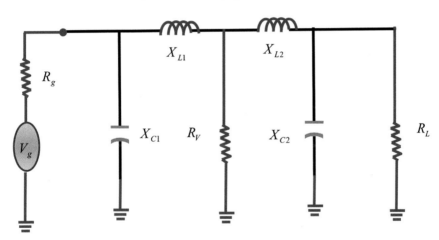

Figure 15.10c π-Network using two L-sections connected back-to-back.

The virtual resistance R_V is estimated from the expression:

$$R_V = \frac{R_H}{(Q^2 + 1)} \qquad (15.20)$$

The resistance R_H is the higher of Figure 15.10c the two design impedances R_g and R_L. The resulting R_V will be lower than either R_g or R_L depending on the desired Q. The range of typical Q values is usually between 5 and 20.

Example 15. 3 (Design of π-Network)

Design a π-network to match a 1000-Ω source to a 100-Ω load at frequency (f) of 50 MHz. You desire a bandwidth (BW) of 10 MHz.

Solution: The value of Q is:

$$Q = \frac{f}{\text{BW}} = \frac{50}{10} = 5$$

Therefore, the intermediate resistance is

$$R_V = \frac{R_H}{(Q^2 + 1)} = \frac{1000}{5^2 + 1} = 38.5\,\Omega$$

The design procedure for the first L-section uses the formulae from the previous section. The values of parameters to use are the desired Q of 5 with an R_L value equal to R_V.

The first inductor may be computed from

$$X_L = Q \cdot R_L = 5 \times 38.5 = 192.5\,\Omega$$

Therefore, the value of the inductor is

$$L_1 = \frac{X_L}{2\pi f} = \frac{192.5}{2\pi \times 50 \times 10^6} = 612.7\,\text{nH}$$

The value of the capacitor C_1 value is:

$$X_C = \frac{R_g}{Q} = \frac{1000}{5} = 200\,\Omega$$

$$C_1 = \frac{1}{2\pi f X_C} = \frac{1}{2\pi \times 50 \times 10^6 \times 200} = 15.9\,\text{pF}$$

The next step is to calculate the values of the inductor and capacitor for the second section. These are L_2 and C_2. For this, we use $R_g = R_V = 38.5\ \Omega$ with the load $R_L = 100\ \Omega$. The value of Q is determined by the L-section relationship:

$$Q = \sqrt{\frac{R_L}{R_g} - 1} = \sqrt{\frac{100}{38.5} - 1} = 1.26$$

The inductance L_2, then, is:

$$X_{L2} = Q \cdot R_g = 1.26 \times 38.5 = 48.51\,\Omega$$

$$L_2 = \frac{X_{L2}}{2\pi f} = \frac{48.51}{2\pi \times 50 \times 10^6} = 154.4\,\text{nH}$$

The capacitance C_2 is:

$$X_{C_2} = \frac{R_L}{Q} = \frac{100}{1.26} = 79.4\,\Omega$$

$$C_2 = \frac{1}{2\pi f X_{C_2}} = \frac{1}{2\pi \times 50 \times 10^6 \times 79.4} = 40.1\,\text{pF}$$

Note that the two inductances are in series so the total is just the sum of the two or:

$$L = L_1 + L_2 = 612.7 + 154.4 = 767.1\,\text{nH}$$

The π-network is shown in Figure 15.11.

15.2.2 T-Networks Design

This section discusses the design of T-networks. The two T-networks discussed are formed from two L-networks in the form of either inductor–capacitor–inductor (LCL) as in Figure 15.13 or inductor–capacitor–capacitor (LCC) connections as in Figure 15.12. The design procedure is given in the following steps.

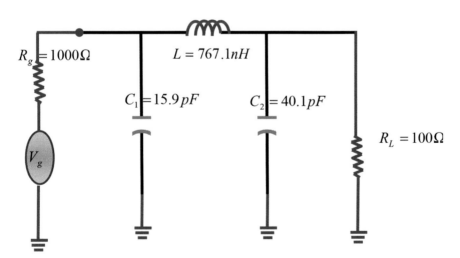

Figure 15.11 Resulting π-network.

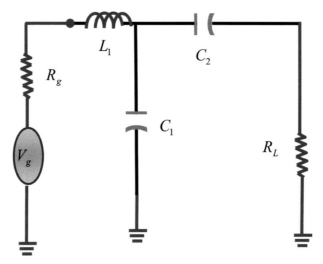

Figure 15.12   LCC matching circuit.

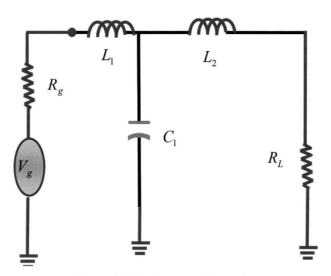

Figure 15.13   Low-pass T-network.

15.2.2.1 LCC design procedure

1. Select the desired bandwidth and calculate $Q(Q = f/\text{BW})$.
2. Calculate $X_L = Q \cdot R_g$
3. Calculate $X_{C_2} = R_L \sqrt{R_g \left(Q^2 + 1\right)/R_L - 1}$

4. Calculate $X_{C_1} = R_g \left(Q^2 + 1\right)/Q\left[QR_L/\left(QR_L - X_{C_2}\right)\right]$
5. Calculate the inductance $L = X_L/2\pi f$
6. Calculate the capacitances $C = 1/2\pi f X_C$

Assume a source or generator resistance of 10 Ω and a load resistance of 50 Ω. Let Q be 10 and the operating frequency be 171 MHz.

$$X_L = Q \cdot R_g = 10 \times 10 = 100\,\Omega$$

$$L = X_L/2\pi f = 100/2\pi \times 171 \times 10^6 = 93.1\,\text{nH}$$

$$X_{C_2} = R_L\sqrt{R_g\left(Q^2 + 1\right)/R_L - 1} = 50\sqrt{\left[(10 \times 101)/50\right] - 1} = 219\,\Omega$$

$$X_{C_1} = R_g\left(Q^2 + 1\right)/Q\left[QR_L/\left(QR_L - X_{C_2}\right)\right]$$
$$= (10 \times 101/10)\left[500/\left(500 - 219\right)\right] = 179\,\Omega$$

$$C_2 = 1/2\pi f X_{C_2} = 1/\left(2\pi \times 171 \times 10^6 \times 219\right) = 4.25\,\text{pF}$$

$$C_1 = 1/2\pi f X_{C_1} = 1/\left(2\pi \times 171 \times 10^6 \times 179\right) = 5.2\,\text{pF}$$

15.2.3 Tunable Impedance Matching Networks

Most of the impedance matching networks discussed so far work over one frequency. Tunable impedance matching networks are designed for operation over a defined frequency bandwidth. In other words, they also match a wider range of impedances. To achieve that, variable inductors and capacitors may be used including microelectromechanical systems (MEMS) switched capacitor and digitally variable capacitors. These are available in IC forms.

15.2.4 Simplified Conjugate Impedance Matching Circuit

The general analysis of impedance matching circuits outlines what happens when a source having complex impedance Z_g transfers power to a complex load Z_L. Maximum power is transferred to the load when the load impedance is matched to the power transfer system impedance. This means that

$$Z_L = Z_g^* \qquad (15.21)$$

The load impedance is the complex conjugate of the system impedance. That makes the total system impedance resistive and is $2R_g$. In other words, both the system and the load have equal resistance, and their reactance should cancel each other out. Effectively, this also means that all the reactive power needs to be reduced to zero to maximise the real power

$$
\begin{aligned}
Z &= Z_L + Z_g = Z_g + Z_g^* = (R_g + jX_g) + (R_g - jX_g) \\
&= 2R_g
\end{aligned}
$$
(15.22)

Most electronic circuits are much more complex than the circuit shown in Figure 15.14. More complex circuits require more elaborate impedance matching considerations, and the rest of this chapter is dedicated to the most prevalent impedance matching cases. Emphasis is given to circuits operating below 1 GHz. This frequency range permits lumped circuit elements to be used for impedance matching.

15.2.4.1 Impedance matching and maximum power transfer consideration

A better understanding of the circuit theory model of wireless power transfer systems can be derived if the transmitting and receiving circuits are treated as two port networks with input and output impedances. The transmitting circuit is connected to the input port which is the output of the voltage source. The load impedance is connected to the output port.

The input impedance looking into the two points to which the input source is connected as shown in Figure 15.15 is influenced by the receiver circuit due to mutual inductance. This influence is modelled by the impedance equation:

$$
\begin{aligned}
Z_{\text{in}} &= Z_1 + \frac{(\omega M)^2}{Z_2} \\
&= R_1 + j\omega L_1 + \frac{1}{j\omega C_1} + \frac{(\omega M)^2}{R_2 + R_L + j\omega L_2 + \frac{1}{j\omega C_2}}
\end{aligned}
$$
(15.23)

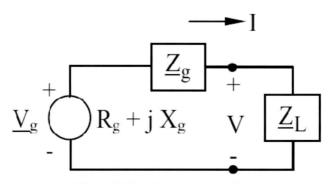

Figure 15.14   Impedance matching.

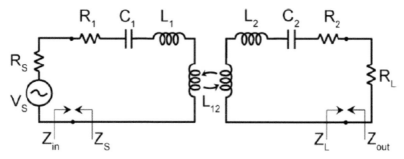

Figure 15.15 Maximum power transfer impedance relationships.

and

$$Z_{\text{out}} = Z_2' + R_1 + j\omega L_1 + \frac{1}{j\omega C_1}$$

$$= R_2 + j\omega L_2 + \frac{1}{j\omega C_2} + \frac{(\omega M)^2}{R_1 + + R_S + j\omega L_1 + \frac{1}{j\omega C_1}} \qquad (15.24)$$

Equation (15.7) which depicts the input impedance of the WPT system includes the load, while Equation (15.8) includes the source resistance. In Figure 15.6, the source will deliver maximum power to the transmitter only when the input impedance is matched to the source impedance. This happens only when the input impedance is equal to the complex conjugate of the source impedance ($Z_{\text{in}} = Z_S^*$). At the output (receiver) circuit, maximum power can be delivered to the load only when the output impedance is also equal to the complex conjugate of the load impedance ($Z_{\text{out}} = Z_L^*$). For any other impedance relationships, the impedance mismatches in the system cause voltage and current reflections in the circuits causing reduced power transfer and hence reduced power transfer efficiency as well. Two impedance matching equations therefore need to be considered as:

$$Z_{\text{in}} = Z_S^*$$
$$Z_{\text{out}} = Z_L^* \qquad (15.25)$$

Hence, at resonance, the system efficiency is

$$\eta = \frac{P_{\text{out}}}{P_{\text{source}}} \times 100\%$$

$$= \frac{\omega^2 M^2 R_L}{\left[R_1 (R_2 + R_L)^2 + \omega^2 M^2 (R_2 + R_L) \right]} \left(1 - |\Gamma_{\text{in}}|^2 \right) \times 100\%$$

$$\qquad (15.26)$$

where the reflection coefficient at the source is given by the expression

$$\Gamma_{\text{in}} = \frac{Z_{\text{in}} - Z_S}{Z_{\text{in}} + Z_S} \qquad (15.27)$$

To compare the efficiency in Equation (15.11) to what was previously obtained, the equation is expressed in terms of the quality factors of the coils given as:

$$\eta = \frac{k^2 Q_1 Q_2 R_2 R_L}{(R_2 + R_L)^2 \left[1 + \frac{k^2 Q_1 Q_2}{(R_2 + R_L)}\right]} \left(1 - |\Gamma_{\text{in}}|^2\right) \times 100\,\% \qquad (15.28)$$

When the following conditions are obeyed

$$Q_2' = Q_2 / (R_2 + R_L),$$
$$R_L' = (R_2 R_L) / (R_2 + R_L),$$
$$\Gamma_{\text{in}} = 0$$

Then, efficiency when the source impedance is considered is given by the following:

$$\eta = \frac{k^2 Q_1 Q_2' R_L'}{[1 + k^2 Q_1 Q_2']} \times 100\,\% \qquad (15.29)$$

References

[1] Bostic, Chris. *RF Circuit Design*. Howard W. Sams (1982).
[2] Frenzel, Louis E. *Principles of Electronic Communication Systems*, McGraw Hill (2008).
[3] Koh Kim Ean, et al. "Impedance Matching and Power Division Algorithm Considering Cross Coupling for Wireless Power Transfer via Magnetic Resonance" (2012).
[4] Nima Soltani, et al. "A High-Gain Power-Matching Technique for Efficient Radio-Frequency Power Harvest of Passive Wireless Microsystems" (2010).
[5] George L. Matthaei. "Microwave Filters, Impedance-Matching Networks, and Coupling Structures". Artech House (1985).
[6] Koh Kim Ean, et al. "Novel Band-Pass Filter Model for Multi-Receiver Wireless Power Transfer via Magnetic Resonance Coupling and Power Division".
[7] Chwei-Sen, Wang, Grant A. Covic. "Power Transfer Capability and Bifurcation Phenomena of Loosely Coupled Inductive Power Transfer Systems" (2004).

16

Design, Analysis, and Optimization of Magnetic Resonant Coupling Wireless Power Transfer Systems Using Bandpass Filter Theory

Henry Mei[1], Dohyuk Ha[2], William J. Chappell[1,2] and Pedro P. Irazoqui[1,2]

[1]Center for Implantable Devices, Weldon School of Biomedical Engineering, Purdue University, West Lafayette, IN 47907 USA
[2]School of Electrical and Computer Engineering, Purdue University, West Lafayette, IN 47907, USA
E-mail:{hmei; pip}@purdue.edu;
dohyuk.ha1.gmail.com; chappell@ecn.purdue.edu

16.1 Introduction

Coupled mode and circuit theory analysis methods have provided great insight regarding the unique behavior of magnetic resonance-coupled (MRC) wireless power transfer systems. However, practical design and optimization methods remain a challenge. This chapter focuses on the development of an alternative MRC model that is derived using bandpass filter (BPF) design theory. Two-port microwave network analysis methods are used to gain insight into system design and optimization. Additionally, a source- and load-included general coupling matrix is synthesized and used to develop new analysis of BPF-modeled MRC wireless power transfer (WPT) systems. Consequently, a unique set of analytical equations are derived which detail the relationships between system parameters and maximum achievable wireless power transfer efficiency (PTE). Also, a figure of merit (FOM) is introduced that explicitly details the extent of tuning controllability as related to system parameters of the BPF-modeled MRC WPT system. The BPF-modeled MRC WPT is realized from the developed theory. Practical system tuning, i.e., critical coupling point

Wireless Power Transfer 2nd Edition, 543–586.

control, is demonstrated through K-inverter networks exhibiting characteristic impedances determined from derived optimization functions.

Wireless power transfer (WPT) technology has fascinated the world ever since Nikola Tesla pioneered the concept at the end of the 19th century [1]. Throughout a significant portion of his life, Tesla envisioned a world where technology could enable energy transfer to far reaches of the world without the need for wires. Unfortunately, due to lacking support and infrastructure, his global reaching WPT endeavors never reached fruition. Instead, wireless communication became the focus of research and development in the wireless technology sector [2].

Consequently, as microwave engineering technology continued to advance, interest in WPT as a means for sourcing power to electronics reemerged. In the 1960s, William C. Brown gained notoriety for his work developing and demonstrating far-field radiofrequency (RF) energy transfer systems, highlighting its useful potential as an alternative method for power delivery [3–5]. Today, the development of WPT technology continues to gain significant interest. Applications utilizing WPT technology include the wireless recharge of consumer electronics including toothbrushes [6], cellular phones [7], and vehicles [8, 9]. Additionally, WPT has been an active research topic as a non-invasive energy delivery method for implantable medical devices [10–12]. Utilizing WPT technology, implantable devices could operate with smaller batteries able to recharge non-invasively from an external source. In devices designed for small implanted spaces, such as the eye or blood vessels [13, 14], the battery could be removed altogether. Overall, integration of WPT technology may enable the development of smaller, safer, and less invasive implantable medical devices [15].

Wireless power transfer can be categorized into two distinct mechanisms: far-field radiative or near-field induction. Each method exhibits a unique set of operating advantages and disadvantages depending on the specific needs, application, and usage model of the applied system. WPT via far-field radiative methods operates from the transmission and capture of propagating electromagnetic (EM) waves. Generally, the wireless power transfer efficiency (PTE) of far-field systems is very low but can operate at very long distances relative to the sizes of the energy transmit and capture antennas. In contrast, near-field inductive methods operate from the transmission and capture of oscillating magnetic fields. Near-field systems typically exhibit extremely high wireless PTE but can only do so at very short distances between source and receiver. In 2007, Kurs et al. [16] advanced the capabilities of traditional near-field WPT through the development of magnetic resonance coupling

(MRC) WPT. Utilizing the phenomena of strongly coupled resonance, the researchers demonstrated WPT at efficiencies much higher than far-field radiative methods at much longer ranges than traditional inductive methods. This first MRC demonstration was conducted using a four-coil magnetically coupled resonant system consisting of two resonator coils, a drive coil, and a load coil (Figure 16.1). The system was theoretically analyzed via coupled mode theory to predict the phenomenon of oscillatory coupled resonance within the system [17]. Although elegant, the physical theories utilized to describe the MRC phenomena did not provide intuitive insight into system behavior and practical design methodologies for engineers. To address these challenges, circuital approaches utilizing voltage and current equations to describe the behavior of four-coil MRC systems have been extensively developed [18–20]. These circuit analysis approaches provided new methods for predicting and tuning the wireless efficiency response of the four-coil MRC system. MRC system tuning methods include coil switching [21], variable coupling method (varying separation distance between drive/load coil and respective resonator coils) [22], and frequency tuning [18]. However, these system tuning methodologies lack capability for practical integration with many applications due to the need for external mechanical control systems and/or complex feedback algorithms. Addressing this, simpler tuning methods via lumped component impedance matching (IM) networks have been demonstrated [23–25]. However, analytical frameworks for achieving optimal IM design still remain elusive.

This chapter focuses on the development of a new framework for MRC system design and analysis that is based on bandpass filter (BPF) theory, microwave network analysis, and general coupling matrix theory. Using BPF modeling, the 4-coil system is simplified into a 2-coil system while still

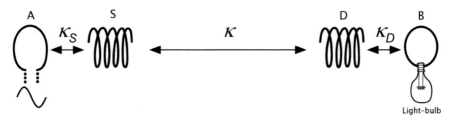

Figure 16.1 Original four-coil MRC system developed by Kurs et al. [16]. Drive coil *A* is magnetically coupled to resonator coil *S* and load coil *B* is magnetically coupled to resonator coil *D*. System of coils *A* and *S* exhibits mid-range resonant WPT to system of coils *B* and *D*. Image reprinted from [16]. Reprinted with permission from AAAS.

maintaining the unique MRC behavior. Additionally, general coupling matrix synthesis and associated analysis methods provide a unique pathway for the derivations of a unique set of analytical equations which describe the parameter relationships for dictating system tuning controllability and optimization. As a result, maximum achievable wireless power transfer efficiency (PTE) can be predicted, controlled, and realized for a given set of initial filter parameters.

16.2 MRC System Equivalent to BPF

The conventional four-coil MRC system can be transformed and modeled as an equivalent BPF network. This transformation is shown in Figure 16.2. The equivalent BPF network (Figure 16.2c) is achieved by creating an equivalent lumped element circuit representation of the coupled resonator circuit. Importantly, the lumped element circuit representations of these couplings are an alternative form of an impedance inverter circuit. The resulting structure is in a form equivalent to that of a generic BPF. The reader is directed to references [26, 27] for more details regarding this BPF equivalence transformation.

16.2.1 Impedance Inverters

Impedance inverters are a form of immittance inverter where immittance is a word derived from the combination of impedance and admittance. Impedance inverters are typically used for the design of narrowband bandpass or bandstop filters. The network representation of a K-inverter is shown in Figure 16.3. Impedance inverters are a two-port network which exhibits the unique property of inverting the load impedance [28]. Thus, the input impedance of the K-inverter is expressed by

$$Z_{\text{in}} = \frac{K^2}{Z_L} \tag{16.1}$$

where Z_{in} is the input impedance looking into the K-inverter network, K is the real-valued characteristic impedance of the impedance inverter, and Z_L is the load impedance. It can be seen from Equation (16.1) that Z_{in} will be inductive/capacitive if Z_L is capacitive/inductive. This condition must hold true in order for K to be real-valued. Thus, K-inverters exhibit a phase shift of $\pm 90^{\circ}$ or some odd multiple thereof [28]. Impedance inverters can be used to transform series-connected elements into shunt-connected elements and vice versa. A simple example of a K-inverter circuit is the quarter-wave transformer transmission line with characteristic impedance, Z_0. Using two-port

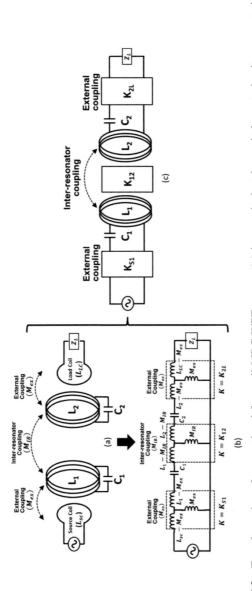

Figure 16.2 Transformation of conventional four-coil MRC WPT system (**a**) into equivalent lumped element circuit representation (**b**). External and inter-resonator couplings are represented with a K-inverter. The resulting circuit can be modeled as a 2-stage BPF (**c**).

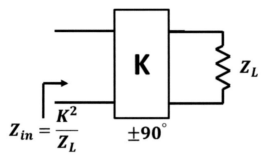

Figure 16.3 Operation of a K-inverter.

network theory, the *ABCD* matrix of a transmission line with length l and wave number $\beta(\beta = 2\pi/\lambda)$ is given by

$$\begin{bmatrix} A & B \\ C & D \end{bmatrix} = \begin{bmatrix} \cos\beta l & jZ_0\sin\beta l \\ j\dfrac{1}{Z_0}\sin\beta l & \cos\beta l \end{bmatrix} \tag{16.2}$$

Observing Figure 16.4 and using Equation (16.2), where $l = \lambda/4$ for a quarter-wave transformer, the *ABCD* matrix of a K-inverter can be determined as

$$\begin{bmatrix} A & B \\ C & D \end{bmatrix} = \begin{bmatrix} 0 & \pm jK \\ \mp\dfrac{j}{K} & 0 \end{bmatrix} \tag{16.3}$$

In addition to quarter-wave transformers, many other circuits can act as impedance inverters including lumped element networks, so long as the

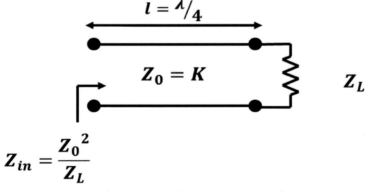

Figure 16.4 Operation of a quarter-wave transformer.

network produces a phase shift of some odd multiple of $\pm90°$ with characteristic impedance that obeys Equations (16.1) and (16.3). As will be shown later, lumped component networks can be utilized to realize an optimum value of characteristic impedance for achieving optimal wireless PTE in a BPF-modeled MRC system.

16.2.2 Two-Stage BPF-Modeled MRC Circuit

Figure 16.5 illustrates a generic bandpass filter prototype utilizing K-inverters and series resonators. The BPF network prototype response can be realized from element, impedance, and frequency scaling transformations of a prototype lowpass filter. From lowpass filter prototypes, a variety of other filter topologies and responses can be designed including bandpass, bandstop, and highpass filters. The reader is directed to references [28, 29–31] for further discussion on the lowpass filter prototypes and the transformation methods for realizing filter topologies with practical impedances and frequency responses. For the remainder of this chapter, the filter theory described will focus only on the principles necessary for the design and analysis related to MRC WPT systems.

The characteristic impedance of the K-inverters can be determined by the equations shown in Figure 16.5 where:

Z_0 = the source impedance

Ω_c = the cutoff angular frequency ($\Omega_c = 1$ rad/s for filter prototype),

FBW = fractional bandwidth

g_i = element values for lowpass prototype filter ($i = 0, 1, 2, ..., n, n+1$)

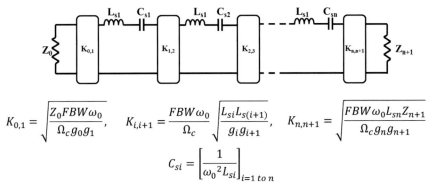

$$K_{0,1} = \sqrt{\frac{Z_0 FBW \omega_0}{\Omega_c g_0 g_1}}, \quad K_{i,i+1} = \frac{FBW \omega_0}{\Omega_c}\sqrt{\frac{L_{si}L_{s(i+1)}}{g_i g_{i+1}}}, \quad K_{n,n+1} = \sqrt{\frac{FBW \omega_0 L_{sn} Z_{n+1}}{\Omega_c g_n g_{n+1}}}$$

$$C_{si} = \left[\frac{1}{\omega_0^2 L_{si}}\right]_{i=1 \text{ to } n}$$

Figure 16.5 Generic BPF prototype with impedance inverter and series resonators. Image adapted and modified from [28].

L_i = series resonator inductor,
C_i = series resonator capacitor,
ω_0 = resonant frequency, and
$Z_{n+1} = Z_L$ = the load impedance.

The element values, g_i, for lowpass filter prototypes can be readily determined from filter design tables of the corresponding filter type (Butterworth, Chebyshev, Elliptic, Gaussian, etc.) and order (*n*) [29]. The characteristic impedances of the inverters, which can be determined by the relationships shown in Figure 16.5, are critically important for realizing a desired BPF network frequency response.

As shown in Figure 16.6, the standard four-coil magnetically coupled MRC system can be represented as a 2-stage BPF with K-inverters. The power source includes a source impedance, Z_s, and the couplings are represented by the K-inverters with characteristic impedance K_{S1}, K_{12}, and K_{2L}. K-inverters K_{S1} and K_{2L} represent external couplings. In standard filter theory, external couplings typically represent couplings which connect the filter to the outside world. In this case, the source impedance and load impedance are connected to series resonator 1 and resonator 2, respectively, and are coupled via K-inverters with characteristic impedance K_{S1} and K_{2L}. Resonators 1 and 2 are made of a series LC circuit. The external couplings are often expressed as external quality (*Q*) factors. The reciprocal of the external *Q* factor represents the external coupling coefficient; for the BPF circuit shown in Figure 16.6, the external coupling coefficients are represented as k_{S1} and k_{2L}. K-inverter K_{12} represents the inter-resonator coupling, and k_{12} represents the magnetic coupling coefficient between resonators 1 and 2. The external coupling K-inverters, K_{S1} and K_{2L}, can be practically realized using lumped elements, whereas inter-resonator coupling K-inverter K_{12} is

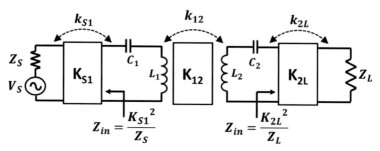

Figure 16.6 BPF-modeled MRC system with K-inverters and series resonators (series *LC* circuit). The series resonators are coupled together through magnetic coupling coefficient, k_{12}.

virtualized through magnetic coupling coefficient, k_{12}. No physical lumped component elements are required for transformation of the inter-resonator coupling between resonators 1 and 2 so as to preserve WPT capability. The coupling coefficients, k, related to the characteristic impedance inverters of the 2-stage BPF ($i = 0, 1, 2, 3$) are given by [26, 32]

$$k_{S1} = \frac{\text{FBW}}{g_0 g_1} = \frac{K_{S1}{}^2}{\omega_0 L_1 Z_s},$$

$$k_{12} = \frac{\text{FBW}}{\sqrt{g_1 g_2}} = \frac{K_{12}}{\omega_0 \sqrt{L_1 L_2}},$$

$$k_{23} = \frac{\text{FBW}}{g_2 g_3} = \frac{K_{2L}{}^2}{\omega_0 L_2 Z_L} \qquad (16.4)$$

The key attribute of the BPF-modeled MRC WPT system filter is to obtain maximum achievable PTE at the resonance frequency of the filter. Other features such as roll-off rate and ripple are not important considerations for the design of the filter. As a consequence, the filter's response type is chosen to exhibit a Butterworth response. The Butterworth response of a 2-stage BPF can be derived from the corresponding Butterworth response lowpass filter prototype. As mentioned previously, the lowpass prototype element values, g_i, can be readily determined from filter tables. Therefore, for a 2-stage ($n = 2$) Butterworth response lowpass to bandpass prototype exhibits element values which are given by

$$g_0 = 1, \quad g_1 = \sqrt{2}, \quad g_2 = \sqrt{2}, \quad g_3 = 1 \qquad (16.5)$$

Substituting Equation (16.5) into Equation (16.4), it is found that

$$k_{S1} = k_{12} = k_{2L} \qquad (16.6)$$

This is known as the matching condition, and achieving this will result in optimum system performance. Coupling coefficient, k_{12}, can be specified based on the expected separation distance, lateral misalignment, or angular misalignment between resonators. Specifically, k_{12} is given by

$$k_{12} = \frac{M_{12}}{\sqrt{L_1 L_2}}, \quad 0 \leq k_{12} \leq 1 \qquad (16.7)$$

where M_{12} is the mutual inductance between coils. Generally, the coupling coefficient of coupled resonators is defined based on the ratio of coupled to stored energy [28]. Thus, a k_{12} value of 0 indicates that no magnetic energy is transferred from the source to receive coil inductor. In contrast, a k_{12} value of 1 indicates that all of the magnetic energy is transferred from the source to the receive coil. Unlike iron core transformers which can exhibit couplings near a value of 1, MRC WPT systems typically exhibit a maximum value of k_{12} that is much less than 1. Therefore, it is critically important to determine the best possible coupling value between resonators and specifying a matched condition coupling value that falls within the range of possible coupling. With k_{12} known, k_{S1} and k_{2L} are also known by virtue of achieving the matched condition described by Equation (16.6). In order for the BPF-modeled MRC system to exhibit this matching condition, the characteristic impedance values of the external coupling K-inverters, K_{S1} and K_{2L}, must be determined. Rearranging Equation (16.4), K_{S1}, K_{12}, and K_{2L} can be calculated by

$$K_{S1} = \sqrt{k_{s1}\omega_0 L_1 Z_S},$$
$$K_{12} = k_{12}\omega_0 \sqrt{L_1 L_2},$$
$$K_{2L} = \sqrt{k_{2L}\omega_0 L_2 Z_L} \qquad (16.8)$$

All parameters for determining the K-inverter impedances in Equation (16.8) are known. Thus, the optimal characteristic impedance of the external coupling K-inverters can be readily determined.

16.2.3 Realization of K-Inverter Circuit and System Matching Conditions

An impedance inverter circuit must be designed such that it exhibits the characteristic impedances, determined from Equation (16.8), for satisfying the matched coupling conditions described in Equation (16.6). Some practical impedance inverter circuits are listed in [28]. K-inverters can be realized using lumped elements, distributed systems (high frequency applications), or a combination of both. Lumped element network impedance inverter circuits represent the most practical for the design of BPF-modeled MRC WPT systems. This is especially true given that near-field WPT systems operate at low frequencies (high kHz to low MHz). An example K-inverter circuit that can be utilized for BPF-modeled MRC system design is shown in Figure 16.7 [32]. The equivalent two-port network representation and associated *ABCD* parameters are shown in Figure 16.8. The characteristic impedance of the K-inverter circuit in Figure 16.7 can be derived by equating the impedance inverter *ABCD* parameters given in Equation (16.3) with the *ABCD* parameters

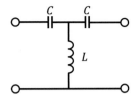

Figure 16.7 Example K-inverter circuit for BPF-modeled MRC WPT system design.

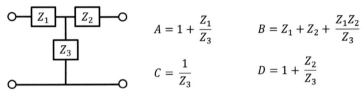

$$A = 1 + \frac{Z_1}{Z_3} \qquad B = Z_1 + Z_2 + \frac{Z_1 Z_2}{Z_3}$$

$$C = \frac{1}{Z_3} \qquad D = 1 + \frac{Z_2}{Z_3}$$

Figure 16.8 Two-port network representation of the K-inverter circuit shown in Figure 16.7 with associated ABCD parameters.

for the two-port K-inverter network representation summarized in Figure 16.8. As a result, it can be found that the characteristic impedance of the K-inverter circuit given in Figure 16.7 is

$$K = \frac{1}{\omega_0 C} = \omega_0 L \qquad (16.9)$$

where ω_0 is the operating resonant frequency. The resonant frequency of the WPT system is determined by the components of the series resonator portion of the BPF-modeled MRC circuit by

$$\omega_0 = 2\pi f_0 = \frac{1}{\sqrt{L_n C_n}} \qquad (16.10)$$

where $n = 1$, 2, for the 2-stage BPF-modeled MRC system. The lumped component values for achieving the matched condition can be determined by equating Equation (16.9) with the corresponding K_{S1} and K_{2L} values that can be calculated from Equation (16.8). Figure 16.9 shows the circuit model of the BPF-modeled MRC WPT system integrated with the lumped component K-inverter circuit.

16.2.4 Example BPF-Modeled MRC WPT System Response

Circuit simulation software can be used to determine the system response of an optimized BPF-modeled MRC system. Some examples include LTSpice© and

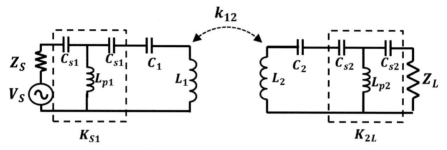

Figure 16.9 Example 2-stage Butterworth BPF-modeled MRC WPT system with integrated K-inverter circuit. BPF circuit adapted and modified from [32].

Advanced Design System©. Here, simulations are performed to demonstrate the optimized design of the BPF-modeled MRC WPT system shown in Figure 16.9. The WPT system parameters for simulation are summarized in Table 16.1. Additionally, a 50-Ω source and load impedance is assumed. Using Equation (16.8), the matched condition for $k_{12} = 0.01$ is achieved when the external coupling characteristic impedance values, K_{S1} and K_{2L}, are 7.75 and 7.89 Ω, respectively. Accordingly, the lumped components of the K-inverter circuits are determined using Equation (16.9) and are given by

$$C_{s1} = \frac{1}{\omega_0 K_{S1}} = 1514.4 \text{ pF} \qquad C_{s2} = \frac{1}{\omega_0 K_{2L}} = 1487.6 \text{ pF}$$

$$L_{p1} = \frac{K_{S1}}{\omega_0} = 91.0 \text{ nH} \qquad L_{p2} = \frac{K_{2L}}{\omega_0} = 92.6 \text{ nH}$$

Figure 16.10 shows the resulting frequency response of the optimized system. Indeed, the frequency response of the filter exhibits peak $|S_{21}|$ and minimum $|S_{11}|$ at the designed resonant frequency of 13.56 MHz (note the dB scale). S-parameter, S_{21}, is known as the transmission coefficient of a two-port network and is related to the PTE by

$$\text{PTE } (\%) = |S_{21}|^2 \times 100 \qquad (16.11)$$

Table 16.1 Initial WPT system parameters for circuit simulation

Parameter	Resonator 1 (Tx)	Resonator 2 (Rx)
f_0	13.56 MHz	13.56 MHz
L_n	1410.0 nH	1460.0 nH
C_n	97.7 pF	94.4 pF
k_{12}	0.01	

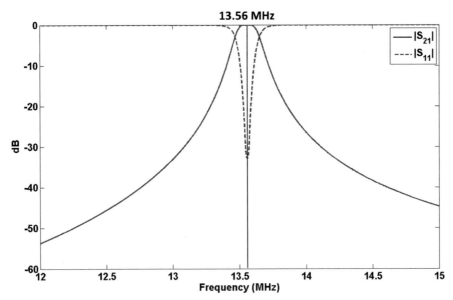

Figure 16.10 Simulated response of BPF-modeled MRC system for system parameters shown in Table 16.1 and corresponding K-inverter lumped component design.

where $|S_{21}|$ is in linear units and not in dB. It is important to highlight that the PTE described by Equation (16.11) denotes the wireless transmission efficiency and not the total system energy efficiency. The total system energy efficiency of a WPT system would include the losses incurred by source impedance, whereas the wireless PTE excludes source impedance and is only determined by the ratio of power delivered at the input of the WPT system and the available power at the output of the WPT system. S-parameter S_{11} is known as the reflection coefficient and can be interpreted as the level of achieved system matching. An $|S_{11}|$ (brackets denote magnitude) value of 0 ($-\infty$ in dB) indicates perfect match, and a value of 1 (0 in dB) indicates no match. As shown in Figure 16.10, a high level of matching is achieved at the resonant frequency and at the specified k_{12} value.

As observed in Figure 16.10, a Butterworth filter response is achieved with peak PTE of 100% occurring at the designed resonant frequency of 13.56 MHz. Indeed, the Butterworth filter response was accurately realized through the filter theory described. The reader may ask how 100% PTE is achieved given that $k_{12} = 0.01$. The answer lies in the fact that so far, the matched condition required for determining the external coupling K-inverter characteristic impedances K_{S1} and K_{2L} assumes ideal (non-lossy) resonator parameters. Such a non-lossy resonator model is not realistic. For example, antenna coils used for WPT typically exhibit significant frequency dependent parasitic losses which deteriorate PTE. In real-life systems, the degradation of wireless PTE is exacerbated as the inter-resonator coupling coefficient between the coils decreases. Consequently, the BPF theory development of MRC WPT systems requires additional optimization methods which account for the characteristics of lossy resonators.

16.3 BPF Model with Lossy Resonator Optimization

As described in the previous section, optimal MRC system design requires matched external and inter-resonator coupling coefficients as described in Equation (16.6). K-inverter characteristic impedances are modified, as described in Equation (16.8), to achieve these matched conditions. Thus far, the derivation of the necessary matched condition assumes ideal filter parameters. To improve upon the BPF model, the incorporation of lossy resonators is necessary.

16.3.1 Lossy Series Resonant Circuit

In the BPF-modeled MRC system, losses can be incorporated into the series resonators as a parasitic series resistance, R_{pn}. Consequently, the series LC resonator circuit is modified as a series RLC circuit as shown in Figure 16.11. The resonant frequency of the resonator is still determined by Equation (16.10). An important parameter of this series resonant circuit is its quality factor, Q, which is given by [31]

$$Q = \omega \frac{\text{average energy stored}}{\text{energy dissipated/cycle}} \qquad (16.12)$$

Thus, resonator Q is a measure of loss of the resonant circuit. For the unloaded series RLC circuit shown in Figure 16.11, the Q of the resonator at the resonant frequency is defined as the unloaded Q, denoted Q_0, and is given by

$$Q_{0n} = \frac{\omega_0 L_n}{R_{pn}} \qquad (16.13)$$

In the design of near-field magnetic WPT systems, R_{pn} is primarily due to the frequency-dependent equivalent series resistance (ESR) of the coil inductors used for WPT. For an inductor, Q_0 typically increases with frequency

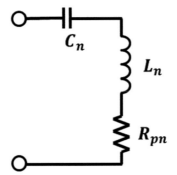

Figure 16.11 Series resonator model which includes series parasitic loss resistance, R_{pn}.

until it reaches its self-resonant frequency at which Q_0 begins to decrease. Consequently, the chosen frequency of operation must be carefully selected such that both resonators used for WPT exhibit sufficiently high Q_0.

As explained in Section 16.1.3, BPF-modeled MRC design requires the determination of optimal external coupling K-inverter characteristic impedances K_{S1} and K_{2L}. For an ideal 2-stage BPF model, the optimal conditions occur when the external and inter-resonator couplings match as described in Equation (16.6). However, in order to account for parasitic losses (finite Q_0) of the series resonators, an alternative determination for optimum K_{S1} and K_{2L} values is required.

16.3.2 Determination of S_{21} Function for Lossy Resonator BPF-Modeled MRC WPT System

As described in Section 16.1.3, determination of K_{S1} and K_{2L} values directly influences the S_{21} frequency response of BPF-modeled MRC system. Analytically then, S_{21} must be a function of K_{S1} and K_{2L}, among other parameters. It is therefore beneficial to derive the S_{21} function of the 2-stage BPF-modeled MRC WPT system. Microwave network analysis methods can be used for the derivation.

In microwave network analysis, Z (Impedance), S (Scattering), and Y (Admittance) parameters and the transmission $ABCD$ matrix can be used to fully characterize a microwave network. In practice, microwave networks can be modeled as a cascade of two-port networks rather than a single one. For characterization, such cascaded networks can be conveniently described using the transmission $ABCD$ matrix [31] of which an equivalent-cascaded $ABCD$ matrix can be derived. The reader is directed to [31] for more information on microwave network analysis.

The BPF-modeled MRC WPT system, shown in Figure 16.6, can be modeled as a series cascade of two-port networks. A total of five two-port networks can be used to represent the BPF model. This includes three K-inverter networks for describing the external and inter-resonator couplings and two series *RLC* networks. Each individual network can be described by an *ABCD* matrix representation where the *ABCD* parameters fully describe each of the individually unique two-port network characteristics. Matrix multiplication between each of the five individual *ABCD* matrices can be conducted resulting in a single equivalent *ABCD* matrix which accurately describes the unique behavior of the entire system. Figure 16.12 shows the series-cascaded two-port network representation and corresponding *ABCD* matrix for each of the five series-cascaded two-port networks which make the 2-stage BPF-modeled MRC WPT system. Specifically, the external and inter-resonator coupling K-inverters (K_{S1}, K_{12}, and K_{2L}) exhibit an *ABCD* matrix described by Equation (16.3). Each of the two-port *RLC* resonator networks can be readily described as network with total series impedance, Z_n. The corresponding *ABCD* matrix for such a network can be easily determined and is usually given in microwave network analysis literature [31]. The total series impedance of a single lossy resonator, at the resonant frequency, is given by

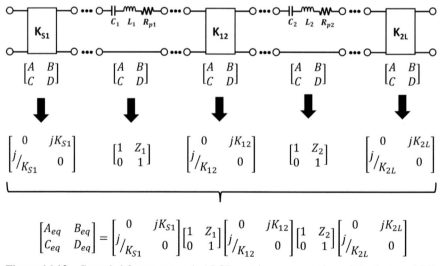

Figure 16.12 Cascaded 2-port network *ABCD* matrix representation of the 2-stage BPF-modeled MRC WPT system.

$$Z_n = R_{pn} - j\frac{1}{\omega_0 C_n} + j\omega_0 L_n,$$

where,

$$R_{pn} = \frac{\omega_0 L_n}{Q_{0n}} \qquad (16.14)$$

where $j = \sqrt{-1}$. Substituting Equation (16.14) into Z_n and using matrix multiplication for each subnetwork shown in Figure 16.12, the equivalent *ABCD* parameters of the whole WPT filter network can be derived to be

$$A_{eq} = \frac{K_{S1}(-1 + C_2\omega_0 (-jR_{p2} + L_2\omega_0))}{C_2 K_{12} K_{2L}\omega_0},$$

$$B_{eq} = -\frac{K_{S1} K_{2L}}{K_{12}},$$

$$C_{eq} = j\frac{\left[-K_{12}{}^2 + \frac{(-1+C_1\omega_0(-jR_{p1}+L_1\omega_0))(-1+C_2\omega_0(-jR_{p2}+L_2\omega_0))}{C_1 C_2\omega_0{}^2}\right]}{K_{S1} K_{12} K_{2L}},$$

$$D_{eq} = \frac{K_{2L}(-1 + C_1\omega_0 (-jR_{p1} + L_1\omega_0))}{\omega_0 C_1 K_{S1} K_{12}} \qquad (16.15)$$

16.3.3 Determination of Optimal K_{S1} and K_{2L} Values for Lossy Resonator System

In microwave network analysis, the Z, S, and Y parameters and *ABCD* parameters of the transmission *ABCD* matrix can be converted into equivalent forms [31]. The S_{21} parameter of the 2-stage BPF-modeled MRC WPT system can be derived from the *ABCD* parameters of the equivalent-cascaded network shown in Equation (16.15). The *ABCD* parameter-derived S_{21} parameter function is given by

$$S_{21} = \frac{2}{A_{eq} + (B_{eq}/Z_0) + (C_{eq}Z_0) + D_{eq}} \qquad (16.16)$$

where $Z_0 = Z_L = Z_S = 50 \ \Omega$ (50 Ω is chosen so as to be convenient for measurement using microwave equipment such as vector network analyzers). Equation (16.16) represents the analytical PTE function of the

2-stage BPF-modeled MRC WPT system. For brevity, the expanded version of Equation (16.16) is not shown. However, it is important to note that Equation (16.16) is a function of the characteristic impedances of the external coupling K-inverters, K_{S1} and K_{2L}, and inter-resonator coupling K-inverter, K_{2L}. Additionally, Equation (16.16) is a function of the source and load impedances (50 Ω), resonant frequency, and series resonator component values, R_{pn}, L_n, and C_n. In the design of the BPF-modeled MRC system, all parameters in Equation (16.16) are known except for the optimum values of the characteristic impedances of the external coupling K-inverters, K_{S1} and K_{2L}. To determine the optimum K_{S1} and K_{2L} impedance values, the magnitude of Equation (16.16) is taken and plotted as a function of both K_{S1} and K_{2L}. The K_{S1} and K_{2L} impedance values which achieve peak $|S_{21}|$, or $|S_{21}|^2$ which relates to wireless PTE as described in Equation (16.11), will be the optimal external coupling impedance values, denoted as K_{S1opt} and K_{2Lopt}. Through Equation (16.16), Figure 16.13 shows the simulated $|S_{21}|^2$ as a function of K_{S1} and K_{2L} for a 2-stage BPF-modeled MRC WPT system with WPT design parameters given in Table 16.2. Note, K_{12} is determined via the corresponding relationship given in Equation (16.4). From Figure 16.13, the peak $|S_{21}|^2$ occurs when the external coupling

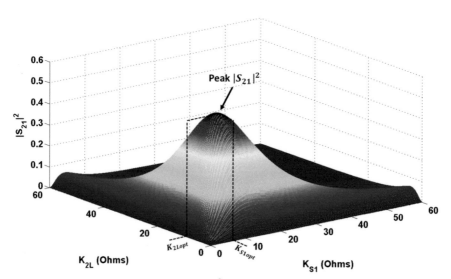

Figure 16.13 Analytically predicted $|S_{21}|^2$ response as a function of external coupling K-inverter characteristic impedance values K_{S1} and K_{2L} of BPF-modeled MRC system with WPT system parameters shown in Table 16.2.

Table 16.2 WPT system parameters used to determine optimal K_{S1} and K_{2L} impedance values

Parameter	Resonator 1 (Tx)	Resonator 2 (Rx)
f_0	13.56 MHz	13.56 MHz
L_n	1410.0 nH	1460.0 nH
C_n	97.7 pF	94.4 pF
R_{pn}	0.3432 Ω	0.3554
Q_{0n}	350	350
k_{12}	0.01	

characteristic impedance values achieve $K_{S1\mathrm{opt}}$ and $K_{2L\mathrm{opt}}$ values of 7.91 and 8.04 Ω, respectively. The corresponding peak $|S_{21}|^2$ is 0.569 ($|S_{21}| = -2.449$ dB). This indicates that given the WPT system parameters summarized in Table 16.2, which includes parasitic loss resistances in resonators 1 and 2, the analytically predicted optimal wireless PTE will be 56.90 %. It is important to point out that this wireless PTE is the analytically predicted maximum achievable wireless PTE for this particular WPT system with parameters given in Table 16.2 and for $k_{12} = 0.01$. Following the determination of the $K_{S1\mathrm{opt}}$ and $K_{2L\mathrm{opt}}$ values, the lumped component values of the K-inverter circuit can be derived using the same procedure conducted in Section X.1.4. The lumped component values of the K-inverter circuit shown in Figure 16.7 can be determined by Equation (16.9) and are given by

$$C_{s1} = \frac{1}{\omega_0 K_{S1\mathrm{opt}}} = 1483.8 \text{ pF} \qquad C_{s2} = \frac{1}{\omega_0 K_{2L\mathrm{opt}}} = 1459.8 \text{ pF}$$

$$L_{p1} = \frac{K_{S1\mathrm{opt}}}{\omega_0} = 92.84 \text{ nH} \qquad L_{p2} = \frac{K_{2L\mathrm{opt}}}{\omega_0} = 94.37 \text{ nH}$$

16.3.4 Circuit Simulation Results and Effect of Q_{0n} on Maximum Achievable PTE

Figure 16.14 shows the resulting circuit-simulated frequency response of the 2-stage BPF-modeled MRC WPT system with resonators exhibiting parasitic series resistance. The simulated circuit consists of the optimally derived K-inverter lumped component circuit values and WPT design parameters summarized in Table 16.2. At an inter-resonator coupling coefficient, k_{12},

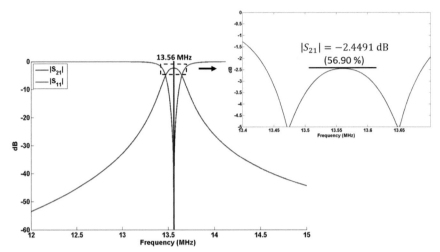

Figure 16.14   Circuit-simulated frequency response for the lossy resonator 2-stage BPF-modeled MRC system with WPT system design parameters summarized in Table 16.2.

of 0.01, the circuit-simulated frequency response shows a peak $|S_{21}|$ with a value of –2.4491 dB (PTE = 56.90%) at the designed resonant frequency of 13.56 MHz. This circuit-simulated response matches exactly with the theoretically predicted response of –2.449 dB (PTE = 56.90%). It is clear that the two-port microwave network analysis approach provides an accurate analytical method for optimizing the design of the 2-stage BPF-modeled MRC WPT system. Indeed, using two-port microwave network analysis, the optimal external coupling K-inverter characteristic impedance values can be derived for a real-life system exhibiting lossy (finite Q_0) resonators.

The finite Q_0 of the series resonators has a significant effect on the maximum achievable WPT efficiency for a given inter-resonator coupling coefficient, k_{12}. Keeping k_{12} = 0.01, Figure 16.15 shows the effect of decreasing resonator Q_0. As resonator Q_0 decreases so too does the maximum achievable PTE. A summary of the analytically predicted and circuit-simulated maximum achievable PTE changes, at $\omega = \omega_0$, due to decreasing resonator Q_0 is summarized in Table 16.3. The maximum achievable wireless PTE drops by ~28 % when the resonator Q_0 (transmit and receive resonators equally) drops from a value of 350–150. It is important to note that $K_{S1\text{opt}}$ and $K_{2L\text{opt}}$ values and corresponding K-inverter circuit lumped component values were updated in the circuit simulation to account for the change in resonator Q_0. Doing this ensures that the analytically predicted and circuit-simulated results exhibit the maximum achievable wireless PTE given the

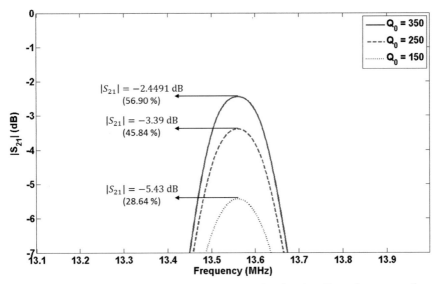

Figure 16.15 Circuit-simulated frequency response showing the effect of resonator Q_0 on the maximum achievable wireless PTE at the resonant frequency. Inter-resonator coupling coefficient, k_{12}, was kept constant with a value of 0.01. K_{S1opt} and K_{2Lopt} values were updated accordingly due to changes in Q_0.

Table 16.3 Summary of maximum achievable wireless PTE for given resonator Q_0 at $\omega = \omega_0$ for $k_{12} = 0.01$, $L_1 = 1410$ nH, $L_2 = 1460$ nH, and 13.56 MHz

Resonator Q_0 ($Q_0 = Q_{01} = Q_{02}$)	K_{S1opt} (Ω)	K_{2Lopt} (Ω)	Analytically Predicted Maximum PTE (%)	Circuit-Simulated Maximum PTE (%)
350	7.91	8.04	56.897	56.90
250	8.04	8.19	45.837	45.84
150	8.50	8.64	28.642	28.64

specific set of resonator Q_0 values. Clearly, in order to obtain high wireless PTE, the resonators of the 2-stage BPF-modeled MRC WPT system must be designed such that Q_0 is as high as possible.

Two-port microwave network analysis methods provide the analytical foundation for optimized design of a 2-stage BPF-modeled MRC WPT system. By developing an optimization model that incorporates parasitic resistance and resonator losses, practical WPT system development can be achieved. However, in the analysis thus far, only the frequency response at the optimized inter-resonator coupling coefficient value was explored. In many WPT applications, however, the coupling between resonators may be dynamic, and thus, the inter-resonator coupling coefficient will deviate from the optimally designed value. In this instance, it is necessary to determine the 2-stage BPF-modeled MRC transfer response not only as a function of frequency but also as a function of changing inter-resonator coupling coefficient.

16.4 BPF Model Analysis Using General Coupling Matrix

The general coupling matrix represents an important and general design technique for synthesizing and analyzing coupled-resonator filters. Introduced in the 1970s, the general coupling matrix provided a new method of filter network synthesis enabling practical filter design, tuning, modeling, and analysis. In addition, real-world effects such as lossy resonators could be accounted for resulting in accurately predicted filter performance characterization [33].

Since then, general coupling matrix synthesis and analysis methods have advanced to accommodate couplings from the source and load to the inter-resonators. Naturally, the source and load general coupling matrix can be synthesized for the 2-stage BPF-modeled MRC WPT system. As a result, appropriate coupling matrix analysis methods can be employed enabling a new level of system optimization as well as the development of closed-form equations for providing the parameter relationships for prediction of system performance and control. The subject of coupling matrix theory and synthesis is extensive. Only the basic theory and analysis techniques necessary for WPT system development will be covered. The reader is directed to references such as [28, 30, 33, 34] for further discussion.

16.4.1 Synthesis of Source and Load Coupling Matrix for BPF-Modeled MRC WPT System

The 2-stage BPF-modeled MRC WPT system with K-inverter and series resonator, as shown in Figure 16.6, can be represented by a general coupling topology. Figure 16.16 shows this general coupling topology representation. The external couplings are represented by the coupling coefficients M_{S1} and M_{2L}, and the inter-resonator coupling is represented as M_{12}. Note, variable M is used instead of k to represent FBW-normalized coupling which will be discussed later. The white circles represent the source and load, while the black circles represent each respective resonator (series RLC circuit previously described in Section 10.2) within the BPF.

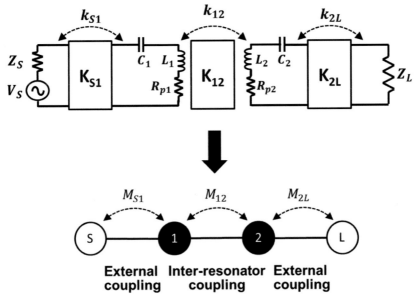

Figure 16.16 General coupling topology (*bottom*) representation of the 2-stage BPF-modeled MRC WPT system (*top*).

	S	1	2	$\cdots$	$n-1$	n	L
S	0	M_{S1}	M_{S2}	$\cdots$	$M_{S,n-1}$	M_{Sn}	M_{SL}
1	M_{1S}	M_{11}	M_{12}	$\cdots$	$M_{1,n-1}$	M_{1n}	M_{1L}
2	M_{2S}	M_{21}	M_{22}	$\cdots$	$M_{2,n-1}$	M_{2n}	M_{2L}
$\vdots$	$\vdots$	$\vdots$	$\vdots$	$\ddots$	$\vdots$	$\vdots$	$\vdots$
$n-1$	$M_{n-1,S}$	$M_{n-1,1}$	$M_{n-1,2}$	$\cdots$	$M_{n-1,n-1}$	$M_{n-1,n}$	$M_{n-1,L}$
n	M_{nS}	M_{n1}	M_{n2}	$\cdots$	$M_{n,n-1}$	M_{nn}	M_{nL}
L	M_{LS}	M_{L1}	M_{L2}	$\cdots$	$M_{L,n-1}$	M_{Ln}	0

Figure 16.17 Source- and load-included general coupling matrix.

Figure 16.17 shows the general coupling matrix for an nth-degree coupled resonator filter. The source- and load-included general coupling matrix is an $(n+2) \times (n+2)$ reciprocal matrix. Non-zero diagonal entries are valid and represent finite Q of the resonators. Using the general coupling topology as a blueprint, the source- and load-included general coupling matrix ($n + 2$ general coupling matrix) for the 2-stage BPF-modeled MRC WPT systems can be synthesized and is given by

$$[M] = \begin{bmatrix} 0 & M_{S1} & 0 & 0 \\ M_{S1} & M_{11} & M_{12} & 0 \\ 0 & M_{12} & M_{22} & M_{2L} \\ 0 & 0 & M_{2L} & 0 \end{bmatrix} \qquad (16.17)$$

The coefficients of the coupling matrix, M_{ij} for i or j = S, 1, 2, ..., n, L, can be calculated by

$$M_{ij} = \frac{1}{\sqrt{g_i g_j}} \qquad (16.18)$$

where g_i and g_j are the element values for the lowpass filter prototype [30]. Substituting Equation (16.18) into Equation (16.4) and rearranging, it can found that the characteristic impedances of the external and inter-resonator K-inverters can be given by

$$K_{S1} = M_{S1}\sqrt{\text{FBW}\omega_0 L_1 Z_S},$$
$$K_{12} = M_{12}\omega_0\sqrt{L_1 L_2},$$
$$K_{2L} = M_{2L}\sqrt{\text{FBW}\omega_0 L_2 Z_L} \qquad (16.19)$$

Therefore, the characteristic impedance of the K-inverters can be determined if the coupling coefficients of the $[M]$-matrix are known. $[M]$-matrix coupling coefficients M_{11}, M_{12}, and M_{22} represent WPT system parameters and are given by [35]

$$M_{11} = \frac{-j}{\text{FBW}Q_{01}},$$
$$M_{12} = \frac{k_{12}}{\text{FBW}},$$
$$M_{22} = \frac{-j}{FBWQ_{02}}, \qquad (16.20)$$

From Equation (16.20), $[M]$-matrix coupling coefficients M_{11} and M_{22} represent the parasitic losses in resonator 1 and resonator 2, respectively. $[M]$-matrix coupling coefficient M_{12} represents the FBW-normalized inter-resonator coupling coefficient. $[M]$-matrix coupling coefficients M_{S1} and

M_{2L} represent the external coupling coefficients that directly modify the characteristic impedances of the external coupling K-inverters. Similar to what was described in Section 10.2., a BPF-modeled MRC WPT system capable of reaching maximum achievable wireless PTE can be realized only if the system is designed with the external coupling K-inverters exhibiting the optimal characteristic impedances K_{S1opt} and K_{2Lopt} for a given a set of WPT system parameters. Note, the WPT system must include parasitic losses of the BPF resonators. From Equation (16.19), it is clear that in order to find K_{S1opt} and K_{2Lopt}, M_{S1opt} and M_{2Lopt} must be determined.

16.4.2 Determination of M_{S1opt} and M_{2Lopt}

The synthesized general coupling matrix for the 2-stage BPF-modeled MRC WPT system, described in Equation (16.17), can be used to compute the filter frequency response in terms of the S-parameters [28]:

$$S_{21} = -2j[A]_{n+2,1}^{-1}$$
$$S_{11} = 1 + 2j[A]_{n+2,1}^{-1} \qquad (16.21)$$

where $[A]_{i,j}^{-1}$ represents the ith row and jth column of matrix $[A]$ which is given by

$$[A] = [M] + \Omega[U] - j[q] \qquad (16.22)$$

where matrix $[U]$ and $[q]$ are $(n+2) \times (n+2)$ matrices denoting the existence of resonators and loads, respectively. For the 2-stage BPF-modeled MRC WPT system, they are given by

$$[U] = \begin{bmatrix} 0 & 0 & 0 & 0 \\ 0 & 1 & 0 & 0 \\ 0 & 0 & 1 & 0 \\ 0 & 0 & 0 & 0 \end{bmatrix}, \ [q] = \begin{bmatrix} 1 & 0 & 0 & 0 \\ 0 & 0 & 0 & 0 \\ 0 & 0 & 0 & 0 \\ 0 & 0 & 0 & 1 \end{bmatrix} \qquad (16.23)$$

Variable Ω is the frequency variable for lowpass filter prototype. Transforming the lowpass response to a bandpass response, Ω becomes

$$\Omega = \frac{1}{FBW}\left(\frac{\omega}{\omega_0} - \frac{\omega_0}{\omega}\right) \qquad (16.24)$$

Using Equations (16.17), (16.22)–(16.24) and substituting the [A]-matrix result into Equation (16.21), the S_{21} function from $[M]$-matrix coupling coefficients is derived to be

$$S_{21} = \frac{2jM_{S1}M_{12}M_{2L}}{M_{12}^2 + (M_{S1}^2 + M_{11} + j\Omega)(M_{22} + M_{2L}^2 + j\Omega)} \qquad (16.25)$$

Using Equation (16.25), the magnitude of S_{21} at the resonant frequency, ω_0, is derived to be

$$|S_{21}|_{\omega=\omega_0} = \frac{2M_{12}M_{2L}M_{S1}}{M_{12}^2 + (M_{22} + M_{2L}^2)(M_{11} + M_{S1}^2)} \qquad (16.26)$$

Equation (16.26) will be used to determine M_{S1opt} and M_{2Lopt} which represent the optimal external coupling values for achieving K_{S1opt} and K_{2Lopt} for a given set of WPT system parameters, i.e., resonator inductance, capacitance, parasitic resistance, and inter-resonator coupling coefficient.

The derivation of M_{S1opt} and M_{2Lopt} from Equation (16.26) begins first with redefining variable M_{12} as M_{12tgt}. Variable M_{12tgt} represents the FBW-normalized targeted inter-resonator coupling coefficient value at which maximum achievable wireless PTE is desired to occur. In Equation (16.26), variables M_{12tgt}, M_{11}, and M_{22} represent known constants defined by the initial WPT system design parameters. The $|S_{21}|_{\omega=\omega_0}$ function will reach a peak value when

$$\frac{\partial |S_{21}|_{\omega=\omega_0}}{\partial M_{S1}} = 0$$

and

$$\frac{\partial |S_{21}|_{\omega=\omega_0}}{\partial M_{2L}} = 0 \qquad (16.27)$$

The optimal values of M_{S1} and M_{2L} (M_{S1opt} and M_{2Lopt}) for reaching maximum achievable PTE at M_{12tgt} are determined by solving the simultaneous equations that are derived from Equation (16.27). As a result, taking only positive values, M_{S1opt} and M_{2Lopt} are given by

$$M_{S1opt} = \frac{M_{11}^{1/4}(M_{12tgt}^2 + M_{11}M_{22})^{1/4}}{M_{22}^{1/4}},$$

$$M_{2Lopt} = \frac{M_{22}^{1/4}(M_{12tgt}^2 + M_{11}M_{22})^{1/4}}{M_{11}^{1/4}} \qquad (16.28)$$

By substituting constitutive relationships from Equation (16.20) into Equation (16.28), the M_{S1opt} and M_{2Lopt} functions can be simplified and determined directly via initial WPT system design parameters. The result is given by

$$M_{S1\text{opt}} = \left(\frac{1 + k_{12\text{tgt}}^2 Q_{01} Q_{02}}{\text{FBW}^2 Q_{01}^2} \right)^{1/4},$$

$$M_{2L\text{opt}} = \left(\frac{1 + k_{12\text{tgt}}^2 Q_{01} Q_{02}}{\text{FBW}^2 Q_{02}^2} \right)^{1/4} \quad (16.29)$$

which provides an exact equation for determining $M_{S1\text{opt}}$ and $M_{2L\text{opt}}$ and ultimately $K_{S1\text{opt}}$ and $K_{2L\text{opt}}$ by way of Equation (16.19). This result indicates two unique properties of BPF-modeled MRC WPT systems: (1) Optimal BPF-modeled MRC system design can be easily achieved given any set of initial WPT system design parameters, and (2) the desired location for maximum achievable PTE can be tuned by parameter $k_{12\text{tgt}}$ which directly modifies $K_{S1\text{opt}}$ and $K_{2L\text{opt}}$. Any K-inverter network can be utilized to achieve optimal wireless PTE so long as it exhibits real characteristic impedances values of $K_{S1\text{opt}}$ and $K_{2L\text{opt}}$. Note, $M_{S1\text{opt}}$ and $M_{2L\text{opt}}$ can also be tuned by the resonator Q_0. However, in practical WPT scenarios, Q_0 is typically set by the coil design and cannot be altered. Additionally, it can be shown that FBW does not modify $K_{S1\text{opt}}$ and $K_{2L\text{opt}}$ as the term cancels out when Equation (16.29) is substituted back into Equation (16.19). As a consequence, $k_{12\text{tgt}}$ $(0 \leq k_{12\text{tgt}} \leq 1)$ is the primary tuning variable. A comparison between $K_{S1\text{opt}}$ and $K_{2L\text{opt}}$ values determined using Equation (16.29) with the graphical method described in Section 10.2 is given in Table 16.4. As shown in Table 16.4, the $K_{S1\text{opt}}$ and $K_{2L\text{opt}}$ values between the graphical and exact equation methods provide similar results indicating the accuracy of coupling matrix analysis. Unlike two-port network analysis, however, the use of coupling matrix analysis methods resulted in the derivations of concise and clear optimization equations detailing unique parameter and parameter relationships for determining $K_{S1\text{opt}}$ and $K_{2L\text{opt}}$ values. In addition to providing an exact method for determining $K_{S1\text{opt}}$ and $K_{2L\text{opt}}$,

Table 16.4 Comparison of $K_{S1\text{opt}}$ and $K_{2L\text{opt}}$ determination for given resonator Q_0 at $\omega = \omega_0$ for $k_{12} = 0.01$, $L_1 = 1410$ nH, $L_2 = 1460$ nH at 13.56 MHz

$Q_0 = Q_{01} = Q_{02}$	Graphical (Two-Port Network Analysis)		Exact Equation (Coupling Matrix Analysis)	
	$K_{S1\text{opt}}$ (Ω)	$K_{2L\text{opt}}$ (Ω)	$K_{S1\text{opt}}$ (Ω)	$K_{2L\text{opt}}$ (Ω)
350	7.9100	8.0400	7.9038	8.0427
250	8.0400	8.1900	8.0432	8.1846
150	8.5000	8.6400	8.4965	8.6458

Equation (16.29) also enables further exploration of optimized system behavior regarding S_{21} as functions of both frequency and inter-resonator coupling.

16.4.3 Examination of M_{S1opt} and M_{2Lopt} on Full S_{21} Response

The functions derived for determining M_{S1opt} and M_{2Lopt} in Equation (16.29) can be substituted back into Equation (16.25) as functions replacing M_{S1} and M_{2L}. As a result, the optimized S_{21} response can be determined both as a function of frequency (lowpass to bandpass frequency transform variable, Ω) and as a function of inter-resonator coupling coefficient, k_{12}, through variable M_{12} as described by Equation (16.20). It is important to emphasize that the M_{12} term in Equation (16.25) is not to be confused with M_{12tgt} which represents a single defined inter-resonator coupling coefficient, k_{12tgt}, at which the BPF-modeled MRC WPT system is optimized for. Variable M_{12} can be varied in a range representative of expected k_{12} values that can occur between the resonator coils. For brevity, the resultant function of the full S_{21} response is not shown, but an example magnitude of S_{21} response, utilizing WPT system design parameters summarized in Table 16.5, is plotted and is shown in Figure 16.18.

Examination of Figure 16.18 reveals the unique resonant coupling $|S_{21}|^2$ behavior specific to MRC systems. In particular, the $|S_{21}|^2$ (PTE) response can be distinguished based on three distinct regions of coupling at the resonant frequency, ω_0: undercoupling, critical coupling, and overcoupling. Undercoupling is a region of rapid PTE decline as k_{12} decreases away from the critical coupling point. Critical coupling is the k_{12} location (k_{12crit}) at which the PTE exhibits a maximum. This represents the maximum achievable wireless PTE given a set of WPT system design parameters. Overcoupling occurs for

Table 16.5   WPT design parameters used to evaluate theoretical model for full $|S_{21}|$ response

Parameters	Resonator 1 (Tx)	Resonator 2 (Rx)
f_0	13.56 MHz	13.56 MHz
FBW	0.043	0.043
L	1410 nH	1460 nH
R_{pn}	0.2529 Ω	0.2619 Ω
Q_{0n}	475	475
k_{12tgt}	0.0338	

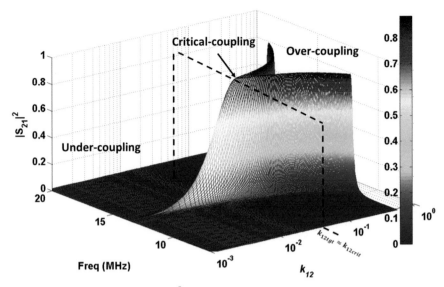

Figure 16.18 Full theoretical $|S_{21}|^2$ (PTE) response as a function of k_{12} and frequency. The WPT design parameters used to evaluate the theoretical model are summarized in Table 16.5.

k_{12} values larger than $k_{12\text{crit}}$. In this region, a phenomenon of frequency splitting occurs resulting in lower and higher modes of resonance within the system. Uniquely, this region can sustain maximum achievable PTE even with varying k_{12} so long as the operating frequency shifts from ω_0 and follows the splitting phenomenon. In contrast, PTE at the resonant frequency in the overcoupling region exhibits a rapid decline similar to that in the undercoupled region. The locations of these three regions can be distinguished by:

1. $k_{12} < k_{12\text{crit}}$: Undercoupled
2. $k_{12} = k_{12\text{crit}}$: Critically Coupled
3. $k_{12} > k_{12\text{crit}}$: Overcoupled

The tuning, or control, of the critical coupling point is of particular interest in MRC system design. A close examination of Figure 16.18 reveals that the PTE at $k_{12\text{tgt}}$ approaches the systems' critical coupling point. This indicates that control of the critical coupling point, in BPF-modeled MRC WPT systems, occurs in the specification of $k_{12\text{tgt}}$ which manifests itself through $M_{S1\text{opt}}$ and $M_{2L\text{opt}}$ and ultimately $K_{S1\text{opt}}$ and $K_{2L\text{opt}}$ values. This phenomenon can be further explored by examining the $M_{S1\text{opt}}$ and $M_{2L\text{opt}}$ functions on the resonant frequency response of $|S_{21}|$.

16.4.4 Examination of M_{S1opt} and M_{2Lopt} on $|S_{21}|_{\omega=\omega_0}$ Response

To determine the optimum $|S_{21}|_{\omega=\omega_0}$ function, the M_{S1opt} *and* M_{2Lopt} functions, described in Equation (16.29), are substituted into Equation (16.26) replacing variables M_{S1} and M_{2L}. Again, as was done in Section 10.3.2, variable M_{12} is left unchanged. The resulting function of optimally tuned $|S_{21}|_{\omega=\omega_0}$ for 2-stage BPF-modeled MRC systems is uniquely derived to be

$$|S_{21}|_{opt,\omega=\omega_0} = \frac{2Q_{01}Q_{02}k_{12}\sqrt{\dfrac{1+k_{12tgt}{}^2Q_{01}Q_{02}}{Q_{01}Q_{02}}}}{2+k_{12tgt}{}^2Q_{01}Q_{02}+k_{12}{}^2Q_{01}Q_{02}+2\sqrt{1+k_{12tgt}{}^2Q_{01}Q_{02}}}$$

$$(16.30)$$

Figure 16.19 shows the plotted results of Equation (16.30) for four different k_{12tgt} optimization points. The resonator Q_0 factors were set to the values specified in Table 16.5. As can be seen, decreasing the k_{12tgt} tuning point causes a corresponding decrease in k_{12crit}. This indicates that lowering k_{12tgt} aids in maximizing range of the 2-stage BPF-modeled MRC system. However, maximizing range results in a trade-off of maximum achievable wireless PTE. Upon closer examination of Figure 16.19, an interesting observation can be made. Specifically, there is an observable discrepancy between the desired location for critical coupling, k_{12tgt} (black diamonds in Figure 16.19), and actual location of critical coupling, k_{12crit} (red dots in Figure 16.19). To investigate this result further, the relationship between k_{12tgt} and k_{12crit} is needed.

16.4.5 Investigation of Relationship between k_{12tgt} and $k_{12\ crit}$

The relationship between k_{12tgt} and k_{12crit} can be determined using the $|S_{21}|_{\omega=\omega_0}$ function described in Equation (16.26) as the foundation for derivation. First, M_{S1opt} *and* M_{2Lopt} are substituted into Equation (16.26) replacing variables M_{S1} and M_{2L}. Second, variable M_{12} is redefined as M_{12crit} to indicate the critical coupling point variable of which is being investigated. Third, the maximum point of the optimized $|S_{21}|_{\omega=\omega_0}$ function is determined by

$$\frac{\partial|S_{21}|_{\omega=\omega_0}}{\partial M_{12crit}} = 0 \qquad (16.31)$$

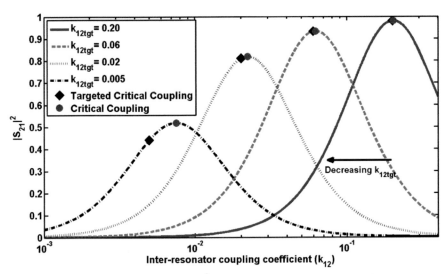

Figure 16.19 Theoretical $|S_{21}|_{opt,\omega=\omega_0}{}^2$ (PTE) response as a function of k_{12} at $\omega = \omega_0$. The black diamonds indicate the PTE at $k_{12} = k_{12tgt}$. The red dot indicates the PTE at $k_{12} = k_{12crit}$. WPT design parameters that are used to evaluate the theoretical model are summarized in Table 16.5.

Solving Equation (16.31) for k_{12crit} and simplifying gives

$$k_{12crit} = \left[\sqrt{M_{S1opt}{}^2 + M_{11}} \times \sqrt{M_{22} + M_{2Lopt}{}^2} \right] \text{FBW} \qquad (16.32)$$

Using the constitutive relationships given in Equations (16.20), (16.32) is simplified to

$$k_{12crit} = \frac{1 + \sqrt{1 + k_{12tgt}{}^2 Q_{01} Q_{02}}}{\sqrt{Q_{01} Q_{02}}} \qquad (16.33)$$

To find the corresponding $|S_{21}|_{\omega=\omega_0}$ value, denoted as $|S_{21}|_{crit}$, Equation (16.32) is renormalized as M_{12crit} and substituted back into the $|S_{21}|_{\omega=\omega_0}$ function described by Equation (16.26). Again, M_{S1} and M_{2L} are replaced with M_{S1opt} and M_{2Lopt}. The final solution of $|S_{21}|_{crit}$ is simplified resulting in

$$|S_{21}|_{crit} = 1 - \frac{1}{k_{12crit} \sqrt{Q_{01} Q_{02}}} = 1 - \frac{1}{1 + \sqrt{1 + k_{12tgt}{}^2 Q_{01} Q_{02}}} \qquad (16.34)$$

An examination of Equation (16.33) indicates that k_{12tgt} and k_{12crit} are not equivalent but in fact scaled by the resonator Q_{0n} factors. This result is an indication of tuning controllability for the BPF-modeled MRC WPT system. Tuning controllability can serve as a useful figure of merit (FOM). The FOM is determined by taking the ratio of k_{12tgt} to Equation (16.33). A perfectly controllable system will exhibit a FOM = 1, indicating that the desired or tuned location of the critical coupling point will be equivalent to the actual location of critical coupling, i.e., $k_{12tgt} = k_{12crit}$. Analytically, this FOM is given by

$$\text{FOM} = \frac{k_{12tgt}}{k_{12crit}} = \frac{k_{12tgt}\sqrt{Q_{01}Q_{02}}}{1 + \sqrt{1 + k_{12tgt}{}^2 Q_{01}Q_{02}}} \tag{16.35}$$

Figure 16.20 shows the effects of varying Q_{01} and k_{12tgt} on the system FOM. Indeed, as k_{12tgt} and/or the transmit resonator quality actor (Q_{01}) decreases so too does the FOM.

It is particularly interesting to note the results when k_{12tgt} is set to equal zero, the minimum value of k_{12tgt} for maximizing system range. Setting $k_{12tgt} = 0$, Equations (16.33) and (16.34) become

$$k_{12crit} = \frac{2}{\sqrt{Q_{01}Q_{02}}},$$

$$|S_{21}|_{crit} = \frac{1}{2} = 0.5,$$

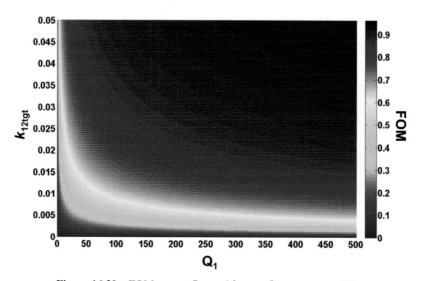

Figure 16.20   FOM versus Q_{01} and k_{12tgt}. Q_{02} was set to 475.

when

$$k_{12\text{tgt}} = 0 \tag{16.36}$$

Interestingly, setting $k_{12\text{tgt}}$ to 0 does not result in a system incapable of WPT. At a tuned optimization point of 0, the location of the critical coupling, $k_{12\text{crit}}$, becomes strictly determined by the resonator Q_{0n} factors. On the other hand, the wireless PTE at the critical coupling point will be 25% ($|S_{21}|_{\text{crit}}^2 \times 100$), independent of the resonator quality. This result is particularly important as it points out the need for high-quality resonators. Specifically, if the resonator Q-factors are too low, then the ability to achieve the critical coupling cannot be achieved. As an example, take a 2-stage BPF-modeled MRC WPT system optimized for $k_{12\text{tgt}} = 0$, with $Q_{01} = Q_{01} = 50$, and antenna coils which can only achieve inter-resonator coupling in the range of $0 \leq k_{12} \leq 0.03$. Using Equation (16.36), $k_{12\text{crit}}$ is calculated to be 0.04. Therefore, the critical coupling point PTE of 25% can never be achieved. This result stresses the importance of having well-designed resonators with high Q-factors. High resonator Q-factors increase not only the magnitude of the maximum achievable PTE but also the control of its location such that critical coupling can occur within the inter-resonator coupling range.

16.5 Experimental Validation

Figure 16.21 describes the 2-stage BPF-modeled MRC circuit used for experimental validation. The corresponding K-inverter circuit is shown in Figure 16.22. This K-inverter circuit consists of series and shunt capacitors, C_{sn} and C_{pn}, respectively. This simple design provides an avenue for practical system design that is amenable for critical coupling point tuning. In addition, this K-inverter design provides the benefit of small size and low loss due to the small surface mount device (SMD) capacitors that can be used. The relationship between the characteristic impedance of the K-inverter, K, and the capacitances, $-C_{sn}$ and C_{pn}, can be determined by equating the real and imaginary input impedance of Figure 16.22 with the K-inverter equation given in Equation (16.1). The series capacitance C_{sn} is negative in order to fulfill the K-inverter *ABCD* matrix given in Equation (16.3). However, this does not represent a problem as $-C_{sn}$ can be conveniently absorbed by the adjacent series resonator capacitance, C_n. All source and load impedances are set to 50 Ω for convenient

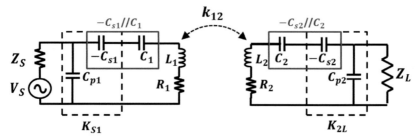

Figure 16.21   2-Stage BPF-modeled MRC system used for experimental validation of coupling matrix analysis and optimization.

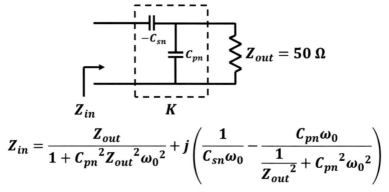

$$Z_{in} = \frac{Z_{out}}{1 + C_{pn}^2 Z_{out}^2 \omega_0^2} + j\left(\frac{1}{C_{sn}\omega_0} - \frac{C_{pn}\omega_0}{\frac{1}{Z_{out}^2} + C_{pn}^2 \omega_0^2}\right)$$

Figure 16.22   K-inverter circuit used for experimental validation. The input impedance, Z_{in}, of the K-inverter circuit is also given.

experimental validation using a vector network analyzer (VNA). Consequently, $-C_{sn}$ and C_{pn} values for the external coupling K-inverters are derived to be

$$C_{pn} = \frac{\sqrt{-K^2 + 50^2}}{50K\omega_0}$$

and

$$-C_{sn} = \frac{R}{K\omega_0\sqrt{-K^2 + 50^2}} \qquad (16.37)$$

Replacing K with the respective K_{S1opt} and K_{2Lopt} values will result in the determination of optimal external coupling K-inverter capacitance values for realizing the maximum achievable PTE response given a set of WPT system parameters and desired tuning point, k_{12tgt}.

16.5.1 Resonator Design and Determination of WPT System Design Parameters

Figure 16.23 shows the experimental setup used to validate the general coupling matrix theoretical design and optimization model. All lumped parameter and S-parameter measurements were conducted using standard 50-Ω one-port and two-port measurements. Two planar spiral transmit (Tx) and receive (Rx) resonator coils were fabricated using 10 AWG insulated copper magnet wire. The coils were designed with an outside diameter of 11.5 cm, 3 turns, and an inter-winding spacing of 6 mm. The parameters of the fabricated coils (L_n, R_{pn}, and Q_{0n}) were measured using a VNA (Agilent E5071B) after two-port calibration. A frequency operation, f_0, of 13.56 MHz was chosen in order to maximize the Q_{0n} factor and also to conform to the industrial, scientific, and medical (ISM) radio band. Both fabricated coils exhibited a Q_0 of 475. Table 16.6 summarizes the experimental WPT design parameter values for resonators 1 and 2. It is important to note that *FBW* is used for convenience

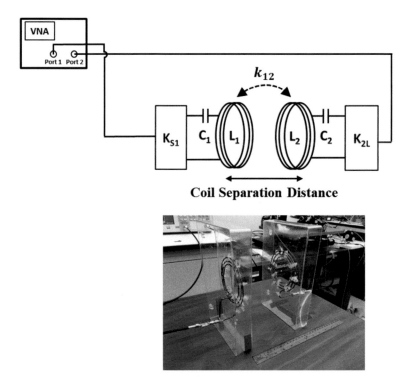

Figure 16.23 Experimental setup for optimized 2-stage BPF-modeled MRC testing.

Table 16.6   Experimental WPT system design parameters

Parameters	Resonator 1 (Tx)	Resonator 2 (Rx)
f_0	13.56 MHz	13.56 MHz
FBW	0.043	0.043
L_n	1410 nH	1460 nH
R_{pn}	0.2529 Ω	0.2619 Ω
Q_{0n}	475	475
C_n^*	97.702 pF	94.356 pF

$^*C_n$ is determined from Equation (16.10) based on the chosen f_0 and series inductance value for the respective resonator ($n = 1, 2$).

and thoroughness but is not a necessary parameter for the determination of K_{S1opt} and K_{2Lopt} values (see Section 16.3.2).

The final parameter needed for system tuning is k_{12tgt}. Figure 16.24 shows the measured k_{12} range between the fabricated planar Tx and Rx coils as a function of separation distance. The range of k_{12} as a function of separation distance was measured by obtaining the S-parameters of only the coils through two-port VNA measurements. The measured two-port S-parameters were converted into Z-parameters, and k_{12} is specifically given by

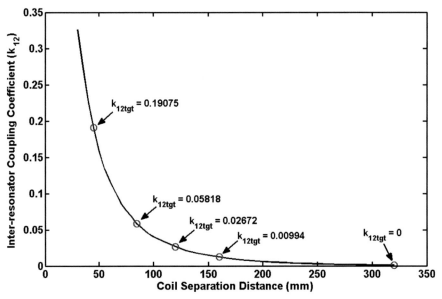

Figure 16.24   Measured k_{12} versus coil separation distance. Red circles indicate the targeted locations for k_{12tgt} optimization tuning points. The minimum coil separation distance is 30 mm due to the plastic enclosure housing the coils.

$$k_{12} = \frac{L_{12}}{\sqrt{L_1 L_2}} = \frac{Im(Z_{12})}{\omega_0 \sqrt{L_1 L_2}} \tag{16.38}$$

where L_{12} is the mutual inductance between the Tx and Rx coils. Mutual impedance, L_{12}, will change as coil separation, angular misalignment, and/or lateral misalignment changes. Figure 16.24 can be interpreted as the range of possible k_{12tgt} tuning points. For this experimental design, the tuning range will exhibit a minimum value of 0 (>320 mm coil separation distance) and a maximum value of 0.33 (30 mm separation distance). Five separate k_{12tgt} tuning points are chosen to reflect the diverse tuning range capability of 2-stage BPF-modeled MRC system design.

Figure 16.23 shows the experimental setup used to validate the general coupling matrix theoretical design and optimization model. All lumped parameter and S-parameter measurements were conducted using standard 50-Ω one-port and two-port measurements.

16.5.2 Optimum Determined K-inverter Capacitance Values

The design parameters in Table 16.6 and the k_{12tgt} points in Figure 16.24 are used to determine K_{S1opt} and K_{2Lopt} values through Equations (16.19) and (16.29). The resulting optimized K-inverter capacitance values for each k_{12tgt} tuning point are given in Table 16.7. SMD capacitors in 0402 package were used for realizing each k_{12tgt} tuning condition. Before placement on the K-inverter printed circuit board (PCB), each SMD capacitor was measured using the VNA at 13.56 MHz to determine actual onboard capacitance. This is necessary due to the 5% to 10% manufacture tolerance typically seen in SMD capacitors.

Table 16.7 Analytically determined capacitance values used for experimental system design at each respective optimization tuning point

	K_{S1opt} Capacitance Values			K_{2Lopt} Capacitance Values		
k_{12tgt}	C_{p1} (pF)	C_{s1} (pF)	$C_{s1}//C_1$ (pF)	C_{p2} (pF)	C_{s2} (pF)	$C_{s2}//C_2$ (pF)
0.19075	255.19	−471.12	123.26	247.00	−470.09	118.05
0.05818	582.10	−676.76	114.19	570.4	−667.00	109.90
0.02672	894.75	−956.35	108.82	878.22	−940.97	104.87
0.00994	1483.95	−1521.10	104.41	1457.68	−1495.50	100.71
0.0	3292.00	−3309.00	100.67	3235.08	−3252.10	97.18

16.5.3 Theoretical versus Measured PTE Response

Figure 16.25 shows the resonant frequency PTE response as a function of k_{12} for the five different k_{12tgt} tuning points. As observed in Figure 16.25, the measured PTE response and critical coupling point converge with the theoretically predicted results. Additionally, a comparison between the theoretically derived and measured $|S_{21}|$ frequency response is made. For brevity, the theoretical $|S_{21}|$ frequency response function is not given, but its derivation is determined by substituting Equation (16.29) into the magnitude of Equation (16.25). Coupling parameter M_{12} was specified at a particular k_{12} value for a k_{12tgt}-optimized system. The comparison results for $k_{12tgt} = 0.00994$ are shown in Figure 16.26. The k_{12} values specified are 0.0012 (300 mm separation distance), 0.1298 (160 mm separation distance), and 0.17296 (50 mm separation distance). As shown, the measured response converges with the theoretical response at each coil separation location. Additionally, undercoupling, critical coupling, and

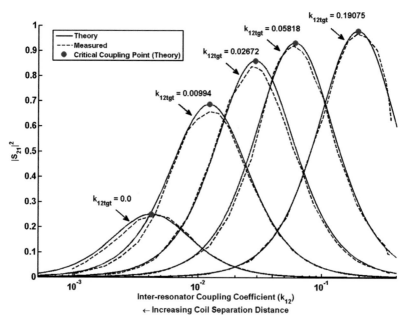

Figure 16.25   Theoretical and measured resonant frequency PTE response plotted as a function of k_{12} for five different k_{12tgt} tuning points. The *arrows* indicate each individual PTE *curve* optimized for the respective k_{12tgt} tuning point. The *red dot* indicates the theoretically predicted critical coupling point.

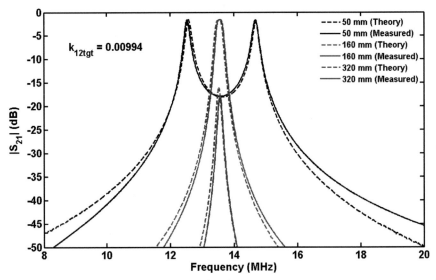

Figure 16.26 Theoretical and measured $|S_{21}|$ transfer response plotted as a function of frequency. The system is tuned for $k_{12tgt} = 0.00994$. Three coil separation distances are plotted to highlight the overcoupling (50 mm), critical coupling (160 mm), and undercoupling (320 mm) regions.

overcoupling are clearly observed. Indeed, the tuning procedure, utilizing general coupling matrix theory and analysis methods, for achieving optimized wireless PTE response is validated.

16.6 Summary of General Coupling Matrix Design Procedure

The procedures for the development and optimization of a 2-Stage BPF-modeled MRC WPT systems are summarized as follows:

Step 1: Determination of WPT System Design Parameters (resonator design). The resonator parameters of the system should first be characterized as a function of frequency. This includes transmit and receive coil inductance, L_n, and unloaded quality factor, Q_{0n}. The operating frequency should be chosen such that both transmit and receive coils exhibit high Q_{0n}. Typically, WPT systems are constrained by system-level requirements. Consequently, the choice of operating frequency may be limited. As a result, significant amount of coil design optimization (coil diameter, number of turns, wire gauge, turn spacing, etc.) should be conducted in an effort to maximize Q_{0n} at the given operating frequency. Once operating frequency and corresponding coil inductances are known, the series resonator capacitance can be calculated.

Step 2: Determination of k_{12} range. In order to determine valid k_{12tgt} points of optimization, the operating inter-resonator coupling coefficient range between transmit and receive coils should be determined. Depending on the physical dimensions, separation distance, and angular misalignment, the k_{12} range may vary considerably. The k_{12tgt} point should be chosen such that range is maximized without sacrificing maximum achievable wireless PTE below the requirements for system operation. At this point, all parameters needed for predication, optimization, and realization of BPF-modeled WPT system design are determined.

Step 3: Determination of K_{S1opt} and K_{2Lopt} Values. K_{S1opt} and K_{2Lopt} should be determined using the parameters determined in steps 1 and 2 with the Equations (16.29) and (16.19).

Step 4: Design K-inverter Circuit. The K-inverter circuit used to realize K_{S1opt} and K_{2Lopt} must be designed such that it achieves the necessary characteristics of a K-inverter network. It is important to specify that in 2-stage BPF-modeled MRC WPT system design, any K-inverter circuit can be utilized, not only the networks specified in this chapter, so long as it achieves the determined K_{S1opt} and K_{2Lopt} impedance. Other impedance inverter circuits can be found in literature [28, 31]. In some K-inverter circuits, negative capacitances and/or inductances may be required. This is acceptable if the negative components can be absorbed by lumped elements of the adjacent resonator.

Step 5: Implementation of 2-stage BPF-modeled MRC WPT System. The K-inverter circuits and resonators can be combined to form the 2-stage BPF-modeled MRC system. It is important to note that the theory discussed in this chapter assumes 50-Ohm source and load impedances. Typically, load impedances such as the input to power conversion and management (rectifier, voltage regulator, boost converter, etc.) exhibit complex load impedances. In such scenarios, impedance matching circuits at the input of power conversion and management circuit should be designed.

16.7 Future Work

The incorporation of resonant coupling matrix analyses for the design of a 2-stage BPF-modeled MRC WPT systems enables a unique path for the derivations of parameters and parameter relationships for predicting and controlling maximum achievable PTE. It is important to consider that up to this point, only a 2-stage system was analyzed and optimized. However, coupling matrix synthesis is extremely versatile and allows for the expansion of additional resonators. Consequently, methods for optimizing multi-resonator and multi-load MRC systems could be developed. This versatility could provide a pathway for the development of optimized and tunable WPT systems with power relaying and selective powering capability using the practical and concise optimization methods derived using BPF theory methods.

16.8 Conclusion

Wireless power technology has developed rapidly since the days of Tesla. The result has been the development of more practical and better performing WPT systems for use with electronics. This chapter has presented the theoretical development for the design and optimization WPT systems by modeling conventional MRC systems as a BPF. As a result, BPF theory analysis was applied resulting in the determination of new parameter relationships and the derivation of simple closed-form design equations to predict and control BPF-modeled MRC WPT system performance. An investigation on system behavior was conducted using two-port network analysis and source and load general coupling matrix analysis methods. Experimental analysis validated the developed theory. Finally, a step-by-step 2-stage BPF-modeled MRC WPT system design procedure is provided.

References

[1] Tesla, N. "The Transmission of Electrical Energy Without Wires as a Means for Furthering Peace". Electrical World and Engineer: Electrical World and Engineer, p. 21 (1905).
[2] Shinohara, N. "Power Without Wires." *IEEE Microwave Magazine* 12: S64–S73 (2011).
[3] Brown, W.C. "Adapting Microwave Techniques to Help Solve Future Energy Problems." *IEEE Transactions on Microwave Theory and Techniques* MT21: 753–763 (1973).

[4] Brown, W.C. "The Technology and Application of Free-Space Power Transmission by Microwave Beam." *Proceedings of the IEEE*: 11–25 (1974).

[5] Brown, W.C. "Optimization of the Efficiency and Other Properties of the Rectenna Element." *Microwave Symposium, 1976 IEEE-MTT-S International*, pp. 142–144 (1976).

[6] Wells, B.P., "Series Resonant Inductive Charging Circuit." US Patent 6,972,543, 6 (2005).

[7] Kim, C.G., D.H. Seo, J.S. You, J.H. Park, and B.H. Cho. "Design of a Contactless Battery Charger for Cellular Phone." *IEEE Transactions on Industrial Electronics* 48: 1238–1247 (2001).

[8] Wang, C.S., O.H. Stielau, and G.A. Covic. "Design Considerations for a Contactless Electric Vehicle Battery Charger." *IEEE Transactions on Industrial Electronics* 52: 1308–1314 (2005).

[9] Yu, X., S. Sandhu, S. Beiker, R. Sassoon, and S. Fan. "Wireless Energy Transfer with the Presence of Metallic Planes." *Applied Physics Letters* 99 (2011).

[10] Lee, S.-Y., C.-H. Hsieh, and C.-M. Yang. "Wireless Front-End with Power Management for an Implantable Cardiac Microstimulator." *IEEE Transactions on Biomedical Circuits and Systems* 6: 28–38 (2012).

[11] Fotopoulou, K., B.W. Flynn, and IEEE. "Wireless Powering of Implanted Sensors Using RF Inductive Coupling." *2006 IEEE Sensors,* vols. 1–3, pp. 765–768 (2006).

[12] Waters, B.H., A.P. Sample, P. Bonde, and J.R. Smith. "Powering a Ventricular Assist Device (VAD) With the Free-Range Resonant Electrical Energy Delivery (FREE-D) System." *Proceedings of the IEEE*, vol. 100, pp. 138–149, January (2012).

[13] Chow, E.Y., A.L. Chlebowski, and P.P. Irazoqui. "A Miniature-Implantable RF-Wireless Active Glaucoma Intraocular Pressure Monitor." *IEEE Transactions on Biomedical Circuits and Systems* 4: 340–349 (2010).

[14] Chow, E.Y., A.L. Chlebowski, S. Chakraborty, W.J. Chappell, and P.P. Irazoqui. "Fully Wireless Implantable Cardiovascular Pressure Monitor Integrated with a Medical Stent." *IEEE Transactions on Biomedical Engineering* 57: 1487–1496 (2010).

[15] Mei, H., and Irazoqui, P.P. "Miniaturizing Wireless Implants." *Nature Biotechnology* 32: 1008–1010 (2014).

[16] Kurs, A., A. Karalis, R. Moffatt, J.D. Joannopoulos, P. Fisher, and M. Soljacic. "Wireless Power Transfer via Strongly Coupled Magnetic Resonances." *Science* 317: 83–86 (2007).

[17] Karalis, A., J.D. Joannopoulos, and M. Soljacic,. "Efficient Wireless Non-Radiative Mid-Range Energy Transfer." *Annals of Physics* 323: 34–48 (2008).

[18] Sample, A.P., D.A. Meyer, and J.R. Smith. "Analysis, Experimental Results, and Range Adaptation of Magnetically Coupled Resonators for Wireless Power Transfer." *IEEE Transactions on Industrial Electronics* 58: 544–554 (2011).

[19] Cannon, B.L., J.F. Hoburg, D.D. Stancil, and S.C. Goldstein. "Magnetic Resonant Coupling as a potential means for wireless power transfer to multiple small receivers." *IEEE Transactions on Power Electronics* 24: 1819–1825 (2009).

[20] Cheon, S., Y.H. Kim, S.Y. Kang, M.L. Lee, J.M. Lee, and T. Zyung. "Circuit-Model-Based Analysis of a Wireless Energy-Transfer System via Coupled Magnetic Resonances." *IEEE Transactions on Industrial Electronics* 58: 2906–2914 (2011).

[21] Kim, J., W.S. Choi, and J. Jeong. "Loop Switching Technique for Wireless Power Transfer Using Magnetic Resonance Coupling." *Progress in Electromagnetics Research-Pier* 138: 197–209 (2013).

[22] Thuc Phi, D., and J.-W. Lee. "Experimental Results of High-Efficiency Resonant Coupling Wireless Power Transfer Using a Variable Coupling Method." *IEEE Microwave and Wireless Components Letters* 21: 442–444 (2011).

[23] Waters, B.H., A.P. Sample, and J.R. Smith. "Adaptive Impedance Matching for Magnetically Coupled Resonators." *Piers 2012 Moscow: Progress in Electromagnetics Research Symposium*, pp. 694–701 (2012).

[24] Lee, W.S., H.L. Lee, K.S. Oh, and J.W. Yu. "Switchable Distance-Based Impedance Matching Networks for a Tunable HF System." *Progress in Electromagnetics Research-Pier* 128: 19–34 (2012).

[25] Beh, T.C., M. Kato, T. Imura, S. Oh, and Y. Hori. "Automated Impedance Matching System for Robust Wireless Power Transfer via Magnetic Resonance Coupling." *IEEE Transactions on Industrial Electronics* 60: 3689–3698 (2013).

[26] Awai, I. "Design Theory of Wireless Power Transfer System Based on Magnetically Coupled Resonators." In Presented at the Wireless

Information Technology and Systems (ICWITS), 2010 IEEE International Conference On, Honolulu, HI (2010).

[27] Awai, I., and T. Ishizaki. "Superiority of BPF Theory for Design of Coupled Resonator WPT Systems." In *Microwave Conference Proceedings (APMC)*, Asia-Pacific, pp. 1889–1892 (2011).

[28] Hong, J.-S. *Microstrip Filters for RF/Microwave Applications*. 2nd ed. John Wiley & Sons, Inc. (2011).

[29] Matthaei, G.L., L. Young, and E.M.T. Jones. *Microwave Filters, Impedance-Matching Networks, and Coupling Structures*. Dedham, MA: Artech House (1980).

[30] Cameron, R.J., C.M. Kudsia, and R.R. Mansour. *Microwave Filters for Communication Systems: Fundamentals, Design, and Application*. Wiley-Interscience (2007).

[31] Pozar, D.M. *Microwave Engineering*. 4 ed. John Wiley & Sons, Inc. (2012).

[32] Ean, K.K., B.T. Chuan, T. Imura, and Y. Hori. "Novel Band-Pass Filter Model for Multi-Receiver Wireless Power Transfer via Magnetic Resonance Coupling and Power Division." In *Presented at the Wireless and Microwave Technology Conference (WAMICON), 2012 IEEE 13th Annual* (2012).

[33] Cameron, R.J. "Advanced Filter Synthesis." *IEEE Microwave Magazine* 12: 42–61 (2011).

[34] Cameron, R.J. "Advanced Coupling Matrix Synthesis Techniques for Microwave Filters." *IEEE Transactions on Microwave Theory and Techniques* 51: 1–10 (2003).

[35] Ha, D., T.-C. Lee, D.J. Webery, and W.J. Chappell. "Power Distribution to Multiple Implanted Sensor Devices Using a Multiport Bandpass Filter (BPF) Approach." In *Presented at the Microwave Symposium (IMS), 2014 IEEE MTT-S International*, Tampa, FL (2014).

17

Multi-Dimensional Wireless Power Transfer Systems

Nagi F. Ali Mohamed[1] and Johnson I. Agbinya

[1]Department of Electrical Engineering,
Faculty of Engineering Technolog, Hoon, Libya

Wireless powering of many devices from a single source is studied. To reduce the rapid decline of inductive power transfer systems, in this chapter, a network of inductive multidimensional wireless power transfer coils is designed and studied in the manner of cellular system. The network coils wirelessly supply devices nearby in the coverage area with electric power. We show that the configuration type and current play important roles in the efficiency of wireless power transfer systems in multiple dimensions. The efficiency is found to be very high at remote locations and acceptable for applications such as biomedical implant devices and lighting of bulbs. The design provides the basis for controlling the power transfer to any direction at any distance within the design.

17.1 Introduction

The ability to gain more wireless power transfer and support a wide area has been studied in various fields such as biomedical applications and charging portable devices. However, users are still required to closely engage these mobile devices, restricting mobility and orientation when charge is applied. Some powering devices are limited to a fixed distance, while their efficiency drops quickly when the receiver is turned around its source. Presently, many enhancing techniques such as relays have been proposed to improve the wireless power connection. Traditionally, the magnetic induction between two-coil system relies on the adjustment of the distance and orientation. Despite these challenges, there is room to improve the efficiency. To overcome

Wireless Power Transfer 2nd Edition, 587–624.

some of these challenges, in particular the orientation, a newly designed coil has been investigated in which the magnetic field can be distributed in multiple directions over an area of seven coils in a cellular network framework [1] and it can be expanded by a repeat of the source in a cluster of seven coils in all directions. The application of such multidirectional wireless power coils includes wireless power supply inside and between houses.

17.2 Related Work

In a related work, [3] shows that by using 3-coil WPT system, the overall efficiency improved with the use of high quality factor. The paper indicates that the efficiency of the 3-coil system has smaller variation than 2-coil system due to the drive and load resistors. In biomedical systems, implants should be designed very carefully in a way that the efficiency of the wireless link never drops; otherwise, it can cause harmful effects such as heating [4, 5]. The team in [4] developed hybrid multilayer coils to overcome heating issues and increase the 'power-efficiency' of the link. Efficiency will easily decay when power is absorbed in the tissue in order to increase the frequency [5, 6] of use. To solve this problem and improve power efficiency, three signals that carry different frequencies and band-pass filter have been used in [5]. Frequencies were divided into three sections: low, medium and high. Both papers approached improvement of the efficiency through their choices of system values. In [7], multidimensional receiver was proposed to attain a more efficient wireless power system by using two-dimensional coil. The antenna was applied and examined for an '*in vivo* robotic capsule' irrespective of the orientation. It was concluded that multicoil design enhances the efficiency drop and 'reduces the effect of coil misalignments' [8]. On the other hand, Han and Wentzloff in [9] reported that efficiency is not improved by using 3×3 array of 60×60 micrometre coils, but it does improve the efficiency when they gather them in one single coil of 200×200 micrometre. Although the single coil has higher efficiency, the array coil received more power wirelessly than the single coil. It seems that efficiency cannot be increased when coils are positioned as an array nearby each other and resonate separately, which allows crosstalk to occur. A 'dead zone' also occurs when array coils resonate with the same frequency as the receiver [10]. In [11], a multiband strongly coupled magnetic resonance was used to deliver power wirelessly to several devices operating at separate frequencies over several bands efficiently. To achieve high wireless power transfer efficiency, the Q-factor must be high and the multiband array loops of the source must match the loops of the receiver side in all parameters in particular the frequency. Jow and Ghovanloo [12] have added coil geometry to the previously described work as an important factor to improve power efficiency. Power distribution is more important than the transfer efficiency in multireceiver system by using a new proposed impedance matching technique in [13]. In [14], multireceiver system was used in the manner of frequency splitting to study the effect of strong coupling between receivers. However, both [13, 14] asserted that receivers must be separated from each other to neglect any effect of coupling except between receivers and the source. Casanova et al. [15] and Seungyoung et al. [23] conclude that multiple receiver coils reduce the interference of coupling efficiency, while multiple transmitter coils increase the power delivery with less impact on the efficiency. Transmitter array coils were examined and found to improve the transfer efficiency in relation to a certain rotation angle of the receiver in [16]. As the method of array coils is differently used, [2, 17–19] developed and studied an array of resonators which coupled to each other by the same frequency and they all receive power from one transmitter. The design effectively extends the power transfer distance and receivers mobility along the array. From their point of view, the array did not experience any interference between the multiple coupled resonators. Besides, in multireceiver systems described in [20] and [21], receivers that are close to the transmitter consume most of the transferred power and the remaining power is consumed by the other receivers.

We recently proposed [22] a multidimensional wireless power transfer systems coil, which was used to improve the link between receiver and transmitter. The authors reported on the study and design of innovative multidimensional inductive resonating structures for precise delivery of flux in multidimensional wireless power transfer and inductive communication systems. Two cooperating resonators were used. A primary transmitter structure supports a secondary transmitter which is architected in the form of six loops forming a hexagonal frame and created from the same inductor. The six loops are individually wound to carry currents either in one direction or in opposing directions without breaking up the resonating loop into separate resonators. A primary cooperating resonator of twice the radius of the individual loops in the hexagonal structure is mounted on top of the secondary hexagonal frame to create a strong flux pointing along the major axis of the hexagonal frame. The design of the multidimensional magnetic field system was described. Experiments were detailed for testing the performance of the multicoil system. The experiments test the flux distribution, the received voltage levels for both linear and networked system of coils. In addition, the relaying configuration and its flux boosting performance were tested. The shape of the magnetic fields created were measured and reported. We demonstrate how to contain the generated field to limit leakage to unwanted areas. The aim of this design was to propose a new multidirectional wireless power transfer with high efficiency. This is in contrast to the simple designs. We used the knowledge of direction of current to design the multidimensional coil systems when the induced flux needs to be focused to different directions. The change in current direction can be achieved by changing the direction of source windings. Figure 17.1 shows an example in which the flux network is the proposed multidimensional coil (MDC) system and in this case a hexagonal coil (HC) system. Each array of equidistantly arranged coils has induced flux in eight directions. The eight directions include the six hexagonal orientations plus the two vertical directions in the array.

Each loop of the transmitter secondary coils produces a magnetic flux directed to the facing loop of the receiver. The flux distribution is controlled by different connections of the secondary coils with respect to the primary coil. The design uses a primary transmitter coil and a secondary transmitter section. The secondary is an array of coils deployed in multiple orthogonal directions. The primary transmitter coil is mounted on top of the secondary transmitter array. Both the primary and the secondary coils are however

Figure 17.1 A network of six MDC receivers placed symmetrically around one MDC transmitter.

connected together in the form which creates the MDC. Each coil of the secondary array has the following parameters. The wire used for the windings has diameter of 1.5 mm. Each coil has 8 turns with equal 7.5 cm diameter. The primary coil has diameter 15 cm wide.

A circular coil compared with over a given area. The design actually provides the ability to increase the short- and long-range performances of wireless power transfer systems by using a primary and secondary transmission involving relay coils. The same values of the voltage received at a receiver of simple coil can be achieved at double the distance by using an MDC. We also showed the shape of the induced flux. Figure 17.2 shows same values of voltages received from different sides of the transmitter supporting the idea of multidirectional flux distribution. Figure 17.3 shows the measured voltages between receivers collected by a relay coil moving along the shared distance between them.

The homogeneity of the flux for both sides introduces a link between each receiver and its neighbour. The magnetic field along the path is strong at the resonator ends and weak at the centre. It is clear that the transmitter delivers voltages (flux) to many directions simultaneously.

A relay was designed to control flux transfer. It was positioned above the primary coil at a distance of 9 cm. In this way, it increases the voltage to about twice for all receivers that were placed around the transmitter.

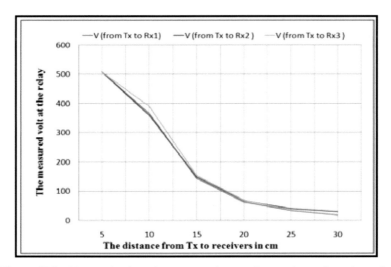

Figure 17.2 Measured voltage between receivers and transmitter using relay coils.

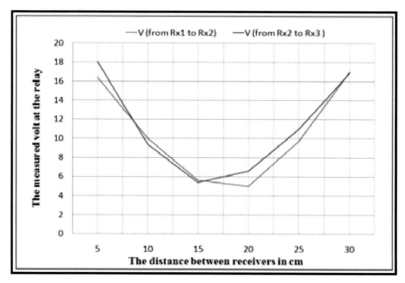

Figure 17.3   Measured voltage between receivers and its neighbour using relay coil.

The performance of the MDC structure was analysed in four scenarios in order to understand the flux pattern of the design. The receiver coil was moved around the transmitter in a dome starting from angle θ equal to zero degree until $90°$ as shown in Figure 17.4. The results are plotted in Figure 17.5. The same results in Figure 17.5 are plotted in three dimensions in Figure 17.6. The pattern shows how the magnetic field strength of the MDC looks like. The red surface illustrates the maximum value of the voltage at a distance of 100 cm from the top, and the green surface shows the minimum value at the

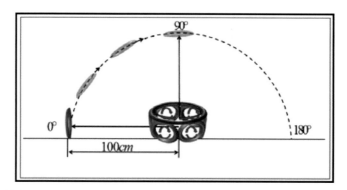

Figure 17.4   Controlling the flux using relay coils on top of the primary transmitter coil.

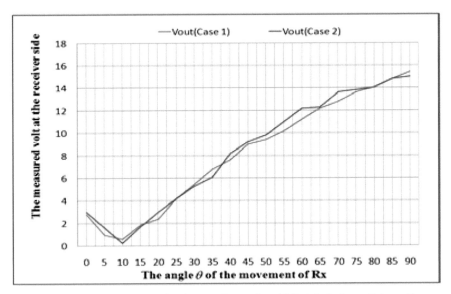

Figure 17.5 The measured voltage at the receiver while it is facing coil (case 1).

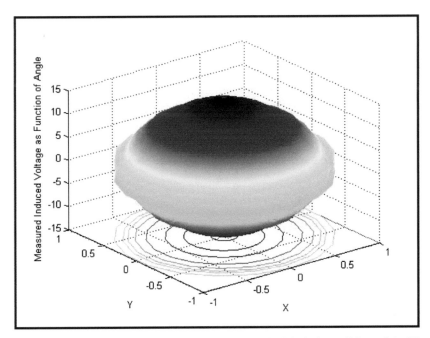

Figure 17.6 The measured voltage at the receiver while it is facing coil (case 1) in 3D.

same distance from the sides. The reverse hemisphere of the pattern shows the negative values or the second half of the magnetic field.

Although the coil was quite small to transfer energy over a network of coils, it does provide a basis for bigger scalable wireless power transfer systems.

17.3 Network of Multidimensional Coils and Radiation Pattern

In this section, we consider extension of previous work, which is plotting the combination of the direct flux of the secondary coils with the orthogonal flux of the primary coil and how they have influences on the receiver. It should be noted that the orientation of the receiver coil in this part is different, but the idea of measuring the voltage is the same. The RX was turned flat on the X-axis at $0°$ so that it will not be facing the transmitter as the one in the previous scenario did. Figure 17.7 illustrates the new orientation of the receiver, which receives most of the flux lines from the small coils while it reaches $90°$. The receiver receives most of the flux lines from the primary coil of the transmitter when it is lying on the same axis.

In Figure 17.8, shows the behaviour of the flux as a function of the change in orientation at 100 cm. These changes illustrate the amount of the field lines crossing the RX while it moves. It seems that most of the flux comes from the primary coil which is taking over the secondary array coils. In view of this, the flux pattern of the design creates two regions. The region close to the MDC is called the near-field region where the secondary coils have more impact over the nearest facing receiver in their sides. The second region starts as the receiver moves far apart from the MDC where the primary coil will

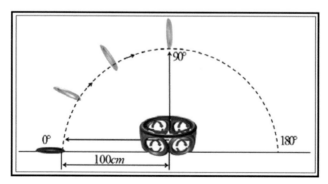

Figure 17.7 The movement of the receiver along a dome when it is lying on $0°$.

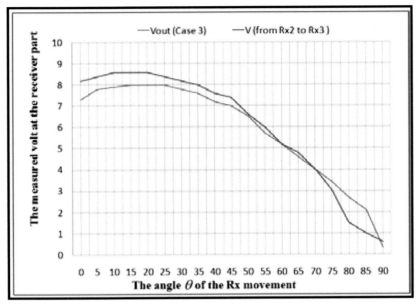

Figure 17.8 The measured voltage of the receiver to a facing coil (case 2).

have the largest impact on receiver. However, the design allows us to extend any of these regions by placing a relay coil in whichever direction we want to. This gives the MDC ability to control the flux.

Figure 17.9 shows the radiation pattern of the design when the receiver coil lies on the same axis as the primary coil (see Figure 17.7). Unlike the previous pattern, the green surface illustrates the minimum value of the voltage at a distance of 100 cm from the top and red surface shows the maximum value at the same distance from the side which means most of the flux lines are received from the primary coil.

Figure 17.10 shows the region covered by the transmitter array. This case starts from 0 to 35 cm where the orthogonal fluxes are about to reach their limit. Each receiver in the diagram received a voltage of about 55 V. How- ever, by placing a relay of radius 3.75 cm in the path 5 cm between the transmitter and each receiver, the voltage increases from 55 V to about 93 V. This means that we can easily extend the short-distance field region. On the other hand, when the relay is replaced on the top of the transmitter primary coil at a distance of 9 cm, the voltage increases from 55 to 114 V. This means that we can have another way of increasing the short and far regions. As the number of relays increases, this region will become wider.

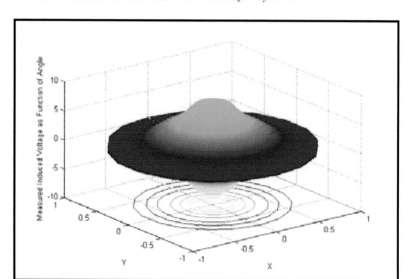

Figure 17.9   The measured voltage of the receiver to a facing coil (case 2) in 3D.

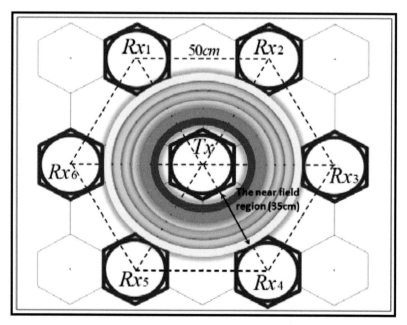

Figure 17.10   The short field region of the transmitter array coils.

The maximum distance reached is about 250 cm. At this distance, the received voltage is 1.2 V. This is where the system efficiency has reduced from around 100 to 0.2 %. Consequently, in this work, we assume that the usable far edge is 250 cm away from the transmitter.

Figure 17.11 shows the electromagnetic field pattern of the MDC where the RX moving along a dome around it at different distances. The distances were measured at different orientation angles of the RX. The required voltage at each point was adjusted to be 1 V in order to plot the edge of the flux. In this way, the 'far field' region was created.

Figure 17.12 shows the polar plot of the radiation pattern measurements listed in Table 17.1. The maximum distance at which the received voltage is 1 V from the side of the MDC is about 100 cm, while the maximum distance from the top (90°) is about 258 cm as shown in the green line.

In contrast, the same measurements have been done after placing a relay over the primary coil by 9 cm to extend the range of receiving the power. The outcome is increasing the maximum distance by nearly 35 cm from both sides as shown in the red line in Figure 17.12.

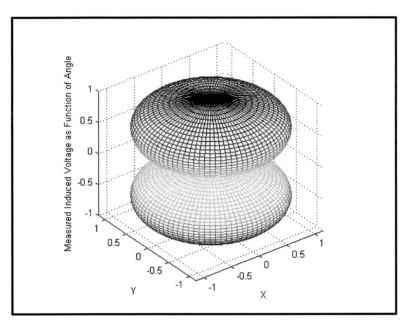

Figure 17.11 The electromagnetic field pattern of the MDC in 3D.

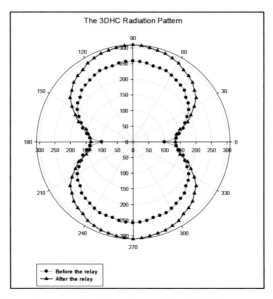

Figure 17.12   The electromagnetic field pattern of the MDC in polar plot before and after placing the relay.

Table 17.1   The maximum distance at which the received voltage is 1 V

Before the Relay		After the Relay	
θ	Distance (cm)	θ	Distance (cm)
0	100	0	135
5	138	5	137
10	150	10	140
15	166	15	155
20	176	20	185
25	190	25	203
30	205	30	225
35	215	35	245
40	220	40	255
45	232	45	268
50	236	50	275
55	240	55	280
60	245	60	290
65	250	65	295
70	250	70	300
75	250	75	302
80	255	80	305
85	257	85	308
90	258	90	310

17.4 Voltage and Current Relation of MDC

In this part, we examine a few of many connection types between coils within the MDC. The connections offer multiple efficiency choices that play an important role in WPT systems and their impact on the communication performance of the MDC network. Changes in the efficiency level will yield different quality of service based on the communication link conditions. For example, if the device that should receive the power wirelessly is positioned within the middle range the connection described in case number 4 will be the best choice for the system because it satisfies the throughput and quality of service objectives. The inductive channel degrades the flux to all devices located around the transmitter. Hence, having control over the efficiency gives us the flexibility to use the right distance with the right connection between (primary and secondary) coils and keep us away from the complexity of the other connections. It should be noted that the electromagnetic link is caused by varying the current flow of the MDC for every (primary and secondary) coils, since the flux amount changes according to the distance between Tx and Rx. In the use of MDC, sometimes a high efficiency such as case number 8 does not necessarily mean a good quality of service because of the current values. Different connections between coils of the MDC lead to different current values. Movement over a distance and changes in current value affect the received wireless power.

When wireless transmission occurs between two MDCs, current is the important factor which determines the type of the connection. For instance, the connection in case number 1 has the lowest voltage received and so the efficiency is very low, but the value of the current is the highest compared to the other connection designs as shown in Table 17.2. This is because this connection type appear to have created the most crosstalk between the coils. In contrast, the connection in case 7 and the connection in case 8 have the highest voltage received and so the efficiency achieved was also high, but the current is low compared to the other connections (see Table 17.2). These two connections create the least crosstalk between the coils. In theory, the efficiency is proportional to the current flows through the transmitter and inversely proportional to the current in the receiver coils as Equation (17.1) shows:

$$\eta = \frac{P_{RX} \times I_{TX}}{P_{TX} \times I_{RX}} \tag{17.1}$$

$$\eta \propto \frac{1}{I_{Rx}}$$

$$\eta \propto I_{Tx}$$

Furthermore, the current values are very important in this system because some applications need higher Amperes than the others. The MDC gives a better use of the power wirelessly in both of the receiver and transmitter sides as long as the right current value is nominated. For example, the current flow in the circuit of case number 7 is slightly low compared to the other cases because of the connection type. On the other hand, the circuit in case number 1 delivers the highest current value of all the other cases because the current distributed in all primary and secondary coils is high. By different connection between coils within the hexagon, we can obtain a unique parameter that impacts the performance of the MDCs network. The power transfer can be delivered with different efficiency depending on the type of device and its power consumption as well as distance. In different operation circuits of MDCs, the current changes cause significant changes in efficiency.

Table 17.2   Current flows through coils (primary and secondary) for different cases of connection

	Case 1	Case 2	Case 3	Case 4
i_{l1}	$1.2 \angle -78.7°$	$0.15 \angle -78.7°$	$0.092 \angle -78.7°$	$0.171 \angle -78.7°$
i_{l2}	$1.2 \angle -78.7°$	$0.15 \angle -78.7°$	$0.092 \angle -78.7°$	$0.171 \angle -78.7°$
i_{l3}	$1.2 \angle -78.7°$	$0.15 \angle -78.7°$	$0.092 \angle -78.7°$	$0.171 \angle -78.7°$
i_{l4}	$1.2 \angle -78.7°$	$0.15 \angle -78.7°$	$0.092 \angle -78.7°$	$0.171 \angle -78.7°$
i_{l5}	$1.2 \angle -78.7°$	$0.15 \angle -78.7°$	$0.092 \angle -78.7°$	$0.171 \angle -78.7°$
i_{l6}	$1.2 \angle -78.7°$	$0.15 \angle -78.7°$	$0.092 \angle -78.7°$	$0.171 \angle -78.7°$
	Case 5	Case 6	Case 7	Case 8
i_{l1}	$0.6 \angle -78.7°$	$0.4 \angle -78.7°$	$0.27 \angle -31.58°$	$0.2 \angle -78.7°$
i_{l2}	$0.6 \angle -78.7°$	$0.4 \angle -78.7°$	$0.27 \angle -31.58°$	$0.2 \angle -78.7°$
i_{l3}	$0.6 \angle -78.7°$	$0.4 \angle -78.7°$	$0.27 \angle -31.58°$	$0.2 \angle -78.7°$
i_{l4}	$0.6 \angle -78.7°$	$0.4 \angle -78.7°$	$0.27 \angle -31.58°$	$0.2 \angle -78.7°$
i_{l5}	$0.6 \angle -78.7°$	$0.4 \angle -78.7°$	$0.27 \angle -31.58°$	$0.2 \angle -78.7°$
i_{l6}	$0.6 \angle -78.7°$	$0.4 \angle -78.7°$	$0.27 \angle -31.58°$	$0.2 \angle -78.7°$

It should be mentioned that the efficiency of the MDC transmitter for most of the cases is higher than that of the conventional coils because of the multidirectional transfer of power. In other words, MDC transfers direct power wirelessly to eight directions, while the normal coils transfer it to two directions. The shape of the MDC determines the coupling coefficient of all its pairs of coils. Orientation also must be identified for the same purpose. An example of a practical implementation of this part is reported in Figure 17.10 where the average voltage received in each RX at distance of 20 cm from the centre of the transmitter coil is approximately 70 V with efficiency of about 14 %. The transmitter was supplied by a class E power amplifier which generates a small current of 200 mA. On the transmitter side, the voltage is assumed to be divided between primary and secondary coils. The voltage of the primary part is the same, and it is lower than the voltage of the secondary coils. Note that the current of the secondary pairs is also divided into six equal values. The resulting voltage at the receiver coil can be obtained directly from one of the secondary resonators. Thus, the 70 V is delivered by the MDC to 8 receivers of the whole network concurrently. On the other hand, the actual values of the elements employed in the network can be described as the sum of the voltage received from all directions, which is about 560 V. This fact provides higher computational performance and shows why we need to have different levels of gain in order to use the MDC network.

17.4.1 Configuration 1

From the circuit in Figure 17.13, the phase currents I_{Z1}, I_{Z2}, I_{Z3}, I_{Z4}, I_{Z5}, and I_{Z6}, represent the secondary currents of the MDC network. Analysis of the circuit consists of finding these currents as a part of showing the relation between currents and efficiency. In constructing the circuit in configuration 1, we made the following calculations:

$$I_{Z1} = \frac{9.8\ \underline{|0}^{\circ}}{8.166\ \underline{|78.7}^{\circ}} = 1.2\ \underline{|-78.7}^{\circ} \tag{17.2}$$

In this type, all coils including the primary coil are supplied directly with 9.8 V, which means they all have the same voltage but the source current is distributed into seven portions.

$$I_{Z1} = I_{Z2} = I_{Z3} = I_{Z4} = I_{Z5} = I_{Z6} \tag{17.3}$$

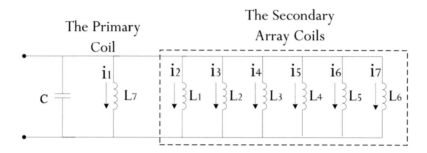

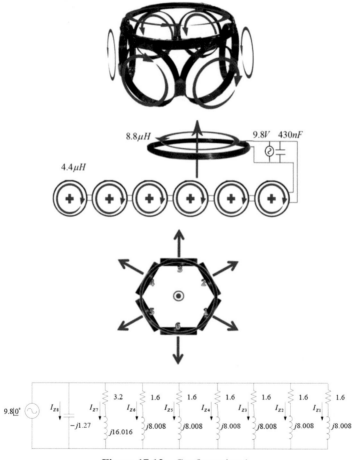

Figure 17.13 Configuration 1.

17.4.2 Configuration 2

From the circuit in Figure 17.14, the impedance seen looking out of the source terminal of the circuit is the impedance of the primary winding plus the impedance of the secondary windings; thus,

$$Z_T = Z_1 + Z_2 + Z_3 + Z_4 + Z_5 + Z_6 + Z_7 \qquad (17.4)$$
$$Z_T = 12.8 + j64.064$$
$$Z_T = 65.33 \,\underline{|78.7}\,^\circ$$

The open circuit value of I_1 is

$$I_1 = \frac{9.8 \,\underline{|0}\,^\circ}{65.33 \,\underline{|78.7}\,^\circ} = 0.15 \,\underline{|-78.7}\,^\circ \qquad (17.5)$$

In similar fashion to Equation (17.3), it can be shown that

$$I_1 = I_{Z1} = I_{Z2} = I_{Z3} = I_{Z4} = I_{Z5} = I_{Z6} \qquad (17.6)$$

In reverse, the source current supplies all coils, which means they all have the same current but the voltage is dissipated into seven portions. Of course, the current of each pier in this configuration is lower than the current of the previous scenario. In case of efficiency, the connection type of configuration 1 is more efficient than connection of configuration 2.

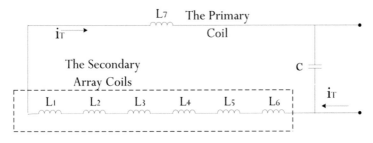

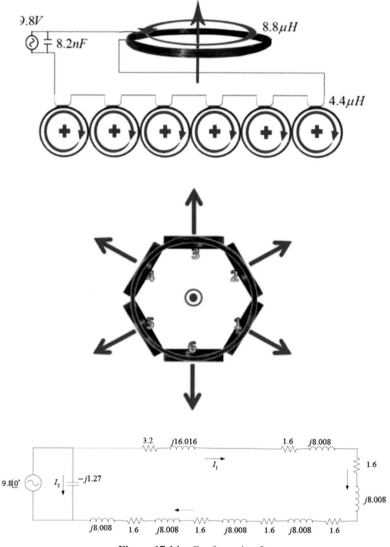

Figure 17.14   Configuration 2.

17.4.3 Configuration 3

In this case in Figure 17.15, the primary coil is connected in series to the secondary coils. The secondary coils are connected together in parallel. Hence, we obtain a slightly different current situation with compared to configuration 1. The impedance seen looking into the terminals of the circuit is the impedance of the primary winding (Z_7) plus the impedance of the parallel-connected secondary windings (Z_T); thus,

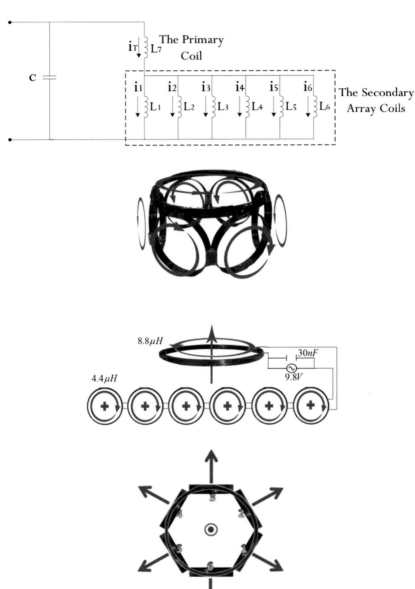

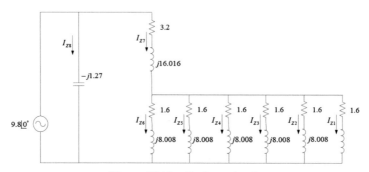

Figure 17.15   Configuration 3.

$$Z_T = (Z_1//Z_2//Z_3//Z_4//Z_5//Z_6)$$

$$Z_T = \frac{Z_1 Z_2 Z_3 Z_4 Z_5 Z_6}{\begin{array}{c} Z_2 Z_3 Z_4 Z_5 Z_6 + Z_1 Z_3 Z_4 Z_5 Z_6 + Z_1 Z_2 Z_4 Z_5 Z_6 + Z_1 Z_2 Z_3 Z_5 Z_6 \\ + Z_1 Z_2 Z_3 Z_4 Z_6 + Z_1 Z_2 Z_3 Z_4 Z_5 \end{array}}$$

$$Z_T = \frac{Z^6}{6Z^5} = \frac{Z}{6} = \frac{8.166 \,\lfloor 78.7^\circ}{6} = 1.361 \,\lfloor 78.7^\circ$$

$$Z_T = 0.266 + j1.335$$

$$Z_T = \frac{Z_1 Z_2 Z_3 Z_4 Z_5 Z_6}{\begin{array}{c} Z_2 Z_3 Z_4 Z_5 Z_6 + Z_1 Z_3 Z_4 Z_5 Z_6 + Z_1 Z_2 Z_4 Z_5 Z_6 + Z_1 Z_2 Z_3 Z_5 Z_6 \\ + Z_1 Z_2 Z_3 Z_4 Z_6 + Z_1 Z_2 Z_3 Z_4 Z_5 \end{array}}$$

$$Z_T = \frac{Z^6}{6Z^5} = \frac{Z}{6} = \frac{8.166 \,\lfloor 78.7^\circ}{6} = 1.361 \,\lfloor 78.7^\circ$$

$$Z_T = 0.266 + j1.335$$

$$Z_T = \frac{Z_1 Z_2 Z_3 Z_4 Z_5 Z_6}{\begin{array}{c} Z_2 Z_3 Z_4 Z_5 Z_6 + Z_1 Z_3 Z_4 Z_5 Z_6 + Z_1 Z_2 Z_4 Z_5 Z_6 + Z_1 Z_2 Z_3 Z_5 Z_6 \\ + Z_1 Z_2 Z_3 Z_4 Z_6 + Z_1 Z_2 Z_3 Z_4 Z_5 \end{array}}$$

$$(17.7)$$

$$Z_T = \frac{Z_1 Z_2 Z_3 Z_4 Z_5 Z_6}{\begin{array}{c} Z_2 Z_3 Z_4 Z_5 Z_6 + Z_1 Z_3 Z_4 Z_5 Z_6 + Z_1 Z_2 Z_4 Z_5 Z_6 + Z_1 Z_2 Z_3 Z_5 Z_6 \\ + Z_1 Z_2 Z_3 Z_4 Z_6 + Z_1 Z_2 Z_3 Z_4 Z_5 \end{array}}$$

$$Z_T = \frac{Z^6}{6Z^5} = \frac{Z}{6} = \frac{8.166 \,\lfloor 78.7^\circ}{6} = 1.361 \,\lfloor 78.7^\circ$$

$$Z_T = 0.266 + j1.335$$

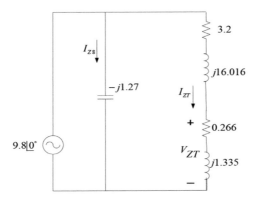

The open circuit value of I_{ZT} is

$$I_{ZT} = \frac{9.8\,\underline{|0}^\circ}{0.266 + j1.335 + 3.2 + j16.016} = 0.554\,\underline{|-78.7}^\circ \qquad (17.8)$$

We now obtain the value of V_{ZT} by multiplying I_{ZT} by the impedance Z_T

$$V_{ZT} = 0.554\,\underline{|-78.7}^\circ \times 1.361\,\underline{|78.7}^\circ = 0.748\,\underline{|0}^\circ \qquad (17.9)$$

Therfore,

$$I_{Z1} = \frac{0.748\,\underline{|0}^\circ}{8.166\,\underline{|78.7}^\circ} = 0.092\,\underline{|-78.7}^\circ \qquad (17.10)$$

$$I_{Z1} = I_{Z2} = I_{Z3} = I_{Z4} = I_{Z5} = I_{Z6} \qquad (17.11)$$

We can describe the circuit in terms of the worst case because it contains the lowest current value flow in each secondary coil.

17.4.4 Configuration 4

The circuit in configuration 4 (Figure 17.16) can be simplified by series–parallel reductions. The impedance of the (Z_1, Z_2 and Z_3) branch is

$$Z_{T1} = (Z_1 + Z_2 + Z_3) \qquad (17.12)$$

$$Z_{T1} = 8.166\,\underline{|78.7}^\circ \times 3 = 24.498\,\underline{|78.7}^\circ$$
$$Z_{T1} = 4.8 + j24.023$$

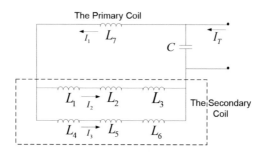

The Primary Coil

I_1 L_7

C

I_T

L_1 I_2 L_2 L_3 The Secondary Coil

L_4 I_3 L_5 L_6

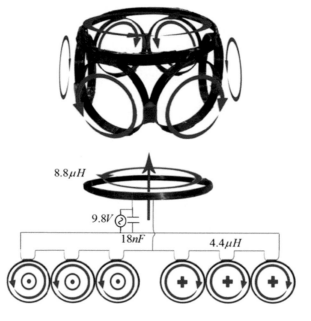

$8.8\mu H$

$9.8V$

$18nF$ $4.4\mu H$

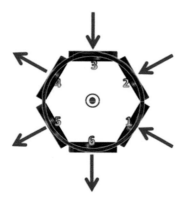

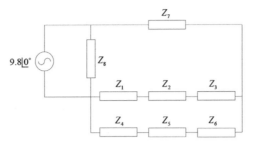

Figure 17.16 Configuration 4.

The impedance of the (Z_4, Z_5 and Z_6) branch is

$$Z_{T2} = (Z_4 + Z_5 + Z_6)$$
$$Z_{T2} = 4.8 + j24.023 \qquad (17.13)$$
$$Z_{T2} = 4.8 + j24.023$$

Note that both branches are in parallel with each other. Therefore, we may replace these two branches with a single branch having an impedance of

$$Z_T = (Z_{T1} // Z_{T2})$$
$$Z_T = \frac{4.8 + j24.023}{2}$$
$$Z_T = 2.4 + j12.012 \qquad (17.14)$$
$$Z_T = \frac{4.8 + j24.023}{2}$$
$$Z_T = 2.4 + j12.012$$

Combining this impedance with the impedance of the primary coil reduces the circuit to the one shown below.

From Figure 17.17, we can find the current

$$I_1 = \frac{9.8\lfloor 0^\circ}{2.4 + j12.012 + 3.2 + j16.016} = \frac{9.8\lfloor 0^\circ}{28.58\lfloor 78.7^\circ}$$
$$I_1 = 0.343 \lfloor -78.7^\circ \qquad (17.15)$$

By knowing the current I_1, we can now work back through the equivalent circuits to find the branch currents in the original circuit by calculating the voltage V_{ZT}.

$$V_{ZT} = 0.343 \lfloor -78.7^\circ \times 12.249 \lfloor 78.7^\circ$$
$$V_{ZT} = 4.2 \lfloor 0^\circ \qquad (17.16)$$

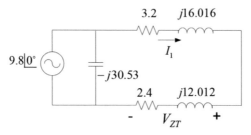

Figure 17.17   Simplified circuit for configuration 4.

We now calculate the current flow through secondary coils

$$I_{Z(1,2,3)} = \frac{V_{ZT}}{Z_{T1}} = \frac{4.2|0^\circ}{24.498\lfloor 78.7^\circ}$$

$$I_{Z(1,2,3)} = 0.171\lfloor -78.7^\circ \qquad\qquad (17.17)$$

$$I_{Z(1,2,3)} = I_{Z(4,5,6)}$$

17.4.5 Configuration 5 (The Simple Coil)

In this configuration (Figure 17.18), we test a conventional coil that has the same parameter as the primary coil to compare it with the MDC. Directly, we can calculate the current flow through it as follows

$$I_T = \frac{V_S}{Z} = \frac{9.8\lfloor 0^\circ}{16.33\lfloor 78.7^\circ}$$

$$I_T = 0.6\lfloor -78.7^\circ \qquad\qquad (17.18)$$

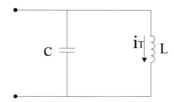

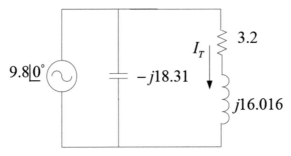

Figure 17.18 Configuration 5.

17.4.6 Configuration 6

In similar fashion to configuration 5, it can be shown that from Figure 17.19,

$$I_{Z(1,2,3)} = \frac{V_S}{Z_{T1}} = \frac{9.8\,\lfloor 0°}{24.498\,\lfloor 78.7°}$$

$$I_{Z(1,2,3)} = 0.4\,\lfloor{-78.7°}$$

$$I_{Z(1,2,3)} = I_{Z(4,5,6)} \tag{17.19}$$

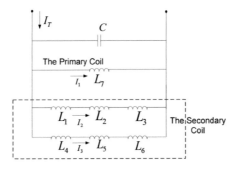

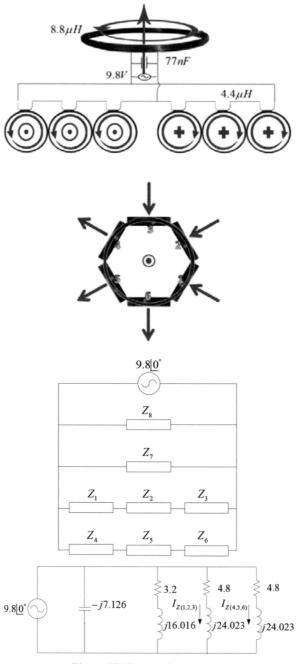

Figure 17.19   Configuration 6.

17.4.7 Configuration 7

In Figure 17.20, the impedances (Z_4, Z_5 and Z_6) connected in parallel (Figure 17.20) can be reduced to a single equivalent impedance (Z_{T1}) as follows

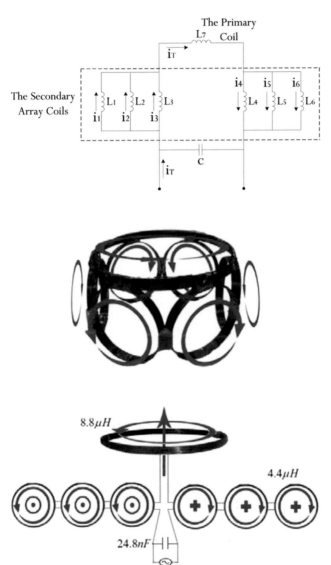

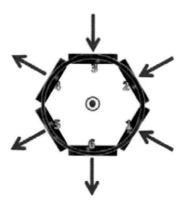

Figure 17.20   Configuration 7.

$$Z_{T1} = Z_{T(4,5,6)} = \frac{Z_4 Z_5 Z_6}{Z_5 Z_6 + Z_4 Z_6 + Z_4 Z_5}$$

$$Z_{T1} = \frac{Z^3}{3Z^2} = \frac{Z}{3} = \frac{1.6 + j8.008}{3}$$

$$Z_{T1} = 2.722\,\lfloor 78.7^\circ$$

$$Z_{T1} = 0.533 + j2.669 \tag{17.20}$$

We may now add this impedance to the impedance of the primary coil (Z_7) as follows:

$$Z_{T2} = (Z_{T1} + Z_7)$$
$$Z_{T2} = 3.2 + j16.016 + 0.533 + j2.669$$
$$Z_{T2} = 3.733 + j18.685 \tag{17.21}$$
$$Z_{T2} = 19.05\,\lfloor 78.7^\circ$$

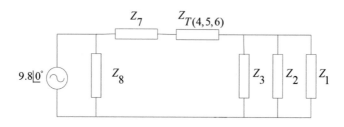

We can replace the parallel combination of the voltage source $(9.8\lfloor 0^{\circ})$ and the impedance of $(-j22.15)$ with the series combination of a current source and the $(3.733-j3.467)$ impedance of Figure 17.21. The source current is

$$I_s = \frac{9.8\lfloor 0^{\circ}}{3.733+j18.685}$$

$$I_s = 0.44\lfloor 90^{\circ} \qquad (17.22)$$

Thus, we can modify the circuit to the one shown below

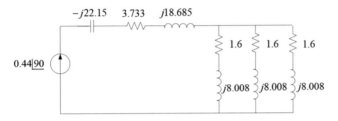

We have reduced the circuit to a simple parallel circuit as shown below
Now we can easily compute the current of each branch which introduces the current flow through secondary coils as follows

$$I_{L1} = \frac{V_s}{Z_1} = \frac{2.23\lfloor 47.12^{\circ}}{8.166\lfloor 78.7^{\circ}}$$

$$I_{L1} = 0.27\lfloor -31.58^{\circ} \qquad (17.23)$$

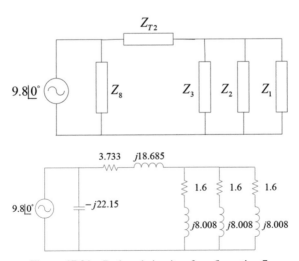

Figure 17.21 Reduced circuits of configuration 7.

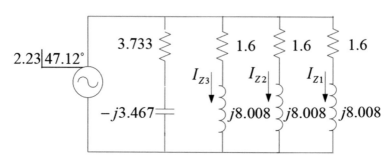

$$I_{Z1} = I_{Z2} = I_{Z3}$$
$$I_{Z1} = I_{Z2} = I_{Z3} = I_{Z4} = I_{Z5} = I_{Z6}$$ (17.24)

17.4.8 Configuration 8

At frequency $w = 1.82$ M rad/s the impedance of the capacitor in Figure 17.22 is

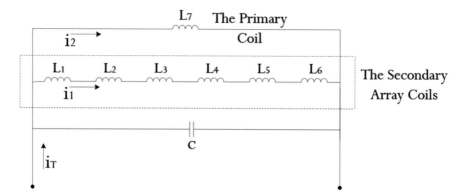

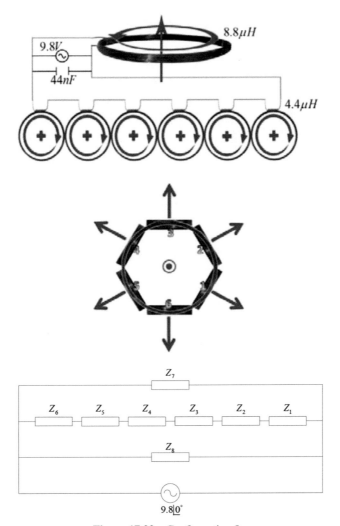

Figure 17.22 Configuration 8.

$$Z_g = \frac{1}{jwc} = \frac{10^9}{18 \times j1.82 \times 10^6} = -j30.53 \qquad (17.25)$$

The impedance of the 8.8 μH inductor is

$$jwL_7 = j1.82 \times 10^6 \times 8.8 \times 10^{-6} = j16.016$$

$$jwL_{(1,2,3,4,5,6)} = j8.008 \qquad (17.26)$$

Summing the impedances (Z_1) through (Z_6) yields

$$Z_1 = Z_2 = Z_3 = Z_4 = Z_5 = Z_6 = 1.6 + j8.008$$

$$Z_T = Z_1 + Z_2 + Z_3 + Z_4 + Z_5 + Z_6$$

$$Z_T = 9.6 + j48.048$$

$$Z_T = 48.99 \underline{|\,78.7^\circ} \tag{17.27}$$

Hence, the branch current is

$$I_T = \frac{9.8 \underline{|\,0^\circ}}{48.99 \underline{|\,78.7^\circ}} = 0.2 \underline{|-78.7^\circ}$$

$$I_T = I_{Z1} = I_{Z2} = I_{Z3} = I_{Z4} = I_{Z5} = I_{Z6} \tag{17.28}$$

Figure 17.23 shows eight curves plotted according to Table 17.3 in 3D. The curves obtained by different connections indicate the voltage received by the

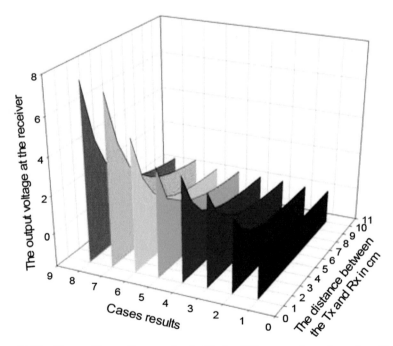

Figure 17.23   The resulting voltage received at Rx for different configurations plotted in 3D.

Table 17.3 Voltage received at RX for different cases of connection

cm	Vo at RX case 1	Vo at RX case 2	Vo at RX case 3	Vo at RX case 4	Vo at RX case 5	Vo at RX case 6	Vo at RX case 7	Vo at RX case 8
1	1	2.2	3.2	3.8	4	5.2	7.2	7.6
2	0.48	1.15	1.8	2.1	2	2.8	4.4	4.4
3	0.28	0.66	1.2	1.2	1.5	1.6	2.8	2.7
4	0.18	0.4	0.8	0.8	1	1	2	1.75
5	0.1	0.27	0.56	0.56	0.78	0.64	1.45	1.25
6	0.08	0.18	0.39	0.4	0.56	0.44	1.08	0.9
7	0.06	0.14	0.3	0.3	0.42	0.32	0.84	0.7
8	0.044	0.1	0.23	0.22	0.32	0.24	0.68	0.54
9	0.036	0.08	0.17	0.2	0.26	0.19	0.54	0.42
10	0.028	0.06	0.145	0.15	0.22	0.15	0.46	0.35

MDC receiver coil at certain distances from MDC transmitter which were classified into eight cases. The maximum distance approaches 10cm away from the transmitter. The results strengthen the belief that current must be chosen carefully for efficient wireless power transfer.

The smallest amount of voltage received describes case number one shown in the violet colour, and it increases with respect to the current changes of the remaining cases. It should be noted that the normal coils case (case number 5) indicates the average value of all measurements in which the other cases evaluated. These results indicate that the flux strength for the MDC network can be higher or less compared with the conventional coil. If the maximum desired power of an application corresponding to maximum magnetic field range is known, we can compute the required power transfer for a shorter magnetic field range. For example, if the maximum far field range of MDC transmitter is 10 cm and if the mid-field range of 5 cm is required for MDC receivers to receive the power accurately, the required power can be delivered with high efficiency by choosing the appropriate connection case. This estimation may not be very useful for mobile applications due to orientation, but it demonstrates that wireless power control is an applicable option.

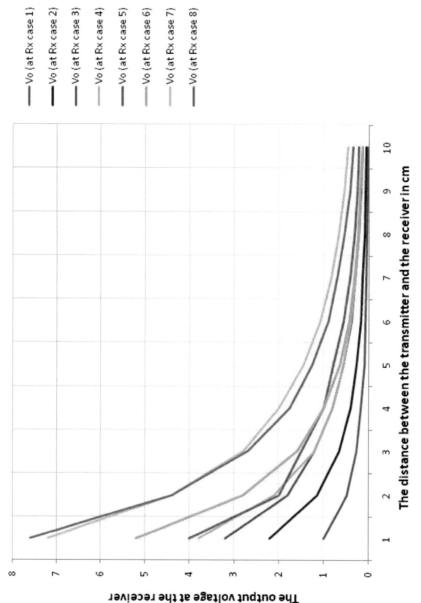

Figure 17.24   The resulting voltage received at RX for different cases of connection.

17.5 Conclusion

This chapter has discussed multidimensional wireless power transfer system to accomplish more efficient performance than conventional coils. It has shown that the configurations of secondary and primary nodes are the major determinants for the overall efficiency of the 3D power system as shown in Figure 17.23 and Table 17.3.

It is shown the electromagnetic field pattern of the design and how it can improve the efficiency depends on the orientation. For the presented connection types, the MDC network system achieves multiple choice efficiencies which depend on the desired current and voltage.

References

[1] J.I. Agbinya, J.I. "A New Framework for Wide Area Networking of Inductive Internet of Things". *Electronics Letters* 47, 21: 1199–1201 (2011).

[2] Hideaki, A., T. Toyohiko, O. Kiyoshi, H. Satoshi, M. Nobuhiro, K. Hiroyasu, and Y. Satoshi. "Equivalent Circuit of Wireless Power Transmission with Coil Array Structures." In *Proceedings of IMWS-IWPT2012*, pp. 115–118 (2012).

[3] RamRakhyani, A.K., and G. Lazzi. "Use of Multi-Coil Telemetry System for High Tolerance Efficient Wireless Power System." In *IEEE International Conference on Wireless Information Technology and Systems*, Hawaii (2012).

[4] Artan, N.S., R.C. Patel, N. Chengzhi, and H.J. Chao. "High-Efficiency Wireless Power Delivery for Medical Implants Using Hybrid Coils." In *2012 IEEE Annual Engineering in Medicine and Biology Society Conference (EMBC)*, pp. 1683–1686, September (2012).

[5] Atluri, S., and M. Ghovanloo. "Design of a Wide-Band Power-Efficient Inductive Wireless Link for Implantable Biomedical Devices Using Multiple Carriers." In *Proceedings of 2nd International IEEE/EMBS Conference Neural Engineering*, pp. 533–537.

[6] Harrison, R.R. "Designing Efficient Inductive Power Links for Implantable Devices." In *Proceedings of 2007 IEEE International Symposium on Circuits and Systems (ISCAS 2007)*, New Orleans, LA, pp. 2080–2083 (2007).

[7] Seon-Woo, L., K. Jong-Dae, S. Ju-Hyun, R. Mun-Ho, and K. Jongwon. "Design of Two-Dimensional Coils for Wireless Power Transmission

to In Vivo Robotic Capsule." In *2005 IEEE 27th Annual Engineering in Medicine and Biology Society Conference (EMBC)*, pp. 1683–1686, September (2005).

[8] Kilinc, E.G., G. Conus, C. Weber, B. Kawkabani, F. Maloberti, and C. Dehollain. "A System for Wireless Power Transfer of Micro-Systems In-Vivo Implantable in Freely Moving Animals." *IEEE Sensors Journal* 14: pp. 522–531 (2014).

[9] Han, S., and D.D. Wentzloff. "Wireless Power Transfer Using Resonant Inductive Coupling for 3D Integrated ICs." In *IEEE 3D System Integration Conference*, pp. 1–5, November (2010).

[10] JinWook, K., S. Hyeon-Chang, K. Do-Hyeon, Y. Jong-Ryul, K. Kwan-Ho, L. Ki-Min, and P. Young-Jin. "Wireless Power Transfer for Free Positioning Using Compact Planar Multiple Self-Resonators." In *IEEE MTT-S International Microwave Workshop Series on Innovative Wireless Power Transmission: Technologies, Systems, and Applications (IMWS)*, pp. 127–130, May (2012).

[11] Jonah, O., S.V. Georgakopoulos, and M.M. Tentzeris. "Multi-Band Wireless Power Transfer via Resonance Magnetic." In *2013 IEEE International Symposium Antennas and Propagation (APSURSI)*, pp. 850–851, July (2013).

[12] Jow, U.M. and M. Ghovanloo. "Design and Optimization of Printed Spiral Coils for Efficient Transcutaneous Inductive Power Transmission." In *IEEE Transactions on Biomedical Circuits and Systems*, 1, pp. 193–202, September (2007).

[13] Koh, K.E., T.C. Beh, T. Imara, and Y. Hori. "Multi-Receiver and Repeater Wireless Power Transfer via Magnetic Resonance Coupling—Impedance Matching and Power Division Utilizing Impedance Inverter." In *15th International Conference on Electrical Machines and Systems (ICEMS)*, 21–24: 1–6 (2012).

[14] Cannon, B.L., F.H. James, D.S. Daniel, and C.G Seth. "Magnetic Resonant Coupling as a Potential Means for Wireless Power Transfer to Multiple Small Receivers." *IEEE Transactions on Magnetics*. 24: 1819–1825 (2009).

[15] Casanova, J.J., Z.N. Low, and J. Lin. "A Loosely Coupled Planar Wireless Power System for Multiple Receivers." *IEEE Transactions on Industrial Electronics* 56: 3060–3068 (2009).

[16] Noriaki, O., O. Kenichiro, K. Hiroki, S. Hiroki, and O. Shuichi. "Efficiency Improvement of Wireless Power Transfer via Magnetic Resonance

Using Transmission Coil Array." In *2011 IEEE International Symposium Antennas and Propagation (APSURSI)*, pp. 1707–1710, July (2011).

[17] Wang, B., W. Yerazunis, and K.H. Teo. "Wireless Power Transfer: Metamaterials and Array of Coupled Resonators." In *Proceedings of the IEEE*, vol. 101, pp. 1359–1368, June (2013).

[18] Lim, H., K. Ishida, M. Takamiya, and T. Sakurai. "Positioning-Free Magnetically Resonant Wireless Power Transmission Board with Staggered Repeater Coil Array (SRCA)." In *IEEE MTT-S International Microwave Workshop Series on Innovative Wireless Power Transmission: Technologies, Systems, Applications (IMWS-IWPT)*, pp. 93–96, May (2012).

[19] Hatanaka, K., F. Sato, H. Matsuki, S. Kikuchi, J. Murakami, M. Kawase, and T. Satoh. "Power Transmission of a Desk with a Cord-Free Power Supply." *IEEE Transactions on Magnetics* 38: 3329–3331 (2002).

[20] Wang, B., D. Ellstein, and K.H. Teo. "Analysis on Wireless Power Transfer to Moving Devices Based on Array of Resonators." *IEEE European Conference on Antennas and Propagation (EUCAP)*, pp. 964–967, March (2012).

[21] Koh, K.E., T.C. Beh, T. Imura, and Y. Hori. "Novel Band-Pass Filter Model for Multi-Receiver Wireless Power Transfer via Magnetic Resonance Coupling and Power Division." In *2012 IEEE 13th Annual Wireless and Microwave Technology Conference (WAMICON)*, pp. 1–6, April (2012).

[22] Nagi, F.A.M., and I.A. Johnson. "Design of Multi-Dimensional Wireless Power Transfer Systems." In *Pan African International Conference on Information Science, Computer and Telecommunication*, pp. 85–91 (2013).

[23] Seungyoung, A., H.P. Hyun, C. Cheol-Seung, K. Jonghoon, S. Eakhwan, B.P. Hark, K. Hongseok, and K. Joungho. "Reduction of Electromagnetic Field (EMF) of Wireless Power Transfer System Using Quadruple Coil for Laptop Applications." In *IEEE MTT-S International Microwave Workshop Series on Innovative Wireless Power Transmission*, pp. 65–68, May (2012).

18

Split Frequencies in Magnetic Induction Systems

Hoang Nguyen[1] and Johnson I. Agbinya

[1]Department of Electronic Engineering, La Trobe University,
Melbourne, Australia

18.1 Introduction

This chapter studies splitting frequency in magnetic coupled resonant systems.
The splitting frequency response in several antenna schemes is presented.
Magnetic resonant coupling system based on strong coupling technique [1]
has recently opened a new research in the wireless power transfer application.
Unlike typical wireless inductive coupling, strong resonant coupling between
two identical coils can allow for more transmission distance and freedom
of position between the transmitter and receiver as well as enhancement
of the performance in the power consumption. Many companies (WiPower,
Powermat, etc.) use this technique to develop wirelessly charging electronic
products such as batteries, smart mobile chargers, and consumer electronics.
Required power of the products can be categorized in three levels as small,
medium, and high power [2]. Two important aspects in the operation of the
wireless power transfer are required to get maximum power transfer and
maximum system efficiency [3].

In the wireless power transfer application, coupling is dependent on
distance between the transmitter and receiver coil. Degree of the coupling
defines the rate of the energy transfer from the transmitter to receiver.
However, the amount of the power delivered to the load not only begins to
fall off with a long distance, but also repeatedly drops at a close distance
and a certain degree of strong couplings [4]. The resultant reductions of
the power in a close distance are dueto splitting phenomenon between

the transmitter and receiver [5]. When the transmitter and receiver are placed closer to where the coupling condition is greater than the critical coupling, the single resonant frequency peak at the receiver load of the receiver quickly changes to double peaks. Therefore, splitting phenomenon negatively impacts the power transfer efficiency and transferring ability. The variation of the gap between the transmitter and receiver leads to the changes of splitting frequencies.

Some works have reported resonant frequency splitting phenomena in magnetic resonant wireless power transfer [3, 4]. The studies on splitting frequency are conducted for two coils [4, 6], three coils [7], four coils [8, 9], and multiple coils [10, 11]. Two concepts coupled mode theory [12–14] or circuit theory [8] are used to analyze the splitting frequency. In coupled coils, comprehensive determination of the splitting frequency is derived using derivative of power transfer with respect to the operation frequency [5, 14]. The relationship between frequency splitting and the splitting point is also well clarified for coupled resonators and low Q factor. Expression of the splitting frequency can be conducted in the voltage ratio, the output power and efficiency [6]. It is shown that splitting frequency can be observed in power transfer analysis. In four-coil resonators, splitting frequency decreases efficiency of the system within a certain distance [9]. Solution to increase the efficiency by reducing the source internal resistance and increasing the mutual inductance of the source coil and the sending coil as well as the mutual inductance of the load coil and the receiving coil, is demonstrated.

The following section will explain the splitting frequency and the power decrease in different antenna configurations. The introduction and method for defining the splitting frequency are to be provided.

18.2 Single Transmitter–Receiver

Equivalent circuit of the two-coil resonant structure is demonstrated in Figure 18.1. The source and destination coils with radii r_1 and r_2 are coupled wirelessly in the distance x. This system can be constructed by the lumped circuit theory using R, L, and C elements. In the circuit, L_1 and L_2 are self-inductances of the two coils, and C_1 and C_2 are capacitors in the first and second circuits, respectively. The internal resistances of the two coils and other losses in the circuit are presented by R_{L1} and R_{L2}. The source and the load resistances are designated R_S and R_L, respectively.

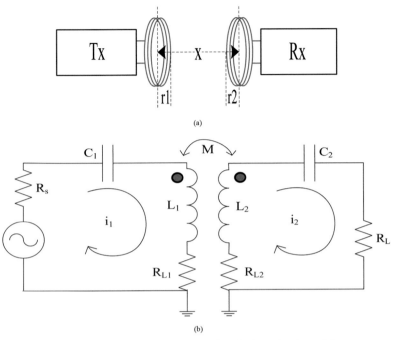

(a)

(b)

Figure 18.1 Single resonator and equivalent circuit model.

Based on the Kirchhoff's voltage law, the circuit at the first and second loops is expressed as the following equation:

$$(R_s + R_{L1} + j\omega L_1 + \frac{1}{j\omega C_1}) I_{Tx} + j\omega M_{12} I_{Rx} = V_{Tx}$$

$$j\omega M_{12} I_{Tx} + (R_L + R_{L2} + j\omega L_2 + \frac{1}{j\omega C_2}) I_{Rx} = 0 \qquad (18.1)$$

where V_{Tx} is the input voltage. The currents induced in the first and second loops are I_{Tx} and I_{Rx}, respectively. The mutual inductance (M_{12}) between two coils is calculated byself-inductance of the coils and coupling coefficient k_{12} as the following equation:

$$M_{12} = k_{12} \sqrt{L_1 L_2} \qquad (18.2)$$

It is seen from Equation (18.2) that the coupling coefficient (k_{12}) measures the strength of the fluxes produced from the transmitter to the receiver, and vice versa.

We simplify Equation (18.1) into the matrix formulas:

$$\begin{bmatrix} Z_{Tx} & j\omega M_{12} \\ j\omega M_{12} & Z_{Rx} \end{bmatrix} * \begin{bmatrix} I_{Tx} \\ I_{Rx} \end{bmatrix} = \begin{bmatrix} V_{Tx} \\ 0 \end{bmatrix} \tag{18.3}$$

where the impedances of the transmitter and receiver are $Z_{Tx} = R_s + R_{L1} + j\omega L_1 + \frac{1}{j\omega C_1}$ and $Z_{Rx} = R_L + R_{L2} + j\omega L_2 + \frac{1}{j\omega C_2}$, respectively. With matrix $A = \begin{bmatrix} Z_{Tx} & j\omega M_{12} \\ j\omega M_{12} & Z_{Rx} \end{bmatrix}$, the current of each loop in Equation (18.4) can be calculated as the following equation:

$$I_i = A^{-1}V = \frac{VA^T}{\det(A)} \quad (i = Tx \text{ or } Rx) \tag{18.4}$$

Therefore, the power transfer between the transmitter and receiver can be derived by the current ratio as follows:

$$\frac{P_{Rx}}{P_{Tx}} = \left| \frac{I_{Rx}}{I_{Tx}} \right|^2 \frac{R_L}{R_S} \tag{18.5}$$

At resonant frequency $\omega_o = \frac{1}{\sqrt{L_1 C_1}} = \frac{1}{\sqrt{L_2 C_2}}$ and loosely coupled, the power transfer is proportional to the efficiency (η_{Tx} and η_{Rx}), quality factor (Q_{Tx} and Q_{Rx}) of the first and second circuits, and coupling coefficient k_{12} as the following equation:

$$\frac{P_{Rx}}{P_{Tx}} = k_{12}^2 \eta_{Tx} \eta_{Rx} Q_{Tx} Q_{Rx} \tag{18.6}$$

where

$$\eta_{Tx} = \frac{R_s}{R_s + R_{L1}} \text{ and } \eta_{Rx} = \frac{R_L}{R_L + R_{L2}}$$

$$Q_{Tx} = \frac{\omega_o L_1}{R_s + R_{L1}} \text{ and } Q_{Rx} = \frac{\omega_o L_2}{R_L + R_{L2}}$$

It is observed from Equation (18.6) that the received power can be improved by the increase of the quality factors. However, as long as circuits have high quality factors, the performance of the power transfer and the efficiency has considerably degraded by the change of the coupling coefficient. To analyze this effect, the power transfer in Equation (18.5) is simulated with both inductors (L_1 and L_2) of 6.3 μH, both capacitors (C_1 and C_2) of 21.9 pF, and source and load resistances of 50 Ω. The internal resistances of the transmitter and receiver are small and negligible. Figure 18.2 shows the curvature of the received power with respect to the change of the frequency and the coupling coefficient k_{12}. It can be seen that when the coupling between the transmitter and receiver is low, the curve maximizes at one peak frequency. However, when the transmitter strongly couples to the receiver, the curve splits into two peaks. Furthermore, the separated gap between two peaks is proportional to the k_{12} increase. The splitting point, which the curve starts to split, is defined by the critical coupling coefficient point as $k_{\text{critical}} = \frac{1}{\sqrt{Q_1 Q_2}}$ [5].

The double peaks in Figure 18.2 refer to two splitting frequencies known as the odd and even frequencies [5, 14]. At a given gap between the transmitter and receiver, the curve of the received power maximizes at two splitting frequencies. If the gap is changed (as $k_{12} = 0.3$, 0.5 and 0.7), the curve will shift to another two splitting frequencies within the strong coupling area.

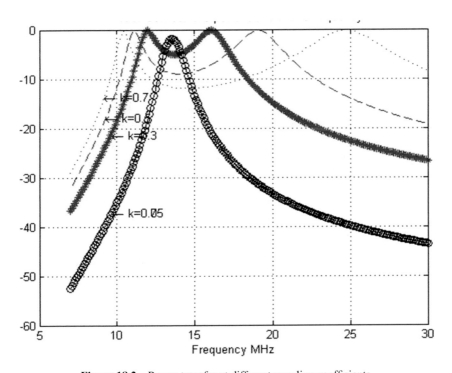

Figure 18.2   Power transfer at different coupling coefficients.

18.3 Determination of Splitting Frequency

Determining the split frequencies requires analysis of the system power transfer matrix using eigenvalue solutions. This starts with defining the system equations. To define splitting frequency, Equation (18.1) is modified as the following expression:

$$[\frac{R_s + R_{L1}}{j\omega L_1} + \left(1 - \frac{\omega_o^2}{\omega^2}\right)] I_{Tx} + k_{12}\sqrt{\frac{L_1}{L_2}} I_{Rx} = \frac{V_{Tx}}{j\omega L_1}$$

$$j\omega M I_{Tx} + [\frac{R_L + R_{L2}}{j\omega L_2} + \left(1 - \frac{\omega_o^2}{\omega^2}\right)] I_{Rx} = 0 \qquad (18.7)$$

If the impedances of the each circuit are $Z'_{Tx} = [\frac{R_s + R_{L1}}{j\omega L_1} + \left(1 - \frac{\omega_o^2}{\omega^2}\right)]$
and $Z'_{Rx} = [\frac{R_L + R_{L2}}{j\omega L_2} + \left(1 - \frac{\omega_o^2}{\omega^2}\right)]$ and the inductor factor $a_1 = \sqrt{\frac{L_1}{L_2}}$,
Equation (18.7) can be presented into the matrix form.

$$\begin{bmatrix} Z'_{Tx} & k_{12}a_1 \\ k_{12}a_1^{-1} & Z'_{Rx} \end{bmatrix} * \begin{bmatrix} I_{Tx} \\ I_{Rx} \end{bmatrix} = \begin{bmatrix} \frac{V_{Tx}}{j\omega L_1} \\ 0 \end{bmatrix} \qquad (18.8)$$

Under strong coupling and high-Q coil of the transmitter and receiver, Equation (18.8) can be simplified as:

$$\begin{bmatrix} 1 & k_{12}a_1 \\ k_{12}a_1^{-1} & 1 \end{bmatrix} \begin{bmatrix} I_{Tx} \\ I_{Rx} \end{bmatrix} = \frac{\omega_o^2}{\omega^2} \begin{bmatrix} I_{Tx} \\ I_{Rx} \end{bmatrix} \qquad (18.9)$$

Solving Equation (18.9), we can obtain the splitting frequencies of the system as $\omega_1 = \frac{\omega_0}{\sqrt{1 + k_{12}}}$ and $\omega_2 = \frac{\omega_0}{\sqrt{1 - k_{12}}}$.

18.3.1 Single Transmitter and Multiple Receiver Configuration

Figure 18.3 represents the structure of power transmission from a transmitter to N receivers. The voltage source V_{Tx} of the transmitter excited the energy to all the receivers with an individual coupling coefficient as k_{Ti} (i is the number of the receiver ranging from 1 to N). The cross-coupling coefficients between two receivers are named as k_{ij} (i and j are the decimal numbers ranging from 1 to N receiver, and $i \neq j$).

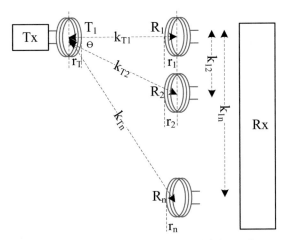

Figure 18.3 Single transmitter and multiple receivers.

Similar to the single transmitter–receiver circuit, the KVL can be normalized as the following matrix:

$$
\begin{bmatrix}
Z'_{Tx} & k_{T1}a_1 & k_{T2}a_2 & \cdots & k_{Tn}a_n \\
k_{T1}a_1^{-1} & Z'_1 & k_{12}a_{12} & \cdots & k_{1n}a_{1n} \\
k_{T2}a_2^{-1} & k_{12}a_{12}^{-1} & Z'_2 & \cdots & k_{2n}a_{2n} \\
\vdots & \vdots & \vdots & 1 & \vdots \\
k_{Tn}a_n^{-1} & k_{1n}a_{1n}^{-1} & k_{2n}a_{2n}^{-1} & \cdots & Z'_n
\end{bmatrix}
\begin{bmatrix}
I_{Tx} \\
I_1 \\
I_2 \\
\cdots \\
I_n
\end{bmatrix}
=
\begin{bmatrix}
\dfrac{V_{Tx}}{j\omega L_{Tx}} \\
0 \\
0 \\
\cdots \\
0
\end{bmatrix}
$$

$$(18.10)$$

where the circuit impedances are $Z'_{Tx} = \dfrac{R_{Tx}}{j\omega L_{Tx}} + 1 - \dfrac{\omega_{Tx}^2}{\omega^2}$, $Z'_i = \dfrac{R_i}{j\omega L_i} + 1 -$

$\dfrac{\omega_1^2}{\omega^2}$ ($i = 1 : N$). The inductor factors are defined as $a_i = \sqrt{\dfrac{L_i}{L_{Tx}}}$ ($i = 1 : N$)

and $a_{ij} = \sqrt{\dfrac{L_i}{L_j}}$ ($i, j = 1 : N, i \neq j$. Under strong coupling and high Q, the

circuit in Equation (18.10) can be transformed as the following equation:

$$
\begin{bmatrix}
1 & k_{T1}a_1 & k_{T2}a_2 & \cdots & k_{Tn}a_n \\
k_{T1}a_1^{-1} & 1 & k_{12}a_{12} & \cdots & k_{1n}a_{1n} \\
k_{T2}a_2^{-1} & k_{12}a_{12}^{-1} & 1 & \cdots & k_{2n}a_{2n} \\
\vdots & \vdots & \vdots & 1 & \vdots \\
k_{Tn}a_n^{-1} & k_{1n}a_{1n}^{-1} & k_{2n}a_{2n}^{-1} & \cdots & 1
\end{bmatrix}
*
\begin{bmatrix}
I_{Tx} \\
I_1 \\
I_2 \\
\cdots \\
I_n
\end{bmatrix}
=
\dfrac{\omega_o^2}{\omega^2}
*
\begin{bmatrix}
I_{Tx} \\
I_1 \\
I_2 \\
\cdots \\
I_n
\end{bmatrix}
$$

$$(18.11)$$

It is shown that the coupling coefficient is the only variable parameter in Equation (18.11). Thus, Equation (18.11) shows the relationship between the frequency response and the coupling coefficients within the system. In other words, it demonstrates the effect of coupling coefficients to the frequency response of the circuit. Therefore, this normalized formula can also predict the splitting frequency for the operating inductive multiple resonators. The next section is categorized in several antenna structures.

18.3.2 A Transmitter and Two Receivers (SI2O)

Using Equation (18.11), the frequency response of SI2O can be obtained as the following formula:

$$\begin{bmatrix} 1 & k_{T1}a_1 & k_{T2}a_2 \\ k_{T1}a_1^{-1} & 1 & k_{12}a_{12} \\ k_{T2}a_2^{-1} & k_{12}a_{12}^{-1} & 1 \end{bmatrix} * \begin{bmatrix} I_{Tx} \\ I_1 \\ I_2 \end{bmatrix} = \frac{\omega_o^2}{\omega^2} \begin{bmatrix} I_{Tx} \\ I_1 \\ I_2 \end{bmatrix} \qquad (18.12)$$

In this matrix, the coupling coefficients from the transmitter to receiver #1 and #2 are k_{T1} and k_{T2}, respectively. The cross-coupling coefficient between the receivers is k_{12}. By solving for the value of the factor $\frac{\omega_0}{\omega}$, Equation (18.12) can give maximum of three eigenvalues corresponding to the three splitting frequencies of the system. However, these values depend on the natural values of coupling coefficients. Thus, we divide the effects of the splitting frequency into two situations without cross-coupling between the receivers and with cross-coupling between the receivers. In the following analysis, we define different splitting cases.

18.3.2.1 Without cross-coupling between receivers
Case 1: When a transmitter is equally coupled to two receivers $k_{T1} = k_{T2}$, three splitting frequencies are obtained from Equation (18.12) as follows:

$$\omega_1 = \omega_o, \ \omega_2 = \frac{\omega_o}{\sqrt{k\sqrt{2}+1}} \ \text{and} \ \omega_3 = \frac{\omega_o}{\sqrt{1-k\sqrt{2}}} \qquad (18.13)$$

Case 2: When the transmitter is coupled unequally to receivers $k_{T1} \neq k_{T2}$, another three frequencies are also provided as follows:

$$\omega_1 = \omega_o, \ \omega_2 = \frac{\omega_o}{\sqrt{1+\sqrt{k_{T1}^2+k_{T2}^2}}} \ \text{and} \ \omega_3 = \frac{\omega_o}{\sqrt{1-\sqrt{k_{T1}^2+k_{T2}^2}}}$$

$$(18.14)$$

Equation (18.14) shows the same results as the work in [15]. However, the splitting frequency is introduced in a different antenna model.

18.3.2.2 With cross-coupling between receivers

Case 1: when all coupling coefficients are constant $k_{T1} = k_{T2} = k_{12} = k$, there are two splitting frequencies as:

$$\omega_1 = \frac{\omega_o}{\sqrt{1 + 2k}} \text{ and } \omega_2 = \frac{\omega_o}{\sqrt{1 - k}} \tag{18.15}$$

Case 2: when the cross-coupling between the transmitter is stronger than coupling from transmitter to receivers $k_{12} = n * k_{T1} = n * k_{T2} = n * k$, we have splitting frequencies as:

$$\omega_1 = \frac{\omega_o}{\sqrt{1 - nk}}, \; \omega_2 = \frac{\omega_o}{\sqrt{\frac{n}{2}k - \frac{\sqrt{n^2+8}}{2}k + 1}}$$

$$\text{and } \omega_3 = \frac{\omega_o}{\sqrt{\frac{n}{2}k + \frac{\sqrt{n^2+8}}{2}k + 1}} \tag{18.16}$$

It is observed in Case 2 (without cross-coupling) that the resonant system creates three splitting frequencies instead of two frequencies when the unequalled cross-coupling is transferred from transmitter to receivers. Hence, the presence of cross-coupling leads to asymmetrical splitting of the frequency.

18.3.2.3 Determination of the power transfer for SI2O

The equivalent circuit relating to the current induced is presented by the following matrix:

$$\begin{bmatrix} Z_{Tx} & j\omega M_{T1} & j\omega M_{T2} \\ j\omega M_{T1} & Z_{Rx1} & j\omega M_{12} \\ j\omega M_{T2} & j\omega M_{12} & Z_{Rx2} \end{bmatrix} * \begin{bmatrix} I_{Tx} \\ I_1 \\ I_2 \end{bmatrix} = \begin{bmatrix} V_s \\ 0 \\ 0 \end{bmatrix} \tag{18.17}$$

where I_{Tx}, I_1, and I_2 are the currents induced in the transmitter and two receivers. Impedances of each loop are $Z_{Tx} = R_{Tx} + j\omega L_{Tx} + \frac{1}{j\omega C_{Tx}}$, $Z_{Rx1} = R_1 + j\omega L_1 + \frac{1}{j\omega C_1}$, and $Z_{Rx2} = R_2 + j\omega L_2 + \frac{1}{j\omega C_2}$. M_{T1}, M_{T2}, and M_{12} are the mutual couplings from the transmitter to two receivers and cross-coupling between the receivers. Solving Equation (18.17), the current induced in the transmitter and two receivers are yielded as the following equations:

$$I_1 = -\frac{V_s\omega(M_{12}M_{T2}\omega + jM_{T1}Z_{Rx2})}{\det(A)}I_2$$

$$I_2 = -\frac{V_s\omega(M_{12}M_{T1}\omega + jM_{T2}Z_{Rx1})}{\det(A)} \qquad (18.18)$$

where the determinant of the matrix A is:

$$\det(A) = Z_{Tx}M_{12}^2\omega^2 - 2jM_{T1}M_{T2}M_{12}\omega^3 + Z_{Rx1}M_{T2}^2\omega^2 + Z_{Rx2}M_{T1}^2\omega^2$$
$$+ Z_{Tx}Z_{Rx1}Z_{Rx2}$$

Consequently, the fractional power efficiency at the load of each receiver is derived as follows:

$$\frac{P_1}{P_{Tx}} = \left[-\frac{\omega(M_{12}M_{T2}\omega + jM_{T1}Z_{Rx2})}{\det(A)}\right]^2 R_{Tx}R_1\frac{P_2}{P_{Tx}}$$

$$\frac{P_2}{P_{Tx}} = \left[-\frac{\omega(M_{12}M_{T1}\omega + jM_{T2}Z_{Rx1})}{\det(A)}\right]^2 R_{Tx}R_2 \qquad (18.19)$$

18.4 A Transmitter and Three Receivers (SI3O)

The similar methodology is applied to SI3O. The splitting frequency is also divided into two situations.

18.4.1 Without Cross-Coupling between the Receivers

Case 1 has three frequencies when there are similar coupling coefficients from the transmitter to receivers.

$$\omega_1 = \omega_o,\ \omega_2 = \frac{\omega_o}{\sqrt{k\sqrt{3}+1}} \text{ and } \omega_3 = \frac{\omega_o}{\sqrt{1-k\sqrt{3}}} \qquad (18.20)$$

Case 2: When we have a different coupling $k_{T1} \neq k_{T2} \neq k_{T3}$, there are also three different frequencies.

$$\omega_1 = \omega_o, \quad \omega_2 = \frac{\omega_o}{\sqrt{1 + \sqrt{k_{T1}^2 + k_{T2}^2 + k_{T3}^2}}} \quad \text{and}$$

$$\omega_3 = \frac{\omega_o}{\sqrt{1 - \sqrt{k_{T1}^2 + k_{T2}^2 + k_{T3}^2}}} \tag{18.21}$$

18.4.2 With the Effect of Cross-Coupling between the Receivers

When the entire coupling coefficients are the same $k_{T1} = k_{T2} = k_{T3} = k_{12} = k_{23} = k$, then the system resolves two splitting frequencies as:

$$\omega_1 = \frac{\omega_o}{\sqrt{1 + 3k}} \quad \text{and} \quad \omega_2 = \frac{\omega_o}{\sqrt{1 - k}} \tag{18.22}$$

Thus, cross-coupling between receivers is detrimental to the system and removes one of the required spectral peaks.

18.5 A Transmitter and N Receivers (SIMO)

Normalizing the analysis in N receiver structure, the splitting frequency is also calculated as the following formula:

18.5.1 Without Cross-Couplings between the Receivers

Case 1: When all the receivers are identical, three splitting frequencies are written as follows:

$$\omega_1 = \omega_o, \quad \omega_2 = \frac{\omega_o}{\sqrt{1 + k\sqrt{N}}} \quad \text{and} \quad \omega_3 = \frac{\omega_o}{\sqrt{1 - k\sqrt{N}}} \tag{18.23}$$

Case 2: Different couplings from a transmitter to receivers and the splitting frequencies are given as follows:

$$\omega_1 = \omega_o, \quad \omega_{2,3} = \frac{\omega_o}{\sqrt{1 \pm \sqrt{\sum_i^N k_{Ti}^2}}} \tag{18.24}$$

18.5.2 With the Effect of Cross-Coupling between the Receivers

When all couplings are equal, the splitting frequencies are obtained as follows:

$$\omega_1 = \frac{\omega_o}{\sqrt{1 + N * k}} \text{ and } \omega_2 = \frac{\omega_o}{\sqrt{1 - k}} \tag{18.25}$$

Again we observe that cross-coupling is detrimental to frequency splitting in this case.

18.5.3 Multiple Transmitter and a Receiver Configuration

Figure 18.4 illustrates the structure of N transmitters and one receiver. The direct couplings from N transmitters to the receiver are k_{Ti} ($i = 1:N$), and the cross-couplings between transmitters are k_{ij} ($i = 1:N, j = 1:N$ and $i \neq j$).

Using the same method in SIMO, the KVL equation of MISO is arranged as the following matrix form:

$$\begin{bmatrix} 1 & k_{12}a_{12} & \cdots & k_{1n}a_{1n} & k_{T1}a_1 \\ k_{12}a_{12}^{-1} & 1 & \cdots & k_{2n}a_{2n} & k_{T2}a_2 \\ k_{13}a_{13}^{-1} & k_{23}a_{23}^{-1} & \cdots & k_{3n}a_{3n} & k_{T3}a_3 \\ \vdots & \vdots & \vdots & \vdots & \vdots \\ k_{1n}a_{1n}^{-1} & k_{2n}a_{2n}^{-1} & \cdots & 1 & k_{Tn}a_n \\ k_{T1}a_1^{-1} & k_{T2}a_2^{-1} & \cdots & k_{Tn}a_n^{-1} & 1 \end{bmatrix} * \begin{bmatrix} I_1 \\ I_2 \\ I_3 \\ \cdots \\ I_n \\ I_{Rx} \end{bmatrix} = \frac{\omega_o^2}{\omega^2} * \begin{bmatrix} I_1 \\ I_2 \\ I_3 \\ \cdots \\ I_n \\ I_{Rx} \end{bmatrix}$$

$$\tag{18.26}$$

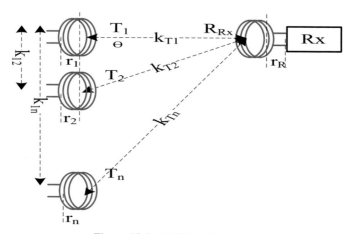

Figure 18.4   MISO configuration.

where the inductor factors are given as $a_i = \sqrt{\dfrac{L_{Rx}}{L_i}}$ $(i = 1 : N)$ and $a_{ij} = \sqrt{\dfrac{L_i}{L_j}}$ $(i, j = 1 : N, i \neq j)$. The value of the splitting frequencies in Equation (18.26) can be solved by the eigenvalues of matrix A ($n \times n$ dimension). It is also shown that Equation (18.26) is similar to Equation (18.11). Therefore, the calculation of the splitting frequency in Equations (18.13–18.16 and 18.20–18.25) can be used for the N transmitters–receiver structure.

18.6 Multiple Transmitters and Multiple Receivers (MIMO)

18.6.1 Determination of Splitting Frequencies (2Tx-2Rx)

The structure of two transmitters and two receivers is illustrated in Figure 18.5, where T_1 and T_2 are two transmitters and R_1 and R_2 are two receivers. The figure also presents the coupling coefficients from transmitters to receivers and cross-couplings between transmitters or between receivers.

Applying the equivalent circuit to the configuration and under high Q, the system can be derived as the matrix equation:

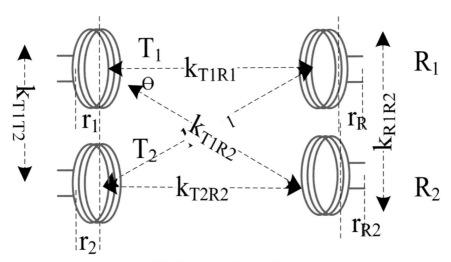

Figure 18.5 Two transmitters and two receivers.

$$\begin{bmatrix} 1 & k_{T1T2}a_{T1T2} & k_{T1R1}a_{T1R1} & k_{T1R2}a_{T1R2} \\ k_{T1T2}a_{T1T2}^{-1} & 1 & k_{T2R1}a_{T2R1} & k_{T2R2}a_{T2R2} \\ k_{T1R1}a_{T1R1}^{-1} & k_{T2R1}a_{T2R1}^{-1} & 1 & k_{R1R2}a_{R1R2} \\ k_{T1R2}a_{T1R2}^{-1} & k_{T2R2}a_{T2R2}^{-1} & k_{R1R2}a_{R1R2}^{-1} & 1 \end{bmatrix} * \begin{bmatrix} I_{T1} \\ I_{T2} \\ I_{R1} \\ I_{R2} \end{bmatrix}$$

$$= \frac{\omega_o^2}{\omega^2} \begin{bmatrix} I_{T1} \\ I_{T2} \\ I_{R1} \\ I_{R2} \end{bmatrix}$$

(18.27)

where the inductor factors stand for $a_{T1T2} = \sqrt{\dfrac{L_{T2}}{L_{T1}}}$, $a_{T1R1} = \sqrt{\dfrac{L_{R1}}{L_{T1}}}$, $a_{T1R2} = \sqrt{\dfrac{L_{R2}}{L_{T1}}}$, $a_{T2R1} = \sqrt{\dfrac{L_{R1}}{L_{T2}}}$, $a_{T2R2} = \sqrt{\dfrac{L_{R2}}{L_{T2}}}$ and $a_{R1R2} = \sqrt{\dfrac{L_{R2}}{L_{R1}}}$. Two situations (with and without cross-coupling) are presented below.

18.6.2 Cross-Couplings Are Ignored

Case 1: When all couplings are similar, Equation (18.24) gives three frequencies as:

$$\omega_1 = \omega_o, \quad \omega_2 = \frac{\omega_o}{\sqrt{1 + 2k}} \quad \text{and} \quad \omega_3 = \frac{\omega_o}{\sqrt{1 - 2k}} \qquad (18.28)$$

Case 2: When there are two different couplings from transmitters to receivers, the split frequencies of the system are calculated as follows:

$$\omega_1 = \omega_o, \quad \omega_2 = \frac{\omega_o}{\sqrt{1 + \sqrt{2}\sqrt{k_1^2 + k_2^2}}} \quad \text{and} \quad \omega_3 = \frac{\omega_o}{\sqrt{1 - \sqrt{2}\sqrt{k_1^2 + k_2^2}}}$$

(18.29)

18.6.3 With Cross-Coupling

Case 1: When all the coupling coefficients are the same, we have two splitting frequencies:

$$\omega_1 = \frac{\omega_o}{\sqrt{1 + 3k}} \quad \text{and} \quad \omega_2 = \frac{\omega_o}{\sqrt{1 - k}} \qquad (18.30)$$

Case 2: When $k_{T1T2} = k_{R1R2} = n * k$, the splitting frequencies are derived as follows:

$$\omega_1 = \frac{\omega_o}{\sqrt{1 - nk}}, \ \omega_2 = \frac{\omega_o}{\sqrt{1 + nk + 2k}} \ \text{and} \ \omega_3 = \frac{\omega_o}{\sqrt{1 + nk - 2k}} \quad (18.31)$$

18.7 Summary

To sum up this chapter, we have presented an overview of the splitting frequency in the wireless power transfer. The explanation of this phenomenon in terms of the power transfer has been illustrated. Three antenna configurations, namely single transmitter–multiple receivers, multiple transmitters–single receiver, and multiple transmitters–multiple receivers, have been classified for deriving eigenvalues and the splitting frequency.

References

[1] Kurs, A., A. Karalis, R. Moffatt, J. D. Joannopoulos, P. Fisher, and M. Soljacic. "Wireless Power Transfer via Strongly Coupled Magnetic Resonances," *Science express* 317: 83–86 (2007).

[2] Ahn, D., M. Kiani, and M. Ghovanloo. "Enhanced Wireless Power Transmission Using Strong Paramagnetic Response." *IEEE Transactions on Magnetics* 50: 96–103 (2014).

[3] Hui, S.Y.R., Z. Wenxing, and C.K. Lee. "A Critical Review of Recent Progress in Mid-Range Wireless Power Transfer. "*IEEE Transactions on Power Electronics* 29: 4500–4511 (2014).

[4] Cannon, B.L., J.F. Hoburg, D.D. Stancil, and S.C. Goldstein. "Magnetic Resonant Coupling as a Potential Means for Wireless Power Transfer to Multiple Small Receivers." *IEEE Transactions on Power Electronics* 24: 1819–1825 (2009).

[5] Wang-Qiang, N., C. Jian-Xin, G. Wei, and S. Ai-Di. "Exact Analysis of Frequency Splitting Phenomena of Contactless Power Transfer Systems." *IEEE Transactions on Circuits and Systems I: Regular Papers* vol. 60: 1670–1677 (2013).

[6] Yiming, Z., and Z. Zhengming. "Frequency Splitting Analysis of Two-Coil Resonant Wireless Power Transfer." *IEEE Antennas and Wireless Propagation Letters* vol. 13: 400–402 (2014).

[7] Runhong, H., Z. Bo, Q. Dongyuan, and Z. Yuqiu. "Frequency Splitting Phenomena of Magnetic Resonant Coupling Wireless Power Transfer." *IEEE Transactions on Magnetics* 50: 1–4 (2014).

[8] Sample, A.P., D.A. Meyer, and J.R. Smith. "Analysis, Experimental Results, and Range Adaptation of Magnetically Coupled Resonators for Wireless Power Transfer." *IEEE Transactions on Industrial Electronics* 58: 544–554 (2011).

[9] Yiming, Z., Z. Zhengming, and C. Kainan. "Frequency-Splitting Analysis of Four-Coil Resonant Wireless Power Transfer." *IEEE Transactions on Industry Applications* 50: 2436–2445 (2014).

[10] Ahn, D., and S. Hong. "Effect of Coupling between Multiple Transmitters or Multiple Receivers on Wireless Power Transfer." *IEEE Transactions on Industrial Electronics*: 1 (2012).

[11] Nguyen, H., Agbinya, J.I., and J. Devlin. "FPGA-Based Implementation of Multiple Modes in Near Field Inductive Communication Using Frequency Splitting and MIMO Configuration." *IEEE Transactions on Circuits and Systems I: Regular Papers.* 62: 302–310 (2015).

[12] Fei, Z., S.A. Hackworth, F. Weinong, L. Chengliu, M. Zhihong, and S. Mingui. "Relay Effect of Wireless Power Transfer Using Strongly Coupled Magnetic Resonances.", *IEEE Transactions on Magnetics* 47: 1478–1481 (2011).

[13] Karalis, A., Joannopoulos, J.D., and M. Soljačić. "Efficient Wireless Non-Radiative Mid-Range Energy Transfer." *Annals of Physics* 323: 34–48 (2008).

[14] Niu, W.Q., W. Gu, J.X. Chu, and A.D. Shen. "Coupled-Mode Analysis of Frequency Splitting Phenomena in CPT Systems." *Electronics Letters* 48: 723–724 (2012).

[15] Dukju, A., and H. Songcheol. "A Study on Magnetic Field Repeater in Wireless Power Transfer." *IEEE Transactions on Industrial Electronics* 60: 360–371 (2013).

19

Recent Advances in Wireless Powering for Medical Applications

Eric Y. Chow[1], David L. Thompson[1] and Xiyao Xin[2]

[1]Cyberonics Inc., Houston, TX 77058 USA
[2]Department of Electrical and Computer Engineering at University of Houston, Houston, TX 77204
E-mail: {eric.chow; david.thompson}@cyberonics.com;
xiyao.xin@gmail.com

An overview of the current state of the art wireless powering technologies is presented in this chapter with a focus on their application to medical implantable applications. A survey of the commercial landscape is also presented, which reviews wireless powering technologies available in marketed medical implantable devices or those in advanced stages of formal product development.

19.1 Introduction

One of the most important goals of the medical device industry is to make better devices for patients. A couple key areas for improvement are device size and lifetime. Circuit fabrication technology has scaled significantly through the years allowing for ultra-miniaturization of the device electronics to the point where the battery is now the dominant limiting factor in most medical devices. Battery technology and density have improved a bit over the years, but a revolutionary advancement in size and overall device lifetime could come from the rapidly advancing wireless power transfer (WPT) technology.

In just the past few years, there has been a surge in the area of wireless powering which is evident by the release of wirelessly rechargeable products and the recent formation of consortiums and standards to address the high-volume consumer electronics market. WPT technology has already begun to show up in common consumer electronics, such as cell phones, as well as in popular venues such as Starbucks. The technologies currently utilized and

Wireless Power Transfer 2nd Edition, 641–680.

available in these applications are primarily inductive coupling or magnetic resonance based.

Wireless powering for implantable medical applications has been around for decades but faces unique challenges due to interactions with the biological environment. The significant benefit of longevity and size reduction encouraged the integration of inductive coupling technology into these applications early on. For this application, however, there are an array of significant challenges including biological tissue effects on electromagnetic propagation, the effects of EM radiation on the body, and packaging requirements of biocompatibility and hermeticity. These significant challenges have resulted in inductive-powering-based technology being the only WPT technique currently in commercially available medical implants.

There are a number of WPT technologies in research stages which may eventually find their way into marketed medical implantable devices. Resonance coupling, which is already commercially available in consumer electronics applications, is continuing to advance on the research and development side. This technology has potential benefits for implantable applications including robustness against misalignment and greater distances of operation, but it faces some of the same challenges as inductive coupling as well as new ones stemming from higher frequencies and/or coil size limitations. Far-field-based powering along with MIMO techniques has recently shown up in the consumer electronics world and could have potential in medical applications due to the potential benefits of miniaturization and operational distance. However, far-field powering utilizes high-frequency EM waves which face significant attenuation and reflections when attempting to propagate through biological tissue making it extremely challenging to achieve adequate efficiencies for implantable applications. Midfield powering has recently been getting a lot of attention in the research world for medical applications as a method which tries to optimize in the area between near-field coupling and far-field propagation. This has the potential for increasing the distance of operation from standard inductive but alleviating some of the tissue-based absorption faced by the far-field technique. Acoustic powering is another wireless power transfer technique that is relatively efficient at transferring power over significant distances through liquid or wet biological tissue, but its performance through air is extremely poor requiring a need for a hybrid approach. Resonance, far-field and MIMO, midfield, and acoustic each have their challenges for biomedical applications, but their potential to revolutionize medical care may result in the necessary research advancements to see their integration into future implantable devices.

19.2 Consortiums, Standards, and WPT in the Consumer Market

The Wireless Power Consortium (WPC) was formed in 2008, and they developed a wireless power standard called "Qi" (after the Chinese word for air or energy flow), which was published in 2009, specifically targeting inductive coupling charging technology [1]. WPC was founded by 8 companies including Philips, Texas Instruments, Logitech, Sanyo, and National Semiconductor and has now expanded to over 200 members including Sony, Samsung, Nokia, Motorola, HTC, LG, Toshiba, and Verizon [2].

More recently, the Power Matters Alliance (PMA) was formed in March 2012 utilizing similar inductive coupling technology as the Qi standard but also has a focus on public environment usability with usage tracking and interoperability [3]. PMA was founded by Procter & Gamble and Powermat Technologies, and currently, they have over 100 members. In October 2012, they were joined by AT&T, Google, and Starbucks, and in June 2014, Starbucks began to rollout wireless charging in their stores [4].

Another consortium was also formed in 2012 under the name Alliance for Wireless Power (A4WP) who promotes the Rezence^{TM} technology and specification which is based on a different type of WPT called magnetic resonance [5]. A4WP was founded by Qualcomm and Samsung leveraging the magnetic resonance technology developed from WiPower (acquired by Qualcomm in 2010) [6]. Intel joined in 2013, and now A4WP has over 140 members including Samsung, WiTricity, Acer, Dell, HP, Lenovo, and Microsoft [7].

Some of the larger players are members of more than one of these consortiums (Samsung is part of all 3), and some phones are developing compatibility across standards [8]. WPC/Qi was established first and has the large number of members and certified products. WPC/Qi and PMA were both established targeting inductive coupling with operational distances of less than a couple inches, but PMA promotes usability in public venues with software- and cloud-based support. A4WP/Rezence was established using a different magnetic resonance technology which allows for greater distances than standard inductive and can support multiple simultaneous devices. In the middle of 2014, WPC added support for resonance wireless charging in their version 1.2 of the Qi standard [9]. On January 5, 2015, PMA and A4WP announced that they will merge creating an organization that also supports both inductive and resonance WPT technologies [10]. Also, in December 2014, Energous Corporation, a company developing a far-field-based

WPT technology called WattUp^{TM}, announced that they would be joining PMA [11].

These recent developments in consortiums and standards highlight the significant attention that WPT is getting for consumer electronics. These standards are valuable for widespread and high-volume applications where compatibility and interoperability are important. The development of WPT in medical applications, particularly in the area of implantable devices, faces significantly different challenges.

19.3 History of Wireless Powering in Medical Implantable Devices

One of the first implantable pacemakers was actually rechargeable. Dr. Elmqvist developed an implantable pulse generator with 2 rechargeable NiCd batteries that was implanted by Dr. Robert Rubio in 1960 in Uruguay. This implant could be recharged wirelessly via inductive coupling and required recharging once a week for 12 h [12]. In the 1970s, Pacesetter Systems, Inc. developed a rechargeable pacemaker, and over 6000 were implanted between 1973 and 1978 [13].

Rechargeable systems, however, typically require the patient to recharge the implant. For these systems, there is the concern for patient compliance because if they fail to recharge their device at the necessary times, it could result in non-operation of the device. For life-sustaining devices, such as pacemakers, this could result in detrimental consequences [14] indicated that placing the "responsibility for recharging in the hands of patients ... is not a good medical practice." According to [15], lawsuits could result if the implant stopped working even though it could have been the patients fault from not recharging as directed.

In addition to the legal implications of rechargeable technology for pacemaker applications, advancement in battery technology, particularly lithium-based primary cells, has lengthened the lifetime of a primary-cell-based device, alleviating the need for rechargeable platforms. Due to the power consumption of pacemaker implants, current primary cell batteries would provide 5–10 years of operation [16]. Substantially longer battery lifetimes may actually not be necessary or desired because other components within the implant may begin to see failures after that duration. Designing the system such that the battery depletion is the first indication for an implant replacement is desirable because battery longevity and end-of-life can be predicted much more accurately than failure of other parts of the implant. Thus, this 5- to 10-year lifespan is currently somewhat of a "sweet spot" [15].

Although the early pacemakers started off as rechargeable platforms, the legal concerns due to patient compliance and the sufficient longevity provided by modern lithium-based battery technology have resulted in only primary-cell-based pacemakers on the market today. This rechargeable technology has found its application in other, non-life-sustaining, implantable applications. One of the most popular rechargeable medical implant applications is pain stimulation. This application utilizes a relatively power-hungry device due to the large voltages and currents required for sufficient stimulation for pain reduction. This application is not life–sustaining, and furthermore, it provides an inherent feedback mechanism to the patient, of returning pain, when the battery charge has been depleted and requires recharging. Other non-life-sustaining applications where rechargeable technologies can be considered include therapies for epilepsy, depression, heart failure, Parkinson's, essential tremor, and dystonia. The platforms for these applications include vagus nerve stimulation (VNS), deep brain stimulations (DBS), and ventricular assist devices (VAD). Implantable sensor-based technology is another area where wireless recharging technology has potential applications.

19.4 Development of a Commercial Rechargeable Active Implantable Medical Device

Developing an active implantable medical device with a secondary, rechargeable battery as opposed to a primary, non-rechargeable battery offers significant opportunities to advance the treatment available to patients using active implantable medical devices. This is because one of the key limitations of all active implants is the power budget, and converting to a rechargeable system allows for a dramatic increase in this power budget. A typical design target for a non-rechargeable active implantable medical device such as the Medtronic Itrel 3^{TM} is an approximately 5-year longevity using a battery with a capacity of a few Amp-hours [17, 18]. This results in a power budget that is quite limiting, keeping certain technologies and options out of reach. In some cases, the use of a rechargeable battery allows an improvement in existing therapy, but certain therapies require significantly more energy and thus basically necessitate the use of rechargeable technology. In addition, the use of a rechargeable battery allows for a decrease in product volume and an increase in the longevity of the implant which will extend the time between surgeries—both of which are significant benefits for the patient. The development of a commercially viable rechargeable implantable device involves a multitude of design trade-offs and system-level decisions that ripple through the entire product development process. Decisions such as the oscillation frequency and the time between recharges have dramatic impacts upon the design and performance of the entire system.

19.4.1 Product Design Implications

The traditional method of wirelessly transferring energy to an implant has been low-frequency near-field inductive coupling; however, this is beginning to change as new, smaller, second- and third-generation rechargeable active implantable medical devices are being developed. There are new products that are in the late stages of development, such as the wireless cardiac stimulation or WiCS system [19] that uses new methods, but near-field, low-frequency inductive coupling is still the most common method used. This is primarily due to the simplicity of this method and the fact that the human tissue is somewhat transparent to this low-frequency electromagnetic energy. In addition, the lower frequency energy also penetrates the housing of the implant relatively well, resulting in less energy lost to the eddy currents generated on the implant case.

The frequency used to recharge the implant has a substantial impact on the overall product design. If the frequency is low enough, and the implant case material is chosen wisely, a significant amount of the energy can pass through the case, and thus, the receiving coil can be placed inside the implant case. This simplifies a number of other issues related to the implant design such as the material biocompatibility of the secondary coil and the need for ceramic or plastic portion of the housing. This also reduces feedthru complexity, an option that is always welcomed by the design team. Some trade-offs, however, are that lower frequencies require larger primary and secondary coils and larger tuning capacitors as well.

It is worth noting that the selection of implant case materials is quite constrained, as a high level of hermeticity is required, along with durability, impact resistance, and of course, biocompatibility. Early implants were simply potted in epoxies, but these were found to be non-hermetic as they eventually became permeated with body fluids. Though new options continue to be tested, at this time there are a rather limited number of viable materials. Now the list of options includes titaniums of grade 1, 2, 5, and 23, ceramics such as Al2O3, glass, and certain grades of stainless steels such as 316L. Each of these materials has a mix of positive and negative attributes, though all are biocompatible and have a proven record of use in permanently implantable devices. The commercially pure titanium grades offer ease of forming and drawing, low price, and reasonable availability. Grades 5 and 23 titanium offer similar biocompatibility to the commercially pure titanium grades but are not as easy to form or draw, which can be limiting for certain packaging designs. These two grades are selected for recharging applications due to their increased electrical resistivity, approximately three to four times the resistivity of the commercially pure titanium grades [20]. The use of these grades has been adopted by many rechargeable active implantable medical device manufacturers because it does improve recharging efficiency at frequencies less than a few hundred kilohertz. Ceramics provide the optimal electromagnetic condition, as they are transparent to the RF energy, allowing the designers to select the RF frequency without regard for the ability to penetrate the housing. Unfortunately, ceramic does come with significant drawbacks. These drawbacks include complexities and expense of manufacturing; difficulties in assembly, brazing, and welding; increased mass; and of course, mechanical toughness. The lack of toughness is the key drawback and may often be deciding factor in a decision against the use of a ceramic case for an implant. The toughness issue, combined with the high density and difficulties of manufacturing, results in a ceramic case that is significantly heavier than a case made of titanium, which may drive a decision to avoid such a design. Finally, glass packaging should be considered. It offers similar attributes to ceramic but is less tough and less hermetic than Al_2O_3, although glass does have the benefit of being easier to manufacture. At this time, the use of ceramics and glass is most practical for smaller implants or for portions of an implant.

The next design decision is the time interval between recharging sessions. Although any interval could be selected, it is common to choose an interval that is convenient for the patient and easy to remember. If the interval is too long, there is a concern for some therapies that the patient may forget to recharge his implant. Note, this is therapy dependent because some therapies, such as pain suppression, have the inherent feedback of pain returning if the implant's battery is depleted. At the other extreme, a very short interval may be overly burdensome to the patient. Often an interval of a few days to a week is selected [17]. As the recharge interval is chosen, the system design of the implant, along with an estimated power budget, should be developed. Then, the interval, combined with the power usage of the implant, will determine the battery capacity and thus the physical size of the battery and the implant itself. In this estimate, the designers should consider including a safety factor to reduce the likelihood of deeply discharging the implant battery, as this may both reduce the implant longevity and expose the patient to loss of therapy, though this comes at a cost in terms of the physical size of the implant. The design of the battery itself is a separate endeavor and is beyond the scope of this chapter.

A recommended overall method of attacking the design of the recharging system is to begin by developing basic system-level and user requirements while researching the thermal parameters of the tissues that surround the implant for all the planned implant locations. The thermal parameters can be used to develop computational models of the implant. The requirements regarding the thermal performance of the system may be internally generated or may be based on compliance to a relevant standard. In any case, the requirements will likely allow for only a small increase in temperature, and as a result, the design must be closely studied to optimize the thermal performance of the system.

19.4.2 Computational Modeling

As discussed above, a logical starting point for the development of computational models is electromagnetics. One of these computational models should focus on the electromagnetics of the system and can provide significant insight into the interaction between the roughly three dozen parameters that impact the electromagnetic efficiency of the system. The team should consider including all the large electrically conductive objects in the system, including the primary coil, secondary coil, the implant housing, the implant circuit board, implant battery, and the external circuit board and battery if near the primary coil. There is a wealth of published information on this topic [21–24], but when reviewing this work, it is important to be certain that the publications include all the key conductors that are needed for the system being developed. The vast majority of the published work does not include a hermetic implant case, which, if made of metal, has a significant impact on the overall system performance.

There are a number of commercially available packages that are capable of performing this modeling. These simulations are often initially simplified by using axisymmetric models. This is a reasonable simplifying assumption especially for initial scoping of the system performance. It is valuable to understand the trade-offs between a highly tuned system and a system that is robust enough to adequately perform across the range of typical patient use cases. It is likely that the system will be tuned for a specific implant depth and the performance will deteriorate when the coils are placed either farther apart or closer together. Obviously, the axisymmetric models will not predict the performance of the system when the primary and secondary coils are misaligned, which is a realistic use case condition and has been shown to have a significant effect upon performance [25]. Because both alignment and depth of implant will vary during actual use, it is recommended that the impact of these variables both be investigated.

A second set of models should incorporate the losses into a thermal model of the implant and the human body. The thermal modeling should incorporate the information regarding implantation location and depth of implant. For some devices, multiple implant locations will need to be modeled. At each implant location, the depth of implantation will need to be varied. Again, the best performance may not be achieved at the most shallow implantation depth.

There are multiple issues to consider when investigating the thermal performance of the system. In regard to depth, there are three basic system features that will change with implant depth: electromagnetic efficiency, the starting temperature of the implant, and finally the influence of the primary coil heating upon the implant temperature. The last item is of particular concern for some designs, as a significant amount of energy is being delivered to the primary coil, causing heating of this coil. Typically, the design allows for a larger increase in temperature of this external component as compared to the implant, but this can have the unintended consequence of heating the tissue around the implant. This occurs because the heat warms the entire area around the implant, reversing the normal condition where the body is able to dissipate heat at the skin surface. Over the long charging time, this can result in an unacceptable increase in heating of the tissue.

The amount of heating that is acceptable is an area of current research [26] and may need to be evaluated for each implant location [27]. In general, the body is quite sensitive to changes in temperature, and a common guideline is that exposing the body to temperatures above $39°C$ for extended amounts of time could be problematic [27]. In addition to the safety issues, designers should consider patient comfort. Investigation into this area has found that a portion of the population is extremely sensitive to changes in temperature and will note discomfort if an object at $37°C$ is placed on their skin [28]. A related point that is important to note is the temperature of the human body is approximately $37°C$ at the core, but it decreases to about $30°C$ at the skin surface [29]. This variation in temperature as a function of depth may be helpful in achieving design objectives for some implants.

The thermal and electromagnetic computational models will allow rapid iteration upon the design concepts with many different variations in a multitude of parameters that control the performance of the rechargeable active implantable medical device. In parallel, the development of new test equipment such as a calorimeter and thermal phantom should be undertaken as they are both needed for physical testing [30]. Then, the development of physical prototypes can begin, with simple bench testing and calorimeter testing to evaluate performance. This work can run in parallel to the development of the rest of the implant, with regular connections early in the process to confirm that the power budget is still on target and to determine any other relationships between the performance of the remainder of the system and the recharging system, such as potential interactions between the recharging frequency and the communication or measurement systems on the implant.

19.5 Comparison of Commercially Available Rechargeable Active Implantable Devices

Medtronic, Boston Scientific, St. Jude Medical, and a number of smaller companies all offer products that utilize a rechargeable power source. For some therapies such as chronic pain, a rechargeable power source is enabling technology, and without the use of such technology, the product longevity is significantly shorter because of the power required for therapy that provides therapeutic benefit [17, 18]. As a result, rechargeable technology is often used for spinal cord stimulation. This is as seen in the Medtronic spinal cord stimulator product line, the majority of which are rechargeable.

The Medtronic product line of rechargeable spinal cord stimulators began in April of 2005 with approval of the Restore^{TM} which was followed by the RestoreUltra^{TM}, RestoreAdvanced^{TM}, and RestoreSensor^{TM}, each with either increased features or reduced volume. These products, shown in Figure 19.1, are recharged using an inductive method. Similarly, Medtronic offers a rechargeable deep brain stimulator that is also recharged using inductive technology. The initial offerings from Medtronic used an innovative approach to the obstacle of the metallic implant housing. They created a front portion of the housing that covered the secondary coil with polysulfone, an RF transparent material [31]. This allows easier inductive transmission of the RF energy from the primary coil to the secondary coil using higher while still containing the secondary coil in biocompatible and durable housing. Both the Restore and RestoreAdvanced feature this approach. Subsequent offerings remove this plastic portion of the housing, presumably to simplify housing and

Figure 19.1 Medtronic restoreultra^{TM} and restoreadvanced^{TM} products.

feedthru designs, reduce costs, and shrink packaging size, but at the expense of adding complexity to the inductive link.

St. Jude Medical has developed a similar product line of spinal cord and deep brain stimulators. Interestingly, they have also followed a similar development path, beginning with an implantable pulse generator, the Eon^{TM} that had a durable, plastic housing for the secondary coil, allowing the use of a much higher frequency of 2 MHz for the inductive power transfer. Subsequent development of the Eon Mini^{TM} included moving the secondary coil inside the titanium case, but with the use of the high-resistivity Ti6Al4V housing and a reduction to 50 MHz, they were also able to simplify the device design and assembly. All the subsequent rechargeable devices that St. Jude has launched are based on this Eon Mini platform. The Eon^{TM} and Eon Mini^{TM} products are shown in Figure 19.2.

Boston Scientific has taken a slightly different approach by placing the secondary coil in the header, allowing for efficient, relatively high-frequency inductive power transfer, while also keeping the secondary coil protected and separated from the metal in the pulse generator. This design has been continued in the Precision Plus^{TM} and Vercise^{TM} implantable pulse generators, shown in Figure 19.3. . . . With the release of the Precision Spectra^{TM}, it appears that Boston Scientific has similarly moved the secondary coil into the titanium case and reduced the frequency into the range of 77 to 90 kHz.

In addition to the big three active implantable medical device manufacturers, there are a handful of smaller companies with inductively rechargeable devices on the market. Although the specific details of each of these devices will not be discussed, it is worth noting that these are all similar to the devices

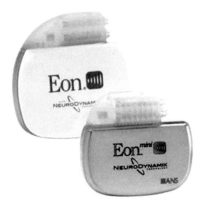

Figure 19.2   St. Jude medical's eon^{TM} and eon mini^{TM}.

Figure 19.3   Boston scientific's precision plus^{TM} and vercise^{TM} products.

from the larger companies. The majority of these devices use a design that is similar to the initial offerings from the larger companies, with a higher recharge frequency and a secondary coil in the header. The remainder of the companies have designed the implant using a lower frequency and secondary coil inside the Ti4Al6V case. Both are viable options that will serve to safely and efficiently recharge the implant.

Finally, there are a few products that are in development, which are not based on inductive power transfer. These products offer the potential to overcome certain problems that are inherit in inductive power transfer. Key among these is the ability to transfer larger amounts of power longer distances. One example of a system in development is Boston Scientific's

Remon ImPressure/Remon CHF sensor which uses acoustic energy to transfer the power deeper into the body. While there is a valid theoretical basis for this approach, it has not yet been proven in a clinical setting.

The technology that makes its way into a commercially available, rechargeable active implantable medical device tends to lag behind the cutting edge technologies, some of which will be discussed in the subsequent sections, with companies choosing to use safe, well-understood methods as a solid foundation upon which to build their product lines. This is for a good reason. The marketplace rewards solid, dependable, products that deliver extremely high reliability performance and have minimal field issues. Related to this concern of reliability, it is important to briefly touch upon the topic of field issues for these products. All of these products have been extremely successful and reliable, improving the lives of hundreds of thousands of patients, but along the way there have been some difficulties related to recharging the implantable pulse generators. In early 2004, Boston Scientific received a handful of complaints related to excessive heating causing discomfort and burns during recharging of their spinal cord stimulator. In late 2008, these complaints, combined with some more serious second- and third-degree burn complaints by approximately 0.3% of the patients, led to a Field Safety Notice and a Class 2 recall for the Boston Scientific SC-5300 charger [32]. Soon after this, St. Jude incurred a limited number of field issues related to their spinal cord stimulator products. A number of these issues were linked to the battery and internal circuitry design of the implantable pulse generator, but in late 2011 and 2012, St. Jude released multiple Field Safety Notices to physicians related to device heating during recharging [33]. It appears that these issues have been resolved, but it is important to note that even with the use of standard inductive technology, there have been difficulties, design issues, and, although rare, injury to patients. These issues are probably a result of a combination of variations in device performance, different use cases, and potential misuses, along with patient use of analgesic medication. For these reasons, the development of a rechargeable implantable active implantable device should include close scrutiny of the thermal performance of the system and should include a variety of patient use cases.

19.6 Resonance Power Transfer

This section begins the more advanced wireless powering techniques that are yet to make it to commercially available medical implantable products but have potential benefits to those applications. Resonance power transfer is available in consumer electronics and has been recently gaining popularity due to recent advances in research, technology development, and the formation of the A4WP. Resonant inductive powering is essentially the wireless transmission of electric power between two magnetically coupled coils operating at resonance. By tuning the transmitter and receiver to the same resonant frequency, a receiver can easily pick up the oscillation from a transmitter, begin oscillating at the same frequency, and continue oscillating. Since the receiver is resonant at the frequency output from the transmitter, it will pick up most of this energy from the transmitter's oscillation. The simplest resonating circuit is an LC resonant circuit. There are two typical LC resonant circuits, which are shown in Figure 19.4:

In both cases, the current on the inductor L will reach a maximum when the circuit reaches resonance with a frequency of:

$$f = \frac{1}{2\pi\sqrt{LC}} \tag{19.1}$$

The inductor L can be realized as coils to generate and/or receive magnetic flux. The capacitor C in Figure 19.4a can be realized with an external tuning capacitor; the capacitor in Figure 19.4b can be an external capacitor parallel to the coil, or the stray parasitic capacitance of the coil itself, or a combination of both. A. Kurs et al. [34] reported the design using two self-resonant coils to transfer power with an efficiency of 40% over a distance of 2 m. A graphic illustration of their model is shown as follow.

In Figure 19.5, both the transmitter and receiver are helix coils with open-circuit ends. In this situation, both coils can be regarded as inductors in parallel with their own parasitic capacitors. When the current on the driver reaches

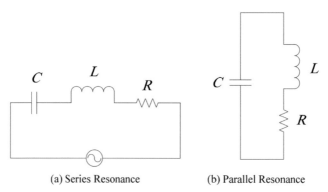

(a) Series Resonance   (b) Parallel Resonance

Figure 19.4   LC resonant circuits.

Figure 19.5 Wireless power transfer with two self-resonant coils.

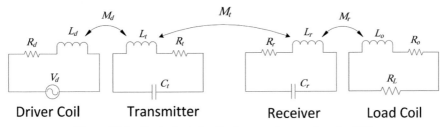

Figure 19.6 Equivalent circuit of resonant power transfer [35].

the self-resonant frequency of the transmitter, the magnetic flux excited in the transmitter reaches a maximum. The receiver is designed to have the same self-resonant frequency as the transmitter so it also excites a resonant magnetic field corresponding to the flux generated by the transmitter. The load coil harvests the flux from receiver and delivers this induced current to the load. The equivalent circuit for the model in Figure 19.5 can be given as:

The driver coil equivalently serves as a voltage source from the transmitter's point of view, and the load coil can be regarded as a load impedance connected to the receiver. Therefore, the four-coil system in Figure 19.6 can be reduced to the circuit shown in Figure 19.7.

The value of the equivalent load resistance can be modified by changing the distance between load coil and the receiver. The optimal power transfer efficiency is

$$e_{opt} = \frac{U^2}{\left(1 + \sqrt{1 + U^2}\right)^2},\qquad(19.2)$$

where $U = \omega M_t / R_t R_r$ [35]. This optimal power transfer efficiency formula indicates that the optimal efficiency can be greatly improved if the parameter U increases. Kesler in [35] also shows that the magnetic coupling coefficient U is given as

$$U = k\sqrt{Q_t Q_r}.\qquad(19.3)$$

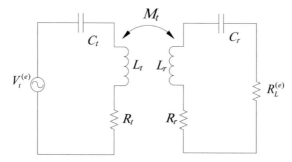

Figure 19.7   Reduced circuit model for resonant power transfer.

where k is the coupling coefficient, and it is given as

$$k = \frac{M.}{\sqrt{L_1 L_2}} \tag{19.4}$$

The range of parameter k is from 0 to 1. When k is close to 0, the system is weakly coupled and there is little or no mutual coupling between the two coils. When k is close to 1, the system is strongly coupled and the mutual inductance of the coils approaches the geometrical average of the self-resonance inductance of both coils.

The power transfer efficiency relies on the quality factors of both the transmitter and receiver and the magnetic flux link between them. From the equation, it is clear that increasing the qualify factor of both coils would improve the efficiency; however, in application, this would also increase the difficulty in tuning. Kurs in [34] pointed out that the receiving power might become almost negligible when the relative frequency shift Δf/f reaches a few times the inverse of the coil quality factor Q. One technique that may help this tuning issue is to add matching networks to the transmitting and receiving coils. Another technique is to add ferrite cores to the coils. The matching network used for resonant power transfer could be as simple as a tuning capacitor in series or in parallel with the coils. In this way, the resonant frequency of both transmitter and receiver can be manually tuned to the same resonant frequency. Waters et al. [36] proposed using a π-shaped matching network for resonant power transfer coils. The π-shaped matching network is illustrated as follow:

This type of network gives a wideband impedance matching to the load. For the circuit in Figure 19.8, the load resistor could have a wider variation while keeping the approximate conjugate matching with the source resistor. Furthermore, the wideband or narrowband property can be tuned with the

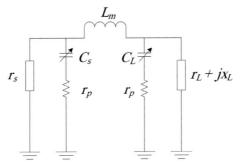

Figure 19.8 π-Shaped network for impedance matching [36].

additional degrees of freedom provided by the two adjustable capacitors. As an example, the tuning algorithm, given that $r_s < r_L$, is based on following formula [36]:

$$Q_m = \sqrt{-1 + \frac{r_L(r_L + r_s)}{L_m^2\omega_0^2} + \frac{2r_L\sqrt{r_Lr_s - L_m^2\omega_0^2}}{L_m^2\omega_0^2}}$$

$$C_s = \frac{\sqrt{-1 + \frac{r_L}{r_s(1+Q_m^2)}}}{r_s\omega_0}$$

$$C_L = \frac{Q_m}{r_L\omega_0} \tag{19.5}$$

The additional degrees of freedom given by the two tuning capacitors (C_s and C_L) are able to tune the quality factor Q within a range given by:

$$0 \leq Q_m \leq \sqrt{\frac{\max(r_s, r_L)}{\min(r_s, r_L)} - 1} \tag{19.6}$$

For the weak coupling case, we can adjust the tuning capacitors to maximize the quality factor to achieve the highest efficiency. In the strong-coupling situation, we may wish to lower the quality factor so that the system is more robust to misalignment and mismatch. Ean et al. [37] proposed a T-shaped network to serve as an impedance inverter in the matching network.

The input impedance of the circuit in Figure 19.9 is given by:

$$Z_{in} = \frac{k^2}{Z_{coil}}, \quad k = \sqrt{\frac{L}{C}} \tag{19.7}$$

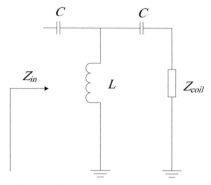

Figure 19.9   T-shaped impedance inverter matching network [37].

when the operating frequency is:

$$\omega_0 = \frac{1}{\sqrt{LC}} \tag{19.8}$$

In [37], the magnetic mutual coupling between transmitter and receiver is also modeled as an impedance inverter connecting both sides. Therefore, the wireless power transfer system becomes a band-pass filter connecting the source and load, which is illustrated in Figure 19.10:

The parameters K_{01} and K_{23} can be determined to achieve conjugate matching after adding the transmitter and receiver to the impedance inverter. For certain resonating field distributions generated by an LC resonator, multiple receivers could be placed within proximity of the transmitter and an optimal impedance inverter network could be designed for each individual receiver [38].

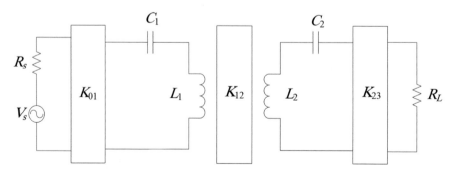

Figure 19.10   Band-pass filter model with impedance inverters [37].

Lee and Lorentz [39] reported a resonant power transfer implementation with a power transfer efficiency of 95% over a 30 cm air gap. They add an LC impedance matching network to the transmitter to ensure that there is an exact match between the resonant frequencies of the transmitter and receiver. Both the transmitter and receiver resonated at a frequency of 3.7 MHz with a Q value of 790 (5.1 uH, 0.15 Ohm). The coupling coefficient of the power delivering distance is about 0.06. Li and Ling [40] implemented a resonant power transfer system of "capacitor loaded coupled loops." When resonant, the coils used for delivering power have a Q value of over 650 and a power transfer efficiency of 60% at a distance of 1 m.

Inserting soft ferrite cores (with high relative permeability μ_r) into the transmitter and receiver enhances the magnetic flux at the proximity of the coils, thus increasing the magnetic coupling coefficient. Park et al. [41] designed a "Dipole Coil Resonant System" (DCRS) that can transfer efficiencies of 29%, 16% and 8% over distances of 3, 4, and 5 m with an operating frequency of 20 kHz. The dipole coils they use are resonant with a tuning capacitor connected in series, and the dipole coils are inserted with 3 m long ferrite cores. The Q value of the transmitter and receiver are 166 and 84. However, the shape of the ferrite core is carefully designed to ensure an approximately uniform flux distribution inside. Such a design enhances the magnetic flux transfer from the transmitter to the receiver and the mutual coupling coefficients are 0.68, 0.39, and 0.26 at distances of 3, 4, and 5 m, which are large enough to compensate for the low Q value of the coils. Therefore, the optimized soft ferrite cores play an important role in improving the power transfer efficiency in this design.

The design in [41] show that it is feasible to transfer power over long distances (5 m as the longest record so far), but it requires a large coil structure, with dimensions comparable to the power transfer distance, and a magnetic core to ensure sufficient flux link over those distances. Using high quality resonating coils is another viable option and helps compensate for the small magnetic coupling coefficient due to the long transfer distances [39, 40]. Tuning networks are utilized to ensure resonance operation at the same frequency for both the transmitter and receiver.

Inductive power transfer typically achieves strong coupling and operates at distances much smaller than the dimension of coils. This large coupling coefficient in inductive coupling makes it difficult for the system to achieve maximum power transfer efficiency at the resonant frequency [42]. This is because the reflected complex impedance from the load coil at the transmitting side will shift the optimal frequency (away from the resonant frequency) for maximum power delivery. Thus, optimal performance for inductive coupling is achieved by shifting the frequency slightly off resonance. Inductive power transfer system also uses tuning capacitors at the transmitter and receiver; however, these capacitors are not used for tuning to resonance, but rather for appropriately compensating the reactance of the coil to obtain maximum inductive coupled power transfer. When the power delivery distance increases (up to several times of the coil dimension), the reflection from the secondary coil is less significant and the mutual coupling coefficient decreases correspondingly. In this case, tuning both coils at exactly the same resonant frequency will maximize the power transfer efficiency. Thus, for close distances, inductive coupling typically achieves the highest efficiency, but for further distances, resonance coupling may achieve higher power transfer.

Medical implantable applications typically use inductive coupling because the greatest absolute efficiency is typically achieved in that use case even though it has limitations in regard to operating distance. Resonance coupling may offer some flexibility in distance and orientation but at the expense of lower efficiency. In currently available medical implants, a titanium case is typically used. Any losses in efficiency will generate eddy currents in the titanium case which is then converted into heat. This heat is also problematic in an implantable setting as it may cause discomfort and possibly harm. Furthermore, to maintain a certain power delivery with this lower efficiency, the flux from the transmitter would have to be increased by increasing the power output which then may encounter electromagnetic interference (EMI) and emission issues. Due to the desire to achieve the highest efficiency possible and minimize any heating and EMI effects, the industry has remained with inductive coupling. However, with advancements in alternative packaging materials and more optimal resonance coupling designs, this technology could find itself in future medical devices due to its benefits including greater operating distances and robustness to movement and orientation.

19.7 Far-Field MIMO

Another wireless powering technique that has recently gotten some public attention is MIMO-based far-field powering, which is being commercially developed by Energous Corporation, who recently joined the PMA. Far-field powering utilizing the multiple-input and multiple-output (MIMO) technique has been developed to perform wireless power transfer via antenna radiation [43]. MIMO utilizes multiple antennas for both the transmitter and receiver, and it has been shown to significantly increase the data rate of wireless systems without increasing the transmitting power or bandwidth ([44] Chapter 10). A typical MIMO system is illustrated as follows:

According to [44], the matrix model of a MIMO system (Figure 19.11) is given by:

$$\mathbf{y} = H\mathbf{x} + \mathbf{n} \tag{19.9}$$

where H is a $n \times m$ matrix of channel gain with each element h_{ij} as zero mean complex circular Gaussian random variables; x is the m-dimensional complex transmitting signal vector; y is the n-dimensional complex receiving signal vector; and n is the dimensional additive white Gaussian noise (AWGN). If we ignore the additive noise term and assume the matrix H is fixed, then the harvested power at the receiver can be calculated as:

$$P_r = tr(E(\mathbf{yy}^T)) = tr(E(H\mathbf{xx}^T H^T)) = tr(HE(\mathbf{xx}^T)H^T)$$
$$= tr(HSH^T), \tag{19.10}$$

where S is called the covariance matrix of the vector x. If we know matrix H perfectly, then it is possible that by adjusting the covariance matrix of the input antenna (a procedure called energy beamforming), one can obtain the maximum harvested power on the receiver side. Zhang and Ho [43] derived the solution for an optimal covariance matrix S for maximizing the harvested power P_r; however, in practice, the matrix H may not be fully known and is subject to changes in the environment.

In order to dynamically adjust the beamforming procedure, Xu et al. [45] proposed a "One-bit Feedback" procedure where the receiver sends a one-bit feedback to the transmitter at each interval indicating an increase or decrease of harvested energy compared to the previous feedback interval. This allows system to estimate matrix H before power transmission. With an estimated H

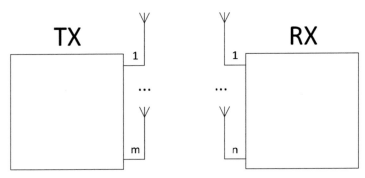

Figure 19.11 Block diagram of MIMO system.

matrix, the system is able to calculate the optimal covariance matrix S and use it to determine the appropriate signal on each transmitting antenna to ensure maximal power harvesting at the receiver(s).

The optimization methods for MIMO wireless power transfer have been developed in [43] and [45]. Energous has recently developed WattUp^{TM} [46], which can transfer power wirelessly to mobile devices using a MIMO system [47]. The WattUp^{TM}-enabled system uses a MIMO transmitter that sends up to 23 dBm across a 120-degree directional span. After determining the optimal direction angle to send the power to receiving devices, the transmitter sends a 3D pocket of energy using the 900-MHz (5.7–5.8 GHz) RF spectrum. This system delivers an average of 2 W to 4 devices simultaneously within a 15 ft. radius. Within 0–5 ft., 4 W of power can be delivered to four devices; if the distance is from 5 to 10 ft., the power delivered to four devices is 2W, and the receiving power will drop to 1 W if the distance increases to 10–15 ft. For mobile devices with low power consumption (less than 5W), this power delivery system may provide a good solution for indoor wireless charging with flexible mobility.

Far-field-based wireless powering techniques are challenging in biomedical applications due to the significant interactions of biological tissue with high-frequency electromagnetic waves. However, this technique may find its application for shallow implants or in external wearable medical applications. The MIMO-based beamforming technique helps to optimize the power transfer link and can account for power transfer to multiple devices and allow for flexibility in movement of the patient while maintaining a charging or powering cycle. Far-field power has the potential benefits of greater operating distances because it is based on propagating waves, but its lower efficiencies, particularly in implantable applications, require significant advancements and optimization techniques, such as MIMO beamforming, to become relevant.

19.8 Midfield Powering

This technology has received a lot of publicity recently particularly with the workout of Professor Ada Poon's Lab at Stanford University where they have developed an incredibly small medical implant that can be powered transcutaneously, through midfield powering, from several centimeters away [48–50]. This technology has the potential to revolutionize patient care by offering implantable devices at significantly smaller scales and wirelessly delivering power to them at relatively greater depths.

The concept of "midfield" power transfer corresponds to a region where the traveling wave cannot yet be approximated by a plane wave, as in far-field, but it is not as burdened by the effects of higher order modes and evanescent modes, as in near-field. Similar to the near-field case, the interaction between the transmitter and receiver sides (antennas/coils) needs to be considered in the analysis for midfield power transfer. For biomedical applications, an implanted device may experience a very inhomogeneous media comprised of skin, muscle, organs, and other biological tissues. A reasonable approximation of this implant environment is a layered media, as is illustrated in Figure 19.12:

For medical implantable transcutaneous applications, particularly when dealing with high frequencies such as those applicable for midfield and far-field powering, dielectric and ohmic losses through biological tissue need to be taken into account in the derivations. In [51], Poon et al. evaluate an inhomogeneous media incorporating the air–muscle interface. In [52], Ho et al., show that for different organ tissues, the optimum frequency for power transfer is different, and approximate evaluations for optimum power transfer frequencies are presented for different human organ tissues including fat, heart, liver, and lung.

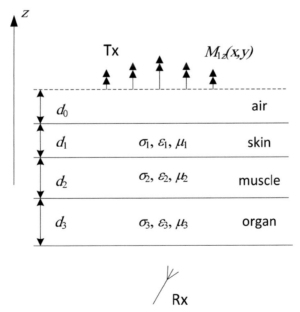

Figure 19.12 Layered media model for deeply implanted biomedical device.

Analysis of a midfield powering system can start with looking at a simple two-port system with a transmitting antenna and a receiving antenna as shown in Figure 19.13.

This simple two-port system has the matrix relationship

$$\begin{bmatrix} V_1 \\ V_2 \end{bmatrix} = \begin{bmatrix} Z_{11} & Z_{12} \\ Z_{21} & Z_{22} \end{bmatrix} \begin{bmatrix} I_1 \\ I_2 \end{bmatrix} \tag{19.11}$$

Based on this simplified model and utilizing the loosely coupled approximation, Kim et al. derive the power transfer efficiency as

$$\eta = \frac{P_r}{P_t} \approx \underbrace{\left(\frac{|Z_{21}|^2}{4R_{11}R_{22}} \right)}_{\eta_c} \underbrace{\left(\frac{4R_{22}R_L}{|Z_{22}+Z_L|^2} \right)}_{\eta_m} \tag{19.12}$$

where power transfer efficiency can be broken up into two main parameters, the coupling efficiency (η_c) and the matching efficiency (η_m) [53]. The matching efficiency (η_m) indicates the percentage of power delivered to the load given the power available from the receiver, and is thus independent of the transmitter. The coupling efficiency (η_c) shows the percentage of power received from transmitter given the total input power from the transmitter, and thus indicates the coupling strength. The goal is to maximize the coupling efficiency (η_c) as that results in the best midfield powering performance.

Kim et al. present a thorough derivation in [53] for a detailed expression of the coupling efficiency presented as a function taking into account the mutual inductances, expressions for all the fields and current distributions, a multilayered medium, and several other key aspects. Incorporating the known quantities (frequency, dielectric properties, etc....) and design criteria (depth, size, etc...), a model can be developed which maps out the current distribution and relates that back to the coupling efficiency. Based on these analytical derivations, Ho et al. developed an optimization model which can focus energy

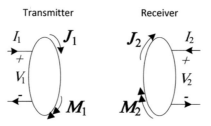

Figure 19.13   Transmit and receive antennas for midfield power transfer.

in a localized target region in biological tissue [54]. This region can then be used as the location of an implant utilizing midfield power harvesting.

On the transmit side, they implement a patterned metal plate, intended for placement close to the body, which can generate this localized and adaptive energy delivery to a midfield depth ($\sim$5 cm or greater) within the body [54]. This plate is essentially a 2×2 slotted antenna array fed by four independent ports and is illustrated in Figure 19.14:

This system can be optimized in real time by varying parameters such as the phase of the currents in the independent loop feeds. These parameters can thus be tuned to generate any kind of optimal current distribution, particularly one with a focused energy-dense region within the tissue at an implant location. The real-time optimization allows the system to adapt to changes such as movement during the power transfer cycle.

A system is developed and evaluated in [54], where they are able to receive about 200 uW of power for both a chest and brain implant from a transmitter positioned about 5 cm away and outputting 500 mW. For lower implant power requirements of around 10 uW, the authors claim that the midfield power transfer distance could be increased to 10 cm. The particular electrical stimulator device developed in [54] had an impressive miniature form factor of 2 mm in diameter and weighed only 70 mg [54].

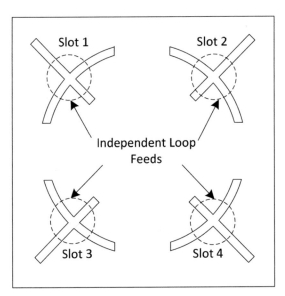

Figure 19.14 Transmitting antenna for midfield power transfer [54].

19.9 Acoustic Powering

Acoustic power transfer is a technique that uses transducers to convert electric power to ultrasonic waves which can propagate through certain media. This wireless powering technique is actually being explored by Boston Scientific as a medical device product after their acquisition of the technology from Remon Medical in 2007 [55]. This technology has found its application in transferring power through biological tissue [56], metal walls [57], and other materials. Acoustic power transfer uses piezoelectric transducers to convert pressure waves to electrical energy. Piezoelectricity, first discovered by the Curie brothers in the late 1800s, is the effect of crystals being able to produce an electric voltage differential under mechanical stresses [58]. From conservation of energy, these crystals could also produce these mechanical characteristics under applied electric fields [59]. To illustrate the piezoelectricity effect, we examine the crystal structure of zincite (ZnO), an increasingly popular material used in thin-film bulk acoustic resonators [60]. The crystal structure of ZnO is shown in Figure 19.15:

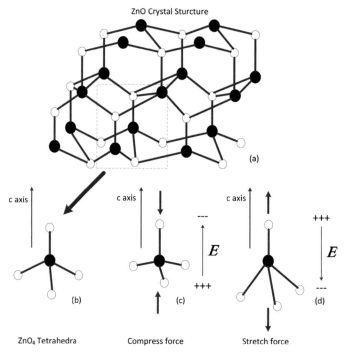

Figure 19.15   Crystal structure of zno and piezoelectricity effect [60].

Figure 19.15a shows all the ZnO_4 tetrahedra having the same orientation, and Figure 19.15b shows that each ZnO_4 tetrahedra has one Zn–O bond, along the c-axis, and three O–Zn–O bonds. When there is a compression (or stretching) force applied along the c-axis, the tetrahedra structure deforms by changing the O–Zn–O bond angle, and the Zn–O bond c-axis becomes increasingly harder to compress (or stretch) [61]. Such deformation causes the displacement of positive and negative charges, i.e., polarization, along the c-axis. The direction of the E-field depends on the direction of the stress force. A compression force produces an E-field in the positive c-axis direction, as shown in Figure 19.15c, and a stretching force produces an E-field in the negative c-axis direction, as shown in Figure 19.15d. Note that this piezoelectricity effect is anisotropic, and thus, stress forces normal to the c-axis will not produce this same polarization effect as a force tangential to the c-axis. If all the ZnO_4 tetrahedra have the same orientation direction and are aligned with the stress force, this would produce the maximum E-field.

If an alternating electric field is applied to the piezoelectric material, the induced alternating stress forces will generate vibrations on the surface. These vibrations could be used to produce pressure waves in a media. These pressure waves could then induce an alternating voltage on a piezoelectric material positioned at a location some distance away. The conversion from electricity to vibrations on a transmitter, vibration transmission through media, and then conversion from vibrations back to electricity on a receiver is essentially the mechanism behind acoustic power transfer.

To help optimize acoustic power transfer, power dissipation can be minimized by maximizing the directivity of the wave propagation. For a circular aperture acoustic antenna, the directivity function $R(\theta)$ is given as [62]:

$$R(\theta) = \left| \frac{2J_1(x)}{x} \right| \tag{19.13}$$

where

$$x = \pi \frac{D}{\lambda} \sin \theta \tag{19.14}$$

and J_1 is the first-order Bessel function, D is the antenna diameter, θ is the angle from the beam center, and λ is the wavelength. These equations show that the antenna beamwidth will become narrower as the ratio of antenna diameter to wavelength increase. For ultrasonic waves with a range of 20 kHz to 200 MHz and approximating the speed of ultrasonic wave propagation in the human body as the speed of sound in water (1484 m/s), the wavelength of the ultrasonic wave will range between 7.4 cm and 7.4 μm. Numerical results from [62] show that D/λ needs to be ≥ 8 to give a beamwidth $\leq 10°$. Given the 7.4 cm to 7.4 μm wavelength range, the dimension of a sufficiently directive acoustic antenna can still be sufficiently small for the application in implanted medical devices. A schematic illustration of an acoustic power transfer system is shown in Figure 19.16:

In Figure 19.16, piezoelectric transducers are used on both the transmitting and receiving sides. Matching layers are inserted between the transducer and media in order to reduce reflections and maximize power transmission into the media. At the transmit side, there is a coupling layer with a larger dimension than the transducer, which is used to increase the effective dimension of the antenna thus increasing its directivity.

Theoretical work has been done for modeling acoustic power transfer. Through analytical analysis, Hu et al. [57] discussed the feasibility of the acoustic power transfer through metal walls. The wave-propagating media in [57] is "sandwiched" by transducers of polarized ceramics PZT-5H as transmitter and receiver. Their results show that the power transfer efficiency reaches peaks at around 300, 600, and 900 kHz with corresponding efficiency of 60%, 80%, and 70% when the load resistance is 10 Ohms, which are not necessarily the fundamental resonant frequency of the media. Such results might be due to the fact that the wave impedance of piezoelectric transducer is not large enough to be regarded as open-circuit end to the wave media. Adding such power source/load with finite impedance might result in the resonating frequency shift off the characteristic frequency of the media. Hsu et al. [64]

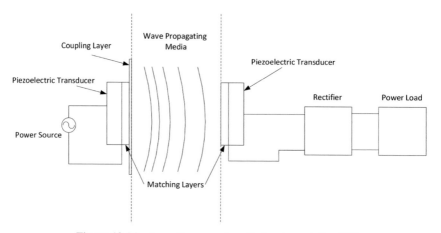

Figure 19.16   Acoustic power transfer implementation [63].

analyzed the piezoelectric transducer power transfer system using a fully coupled field equation. The constitutive equation of piezoelectric transducer is given as:

$$T_{ij} = c_{ijkl}^{E} S_{kl} - e_{kij} E_k$$
$$D_j = e_{jkl} S_{kl} + \varepsilon_{jk}^{S} E_k \qquad (19.15)$$

where $i, j, k, l = 1, 2, 3$. The first formula shows that both mechanical strain S and electric field E can induce a mechanic stress T through elastic constant c and piezoelectric stress constant e; the second formula shows that both mechanical strain S and electric field E can also produce an electrical displacement D. By applying mechanical equations of stress and Maxwell equation of the electrical displacement, theoretically we can calculate the transfer function of the entire wireless power transfer system and the optimal load impedance given the certain wave-propagating media [64]. However, for biomedical applications, the wave-propagating media is complex with various tissue types such as skin, muscle, and organ tissue. Such wave-propagating media is complicated and unpredictable. It might be more reliable if we base the design on actual experiments and measurements. An acoustic power receiver is described in [65] where they seal the receiver into a package of cohesive gel to see whether the shape of the package might affect the receiving power. The receiver is a circular shape metal plate (with a diameter of 7 mm) connecting the piezoelectric transducer with an ethyl cyanoacrylate adhesive. The dominant frequency is 35 kHz, and streaky pork was used to mimic muscle and fat tissue. The measurement of their system shows a power transfer efficiency of –38.16 dB (cubic seal) and –55.28 dB (sphere seal) in fatty tissue and –51.21 dB (cubic seal) and –47.27 dB (sphere seal) in muscle tissue, when the distance between the transmitter and receiver is 15 mm. Though the work does implement exact impedance matching, analysis was on the matching and reflection behaviors that occur in the gel-tissue–air interface [65].

To maximize the performance of the acoustic power transfer system, impedance matching should be used to minimize the wave reflection at all the interfaces between the power source at the transmitter side and the load at the receiver side. A typical value of characteristic impedance of conventional piezoelectric ceramics (lead zirconate-titanate, lead metaniobate, and modified lead titanate) is around 20~30 MRayl [66]. The impedance biological tissue will vary depending on the type. But most kinds of human tissue have the acoustic impedance close to water (1.48 MRayl), such as fat (1.33 MRayl), blood (1.66 MRayl), and muscle (1.65~1.74 MRayl) [67, 68]. The reflection coefficient of acoustic waves over a two-media boundary is given as:

$$\Gamma = \frac{Z_{\text{tissue}} - Z_{\text{pzt}}}{Z_{\text{tissue}} + Z_{\text{pzt}}} \qquad (19.16)$$

Using the approximate typical wave impedance values for piezoelectric transducers and biological tissue, the magnitude of the reflection coefficient can reach 0.91 indicating that almost all of the signal would be reflected at the interface boundary. A matching layer can be added to significantly reduce this reflection by impedance matching between the two layers. A common design of a matching layer is utilizing a quarter-wavelength layer with a characteristic impedance of:

$$Z_m = \sqrt{Z_{\text{tissue}} Z_{\text{pzt}}} \qquad (19.17)$$

Using this equation, we determine that the characteristic impedance of an optimal quarter-wavelength matching layer between the piezoelectric transducer and biological tissue should be around 7 MRayl. There are only a few types of biocompatible materials that have this characteristic impedance, but a matching layer consisting of multiple layers of different materials can be used to also achieve this impedance. Multiple layers with their own characteristic impedance and thicknesses offer many degrees of freedom to help achieve the optimal impedance matching interface. Callens et al. [69] reported a two-layer structure with one of the layer is a type of glue (Araldite) with known acoustic impedance (3.04 MRayl). The impedance of the glue is between that of water and piezoelectric transducer, so it adds degree of freedom for choosing the second matching layer by adjusting the glue layer thickness. The second layer in [69] is Schott glass with acoustic impedance of 11.1 MRayl. Such design reduces the reflection to minimum while broadening the selection of matching layer, so one does not need to seek for quarter-wavelength matching layer with a certain value of acoustic impedance.

Numerous research groups have successfully implemented acoustic powering in implanted medical applications. Maleki et al. [70] reported the design of a micro-oxygen generator (IMOG) powered by an ultrasonic wave. IMOG is used for tumor treatment by implanting the device within a tumor to create a concentrated oxygen distribution which enhances the ability of the radiation treatment to permanently damage the tumor DNA. Acoustic powering alleviates the need for a battery on IMOG allowing for the smaller form factor facilitating implantation within a wide array of tumors. The calculated power efficiency of the system is 0.1% over a distance of 3 cm with a received power transfer of 300 μA, which is sufficient for the IMOG device. Sanni et al. [71] discuss a combined inductive and acoustic power transfer system. Inductive coupling is used in the first subsystem to transfer power a short distance through the skin and then an ultrasonic subsystem transfer power through a longer distance deep into the tissue. A subcutaneous transponder is utilized to connect these two subsystems. The inductive power transfer subsystem is empirically validated with a 10 mm air gap between the transmitter and receiver achieving an efficiency of 83% with 6W power input at the primary coil and 5W of power received on the secondary coil. The ultrasonic powering subsystem transfers power over a distance of 70 mm through a tissue phantom and achieved an efficiency of 1% with 0.8 W input at the transmitter and 8 mW delivered to the load at the receiver. Arra et al. utilized an impedance matching technique to achieve acoustic powering with efficiencies of 25% at a distance of 100 mm [72]. Hu et al. [73] used a double-layer matching design for both the transmitter and implanted receiver which are piezoelectric transducers with diameters of 16 mm. The matching layer consists of aluminum alloy 6061T6 and polymethylmethacrylate (PMMA) sub-layers with thickness of 1.6 and 0.68 mm. The experimental results show that the matched acoustic powering system can reach efficiencies of 12% (−9.28 dB) through 10 mm of tissue phantom. Without the matching layer, Hu et al. measured lower efficiencies of about 6% (−12.41 dB).

Through tissue, acoustic power transfer could have advantages over electromagnetic propagation techniques which face significant tissue-induced attenuation; however, acoustic powering may have limitations in performance through air media. In tissue, electromagnetic radiation decays exponentially due to the electric conductivity and dielectric loss resulting in significant attenuation which further and rapidly increases with frequency. For comparison, an optimized design for midfield powering implanted neurostimulator in the brain achieved an efficiency of 0.04% (500 mW input, 200 uW output) through a distance of 5 cm [52]. Using acoustic power transfer, the IMOG system in [70] achieved an efficiency of 0.1% through 3 cm of tissue. Although this is not a 1-to-1 comparison, it can be seen that electromagnetic propagation-based techniques experience a much stronger decay through tissue. However, when evaluating acoustic powering through air, the situation is different. Acoustic wave attenuation follows Stoke's law [74]:

$$A(d) = A_0 e^{-\alpha d}, \tag{19.18}$$

where

$$\alpha = \frac{2\eta\omega^2}{3\rho V^3}, \tag{19.19}$$

and the parameter η is the dynamic viscosity coefficient, ω is the angular frequency, ρ is the fluid (liquid or gas) density, and V is the speed of sound in the medium. Compared to human tissue and water, the density of sound in air is much smaller and the speed is much slower. This results in a much larger decay coefficient for acoustic power transfer through air resulting in relatively low efficiencies. Similar to attenuation of EM waves, increasing the frequency of the sound waves also increases the decay, resulting in a similar transducer size to frequency to decay rate trade-off. Overall, acoustic power transfer may have improved performance over EM wave-based techniques when transmitting significant distances through biological tissue but poorer performance when transmitting through air media. A hybrid EM–acoustic approach could be used to achieve performance through both air and biological tissue, but the optimal solution will ultimately be application dependent.

19.10 Conclusions

Wireless powering and recharging has made significant recent advances in the consumer electronics market. The medical device industry has yet to move away from traditional inductive coupling which saw its first use several decades ago. With the recent ramp-up of wireless powering in consumer electronic products, the implantable device industry may be able to leverage some of these more mature technologies and standards if the benefits outweigh the risks. Research in wireless powering techniques is continuing to advance, particularly in its utilization in medical applications. With the recent advancements on the commercial and standards side and the progress on medical application-targeted research development, implantable devices may be due for a long-awaited advancement in wireless power transfer.

References

[1] "Wireless Power Consortium." [Online]. Available: http://www.wireless powerconsortium.com/ Geopend 17 1 (2015).
[2] "Wireless Power Consortium | IEEE-ISTO." [Online]. Available: http:// www.ieee-isto.org/member-programs/wireless-power-consortium Ge-opend 17 1 (2015).
[3] "http://www.powermatters.org/." [Online]. Available: http://www.power matters.org/ Geopend 17 1 (2015).

[4] Starbucks, "National Rollout of Wireless Charging, by Duracell Powermat, Begins in Starbucks Stores | Starbucks Newsroom." Starbucks, 11 6 2014. [Online]. Available: http://news.starbucks.com/news/national-rollout-of-wireless-charging-by-duracell-powermat-begins-in-starbu Geopend 17 1 (2015).

[5] "Rezence Technology." [Online]. Available: http://www.rezence.com/technology/meet-rezence Geopend 17 1 (2015).

[6] "WiPower |Qualcomm." [Online]. Available: https://www.qualcomm .com /products/wipower Geopend 17 1 (2015).

[7] "Alliance for Wireless Power Membership Doubles on Expanded Application of Rezence Technical Specification to Tablets, Notebooks and Laptops." Alliance For Wireless Power, 7 1 2014. [Online]. Available: http://www.rezence.com/media/news/alliance-wireless-power-membership-doubles-expanded-application-rezence-technical Geopend 17 1 (2015).

[8] Rubino, D. "AT&T Lumia 830 comes with two dual-use Qi/PMA wireless charging covers in green, black." 16 10 2014. [Online]. Available: http://www.windowscentral.com/att-lumia-830-comes-two-dual-pma-qi-covers Geopend 17 1 (2015).

[9] Francesca, T. "Major Milestones for V1.2 Resonance Specification." Wireless Power Consortium, 29 7 2014. [Online]. Available: http://www.wirelesspowerconsortium.com/blog/86/major-milestones-for-v12-resonant-specification Geopend 17 1 (2015).

[10] Cheng, R. "Key wireless charging groups A4WP, PMA agree to merge." CNET, 5 1 2015. [Online]. Available: http://www.cnet.com/news/key-wireless-charging-groups-a4wp-pma-agree-to-merge/ Geopend 17 1 (2015).

[11] "Energous Corporation Joins Power Matters Alliance—Yahoo Finance." Energous Corporation, 2014 3 12. [Online]. Available: http://finance.yahoo.com/news/energous-corporation-joins-power-matters-150000637.html Geopend 18 1 (2015).

[12] Fiandra, O. "The First Pacemaker Implant in America." *PACE* 11: 1234–1238 (1988).

[13] Greatbatchen, W. and C. F. Holmes, "History of Implantable Devices." *IEEE Engineering in Medicine and Biology*: 38–49 (1991).

[14] Mallela, V.S., V. Ilankumaranen, and S. N. Rao, "Trends in Cardiac Pacemaker Batteries." *Indian Pacing Electrophysiol Journal* 4, no. 4: 201–212 (2004).

[15] Fogoros, R.N. "My Battery Is Low—So Why Does My Whole Pacemaker Need To Be Replaced?." About.com, 4 12 2014. [Online]. Available:

http://heartdisease.about.com/od/pacemakersdefibrillators/f/My-Battery
-Is-Low-So-Why-Does-My-Whole-Pacemaker-Need-To-Be-Replaced
.htm Geopend 17 1 (2015).

[16] Staff, M.C. "Pacemaker Results—Tests and Procedures." Mayo
Clinic, 10 4 2013. [Online]. Available: http://www.mayoclinic.org/tests-
procedures/pacemaker/basics/results/prc-20014279 Geopend 17 1 (2015).

[17] Medtronic, [Online]. Available: http://professional.medtronic.com/wcm/
groups/mdtcom_sg/@mdt/@neuro/documents/documents/scs-bat-logev-
specs.pdf Geopend 14 January (2015).

[18] Young Hoon Jeon, M. "Spinal Cord Stimulation in Pain Management: A
Review".

[19] ERB Systems, Inc., [Online]. Available: http://www.ebrsystemsinc.com/
news/first_in_man Geopend 15 January (2015).

[20] Metals, A.S.f. *ASM Handbook Volume 2: Properties and Selection:
Nonferrous Alloys and Special-Purpose Materials.* Cleveland: ASM
International (1990).

[21] Pengfei Li, B.R. "A Wireless Power Interface for Rechargeable Battery
Operated Medical Implants." *Circuits and Systems II: Express Briefs,
IEEE Transactions on* 54, no. 10: 912–916 (2007).

[22] Wang, P., Z.B. Liang, X. Yeen and W. Ko. "A Simple Novel Wireless
Integrated Power Management Unit (PMU) for Rechargeable Battery-
Operated Implantable Biomedical Telemetry Systems." *Bioinformatics
and Biomedical Engineering (iCBBE), 2010 4th International Conference on*, pp. 1–4 (2010).

[23] Jeutter, D. "A Transcutaneous Implanted Battery Recharging and
Biotelemeter Power Switching System." *Biomedical Engineering, IEEE
Transactions on* 29, no. 5, pp. 314–321 (1982).

[24] Ghovanloo, M.S. "A Wide-Band Power-Efficient Inductive Wireless
Link for Implantable Microelectronic Devices Using Multiple Carriers."
Circuits and Systems I: Regular Papers, IEEE Transactions on 54, no.
10: 2211–2221 (2007).

[25] Lovik, R.D.e.a. "Surrogate Human Tissue temperatures Resulting From
Misalignment of Antenna and Implant During Recharging of a Neu-
romodulation Device." *Neuromodulation: Technology at the Neurla
Interface*: 501–511 (2011).

[26] Yarmolenko PS, E.J. Moon, C. Landon, A. Manzoor, D.W. Hochman,
B.L. Viglianti, and M.W. Dewhirst. "Thresholds for thermal damage to
normal tissues." *International Journal of Hyperthermia:* 1–26 (2011).

[27] Rhoon, G.v., T. Samaras, P. Yarmolenko, M. Dewhirst, E. Neufelden, and N. Kuster. "CEM43°C Thermal Dose Thresholds: A Potential Guide for Magnetic Resonance Radiofrequency Exposure Levels?" *European Radiology*: 2215–2227 (2013).

[28] Kuhtz-Buschbeck, J.P., W. Andresen, S. Göbel, R. Gilsteren, and C. Stick. "Thermoreception and Nociception of the Skin: A Classic Paper of Bessou and Perl and Analyses of Thermal Sensitivity During a Student Laboratory Exercise." *AJP Advances in Physiology Education* 34, no. 2: 25–34 (2010).

[29] Yang Yang E.A. "Suitability of a Thermoelectric Power Generator for Implantable Medical Electronic Devices." *Journal of Physics D: Applied Physics*: 5790–5800 (2007).

[30] Weinmann, J.J.e.a. "Heat Flow from Rechargeable Neuromodulation Systems into Surrounding Media." *Neuromodulation: Technology at the Neural Interface*: 114–121 (2009).

[31] Medtronic, [Online]. Available: http://manuals.medtronic.com/wcm/groups/mdtcom_sg/@emanuals/@era/@neuro/documents/documents/wcm_prod042265.pdf Geopend 14 January (2015).

[32] FDA, [Online]. Available: http://www.accessdata.fda.gov/scripts/cdrh/cfdocs/cfRES/res.cfm?id=73737 Geopend 15 January (2015).

[33] FDA, [Online]. Available: http://www.accessdata.fda.gov/scripts/cdrh/cfdocs/cfRES/res.cfm?id=115091 Geopend 15 January (2015).

[34] Kurs, A., A. Karalis, R. Moffatt, J.D. Joannopoulos, P. Fisheren, and M. Soljacic. "Wireless Power Transfer via Strongly Coupled Magnetic Resonance." *Science* 317, no. 5834: 83–86 (2007).

[35] Kesler, M. "Highly Resonant Wireless Power Transfer: Safe, Efficient and over Distance." WiTricity Corporation: Watertown, MA (2013).

[36] Waters, B.H., A.P. Sampleen, and J.R. Smith. "Adaptive Impedance Matching for Magnetically Coupled Resonators." in *PIERS Proceedings*, Moscow, Russia (2012).

[37] Ean, K.K., B.T. Chuan, T. Imuraen, and Y. Hori. "Novel Band-Pass Filter Model for Multi-Receiver Wireless Power Transfer via Magnetic Resonance Coupling and Power Division." in *2012 IEEE 13th Annual Wireless and Microwave Technology Conference (WAMICON)*, Cocoa Beach (2012).

[38] Koh, K.K., T.C. Beh, T. Imuraen, and Y. Hori. "Impedance Matching and Power Division Using Impedance Inverter or Wireless Power Transfer via Magnetic Resonant Coupling." *IEEE Transactions on Industry Applications* 50, no. 3, pp. 2061–2070 (2014).

[39] Leeen, S., R.D. Lorentz. "Development and Validation of Model for 95%-Efficiency 220-W Wireless Power Transfer Over a 30-cm Air Gap." *IEEE Transactions on Industry Applications* 47, no. 6: 885–892 (2011).

[40] Lien, C., and H. Ling. "Investigation of Wireless Power Transfer Using Planarized, Capacitor-Loaded Coupled Loops." *Progress In Electromagnetics Research* 148: 223–231 (2014).

[41] Park, C., S. Lee, G. Choen, and C.T. Rim. "Innovative 5-m-off-Distance Inductive Power Transfer Systems With Optimally Shaped Dipole Coils." *IEEE Transactions on Power Electronics* 30, no. 2: 817–827 (2015).

[42] "Magnetic Resonance and Magnetic Induction—What is the best choice for my application." Wireless Power Consortium, [Online]. Available: http://www.wirelesspowerconsortium.com/technology/magnetic-resonance-and-magnetic-induction-making-the-right-choice-for-your-application.html.

[43] Zhangen, R., C.K. Ho. "MIMO Broadcasting for Simultaneous Wireless Information and Power Transfer." in *2011 IEEE Global-break Telecommunications Conference (GLOBECOM 2011)*, Houston (2011).

[44] Goldsmith, A. "Wireless Communication." New York: Cambridge University Press (2005).

[45] Xuen, J., and R. Zhang. "Energy Beamforming With One-Bit Feedback." *IEEE Transactions on Signal Processing* 62, no. 20: 5370–5381 (2014).

[46] Souppouris, A. "This Router Can Power Your Devices Wirelessly from 15 feet Away." Engaget, [Online]. Available: http://www.engadget.com/2015/01/05/energous-wattup-wireless-charging-demo/.

[47] "Product Overview." Energous Corporation, [Online]. Available: http://www.energous.com/overview/.

[48] Statt, N. "The Future of 'Microimplants' in Your Body: Wireless charging—CNET." CNET, 19 5 2014. [Online]. Available: http://www.cnet.com/news/the-future-of-microimplants-in-your-body-wireless-charging/ Geopend 18 1 (2015).

[49] "Wirelessly Powering Medical Chips Inside Your Body." Forbes, 26 5 2014. [Online]. Available: http://www.forbes.com/sites/jenniferhicks/2014/05/26/wirelessly-powering-medical-chips-inside-your-body/ Geopend 18 1 (2015).

[50] Jacobs, S. "Magnetic Fields That Could Power Tiny Implants |MIT Technology Review." MIT Technology Review, 21 8 2014. [Online]. Available: http://www.technologyreview.com/news/530006/ wireless-power-for-minuscule-medical-implants/ Geopend 18 1 (2015).

[51] Poon, A.S.Y., S. O'Driscollen, and T. Meng. "Optimal Operating Frequency in Wireless Power Transmission for Implantable Devices." in *Proceedings of the 29th Annual International Conference of the IEEE EMBS Cite Internationale*, Lyon, France (2007).

[52] Ho, J.S., S. Kimen, and A.S.Y. Poon. "Midfield Wireless Powering for Implantable Systems." *Proceedings of IEEE* 101, no. 6: 1369–1378 (2013).

[53] Kim, S., J.S. Hoen, and A.S.Y. Poon. "Wireless Power Transfer to Miniature Implants: Transmitter Optimization." *IEEE Transactions on Antennas and Propagation* 60, no. 10: 4838–4845 (2012).

[54] Ho, J.S., A.J. Yeh, E. Neofytou, S. Kim, Y. Tanabe, B. Patlolla, R.E. Beyguien and A.S.Y. Poon. "Wireless Power Transfer to Deep-Tissue Microimplants." *Proceedings of National Academy of Science* 111, no. 22: 7974–7979 (2014).

[55] "Boston Scientific Corporation (BSX) Announces Agreement to Acquire Remon Medical Technologies Ltd." DeviceSpace, 26 6 2007. [Online]. Available: http://www.devicespace.com/News/boston-scientific-corporation-announces-agreement/61182 Geopend 18 1 (2015).

[56] Ozeri, S., D. Shmilovitz, S. Singeren, and C. Wang, "Ultrasonic Transcutaneous Energy Transfer Using a Continuous Wave 650 kHz Gaussian Shaded Transmitter." *Ultrasonics* 50, no. 7: 666–674 (2010).

[57] Hu, Y., X. Zhang, J. Yangen, and Q. Jiang. "Transmitting Electric Energy Through a Metal Wall by Acoustic Waves Using Piezoelectric Transducers." *IEEE Transactions on Ultrasonics, Erroelectrics and Frequency Control* 50, no. 7: 773–781 (2003).

[58] Curieen, J., P. Curie. "Development, Via Compression, of electric polarization in hemihedral crystals with inclined faces." *Bulletin de la Societe de Minerologique de France* 3: 90–93 (1880).

[59] Lippmann, G.J. "Principal of the Conservation of Electricity." *Annales de chimie et de physique* 24: 145–177 (1881).

[60] Trolier-McKinstry, S. "Chapter 3: Crystal Chemistry of Piezoelectric Materials." in *Piezoelectric and Acoustic Materials for Transducer Applications*, Berlin: Springer (2008).

[61] Newnham, R.E. *Properties of Materials: Anisotropy, Symmetry and Structure*. Oxford: Oxford Press (2005).

[62] Singal, S.P. *Acoustic Remote Sensing Applications*. Springer Press (1997).

[63] Ozerien, S., D. Shmilovitz. "Ultrasonic Transcutaneous Energy Transfer for Powering Implanted Devices." *Ultrasonics* 50: 556–566 (2010).

[64] Hsu, Y., C. Leeen, and W. Hsiao. "Optimizing Piezoelectric Transformer for Maximum Power Transfer." *Smart Materials and Structures* 12: 373–383 (2003).

[65] Shihen, P., and W. Shih. "Design, Fabrication, and Application of Bio-Implantable Acoustic Power Transmission." *Journal of Microelectromechanical Systems* 19, no. 3: 494–502 (2010).

[66] Smith, W.A. "Composite Piezoelectric Materials for Medical Ultrasonic Imaging Transducers—A Review." *IEEE CH2358-0.86/0000-0249*, pp. 249–256 (1986).

[67] Goss, S.A., R.L. Johnstonen, and F. Dunn. "Compilation of Empirical Ultrasonic Properties of Mammalian Tissues." *The Journal of the Acoustical Society of America* 64, no. 2: 423–457 (1978).

[68] Goss, S.A., R.L. Johnstonen, and F. Dunn. "Compilation of Empirical Ultrasonic Properties of Mammalian Tissues." *The Journal of the Acoustical Society of America* 68, no. 1: 93–108 (1980).

[69] Callens, D., C. Bruneelen, and J. Assaad, "Matching Ultrasonic Transducer Using two Matching Layers where One of them is Glue." *NDT&E International* 37: 591–596 (2004).

[70] Maleki, T., N. Cao, S.H. Song, C. Kao, S. Koen, and B. Ziaie. "An Ultrasonically Powered Implantable Micro-Oxygen Generator (IMOG)." *IEEE Transactions on Biomedical Engineering* 58, no. 11: 3104–3111 (2011).

[71] Sanni, A., A. Vilchesen, and C. Toumazou. "Inductive and Ultrasonic Multi-Tier Interface for Low-Power Deeply Implantable Medical Devices." *IEEE Transactions on Biomedical Circuits and Systems* 6, no. 4: 297–308 (2012).

[72] Arra, S., J. Leskinen, J. Heikkilaen, and J. Vanhala, "Ultrasonic Power and Data Link for Wireless Implantable Applications." in *2th International Symposium on Wireless Pervasive Computing*, San Juan (2007).

[73] Hu, Y., P. Liao, W. Shih, X. Wangen P. Chang, "Study on Acoustic Impedance Matching of Human Tissue for Power Transmitting/Charging System of Implanted Biochip." in *Proceedings of 2009 IEEE 3rd International Conference on Nano/Molecular Medicine and Engineering*, Tainan (2009).

[74] Stokes, G.G. "On the Theories of Internal Friction in Fluids in Motion, and of the Equilibrium and Motion of Elastic Solids." *Transactions of the Cambridge Philosophical Society* 8, no. 22: 287–342 (1845).

[75] Medtronic 1, [Online]. Available: http://professional.medtronic.com/ wcm/groups/mdtcom_sg/@mdt/@neuro/documents/documents/scs-bat-logev-specs.pdf Geopend 14 January (2015).

[76] Medtronic 2, [Online]. Available: http://professional.medtronic.com/pt/ neuro/scs/prod/index.htm#.VLaMcCvF98E Geopend 14 January (2015).

[77] Medtronic 3, [Online]. Available: http://manuals.medtronic.com/wcm/ groups/mdtcom_sg/@emanuals/@era/@neuro/documents/documents/ wcm_prod042265.pdf Geopend 14 January (2015).

[78] FDA 2, [Online]. Available: http://www.accessdata.fda.gov/scripts/cdrh/ cfdocs/cfRES/res.cfm?id=73737 Geopend 15 January (2015).

[79] FDA 1, [Online]. Available: http://www.accessdata.fda.gov/scripts/cdrh/ cfdocs/cfRES/res.cfm?id=115091 Geopend 15 January (2015).

[80] Yangen, Y. et al. "Suitability of a Thermoelectric Power Generator for Implantable Medical Electronic Devices." *Journal of Physics D: Applied Physics*: 5790–5800 (2007).

[81] Jeon, Y.H. "Spinal Cord Stimulation in Pain Management: A Review".

[82] Loviken, R.D., et. al. "Surrogate Human Tissue temperatures Resulting From Misalignment of Antenna and Implant During Recharging of a Neuromodulation Device." *Neuromodulation: Technology at the Neurla Interface*: 501–511 (2011).

[83] Chew, W.C. "Waves and Fields in Inhomogeneous Media. New York: IEEE Press (1995).

20

Induction Cooking and Heating

20.1 Introduction

Over the last decade, domestic applications of induction heating have increased. This chapter is dedicated to one of the applications – induction cookers. The working principles of induction heating are historically well known. When an oscillating current is passed through an inductor, a magnetic field is created in a small region surrounding the inductor. The magnetic field may also be created by moving the inductor inside a magnetic field or an inductor inside a magnetic field. These principles have been known for centuries and have formed the basis for many of the industrial and household equipment and devices such as electric motors, electric generators and electric power systems and in modern times induction cookers. Other applications of magnetic induction include wireless power chargers, wireless power transfer and magnetic levitation trains. When an oscillating magnetic field is created in the neighbourhood of magnetic materials, induced magnetic fields are created in the magnetic material (load) and can result in induction heating. The induced field is proportional to the magnetic properties of the source and load. If however the load is highly susceptible to magnetism, large magnetic fields will also be developed in the load. This principle is useful for creating significant heating effect in the load. The heating is due to the resistive losses in the load. Induction heating is used in furnaces, welding machines and for applications that require heating up and melting purposes. Since the induced magnetic field is localised around the inductor, the heat developed is also localised to a small region. This localisation of heat is highly desirable in induction cookers and other similar applications.

The rate of heating is proportional to the intensity of the current passing through the inductor, the specific heat of the material, the frequency of the induced current, the permeability of the material and the resistance of the material. Modern induction cookers operate around the 20 kHz to 100 kHz frequency range.

20.2 Advantages of Induction Cooking

The present dominance of electric and gas cookers as well as microwave ovens has meant that induction cooking has not gained traction as is expected. This is despite the numerous advantages of induction cooking. Some of the benefits of induction cooking include speed, efficiency, accuracy, specificity and its green energy credentials. The latency between when an induction cooking is turned on and when heat is created in the pan is extremely small and negligible. This means that when the switch is turned on, heat should be almost instantly expected in the cooking pot. When the cooking pot is removed from the top of the cooker, the heating effect is also removed because there is no magnetic pot to create the heat. This creates an easy but effective means for avoiding being burned by inductive cookers. They can thus be quickly turned on and kept inactive until an induction pot is placed on them.

Induction cookers have the highest efficiency when compared with gas and electric cookers. Gas cooking is only 35% efficient, while induction cooking is nearly 90% efficient [1]. In other words, there is a factor of two and half improvement in induction cooking above gas cooking. This also means that energy is saved with the use of induction cooking. Apart from these benefits, induction cooking is clean, neat, exact and accurate. In other words, energy spillage from induction systems is predominantly almost zero provided the induction pot is well positioned on the cooking top.

20.3 Theory of Induction Heating

When a current passes through a conductor, it creates a magnetic field. The magnetic field normally surrounds the conductor. To understand how magnetic fields can be created, it is essential to observe the relationship between a current, the permeability of the medium in which the conductor is placed and the resulting magnetic field. Jean Pierre Ampere, a French scientist, was first to present this relationship which has rightly become known in electrical engineering as Ampere's law. Consider a conductor carrying the current I amperes. If the conductor is located in free space, Ampere's law states that the magnetic field created by this current is given by the product of the magnitude of the current times the permeability of free space. It is logical that the permeability of free space comes into this equation because it characterises the magnetic behaviour of this medium (free space). Since the magnetic field surrounds the conductor, by taking the sum of small elements of the magnetic field around a closed contour, the result should be the same as the product of the permeability of free space and the current. Ampere wrote an equation for this as a circulation (line integral) Equation (20.1):

$$\oint B_{||}.dS = \mu_0 I \qquad (20.1)$$

In discrete form, Ampere's law is given by Equation (20.2):

$$\sum B_{||}\Delta l = \mu_0 I \qquad (20.2)$$

The sum of the magnetic field is taken round the closed loop over small increments of distance as in Figure 20.1. The magnetic field is concentric with the current-carrying conductor. Its strength decreases with distance r from the conductor. Thus, over a large distance *r* from the conductor, the loop subtends a larger circumference, and hence, more components (weaker field components) of the field are required to be summed to give the same result as when the circumference is small. If the conductor is reshaped into a coil with N turns, the magnetic field is strengthened N-fold. Therefore, the magnetic field inside and outside the coil is a function of the number of turns N, the magnitude of the current I and the radius of the loop. This is written with the following relationship in Equation (20.3):

$$\mathbf{H} = fn(I, N, r) = \mathbf{H}(I, N, r) \qquad (20.3)$$

The magnetic field strength H has units of amperes per metre (A/m). Often the presence of magnetic field around a wire is a puzzle to many people.

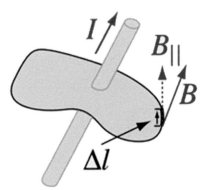

Figure 20.1 Line integral.

However, the current flowing in the conductor is the cause of the magnetic field (the effect). The density of the magnetic field or the flux density is proportional to the permeability of the medium in which the current-carrying conductor is in. In other words, the magnetic field density is given as in Equation (20.4):

$$\mathbf{B} = \mu\mathbf{H}(I, N, r) \tag{20.4}$$

For a practical medium, the permeability can be compared with the permeability of free space via the relationship $\mu = \mu_r\mu_0$, and μ_r is called the relative permeability of the medium. Its value describes how magnetic this medium in in comparison with free space. Since the permeability of free space is $\mu_0 = 4\pi \times 10^{-7}$ (Tesla$/A/m$), the relative permeability allows us to quantify how magnetic materials are. The relative permeability of a diamagnetic or paramagnetic material is unity. Ferromagnetic materials such as nickel and iron have much higher relative permeability. Alloys of nickel and iron can have relative permeability up to 20,000. These large values mean that the material is able to convert (amplify) a magnetic field to very high values. This behaviour is essential for materials that are used for inductive heating because they are able to concentrate the magnetic field $\mathbf{H}$ (or flux density $\mathbf{B}$) in the material.

Until now, we are yet to explain how the magnetic field results in heating. Historically, a notable explanation of how a magnetic field results in heating was first given in Faraday's law of induction stated as Equation (20.5):

$$\nabla \times \mathbf{E} = -\frac{\partial\mathbf{B}}{\partial t} \tag{20.5}$$

Although Faraday's law of induction has formed part of Maxwell's equations, the real honour should go to Faraday. The law establishes that when and where there is a time-varying magnetic field, there will also arise a varying electric field and that the electric field will curl or surround the magnetic field lines. The two fields are at right angles to each other (Figure 20.2) and propagate together in tandem.

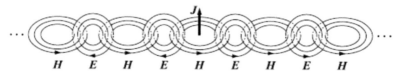

Figure 20.2   Coupling of electric and magnetic fields.

The expression also inherently shows that the electric field is a vector with components in the x, y, and z directions at every point in space. In other words, at every point in space surrounding a time-varying magnetic field, there will exist also an electric field with components in the three spatial directions. If the medium in which the magnetic field exists has variations in permeability as a function of location, the above equation also suggests that the excited electric fields will also depend on that variation. Indeed in an induction cooking set-up, the variation of permeability exists in the spaces around the various components, and hence, varying levels of magnetic field strengths can also be measured in those places [2]. Also inherent in Equation (20.5) is the fact that the magnetic field should be varying. The cause of the variation may be a conductor moving inside a static magnetic field, or a current-carrying conductor moving inside a static magnetic field, or a magnet moving towards a stationary current-carrying coil or a rotating conductor (armature) between two poles of stationary magnets. The point being made is that there has to be some dynamics or motion to excite the induced current (voltage). Hence, the induced voltage at a point r is proportional to the rate of change of the magnetic field as Equation (20.6):

$$\frac{\mathbf{E(r)}}{r} = \text{constan}\,t \times \frac{\partial \mathbf{B}}{\partial t} \qquad (20.6)$$

Faraday actually stated the law in terms of induced voltage with Equation (20.7):

$$V_{\text{ind}} = \mathbf{A}\gamma\frac{\partial \mathbf{B}}{\partial t} \qquad (20.7)$$

The area A is perpendicular to the direction of the magnetic field **B**. The induced voltage is reduced in magnitude if the area A subtends a non-perpendicular angle with the direction of the magnetic field (Figure 20.3). If that happens, the induced power is less than optimum due to existence of a reactive power in the system. In an induction cooker, the existence of large relative permeability in Equation (20.7) is that due to the ferromagnetic bottom, the induced electric field is also very strong. This strong field also causes eddy currents (induced currents) in the bottom of the pot. This in turn creates an opposing magnetic field to that at the top of the induction cooktop coil. Thus, the possibility of existence of stray fields is strongly minimised due to the opposing magnetic fields [3].

Figure 20.4 also shows the arrangement of the ferromagnetic pot, coil and base.

Figure 20.3   Induction cooker [4].

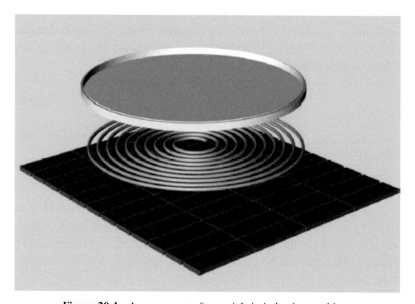

Figure 20.4   Arrangement of materials in induction cooking.

The induced voltage due to a coil of N turns is given by Equation (20.8)

$$V_{\text{ind}} = -\text{N.A.}\frac{\partial \mathbf{B}}{\partial t} = -N\frac{\Delta(AB.A)}{\Delta t} \text{ (volts)} \qquad (20.8)$$

The negative sign shows that the induced magnetic field opposes the field creating it.

20.4 Building Blocks of Induction Cooker

The building blocks of a typical induction cooking system are given in Figure 20.5. The power source is typically a 50-Hz (60 Hz US) AC source. The rectification process may involve the use of high-current diodes, SCRs and insulated gate bipolar transistor (IGBT). The rectifier converts the alternating power source to a direct current (DC). The rectifier is followed by a high-frequency inverter. Its main roles are to convert the acquired DC signals with ripples into a high-frequency AC signal for application to the receiver or load. The IGBT combines the low conduction loss of a bipolar junction transistor with the fast switching speed of a power MOSFET.

An IGBT is ideal for the power electronics application provided in induction cooking. Since this chapter is more focused on the general principles of induction cooking (IC), we will not go further in discussing the working principles of IGBT which can be found in most power electronics text books. The junction temperature of an IGBT is one of its most important parameters because it sets limits on the heating capacity of an induction cooker. Once its maximum temperature is reached, the cooking system must decrease the operating temperature. This temperature depends on its power losses. A method for choosing appropriate IGBT for induction heating is presented in [4].

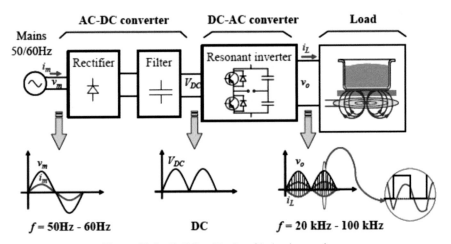

Figure 20.5 Building blocks of induction cookers.

20.4.1 Rectifiers

Most homes worldwide are supplied with a single-phase power. Therefore, it is essential to look at single-phase rectifiers. Two types of single-phase rectifiers are described in this section using power diodes and silicon controlled rectifiers (SCR).

For the diode rectifiers, a bridge rectifier (Figure 20.6) is often used and is popular for power conversion from AC to DC for application in home appliances.

The output voltage of a bridge rectifier is calculated starting from the expression for the input voltage as a function of angle given as Equation (20.9):

$$V(\theta) = \sqrt{2}\,V_m \sin\theta \qquad (20.9)$$

The average value of this voltage is obtained by integrating this function over one quadrant using Equation (20.10)

$$V_{av} = \frac{1}{\pi/2}\int_0^{\pi/2} \sqrt{2}\,V_m \sin\theta.d\theta = \frac{2\sqrt{2}\,V_m}{\pi} \qquad (20.10)$$

This output voltage from the rectifier is 'almost' a steady DC voltage but has ripples. To ensure the ripples are reduced as much as possible, a filter is used. The filter is either a series RL connection or a parallel RC connection.

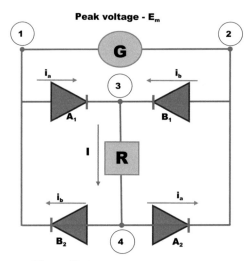

Figure 20.6   Full-wave diode rectifier.

20.4.1.1 SCR rectifiers

The silicon controlled rectifier is a semiconductor device with three terminals. It is traditionally used as a switching device and has found applications in voltage rectifiers, inverters and power flow. Its capability to hand large currents up to thousands of amperes and voltages up to 1 kV has made SCR an extremely versatile device in power electronics. It combines the features of a transistor and rectifier.

20.4.1.2 Half-wave scr rectifier

Figure 20.7 depicts a SCR half-wave rectifier. The input to the rectifier comes from the secondary of a transformer. The SCR is triggered using the variable resistor r. The triggering angle for the SCR is α. This means the SCR will conduct from α to 180° during the positive half-cycles. Let the input to the rectifier be given by the Equation (20.11)

$$V(\theta) = \sqrt{2}\,V_m \sin\theta \qquad (20.11)$$

where $V(\theta)$ is the forward breakdown voltage of the SCR. The firing angle is $\theta = \alpha$. The conduction angle is $180° - \alpha$. The average output voltage from the rectifier is then given by the integral in Equation (20.12)

$$V_{av} = \frac{1}{2\pi} \int_{\alpha}^{180°} \sqrt{2}\,V_m \sin\theta.d\theta$$

$$= \frac{V_m}{\sqrt{2\pi}} \left[-\cos\theta\right]_{\alpha}^{180°} = \frac{V_m}{\sqrt{2\pi}}(1 + \cos\alpha) \qquad (20.12)$$

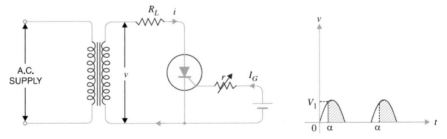

Figure 20.7   Half-wave scr rectifier.

Therefore, the average current which flows during rectification is determined with Equation (20.13):

$$I_{av} = \frac{V_{av}}{R_L} = \frac{V_m}{\sqrt{2}\pi R_L}(1 + \cos\alpha) \qquad (20.13)$$

Maximum current will flow through the load when the triggering angle is zero ($\alpha = 0°$). This current is given by Equation (20.13a)

$$I_{av} = \frac{\sqrt{2}V_m}{\pi R_L} \qquad (20.13a)$$

This current is equal to that of an ordinary half-wave rectifier. As the triggering angle increases, the average current available at the load decreases until it reaches zero when the triggering angle is 180°.

20.4.1.3 Full-wave scr rectifier

Figure 20.8 shows a full-wave SCR rectifier. The circuit is a combination of two half-wave rectifiers feeding the same load. The transformer is centre-tapped. Two gate-controlled power supply sources are used.

Each half-wave rectifier is supplied with equal input voltages as in the half-wave rectifier. Therefore, the average voltage supplied to the load is given by Equation (20.14)

$$V_{av} = 2 \times \left[\frac{1}{2\pi}\int_{\alpha}^{180°}\sqrt{2}\,V_m\sin\theta.d\theta\right]$$

$$= \frac{2\,V_m}{\sqrt{2}\pi}[-\cos\theta]_{\alpha}^{180°} = \frac{\sqrt{2}V_m}{\pi}(1 + \cos\alpha) \qquad (20.14)$$

Thus, the average voltage is twice the average voltage from the half-wave rectifier. Therefore, the average current from the full-wave SCR rectifier is given by the Equation (20.15)

$$I_{av} = \frac{V_{av}}{R_L} = \frac{\sqrt{2}\,V_m}{\pi R_L}(1 + \cos\alpha) \qquad (20.15)$$

Now that we have described the rectifiers fully, the design of the inverter will be described in the next section.

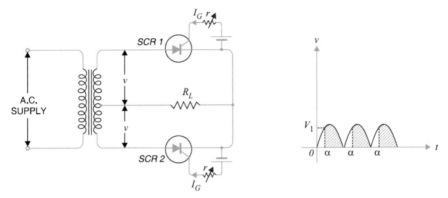

Figure 20.8 Full-wave scr rectifier.

20.4.2 Inverters

A range of electronic devices and circuits are used to create power inverters (DC to AC). Popular devices include SCR [4], MOSFET [5] and IGBT [1, 4]. Inverters also come in different forms and are also designed for different purposes. There are three forms of DC to AC inverters depending on their output waveforms. Inverters which produce square wave outputs have very poor performance with total harmonic distortions of up to 45%. This means that the signal applied to the load from the inverter contains a lot of unfiltered harmonics which distort the signal the load sees. The modified sine wave inverters produce output waveforms which are modified square waves with dead zones between the positive and negative transitions. Although better than the square wave inverters because of reduced harmonic components, they still perform poorly with almost 24% total harmonic distortion (THD). Pure sine wave inverters have the least THD of about 3% and hence much more expensive than the previous two types. The harmonics originate from the decomposition of the square wave inputs to the inverters. When expressed in Fourier series, often only the fundamental component of the Fourier terms is required by the resonant inverter. The remaining components appear as harmonic distortion. The pure sine wave inverters are suitable for running sensitive equipment such as laptop computers, medical equipment, printers, digital clocks and power tools that could be damaged if operated with square and modified sine wave inverters.

The inverter for an induction cooker is typically a single-phase SCR inverter. The role of the inverter is to transform the DC input into a pulsed (bipolar square) or alternating current or output voltage. The SCR acts as a controlled switch which opens and closes a DC circuit alternatively. To illustrate how a DC voltage is converted to an alternating voltage, we first show the circuit with two switches (Figure 20.9) and later replace the two switches with SCRs.

By replacing the switches S1 and S2 with SCRs, the static circuit is transformed into the following switching circuit of Figure 20.10.

The circuit arrangement is such that when one SCR is being turned on, the other one is turned off as well. The two SCRs thereby alternate between being on and off. Both SCRs are never on at the same time and also never both off at the same time.

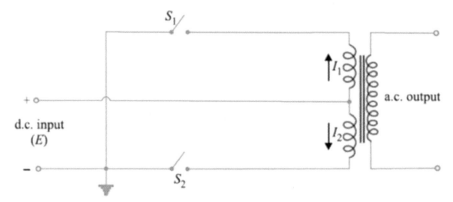

Figure 20.9   Voltage switching.

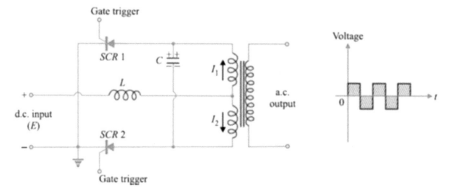

Figure 20.10   SCR voltage switching.

20.4.2.1 Fourier series of output voltage

A periodic signal can be represented as a sum of sinusoids using Fourier series. The Fourier series of a periodic function is defined by Equation (20.16).

$$v(t) = \sum_{n=0}^{N} a_n \cos n\omega_o t + b_n \sin n\omega_o t \qquad (20.16)$$

The value of N is infinitely large. This expression provides the basis for decomposing a signal into its constituent harmonics of the fundamental frequency. It also means that the same formula can be used to create a waveform once the fundamental frequency is given. To obtain a simple expression for the Fourier series, the coefficients need to be determined. The coefficients are determined by integrating the periodic waveform over one period of the signal with Equations (20.17a) to (20.17c) [6]:

$$a_0 = \frac{1}{T} \int_{t_0}^{T+t_0} v(t)\, dt \qquad (20.17a)$$

$$a_n = \frac{1}{T} \int_{0}^{T} v(t) \cos n\omega_0 t\, dt; \quad n \neq 0 \qquad (20.17b)$$

and

$$b_n = \frac{1}{T} \int_{0}^{T} v(t) \sin n\omega_0 t\, dt \qquad (20.17c)$$

Some of the regular waveforms of interest in inverter design include the rectangular waveform given in Figure 20.11. This waveform has peak voltages $\pm V$ volts at half-period intervals.

The Fourier series of this waveform can be shown to be given by the superposition Equation (20.18):

$$v(t) = \frac{4V}{\pi} \sum_{n=0}^{\infty} \frac{1}{2n+1} \sin(2n+1)\omega_0 t \qquad (20.18)$$

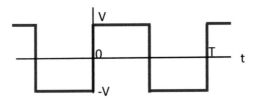

Figure 20.11   Rectangular voltage pulse.

More than 93% of the energy in the signal occurs within the first five harmonics of the waveform. The rest contribute to smoothing the shape of the rectangular pulse. In practice, it is impossible to sum all the infinite components. Hence, the summation is normally taken over N harmonics where N is reasonably large for the application.

Consider the same waveform but shifted in time as in Figure 20.12. The axes cut through the width of the rectangular pulse. The Fourier series of the waveform can be shown to be given by Equation (20.19).

$$v(t) = \frac{4V}{\pi} \sum_{n=odd}^{\infty} \frac{(-1)^q}{n} \cos(n\omega_0 t)$$

$$q = (n-1)/2 \tag{20.19}$$

The superposition Equations (20.18) and (20.19) are used as inputs to the resonant tank circuits. The resonant inverters are tuned to the fundamental components when $n = 1$. The remaining harmonics appear as the total harmonic distortions (THD).

20.4.2.2 IGBT inverters

Use of insulated gate bipolar transistor (IGBT) is popular in the design of high-power inverters. Often how the inverter output is coupled to the load is used in naming the type of inverters. If the inductor is coupled to the load with a T-network consisting of 'inductor, capacitor, inductor', the inverter is called an LCL inverter. The inverter is thus an inductor-coupled inverter to the load. In an LCC, the inverter output is coupled to the load through a capacitor, the so-called capacitive coupling. Other forms of coupling including LLC [7–9] have been reported. Series connection and coupling using RLC and LC circuits have been used for induction cooking [8]. In this section, we discuss two inverters using IGBT as the active device.

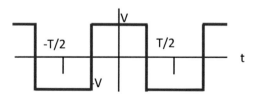

Figure 20.12   Shifted rectangular voltage pulse.

20.4.2.2.1 LCL configuration

In the voltage-fed inverter, the output of the inverter is a rectangular pulse and the output current is sinusoidal. The rectangular pulse is used as the input to the resonant tank circuit. The tank circuit is an LCL configuration (Figure 20.13).

The impedance of the tank circuit is given by Equation (20.20).

$$Z = j\omega L_1 + \frac{(R + j\omega L_2)\,(1/j\omega C)}{R + j\omega L_2 + \frac{1}{j\omega C}}$$

$$= \frac{R + j\omega(L_1 + L_2) - \omega^2 L_1 C(R + j\omega L_2)}{1 + j\omega C(R + j\omega L_2)} \tag{20.20}$$

At resonance, the imaginary part of this impedance is zero, or the impedance is purely resistive. This means that we can write the imaginary part of the impedance as Equation (20.21)

$$Z_i = \omega(L_1 + L_2) - \omega^3 L_1 L_2 C = 0 \tag{20.21}$$

This yields the expression when the impedance tends to zero as Equation (20.22)

$$Z \to 0 \quad \omega_{01} = \sqrt{\frac{(L_1 + L_2)}{C(L_1 L_2)}} = \frac{1}{\sqrt{C(L_1 L_2)/(L_1 + L_2)}} \tag{20.22}$$

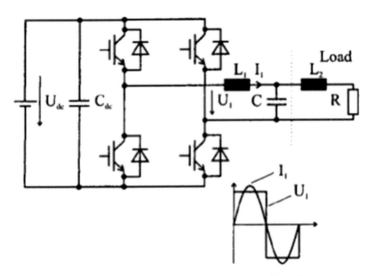

Figure 20.13   LLC inverter tank circuit configuration.

When $L_2/L_1 \ll 1$, the impedance tends to infinity and we have a second resonant frequency and that is given as Equation (20.23):

$$Z \to \infty \quad \omega_{02} = \sqrt{\frac{(L_1 + L_2)}{C(L_1 L_2)}} = \frac{1}{\sqrt{C.L_2}} \qquad (20.23)$$

The LCL circuit operates at a resonant frequency equal to the resonant frequency of a series LC circuit when the impedance approaches zero.

20.4.2.2.2 CCL configuration

In the current-fed inverter, the output voltage of the inverter is a sinusoidal pulse and the output current is rectangular. The rectangular current pulse is used as the input to the resonant tank circuit. The tank circuit has CCL configuration (Figure 20.14) with impedance as in Equation (20.24):

$$Z = \frac{1/j\omega C_1 \times (j\omega L + 1/j\omega C_2 + R)}{1/j\omega C_1 + (j\omega L + 1/j\omega C_2 + R)} = \frac{(j\omega L + 1/j\omega C_2 + R)}{1 + j\omega C_1 R - \omega^2 L C_1 + \frac{C_1}{C_2}}$$

$$(20.24)$$

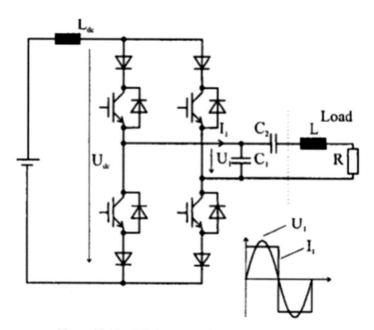

Figure 20.14   CCL inverter tank circuit configuration.

We can show that this expression also leads to two resonant frequencies given by Equations (20.25a) and (20.25b).

$$Z \to 0 \quad \omega_{01} = \frac{1}{\sqrt{L.C_2}} \tag{20.25a}$$

and

$$Z \to \infty \quad \omega_{02} = \frac{1}{\sqrt{LC_1 C_2 / (C_1 + C_2)}} \tag{20.25b}$$

This circuit operates at a resonant frequency when $Z \to \infty$ or at the resonant frequency of an equivalent parallel resonant circuit.

20.4.3 Half-Bridge Inverter Design

Normally for an inverter heating to maximise power supplied to the load and reduce the reactive power, an induction heater will use a resonant tank circuit. The resonant circuit consists of the work coil on which the object to be heated (induction pot) is placed. The tank circuit is driven at resonance or close to resonance which helps to remove or limit the reactive power component.

Figure 20.15 shows a half-bridge inverter using an LCL parallel tank circuit.

A transistor half-bridge is shown. The half-bridge inverter produces a square wave output which is applied to a parallel-tank LC circuit. A matching inductor is used. A DC blocking capacitor is also connected in series with the

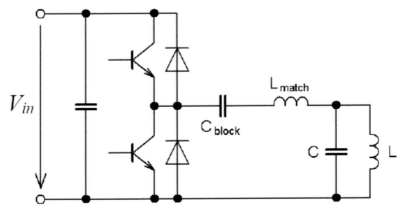

Figure 20.15   Half-bridge inverter design circuit.

matching inductor. The parallel tank circuit is chosen with the objective of producing a steady working frequency at resonance. The tank circuit and its accompanying matching and DC blocking produce low-Q systems typically in the range $Q < 40$.

A series form of this circuit is much more simplified and disposes of the blocking capacitor and matching inductor (Figure 20.16). The resonant circuit again maintains a relatively steady working frequency and also helps to limit reactive power being delivered to the load.

The design example in this section follows the design example in [9] by assuming a total required power from the mains supply. We assume this power is 10 kW. Since the inverter produces square waveforms, we assume that for each half-cycle, the same power is delivered to the load. Therefore, the calculations may be done for just one half-cycle and be applicable for the second half-cycle.

At DC, the power available is known and is given by the product of the DC voltage V_{dv} and the average current I_{avg} in Equation (20.26)

$$P_{dc} = V_{dv} I_{avg} \qquad (20.26)$$

Since the tank circuit receives a near-perfect sine wave, the average current delivered to the tank circuit is given by the rms current in Equation (20.27).

$$I_{av} = I_{rms} \frac{2\sqrt{2}}{\pi} \qquad (20.27)$$

The constant multiplier $2\sqrt{2}/\pi$ is obtained by first expressing the square wave at the output of the inverter and taking only the fundamental component. By

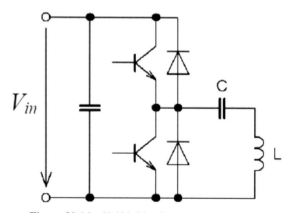

Figure 20.16   Half-bridge inverter with LC load.

inserting the expression for the average current in Equation (20.27), we have the expression for the maximum rms current to be as in Equation (20.28):

$$I_{rms} = \frac{P}{V_{dc}} \frac{\pi}{2\sqrt{2}} \qquad (20.28)$$

For a mains supply in Australian homes of 230 volts, the peak DC value will be $230\sqrt{2} = 325.2$ volts. Using a half-bridge, there will be only half (162.6 volts) of this voltage available.

We now have all the required expressions to use in designing the tank circuit. Let the operating frequency be 50 kHz. Assume the loaded Q-factor of the work coil can be measured and is 30. Generally, if the same current flows through the resistive and inductive parts of the work coil, which is true, then we can express the Q-factor as a ratio of the real and imaginary powers in the system which is given by Equation (20.29):

$$Q = \frac{P_Q}{P} = \frac{I_{rms} X_L}{I_{rms} R} = \frac{\omega L}{R} \qquad (20.29)$$

10 The reactive power in the work coil is shown in Equation (20.30):

$$P_Q = I_{rms}^2 X_L = I_{rms}^2 2\pi f L \qquad (20.30)$$

Since the rms current is

$$I_{rm} = \frac{10,000\,W}{162.6} \frac{\pi}{2\sqrt{2}} = 68.3\,A$$

$$L = \frac{P_Q}{I_{rms}^2 2\pi.f} = \frac{Q.P}{I_{rms}^2 2\pi.f} = \frac{30 \times 10,000}{(68.3)^2 \times 2\pi \times 50,000}$$

$$= \frac{3}{(68.3)^2 \times \pi} = 204.7\,\mu H$$

Hence, the capacitor required for resonance is

$$C = \frac{1}{(2\pi.f)^2} = \frac{1}{(100,000 \times \pi)^2} = 101.3\,nF$$

The number of turns required to obtain the required inductance is estimated using Equation (20.31):

$$L = \frac{a^2 N^2}{9a + 10b}\,\mu H \qquad (20.31)$$

where

- a is the diameter of the coil in inches,
- b is the height of the coil also in inches and
- N is the number of turns.

Observe that the unit for the variables a and b is inches rather than centimetre. Conversion to centimetre is also possible.

References

[1] Pathak, Abhijit D., and Kyoung-Wook Seok. "New Optimised IGBTs for Induction Cooking." *Power Electronics Europe* 1: 37–39 (2007).

[2] Mariani Primiani, V., S. Kovyryalov, and G. Cerri. "Rigorous Electromagnetic Model of An Induction Cooking System." *IET Science Meas. Technolology* 6, no. 4: 238–246 (2012).

[3] Irnich, Werner, and Alan D. Bernstein. "Do Induction Cooktops Interfere with Cardiac Pacemakers?" *Europace* 8: 377–384 (2006).

[4] I. Millán, D. Puyal, J.M. Burdío, O. Lucía, and D. Palacios. "IGBT Selection Method for the Design of Resonant Inverters for Domestic Induction Heating." *Proceedings of the 13th European Conference on Power Electronics and applications, EPE'09*, pp. 1–7 (2009).

[5] Esteve, Vicente, Jose Jordan, Esteban Sanchis-Kilders, Enrique J. Dede, Enrique Maset, Juan B. Ejea and Augustin Ferreres. "Comparative Study of a Single Inverter bridge for Dual-frequency Induction heating Using Si and SiC MOSFETs." *IEEE Transaction on Industrial Electronics* 62, no. 3, 1440–1450, March (2015).

[6] Dorf, Richard C. *Introduction to Electric Circuits*, 2nd edition. John Wiley & Sons, Inc.: New York (1993).

[7] Espí Huerta, José M., Enrique J. Dede García Santamaría, Rafael García Gil, and Jaime Castelló Moreno. "Design of the *L-LC* Resonant Inverter for Induction Heating Based on Its Equivalent SRI." *IEEE Transaction on Industrial Electronics* 54, no. 6: 3178–3187 (2007).

[8] Espi, J.M., E.J. Dede, A. Ferreres, and R. Garcia. "Steady-State Frequency Analysis of the "LLC" Resonant Inverter for Induction Heating." pp. 22–28.

[9] Martis, Jan, and Pavel Vorel. "Apparatus for Induction Heating 2.5 kW Using a Series Resonant Circuit."

[10] Dieckerhoff, Sibylle, Michael J. Ryan and Rik W. De Doncker. "Design of an IGBT-based LCL-Resonant Inverter for High-Frequency Induction Heating.": 2039–2045 (1999).

[11] Saoudi, Magdy, Diego Puyal, Héctor Sarnago, Daniel Antón and Arturo Mediano. "A New Multiple Coils Topology For Domestic Induction Cooking System." pp. 1–7.

Index

Editor's Biography

Johnson I. Agbinya graduated from La Trobe University with a PhD in microwave radar remote sensing (MSc Research University of Strathclyde, Glasgow, Scotland (1982) in microprocessor techniques in digital control systems) and BSc (Electronic/Electrical Engineering, Obafemi Awolowo University, Ile Ife, Nigeria. He is currently Head of School of Information Technology and Engineering at Melbourne Institute of Technology (Australia). Prior to this he was Associate Professor (remote sensing systems engineering) in the department of electronic engineering. He is also Adjunct Professor at Nelson Mandela African Institute of Science and Technology (NM-AIST) Arusha Tanzania and PhD Research supervising Professor at Sudan University of Science and Technology Khartoum, Sudan.

He was Extraordinary Professor in telecommunications at Tshwane University of Technology/French South African Technical Institute in Pretoria, South Africa. He was also a Senior Research Scientist at CSIRO Telecommunications and Industrial Physics (1994–2000; renamed CSIRO ICT) in biometrics and remote sensing and Principal Engineer Research at Vodafone Australia research from 2000 to 2003. He is the author of six technical books in electronic communications including Principles of Inductive Near Field Communications for Internet of Things (River Publishers, Aalborg Postkontor, Denmark, 2011); IP Communications and Services for NGN (Auerbach Publications, Taylor & Francis Group, USA, 2010) and Planning and Optimisation of 3G and 4G Wireless Networks (River Publishers, Aalborg Postkontor, Denmark, 2009).

He is Consulting Editor for River Publishers Denmark on new areas in Telecommunications and Science and also the founder and editor-in-chief of the African Journal of ICT (AJICT) and founder of the International Conference on Broadband Communications and Biomedical Applications (IB2COM). His current research interests include remote and short range sensing, inductive communications and wireless power transfer, Machine to Machine communications (M2), Internet of Things, wireless and mobile communications and biometric systems.

Dr. Agbinya is a member of IEEE, ACS and African Institute of Mathematics (AIMS). He has published extensively on broadband wireless communications, sensors, inductive communications, biometrics, vehicular networks, video and speech compression and coding, contributing to the development of voice over IP, intelligent multimedia sub-system and design and optimisation of 3G networks. He was recipient of research and best paper awards and has held several advisory roles including the Nigerian National ICT Policy initiative.